Lecture Notes in Networks and Systems

Volume 281

AF862950

The series "Lecture Notes in Networks and Systems" publishes the latest developments in Networks and Systems—quickly, informally and with high quality. Original research reported in proceedings and post-proceedings represents the core of LNNS.

Volumes published in LNNS embrace all aspects and subfields of, as well as new challenges in, Networks and Systems.

The series contains proceedings and edited volumes in systems and networks, spanning the areas of Cyber-Physical Systems, Autonomous Systems, Sensor Networks, Control Systems, Energy Systems, Automotive Systems, Biological Systems, Vehicular Networking and Connected Vehicles, Aerospace Systems, Automation, Manufacturing, Smart Grids, Nonlinear Systems, Power Systems, Robotics, Social Systems, Economic Systems and other. Of particular value to both the contributors and the readership are the short publication timeframe and the world-wide distribution and exposure which enable both a wide and rapid dissemination of research output.

The series covers the theory, applications, and perspectives on the state of the art and future developments relevant to systems and networks, decision making, control, complex processes and related areas, as embedded in the fields of interdisciplinary and applied sciences, engineering, computer science, physics, economics, social, and life sciences, as well as the paradigms and methodologies behind them.

Indexed by SCOPUS, INSPEC, WTI Frankfurt eG, zbMATH, SCImago.

All books published in the series are submitted for consideration in Web of Science.

More information about this series at https://link.springer.com/bookseries/15179

Hiren Kumar Deva Sarma · Valentina Emilia Balas ·
Bhaskar Bhuyan · Nitul Dutta
Editors

Contemporary Issues in Communication, Cloud and Big Data Analytics

Proceedings of CCB 2020

Springer

Editors
Hiren Kumar Deva Sarma
Department of Information Technology
Sikkim Manipal Institute of Technology
Majitar, Sikkim, India

Bhaskar Bhuyan
Department of Information Technology
Sikkim Manipal Institute of Technology
Majitar, Sikkim, India

Valentina Emilia Balas
Department of Automatics and Applied Software
Aurel Vlaicu University of Arad
Arad, Romania

Nitul Dutta
Department of Computer Science and Engineering
Marwadi University
Rajkot, Gujarat, India

ISSN 2367-3370 ISSN 2367-3389 (electronic)
Lecture Notes in Networks and Systems
ISBN 978-981-16-4243-2 ISBN 978-981-16-4244-9 (eBook)
https://doi.org/10.1007/978-981-16-4244-9

This Springer imprint is published by the registered company Springer Nature Singapore Pte Ltd.
The registered company address is: 152 Beach Road, #21-01/04 Gateway East, Singapore 189721, Singapore

Preface

Communication, cloud and big data have been three highly popular researched domains in the past decade. Society has witnessed significant innovation in communication paradigms. Communication methods and associated technologies have witnessed huge changes and tremendous growths. This journey is still on. Due to significant penetration of the Internet and mobile technologies in the human society, there has been shift in the communication means and methods, and ultimate goal is perhaps to achieve easy-to-use and low-cost technologies as the user-friendly communication systems.

Cloud technology has supported information technology industries significantly. Cloud is being relied for data storage, computing and even for providing different services to the end-users. Enterprises have become dependent heavily on cloud technologies. Although there are many research issues as well as policy-level issues, cloud technology has become indispensable for the world of information technology.

At present scenario, it is visible that the human society is gradually becoming heavily dependent on data. It will probably be not wrong if it is said that present form of society is driven by data. Here, the technology named as big data analytics plays a vast role. Big data analytics has tremendous potential that can impact the way enterprises, business houses, organizations, even the nations across the globe can function. There has been huge amount of data getting generated digitally, everyday due to the usage of the Internet as a whole, and the technologies like social networks, Internet of Things (IoT), and cyberphysical systems (CPS) in specific. Processing, storage, distribution, and security of these data are major concerns. There is a need of innovation of tools and technologies for these purposes.

These three domains have tremendous research scopes in isolation as well as when looked as connected technologies to each other. The purpose of this book is to share the latest research findings in the areas of communication, cloud technologies and big data analytics.

The book is comprised of 38 research articles. The articles address various issues in communication, cloud and big data analytics. There are few articles in the general area of computing and healthcare informatics as well. The articles have been presented by the respective authors during the first International Conference on Communication,

Cloud and Big Data 2020 (CCB 2020), organized by the Department of Information Technology at Sikkim Manipal Institute of Technology (SMIT), Sikkim, during December 18–19, 2020. CCB 2020 is the fourth one in the series.

We are thankful to all the contributing authors of CCB 2020. We thank Sikkim Manipal Institute of Technology for extending all support in organizing CCB 2020. We are highly grateful to Prof. Janusz Kacprzyk for his guidance and encouragement throughout the journey. We thank the editorial team of Springer Nature for extending great support to us. We thank Springer Nature for supporting us as the publication partner and all the help that have been extended throughout the process of publishing this book.

Gangtok, India — Dr. Hiren Kumar Deva Sarma
Arad, Romania — Dr. Valentina Emilia Balas
Gangtok, India — Dr. Bhaskar Bhuyan
Rajkot, India — Dr. Nitul Dutta

Contents

Editors and contributors

About the Editors

Dr. Hiren Kumar Deva Sarma is Professor in the Department of Information Technology, Sikkim Manipal Institute of Technology, Sikkim. He received Bachelor of Engineering in Mechanical Engineering from Assam Engineering College, Guwahati, Assam (1998). He completed Master of Technology in Information Technology from Tezpur University, Assam (2000). He received Doctor of Philosophy (in Computer Science and Engineering) from Jadavpur University, West Bengal (2013). He has co-authored two books, edited three book volumes, and published more than 70 research papers in different International Journals and referred International and National Conferences of repute. He is the recipient of Young Scientist Award from International Union of Radio Science (URSI) in the XVIII General Assembly 2005, held at New Delhi, India, and has received IEEE Early Adopter Award in 2014. His current research interests are networks, network security, robotics, and big data analytics.

Dr. Valentina Emilia Balas is currently Full Professor in the Department of Automatics and Applied Software at the Faculty of Engineering, "Aurel Vlaicu" University of Arad, Romania. She holds a Ph.D. in Applied Electronics and Telecommunications from Polytechnic University of Timisoara. Dr. Balas is Author of more than 350 research papers in refereed journals and International Conferences. She is Editor-in Chief to *International Journal of Advanced Intelligence Paradigms* (IJAIP) and to *International Journal of Computational Systems Engineering* (IJCSysE), Member in Editorial Board member of several national and international journals, and is Evaluator Expert for national, international projects and Ph.D. thesis. She served as General Chair of the International Workshop Soft Computing and Applications (SOFA) in eight editions 2005–2020 held in Romania and Hungary. Her research interests are in intelligent systems, fuzzy control, soft computing, smart sensors, information fusion, modeling, and simulation.

Dr. Bhaskar Bhuyan is presently working as Associate Professor in the Department of Information Technology, Sikkim Manipal Institute of Technology affiliated to Sikkim Manipal University, Sikkim, India. He did his B.E. (1997) in Computer Science & Engineering from Motilal Nehru Regional Engineering College (now NIT), Allahabad, India. He did his M.Tech. (2000) in Information Technology and Ph.D. (2017) in Computer Science & Engineering from Tezpur University, Assam, India. He has 18+ years of professional experience in teaching as well as in industry. He has published several research papers in various conferences and journals of repute, and co-edited one book (conference proceedings). His research interests include computer networks, wireless sensor networks, mobile ad hoc networks, Internet of things, and cloud computing.

Dr. Nitul Dutta is Professor in the Department of Computer Engineering, Faculty of Engineering (FoE), Marwadi University, Rajkot, Gujarat since 2014. He has more than 20 years of experience in teaching and research. He received B.E. degree in Computer Science and Engineering from Jorhat Engineering College, Assam (1995) and M. Tech. degree in Information Technology from Tezpur University, Assam (2002). He received Ph.D. (Engineering), in the field of Mobile IPv6 from Jadavpur University, West Bengal (2013). He has published 15 research papers in various journals and 30 research papers in various conferences of repute. He has co-edited three books (two conference proceedings and one edited volume). He has successfully completed two sponsored research projects funded by AICTE, Government of India. His current research interests are wireless communication, mobility management in IPv6 based network, cognitive radio networks, and cyber security.

Contributors

Ashwin Adarsh Department of Computer Science & Engineering, Sikkim Manipal Institute of Technology, Sikkim Manipal University, Bangalore, India

Ayush Agarwal Amity School of Engineering and Technology, Amity University Noida, Noida, Uttar Pradesh, India

Ali Nadim Alhaj Department of Computer Engineering, Marwadi University, Rajkot, Gujarat, India

Faraaz Ali Amity School of Engineering and Technology, Amity University Noida, Noida, Uttar Pradesh, India

P. A. Alvi Department of Physics, Banasthali Vidyapith, Jaipur, Rajasthan, India

B. Anand Department of EIE, Hindusthan College of Engineering and Technology, Coimbatore, Tamil Nadu, India

Ravi Anand Department of Computer Science and Engineering, Sikkim Manipal Institute of Technology, Sikkim Manipal University, Majitar, India

Vivek Kumar Anand Marwadi University, Rajkot, India

T. Ananth Kumar Department of CSE, IFET College of Engineering, Villupuram, Tamil Nadu, India

T. S. Arun Samuel Department of ECE, National Engineering College, Kovilpatti, Tamil Nadu, India

Rishi Asthana Goel Institute of Technology and Management, Lucknow, Uttar Pradesh, India

Polash Banerjee Department of CSE, SMIT, SMU, Majitar, Sikkim, India

Udayan Baruah Department of Information Technology, Sikkim Manipal Institute of Technology, Sikkim Manipal University, Majitar, Sikkim, India

Shreedevi Subrahmanya Bhat Department of Information Science and Engineering, SDM College of Engineering and Technology, Dharwad, India

Anushka Bhattacharya Department of Computer Science, Bengal Institute of Technology, Kolkata, West Bengal, India

S. R. Biradar Department of Information Science and Engineering, SDM College of Engineering & Technology, Dharwad, Karnataka, India;
Affliated to Visvesvaraya Technological University (VTU), Belagavi, India

D. Boopathi Department of EEE, Paavai Engineering College, Pachal, Tamil Nadu, India

Naiwrita Borah Department of Information Technology, Sikkim Manipal Institute of Technology, Sikkim Manipal University, Majitar, Sikkim, India

Sudipta Chakrabarty Department of Computer Applications, Techno India, Kolkata, West Bengal, India

Madhura Chakraborty Department of Electronics and Communication Engineering, JIS College of Engineering, Kalyani, Nadia, West Bengal, India

Kamlesh Chandravanshi Department of Information Technology, LNCT, Bhopal, India

Niranjan Chavan Health Data Analytics & Visualization Environment, Amity Institute of Public Health, Amity University Uttar Pradesh, Noida, India

Tejbanta S. Chingtham Department of Computer Science and Engineering, Sikkim Manipal Institute of Technology, Sikkim Manipal University, Majitar, Sikkim, India

Tejbanta Singh Chingtham Department of CSE, Sikkim Manipal Institute of Technology, SMU, Majitar, Sikkim, India

Prasenjit Das Chitkara University School of Computer Applications, Chitkara University, Baddi, Himachal Pradesh, India

Parul Datta Chitkara University School of Engineering and Technology, Chitkara University, Baddi, Himachal Pradesh, India

Shankar Debnath Department of Computer Science and Engineering, JIS College of Engineering, Kalyani, Nadia, West Bengal, India

Krishna Delvadia Department of Information Technology, Uka Tarsadia University, Surat, Gujarat, India;
Department of Computer Engineering, Marwadi University, Rajkot, Gujarat, India

Amlan Jyoti Dey Technology Specialist, NIIT Technologies Ltd, Noida, India

Sudipta Dey Sikkim Manipal Institute of Technology, Majitar, Rangpo, Sikkim, India

Prantik Dey Sikkim Manipal Institute of Technology, Majitar, Rangpo, Sikkim, India

Dependra Dhakal Sikkim Manipal Institute of Technology, Majitar, Rangpo, Sikkim, India

Ambarish Dutta Department of Computer Science and Engineering, Sikkim Manipal Institute of Technology, Sikkim Manipal University, Majitar, India

Nitul Dutta Department of Computer Engineering, Marwadi University, Rajkot, Gujarat, India

Sushanta Kabir Dutta Department of Electronics and Communication Engineering, North Eastern Hill University, Shillong, India

Sourav Ghosh Department of Computer Science and Engineering, JIS College of Engineering, Kalyani, Nadia, West Bengal, India

Angana Goswami Department of Computer Science & Engineering, Sikkim Manipal Institute of Technology, Sikkim Manipal University, Sikkim, India

Sandeep Gurung Sikkim Manipal Institute of Technology, Majitar, Sikkim, India

Mousum Handique Assam University, Silchar, Assam, India

Priyanka Hazowary Department of Electronics and Communication Engineering, North Eastern Hill University, Shillong, India

Shrikrishna Sharad Huilgol Department of Information Science and Engineering, SDM College of Engineering & Technology, Dharwad, Karnataka, India;
Affliated to Visvesvaraya Technological University (VTU), Belagavi, India

Md. Ruhul Islam Department of Computer Science and Engineering, Sikkim Manipal Institute of Technology, Sikkim Manipal University, Majitar, India

K. S. Jagadeeshgowda Department of Computer Science, Sri Krishna Institute of Technology, Bengaluru, India

K. Jagatheesan Department of EEE, Paavai Engineering College, Pachal, Tamil Nadu, India

Rajiv Janardhanan Health Data Analytics & Visualization Environment, Amity Institute of Public Health, Amity University Uttar Pradesh, Noida, India;
Laboratory of Disease Dynamics & Molecular Epidemiology, Amity Institute of Public Health, Amity University Uttar Pradesh, Noida, India

A. Jayakumar Electronics and Communication Engineering, IFET College of Engineering, Villupuram, India

Ajeya Jha Department of Management Studies, SMIT, SMU, Majitar, India

Aranya Jha Department of CSE, SMIT, SMU, Majitar, Sikkim, India

Mahendra Ku. Jhariya Department of Computer Science and Engineering, MANET, Bhopal, Madhya Pradesh, India

Swati Kamat Department of Information Science and Engineering, SDM College of Engineering and Technology, Dharwad, India

Priyanka Kamath Department of Information Science and Engineering, SDM College of Engineering and Technology, Dharwad, India

V. Kanendra Naidu School of Electrical Engineering, College of Engineering, Universiti Teknologi MARA, Shah Alam, Selangor, Malaysia

Vijeta V. Kerur Department of Information Science and Engineering, SDM College of Engineering & Technology, Dharwad, Karnataka, India;
Affliated to Visvesvaraya Technological University (VTU), Belagavi, India

Aditya Kopparthi Amity School of Engineering and Technology, Amity University Noida, Noida, Uttar Pradesh, India

Madhuri Kulkarni Department of Information Science and Engineering, SDM College of Engineering and Technology, Dharwad, India

Abhishek Kumar Chitkara University School of Engineering and Technology, Chitkara University, Baddi, Himachal Pradesh, India

Uday Kumar Delhi State Cancer Institute, Dilshad Garden, Delhi, India

Utsav Kumar Department of Computer Science and Engineering, Sikkim Manipal Institute of Technology, Sikkim Manipal University, Majitar, India

Venktesh Kumar Department of Computer Science and Engineering, Sikkim Manipal Institute of Technology, Sikkim Manipal University, Majitar, India

V. Kumarakrishnan Department of EEE, Paavai Engineering College, Pachal, Tamil Nadu, India

Rinika Manna Department of Computer Science and Engineering, JIS College of Engineering, Kalyani, Nadia, West Bengal, India

Suman Mohanty Department of Computer Science and Engineering, Sikkim Manipal Institute of Technology, Sikkim Manipal University, Majitar, India

Santanu Mohapatra Department of Computer Science & Engineering, Sikkim Manipal Institute of Technology, Sikkim Manipal University, Sikkim, India

M. Navaneetha Velammal Department of ECE, Francis Xavier Engineering College, Tirunelveli, Tamil Nadu, India

Pushpalatha S. Nikkam Department of Information Science and Engineering, SDM College of Engineering and Technology, Dharwad, India

Dhruba Ningombam Department of Computer Science and Engineering, Sikkim Manipal Institute of Technology, Sikkim Manipal University, Majitar, India

Chitrapriya Ningthoujam Department of Computer Science and Engineering, Sikkim Manipal Institute of Technology, Sikkim Manipal University, Majitar, Sikkim, India

V. K. Parvati Department of Information Science and Engineering, SDM College of Engineering & Technology, Dharwad, Karnataka, India;
Affliated to Visvesvaraya Technological University (VTU), Belagavi, India

Sanjit Paul Department of Computer Science and Engineering, JIS College of Engineering, Kalyani, Nadia, West Bengal, India

Sayan Paul Department of Computer Science and Engineering, JIS College of Engineering, Kalyani, Nadia, West Bengal, India

Payash Pradhan Department of Computer Science & Engineering, Sikkim Manipal Institute of Technology, Sikkim Manipal University, Bangalore, India

Praveen Kumar Pradhan Centre for Computers and Communication Technology, Chisopani, South Sikkim, India

Amrit Prasad Assam University, Silchar, Assam, India

P. Pugazhendiran Electrical and Electronics Engineering, IFET College of Engineering, Villupuram, India

Prativa Rai Sikkim Manipal Institute of Technology, Majitar, Sikkim, India

Prerna Rai Department of IT, SMIT, East Sikkim, Gangtok, Sikkim, India

Arjun Rajput Department of Computer Science and Engineering, Lovely Professional University, Phagwara, Punjab, India

Raja Siddharth Raju Sikkim Manipal Institute of Technology, Majitar, Sikkim, India

Pranati Rakshit Department of Computer Science and Engineering, JIS College of Engineering, Kalyani, Nadia, West Bengal, India

Priya Ranjan SRM University, Mangalagiri, Andhra Pradesh, India

Mohit Rathor Sikkim Manipal Institute of Technology, Majitar, Rangpo, Sikkim, India

Debasmita Ghosh Roy School of Automation, Banasthali Vidyapith, Jaipur, Rajasthan, India

Dwaipayan Roy Department of Computer Science and Engineering, JIS College of Engineering, Kalyani, Nadia, West Bengal, India

Leena I. Sakri Department of Information Science and Engineering, SDM College of Engineering and Technology, Dharwad, India

Kamal Sapkota Department of Electrical and Electronics Engineering, Sikkim Manipal Institute of Technology, Sikkim Manipal University, Majitar, Sikkim, India

Hiren Kumar Deva Sarma Department of Information Technology, Sikkim Manipal Institute of Technology, Majitar, Sikkim, India

Biswaraj Sen Department of Computer Science & Engineering, Sikkim Manipal Institute of Technology, Sikkim Manipal University, Bangalore, India

Rishav Sen Department of Computer Science and Engineering, JIS College of Engineering, Kalyani, Nadia, West Bengal, India

Shabbiruddin Department of Electrical and Electronics Engineering, Sikkim Manipal Institute of Technology, Sikkim Manipal University, Majitar, Sikkim, India

Anand Sharma Computer Science and Engineering Department, MUST, Lakshmangarh, Sikar, India

Anurag Sharma Department of Computer Science and Engineering, Sikkim Manipal Institute of Technology, Sikkim Manipal University, Majitar, India

Kalpana Sharma Department of Computer Science & Engineering, Sikkim Manipal Institute of Technology, Sikkim Manipal University, Bangalore, India; Department of Computer Science & Engineering, Sikkim Manipal Institute of Technology, Sikkim Manipal University, Sikkim, India

Sandeep Shaw Department of Computer Science and Engineering, JIS College of Engineering, Kalyani, Nadia, West Bengal, India

Karma Sonam Sherpa Sikkim Manipal University, Gangtok, Sikkim, India

Kumar Dron Shrivastav Health Data Analytics & Visualization Environment, Amity Institute of Public Health, Amity University Uttar Pradesh, Noida, India

Ashi Singh Department of Computer Science & Engineering, Sikkim Manipal Institute of Technology, Sikkim Manipal University, Sikkim, India

Balwant Singh Amity School of Engineering and Technology (ASET), Noida, Uttar Pradesh, India

Vikash Kumar Singh Department of Computer Science & Engineering, Sikkim Manipal Institute of Technology, Sikkim Manipal University, Bangalore, India; Department of Computer Science & Engineering, Sikkim Manipal Institute of Technology, Sikkim Manipal University, Sikkim, India

Aaditaa Soni MUST, Lakshmangarh, Sikar, India

Gaurav Soni Department of Computer Science and Engineering, Lovely Professional University, Phagwara, Punjab, India

Rupaban Subadar Department of Electronics and Communication Engineering, North Eastern Hill University, Shillong, India

Tanuja Subba Department of CSE, Sikkim Manipal Institute of Technology, SMU, Majitar, Sikkim, India

R. Sujay Department of Information Science and Engineering, SDM College of Engineering & Technology, Dharwad, Karnataka, India;
Affliated to Visvesvaraya Technological University (VTU), Belagavi, India

S. Sundaresan Electronics and Communication Engineering, National Institute of Technology Puducherry, Puducherry, India

K. Suresh Kumar Electronics and Communication Engineering, IFET College of Engineering, Villupuram, India

M. Suryaganesh Department of ECE, National Engineering College, Kovilpatti, Tamil Nadu, India

Tshering Lhamu Tamang Department of Computer Science & Engineering, VMWARE Technologies Bangalore, Bangalore, India

Dereje Tirfe Marwadi University, Rajkot, India

Malay Ranjan Tripathy Amity School of Engineering and Technology (ASET), Noida, Uttar Pradesh, India

G. Vijayakumar Department of EEE, Muthayammal Engineering College, Rasipuram, Tamil Nadu, India

Communication

Reliable Data Delivery in Software-Defined Networking: A Survey

Prerna Rai and Hiren Kumar Deva Sarma

Abstract Software-defined networking is a network paradigm which has shifted the intelligence of the underlying network interface to an independent control plane. The controller acts independently from the data plane. This isolation of the plane brings about greater opportunities but with a challenge of attaining higher reliability and optimum energy consumption in packet delivery. This paper aims at identifying approaches that improve the performance of the software-defined networking (SDN) with increase in reliability, specially required for real-time data delivery.

Keywords Reliability · Software-defined networking · Data delivery · Quality of services · Real-time data delivery

1 Introduction

If we see around today and the prevailing pandemic situation, we can witness the growing needs for data communication. The requirement increases the need for the number of devices to be connected. The emerging new technology of IoT, cloud computing, and the Internet have also added to the increase of network users today. Every single user may be connected to multiple devices that are accessed from any place using an Internet. The increase in number of users and the number of devices has increased the size of the network [1]. The existing traditional network may not be suitable to handle such huge growth in the number of user and size of the network. A traditional type of network requires manual configuration and are not scalable. Neither it is flexible. Handling such huge growth in the network can create a bottleneck. The network devices that it uses require manual configuration and it lacks automation. The hardware devices are custom made and are vendor specific. So, this huge change in the network may be susceptible to changes. Moreover, the manual configuration is at the risk of being error prone. The vertical integration of the layers restricts the network from growing. In the previous traditional IP network, decentralization fails to tackle with the growing data size. It has a fixed characteristic

P. Rai · H. K. D. Sarma (✉)
Department of IT, SMIT, East Sikkim, Gangtok, Sikkim 737136, India

H. K. D. Sarma et al. (eds.), *Contemporary Issues in Communication, Cloud and Big Data Analytics*, Lecture Notes in Networks and Systems 281,
https://doi.org/10.1007/978-981-16-4244-9_1

[2]. With time and the date, software-defined networking (SDN) has become the need of an hour. It came up to solve the issues faced by the traditional network. It breaks the vertical integration and increases flexibility. The SDN is a new approach in the field of the network. The SDN has adopted many features which makes it better than the traditional network. The features of SDN allow it to handle the big data coming out from sources such as the Internet, the Internet of things, and cloud computing. SDN is scalable, programmable, flexible, and reliable in terms of data communication. SDN architecture has three distinct layers. They are data plane, control plane, and management plane or an application plane. The central point of the whole architecture is the control plane. It is considered to be the heart of the network. The data plane is considered only to be forwarding device and do not make any decisions. The decisions for routing data in the network are taken up by the control plane. The forwarding devices are a dummy device which forwards the packet based on the decision made by the controller.

A Brief History of SDN

SDN is said to be the new approach in the sense that it is now gaining popularity. Today companies such as Google data center is based on SDN. The advent of SDN started long time back. Researches had already understood that the amount of data is soon going to increase tremendously. Due to the increasing use of the Internet, the data was sure to increase. Moreover, devices getting connected to the network also would increase. And by the end of 2020, the increase in data is seen to be quite a lot. Many studies were undertaken to come up with the technology that can handle the data as is today. In 1996 [3], IPsilon was developed. Followed by the Tempest in 1998, by the end of the year 2000, the breaking of vertical integration had started such that the decision-making capability of forwarding devices was shifted to the control plane. In 2007, networking technique names as Ethane came into being. It was developed for continuous communication between the different layers and was a great success. SDN had already started its entry into the networking field, but in the year 2008, it saw its real new dawn. It was due to the introduction of OpenFlow protocol, which was introduced by Stanford university [4] and later standardized by the open network foundation (ONF) in the year 2011 [*ONF, "ONF Overview," 2014.*]. The major foundation behind the SDN evolution is network virtualization [5]. With this new idea, there has been a tremendous transformation in the nature, design, and management of a network.

In this paper, various approaches used for achieving reliability in SDN are outlined. The rest of the paper is organized as follows: Sect. 2 details architecture of SDN, followed by Sect. 3 in which traditional networks are compared with SDN. Section 4 outlines challenges faced in SDN followed by Sect. 5, in which the state of the art in reliable data delivery in SDN is presented. Section 6 presents various approaches used for reliable data delivery in SDN followed by Sect. 7 in which a comparison between different approached of reliable data delivery in SDN has been presented. The paper is concluded in Sect. 8.

2 Architecture of SDN

With the introduction of SDN, the networking concept and design is revolutionary [5, 6]. When compared to the traditional network, this new idea has brought about great changes in the network design, concept, and management. [7] SDN architecture comprises of three basic components. They are control layer, data layer, and application or management layer [7, 8]. It has been standardized by the open network foundation (ONF 2014). The major idea behind the evolution of SDN lies in the fact that the brain of the network is to be removed from the lowest layer and shift it to the central controller. The distributed nature of the traditional network was the major setback for it to handle larger data. In SDN architecture, there is a separate layer that fully controls the entire data communication in central mode. The overhead on the forwarding device is reduced. Moreover, the central controller provides global network view, scalable, agile, and they can be programmed. Network management and configuring is also ease out with reduction, in overall cost.

The block diagram for SDN architecture is depicted as in Fig. 1. The various components in the SDN layered architecture comprises of:

(i) Forwarding/data plane: The layer includes switching or forwarding devices. These devices do not make decisions on their own but depends upon the controller. It uses flow-based routing that forwards the packet according to match and action field of the routing table. The routing table is maintained in the router hardware memory known as the ternary content address memory (TCAM). These memories are very expensive but limited to its storage size. It needs to be used cautiously.

(ii) Control plane: The layer is the heart or brain of SDN architecture. It contains network controller. These controller makes all the necessary routing decisions for the packets to have end-to-end delivery. It is logically centralized, so that it can be the central point of collecting network status and information. It has a global view.

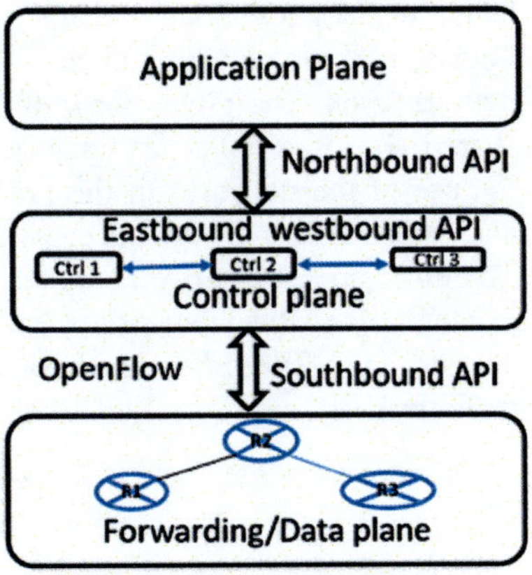

Fig. 1 SDN architecture

(iii) Application plane: This layer uses REST API' s for applications to be forwarded in the network. It includes application such as firewall application and many others.

(iv) Northbound API: This is an interface between application and control plane. It makes use of API for communication.

(v) Southbound API: This is an interface between the control plane and the data plane. The requested route for data delivery is made available to the controller using this interface. The controller also makes use of this interface to send the routes that are to be used for packet forwarding. The most common protocol used in this interface for communication is OpenFlow ver 1.0, ver 1.1, ver 1.2, etc. [9].

(vi) Southwest bound API: In certain architecture where there are multiple controllers, there too exists southwest bound interface. This interface allows the controllers to communicate among each other for decision making.

SDN evolution has been a boon to the existing network infrastructure. The layered architecture also improves level of abstractions [7]. Such that the forwarding of data is done hiding the underlying details about the existing hardware. Distributed abstraction does not need to show the way of distributing data. Specification abstraction allows network application that work without having to know the characteristics of the network. Having said all that SDN still faces lots of challenges in terms of centralization, reliability, and scalability. The aim of the study is to bring out techniques and methods used by different researchers for reliable data delivery in SDN. It also provides a comparative study between these techniques based on handling reliability of the network in data transfer.

3 The SDN Versus Traditional Network

The main aim of SDN is to simplify the network functioning and also to optimize management of the network [10] by providing network automation, while on the other hand, legacy networks are difficult to automate because it is distributed in nature. Each routing or switching devices have an individual controller, which makes their own decision. The global view of the network is not known to each device unless shared. The traditional network cannot meet the growing demand for service and the use of the network. In the present-day scenario, real-time data such as audio data, video data, and big data centers provide major challenges to network operators. Therefore, in order to overcome the drawback of traditional network SDN came into existence. It enabled network operators with an efficient, flexible, agile, and scalable network [11]. The difference between the traditional network and SDN is as given in Table 1.

Table 1 SDN versus legacy network

Sl. no	Parameter	Software-defined network	Traditional network
1	Network control	Logically controlled centrally	Distributed in nature
2	Routing	Flow-based forwarding	Destination based
3	Flexibility	Flexible	fixed
4	Integration	Separation of control plane from data plane	Forwarding devices can control and forward
5	Program	Can be programmed	Cannot programmed easily
6	Cost of management and hardware	Cost effective	Expensive
7	Vendor architecture	Vendor Neutral	Vendor specific
8	Network view	Global view	Partial view

4 Challenges in SDN

SDN is a network technology that is emerging today. It can dynamically handle the network in simple and cost-effective manner [3]. As stated in the above Table 1, SDN provides many advantages over the traditional networking. Even though SDN has created a great interest in the area of networking, its utilization in the industrial field is yet to flourish [3]. Despite many abilities and application of SDN, there still exist many challenges. These challenges are to be taken care of. Some of the challenges as stated in [10–12] are:

i **Placement of Controller**: Logically centralized controllers are susceptible to single-point failure. Using multiple controllers can solve this problem, but the controller needs to be intelligently placed, such that it reduces controller overhead. Yet today placement of the controller in SDN is a real challenge [11].

ii **Scalability**: SDN has evolved to improve network scalability. As the number of data increases, the amount of data to be collected from the global network also increases. The centralized controller may be overloaded. The major issues that might arise are such as induced latency and delay. This is due to the exchanging of information between data plane and control plan and also due to communication between multiple controllers in flat or mesh architecture.

iii **Reliability** [13]: Reliability of data delivery relies on delivery of data that was expected to be delivered. Real-time data such as audio/video streaming and audio/video conferencing require higher reliability with timely delivery. There exist many research studies that aim at improving reliability in SDN. Initiatives have been taken to improve over fault handling, detection, and prediction. But reliability is mostly achieved with higher complexity and cost.

iv **Consistency**: Compared to the centralized controller, distributed controller platform face consistency problem. Achieving good consistency with a better performance is a real challenge for SDN.

v **Security** [14]: Controllers in SDN are the central point of decision making. But the controller being centralized is susceptible to attacks. The central body is easily seen by the external source. The flexibility and visibility of the network tend to make the network more vulnerable.

5 State of the Art in Reliable Data Delivery in SDN

The area of application of SDN, for example, audio conferencing as shown in Fig. 2 below, is very promising area of research. In SDN, the data delivery should be very reliable such that what needs to be received is what is sent and also within a limited time frame. There are certain data's that are within the time frame, and their meaning will only last until they are delivered timely. For example, in audio/video conferencing or any other real-time data delivery, if the data delivery is not reliable, then due to slow convergence and packet loss, the data that is delivered to the destination may not be meaningful. Loss of packets and timely delivery of data is of utmost importance, especially in today's era when the data size is increasing tremendously.

Studies and research undertaken in SDN highly relates to the requirement for scalability, reliability, and flexibility. Data delivery in the traditional network is IP based, and flexibility, scalability, and reliability are a real concern. The data travels from source to destination. SDN uses flow-based delivery rather than destination.

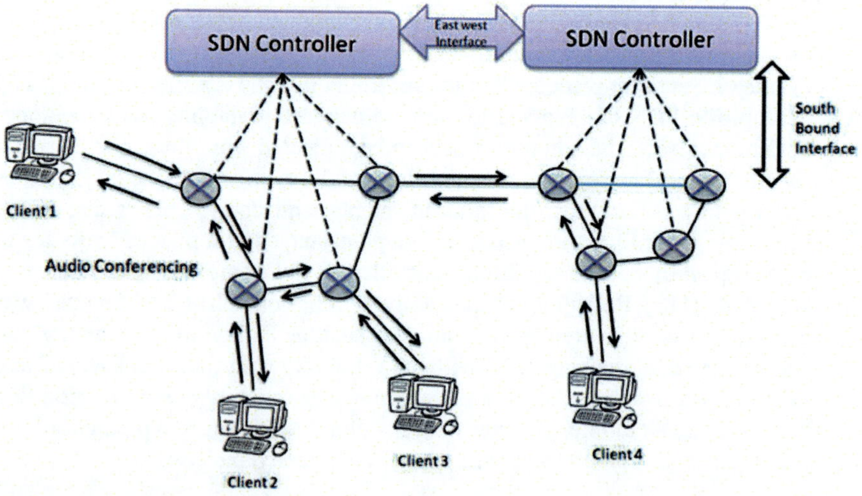

Fig. 2 Audio conferencing in SDN

Flow-based data delivery routes the data looking at the destination address. The SDN uses controller to decide upon the routes and store it in the routing table. The major concern about the data delivery is reliability. In the network, when the amount of data increases, congestion can occur. The congested network causes loss of packets and delay in delivery. This causes unreliable data delivery in the network. If any fault occurs in the path, the data need to be rerouted such that delivery of data is done. But the path fault may cause the data delivery to fail. This all causes the network data delivery to be unreliable.

The main motivation of this paper is to identify the various studies undertaken for reliable data delivery in SDN. This paper brings about various techniques used by researchers who have worked in improving the reliability of SDN. There are many types of data such as text, audio, and video. Among many different types of traffic, the major concern in data delivery is real-time data. Real-time data are time sensitive. They are to be delivered as they are presented, e.g., audio/video streaming, audio/video conferencing, vehicle tracking using GPS, weather forecasting, etc. [15]. SDN aims in providing a reliable transmission of time-based data without compromising the quality of the network. There are three different types of real-time data [16]. They are:

(i) Hard real-time data: It should meet the specified deadline else the system fails such as nuclear system, pacemaker, etc.
(ii) Firm real-time data: It can miss deadlines, but the performance degrades, and the value of the task may be of no value after the deadline such as weather forecast.
(iii) Soft real-time data: It can frequently miss the deadlines, but the value of the data will reduce considerably such as audio/ video data, Web browsers.

Reliability: Reliability of a network in general terms can be represented using $R(t)$. $R(t)$ is a network reliability at time t. It is the probability that all the nodes in the network are functioning and communicating with each other over the time interval $[0, t]$ [17].

Failure rate determines the prediction of reliability. There are three common basic categories of failure rates:

(i) Mean time between failure (MTBF): It is the measure of reliability used for systems that can be repaired. MTBF can be stated as the time passed before the failure of system under the condition of a constant failure rate. It can also be defined as the expected time between the two-consecutive failure for the system that is repairable.
(ii) Mean time to failure: This is for non-repairable system. It is the expected mean time until the first failure of any device. It is a mean over a large period of time with large number of units.
(iii) Mean time to repair: It is the total time taken to take corrective and preventive repair divide by the total number of repairs. It is the expected span of time from a failure to the completion of repair or maintenance. This is used with repairable systems.

As per [15], a reliable network is a network such that it informs about the failure in delivering the data and do provide mechanism to handle the failure, so that the network convergence time is short. Moreover, with real-time data delivery, it should be timely delivery with reduced latency and delay. Not only delivery of data but routing convergence time also plays major role in improving the network reliability as rightly proven in [12]. The main aspect of achieving reliable data delivery is routing and controller placement. The purpose of this study is to bring about various reliability-based approaches in SDN such that the network delivers the data in time with lowered latency and convergence time [12]. Network convergence time depends on routing protocols implemented by the routers. It is the time taken by the router to resume communication or transfer the information. The main task of the routing protocol is to detect link failure at a faster rate and to search for an alternative route for the data to be delivered to its destination on time. When failure occurs if the routing protocols are faster in finding an alternative route, the network can converge at a faster rate and hence improve over the network reliability and can be used in real-time application. The other main area that influences reliability aspect of SDN is the placement of controller [18]. Controller placement affects flow setup latency, fault tolerance, and also performance metrics. [19]. The paper brings about a varied techniques and methodology used by different researchers as in section below.

Application Area of SDN: An Example
The transfer of audio data can be routed in scenario such as shown in Fig. 2. The outcome can be applicable for reliable transfer of real-time data in software-defined network.

6 SDN and Its Reliability in Data Delivery

SDN initiated its evolution due to the rigid nature of traditional network. In order to handle the huge data, SDN comes into its role at ease. But despite having many features and characteristics that can solve many existing issues, it is still in its preliminary stage and quite young to include many features to solve n numbers of networking issues. The dawn of SDN is seen but yet not matured enough. SDN still have challenges to be solved as stated above. Day and night researchers have been trying hard, and many of them have come up with many approaches to provide light toward solving these challenges. Among many SDN challenges, this paper discusses about the different approaches that can be used to improve SDN reliability. In this paper, the term SDN reliability is discussed in general, based on controller and data plane using OpenFlow protocols as southbound API as interface between control plane and data plane.

Reliability is directly proportional to network performance. Reliable is the network, better is the network performance.

When a data is to be delivered in the network, the major concern is its end-to-end delivery. But during the data transmission, the data needs to face a real challenge of passing through the routes that might not be efficient or congested. Having inefficient routes may allow the data packet to be lost on its way and may never be delivered. Moreover, the SDN has a central controller that controls the forwarding of data packet over the network. The routes to be taken by the data packets are decided by the controller. It can collect network information and network status such that the decisions are real time. But the major issue arises when the single controller is down. It is at a risk of single-point failure. And if such failure occurs, then the data delivery will completely come to halt. Therefore, rather than single central controller, a distributed controller can be used. For improving network reliability of a network, the study in [13] proposes an architecture that is distributed in nature. It uses controller in distributed nature. The main approach used is the master–slave controller and aims at achieving fault-tolerant network and improved recovery time. The configuration uses a coordinator finder algorithm that selects a coordinator controller based on the rate of reliability for each subnet. This approach is very good in improving the reliability based on fault tolerance, but the distributed nature of controller increases latency between the controllers. For reliable data delivery in SDN, routing the routes taken by the data packets is to be reliable. If any routes are congested, the paths need to be rerouted dynamically such that the network is able to cope up with the route change dynamically as studied by Martini et al. [20]. The dynamic routing is beneficial for better end-to-end delivery, but the approach used in this study is not effective for smaller network. In the study done by Guan et al. [21], it states that with the increase data traffic, complexity of network also increases with increase in CPU utilization and greater chances of loss of packets in SDN. But the paper does not provide solution toward achieving reliability and scalability issues. The inflow of data is a great challenge for delivering the data reliably in SDN. One of the studies provided by Comer et al. [22] aims at handling network failure before it occurs, which means fault-avoidance mechanism. In this mechanism, multiple copies of paths are made. It uses redundant controller for managing these multiple copies. The main idea of using redundancy is to ensure that the packets reach its destination on time. The copies of packets are sent over multiple paths, and it ensures that one of the paths with at least one copy of packets is delivered. The failure of one path will not affect the packet delivery. It uses proactive approach for avoiding fault from occurring. But due to redundancy, the overhead caused in the network and controller is huge with huge latency. Another study by Mendiratta et al.[23] focuses on failure of hardware components. It examines the network reliability over the data flow when any hardware component fails. It framework discusses about the hardware fault tolerant. The study undertaken is only applicable for smaller network and not discussed for the network with larger size. In another paper by Yu et al. [24], it provides an approach based on OpenFlow protocol in SDN. It aims at improving the protocol. It uses an independent platform and aims at making the network flexible and reliable with better forwarding behavior. Despite of many feature, the approach does not consider link load, end-to-end data delivery, etc. Network congestion is one very important criterion for reliable data delivery in SDN. If the SDN network is congested, packet loss and delay would be incurred. Therefore, handling network congestion is an utmost priority for SDN network for it to delivery data reliably. The study discussed in [25] provides a mechanism to control congestion in SDN. It uses OpenFlow protocol. The proposed mechanism detects congestion and thereafter reroutes the packets using a controller. The reliability in this mechanism improves but only limited to certain no of packet size. In the study discussed in [26], an algorithm is designed to find an optimal traffic path using a nonlinear optimization framework for individual switch. The study uses centralized controller and fault-handling mechanism. The approach used in this study solves the issue of reliability to certain extent. In [27], a fault tolerance SDN is discussed where it uses master–slave SDN controller. This improves network reliability. In [28], the proposed mechanism aims at optimizing energy consumption required in routing data. In [29], a reliable transmission framework is proposed which converting TCP packet to UDP packet. UDP packets create lesser overhead than TCP; therefore, use of UCP for transmission is efficient while network reliability is due to TCP which can retransmit when there is packet loss. The approach improves the network reliability, but the use of predefined routes might create scalability issues. The use of a single centralized controller acts as single-point failure. For network reliability, many network resources are utilized among which energy consumption is one of them in [30–32], and it focuses on efficient and optimum utilization of energy. The study in [33] proposes a framework that checks the network load and dynamic path flows. It uses an algorithm that is able to control network topology with its traffic load. An "application-based network operations (ABNO)" [34] is another technology toward network operators. This approach is vulnerable to attacks.

One of the important parameters that measures network performance is convergence time. Network reliability is concern with it. The research provided in paper [12] provides an experimental study on "routing convergence time" and is depicted as the sum of:

Failure detection time+ flooding of information time+ processing the routing updates time+ computation paths time+ alternative path installation time.

The proposed methodology depicts that convergence time in SDN is consistently same for different topology unlike traditional network. It shows that SDN has timely data delivery compared to the traditional network. The paper [35] proposes "Route-Flow Control Platform (RFCP)." It provides effective load balancing and path selection. This methodology removes the need for processing, storing, and maintenance across a multiple-edge environment. The summary of a research gap based on works mentioned above is listed in Table 2.

7 Relevant Findings from the Survey

From the various study undertaken, certain relevant findings have been compared and are highlighted in Table 3. Different researcher has put forward different approaches and methods to provide reliable end-to-end data delivery. The comparison of the different study is based on the numbers of controller, controller placement, fault tolerance, packet delay, and jitter. From the comparative study, it is seen that every

Table 2 Features and limitations in various study

Sl. no	Title of the paper	Features/approach used	Limitations
i	[13] "On reliability improvement of software-defined networks"	• It uses "RDSDN" • Use algorithm known as "coordinator finder algorithm" to select a master controller • Improves reliability	• The use of multiple controllers induces latency and delay of packets • It does not consider network element failure
ii	[20] "SDN controller for context-aware data delivery in dynamic service chaining"	• Service-based end-to-end SDN controller • It addresses context-based data delivery from one end to other • Optimize rules of forwarding	• Only applicable for smaller scale • Link load and packet inflow is not considered • Size of network topology is not considered
iii	[21] "Reliability and scalability issues in software-defined network frameworks"	• It addresses reliability and scalability issues of SDN • Verifies CPU utilization that increases with packet inflows	• It does not consider link load balancing for an end-to-end delivery
iv	[22]"Resilient packet delivery through software-defined redundancy: An experimental study"	• Creates a predefined route • It proactively creates a redundant • It uses redundant controller that improves reliability	• Creates extra overhead on the path • Increases route congestion
v	[24] "Forwarding programming in protocol-oblivious instruction set"	• Uses an approach known as "POF Flow Instruction Set (POF-FIS)"	• It does not consider any placement of controller and link load
vi	[25]"Congestion control in software-defined data center networks through flow rerouting"	• Congestion control and management approach • It calculates link load on various links and route packet along lightly loaded shortest path • Improve throughput and reduced delay	• Rerouting of path is not possible • It also does not consider fault-avoidance mechanism
vii	[12] "Comparative analysis of SDN and conventional networks using routing protocols"	• Shows that SDN converge faster than traditional network	• Performance degrades with increase in size

(continued)

Table 2 (continued)

Sl. no	Title of the paper	Features/approach used	Limitations
viii	[29] "SDUDP: A reliable UDP-based transmission protocol over SDN"	• Retransmission guarantees reliability • Uses an approach that transmits TCP packets as UDP packets and vice versa	• Increase overhead due to UDP packets • Cannot handle growing network size
ix	[23] "How reliable is my software-defined network? models and failure impacts"	• Use continuous–time Markov chain (CTMC) model • Assess how tolerant are both switches and controller toward fault • Considers both switch and controller failure	• Not applicable for larger network • Do not consider link loss probability
x	[36]"Better Safe than Sorry: Modeling reliability and security in replicated SDN controllers"	• Combine reliability and security issue • Use byzantine fault-tolerance algorithm	• It considers reliability in limited scenario • The approach requires more refined approach to improve SDN reliability

different approach has their own advantages and disadvantages. Some have focused reliability on controller placement, while some other researchers have focused on fault-handling mechanism.

8 Conclusion

The paper briefly describes the need for reliable SDN network. Various studies undertaken by different researchers show that many works have been done in the area of SDN in order to improve reliability in SDN. SDN being the present and future of networking, reliability is one of the most important challenges that needs to be resolved. The study also highlights the comparative study on different approaches used to improve SDN reliability. The study does not include quality of service (QoS) required for effective and reliable end-to-end data delivery. Further we can also study the QoS for real-time data delivery.

Table 3 Comparison based on reliability measuring parameters

Attributes → author/paper	Single controller	Distributed controller	Convergence time	Quality of service	Approach used	Fault tolerance	Delay & jitter
Moazzeni Shadi et al. [13]	No	Yes	High	High	Proactive	High	Propagation delay incurred
Martini et al. [20]	Yes	No	Low	Extended QOS	Service-oriented end-to-end delivery	Medium	Reduced
Comer et al. [22]	Yes, but redundant controller	No	Very high	Guaranteed delivery	Proactive	High	Minimum
Mendiratta et al. [23]	No	Yes	Not known	Degrade with control and data plane failure	Reactive	Low	Not known
Gholami et al. [25]	No	Yes	High	Improved QoS	Rerouting of selected flows	High	Minimum with low packet loss
Lin et al. [26]	Yes	No	Fast convergence	Not known	Polynomial time approximation (PTAA)	Minimum	80% delay reduction
Gonzalez et al. [27]	No	Yes master—slave	Slower convergence	Not known	Uses replication scheme	High	Minimum
Wang et al. [29]	Yes	No	Slower convergence	Improved QoS	SDUDP	Not known	Minimum
Amir et. al. [37]	No	Yes	Fast	QoS based on improved latency	Network clustering	Good	Minimum

References

1. Nunes, B.A.A., Mendonca, M., Nguyen, X.-N., Obraczka, K., Turletti, T.: A survey of software-defined networking: past, present, and future of programmable networks. IEEE Commun. Surv. Tutor. **16**(3), 1617–1634 (2014)
2. Prajapati, A., Sakadasariya, A., Patel, J.: Software defined network: future of networking. In: 2018 2nd International Conference on Inventive Systems and Control (ICISC), pp. 1351–1354. IEEE (2018)
3. Sezer, S., Scott-Hayward, S., Chouhan, P.K., Fraser, B., Lake, D., Finnegan, J., Viljoen, N., Miller, M., Rao, N.: Are we ready for SDN? Implementation challenges for software-defined networks. IEEE Commun. Mag. **51**(7), 36–43 (2013)
4. Naous, J., Erickson, D., Covington, G.A., Appenzeller, G., McKeown, N.: Implementing an OpenFlow switch on the NetFPGA platform. In: Proceedings of 4th ACM/IEEE Symposium Architecture for Networking and Communications Systems, (ANCS '08), p. 1 (2008)
5. Feamster, N., Rexford, J., Zegura, E.: The road to SDN: an intellectual history of programmable networks. ACM SIGCOMM Comput. Commun. Rev. **44**(2), 87–98 (2014)
6. Kobayashi, M., Seetharaman, S., Parulkar, G., Appenzeller, G., Little, J., Reijendam, J.V., Weissmann, P., Mckeown, N.: Maturing of OpenFlow and software-defined networking through deployments. Comput. Netw. **61**, 151–175 (2014)
7. Kreutz, D., Ramos, F., Verissimo, P., Rothenberg, C.E., Azodolmolky, S., Uhlig, S.: Software-defined networking: a comprehensive survey. arXiv preprint arXiv:1406.0440 (2014)
8. Jarraya, Y., Madi, T., Debbabi, M.: A survey and a layered taxonomy of software-defined networking. IEEE Commun. Surv. Tutor. **16**(4), 1955–1980 (2014)
9. Lara, A., Kolasani, A., Ramamurthy, B.: Network innovation using openflow: a survey. IEEE Commun. Surv. Tutor. **16**(1), 493–512 (2013)
10. Bannour, F., Souihi, S., Mellouk, A.: Distributed SDN control: survey, taxonomy, and challenges. IEEE Commun. Surv. Tutor. **20**(1), 333–354 (2018)
11. Shamugam, V., Murray, I., Leong, J.A., Sidhu, A.S.: Software Defined Networking challenges and future direction: a case study of implementing SDN features on OpenStack private cloud. In: IOP Conference Series: Materials Science and Engineering, vol. 121, no. 1 (2016)
12. Gopi, D., Cheng, S., Huck, R.: Comparative analysis of SDN and conventional networks using routing protocols. In: 2017 International Conference on Computer, Information and Telecommunication Systems (CITS), pp. 108–112. IEEE (2017)
13. Moazzeni, S., Khayyambashi, M.R., Movahhedinia, N., Callegati, F.: On reliability improvement of software-defined networks. Comput. Netw. **133**, 195–211 (2018)
14. Rawat, D.B., Reddy, S.R.: Software defined networking architecture, security and energy efficiency: a survey. IEEE Commun. Surv. Tutor. **19**(1), 325–346 (2016)
15. Lee, M.-C., Sheu, J.-P.: An efficient routing algorithm based on segment routing in software-defined networking. Comput. Netw. **103**, 44–55 (2016)
16. Martin, J.: Programming Real-time Computer Systems, p. 4. Prentice-Hall Inc.,, Englewood Cliffs, NJ (1965). ISBN 978-0-13-730507-0
17. Large, D., Farmerm, J.: Chapter 12: network reliability and availability. In: Broadband Cable Access Networks, 1st edn (2008). ISBN: 9780123744012
18. Hu, Y.-N., Wang, W.-D., Gong, X.-Y., Que, X.-R., Cheng, S.-D.: On the placement of controllers in software-defined networks. J. China Univ. Posts Telecommun. **19**, 92–171 (2012)
19. Jammal, M., Singh, T., Shami, A., Asal, R., Li, Y.: Software defined networking: state of the art and research challenges. Comput. Netw. **72**, 74–98 (2014)
20. Martini, B., Paganelli, F., Mohammed, A.A., Gharbaoui, M., Sgambelluri, A., Castoldi, P.: SDN controller for context-aware data delivery in dynamic service chaining. In: Proceedings of the 2015 1st IEEE Conference on Network Softwarization (NetSoft), pp. 1–5 (2015)
21. Guan, X., Choi, B.-Y., Song, S.: Reliability and scalability issues in software defined network frameworks. In: 2013 Second GENI Research and Educational Experiment Workshop, pp. 102–103. IEEE (2013)

22. Comer, Douglas, Karandikar, R.H., Rastegarnia, A.: Resilient packet delivery through software defined redundancy: an experimental study. In: 2017 IEEE 7th Annual Computing and Communication Workshop and Conference (CCWC), pp. 1–6. IEEE (2017)
23. Mendiratta, V.B., Jagadeesan, L.J., Hanmer, R., Rahman, M.R.: How reliable is my software-defined network? Models and failure impacts. In: 2018 IEEE International Symposium on Software Reliability Engineering Workshops (ISSREW), pp. 83–88 (2018)
24. Yu, J., Wang, X., Song, J., Zheng, Y., Song, H.: Forwarding programming in protocol-oblivious instruction set. In: 2014 IEEE 22nd International Conference on Network Protocols, pp. 577–582 (2014)
25. Gholami, M., Akbari, B.: Congestion control in software defined data center networks through flow rerouting. In: 2015 23rd Iranian Conference on Electrical Engineering, pp. 654–657. IEEE (2015)
26. Lin, S.-C., Wang, Pu., Luo, M.: Control traffic balancing in software defined networks. Comput. Netw. **106**, 260–271 (2016)
27. Gonzalez, A.J., Nencioni, G., Helvik, B.E., Kamisinski, A.: A fault-tolerant and consistent SDN controller. In: 2016 IEEE Global Communications Conference (GLOBECOM), pp. 1–6. IEEE (2016)
28. Xie, K., Huang, X., Zhang, P., Hao, S.: Elastic Multi-Controller SDN in Data Center Networks
29. Wang, M.-H., Chen, L.-W., Chi, P.-W., Lei, C.-L.: SDUDP: a reliable UDP-based transmission protocol over SDN. IEEE Access **5**, 5904–5916 (2017)
30. Fernández-Fernández, A., Cervelló-Pastor, C., Ochoa-Aday, L.: Energy efficiency and network performance: a reality check in SDN-based 5G systems. Energies **10**(12), 2132 (2017)
31. Priyadarsini, M., Bera, P., Rahman, M.A., A new approach for energy efficiency in software defined network. In: 2018 Fifth International Conference on Software Defined Systems (SDS), pp. 67–73. IEEE (2018)
32. Tuysuz, M.F., Ankarali, Z.K., Gözüpek, D.: A survey on energy efficiency in software defined networks. Comput. Netw. **113**, 188–204 (2017)
33. Li, T., Sun, Y., Sobe, M., Strufe, T., Santini, S.: Energy-efficient SDN control and visualization. In: NOMS 2018–2018 IEEE/IFIP Network Operations and Management Symposium, pp. 1–2 (2018)
34. King, D., Rotsos, C., Aguado, A., Georgalas, N., Lopez, V.: The software defined transport network: fundamentals, findings and futures. In: 2016 18th IEEE International Conference on Transparent Optical Networks (ICTON), pp. 1–4 (2016)
35. Rothenberg, C.E., Nascimento, M.R., Salvador, M.R., Corrêa, C.N.A., de Lucena, S.C., Raszuk, R.:"Revisiting routing control platforms with the eyes and muscles of software-defined networking. In: Proceedings of the First Workshop on Hot Topics in Software Defined Networks, pp. 13–18. ACM (2012)
36. Kriaa, S., Papillon, S., Jagadeesan, L., Mendiratta, V.: Better safe than sorry: modeling reliability and security in replicated SDN controllers. In: 2020 16th International Conference on the Design of Reliable Communication Networks DRCN 2020, pp. 1–6. IEEE (2020)
37. Javadpour, A.: Providing a way to create balance between reliability and delays in SDN networks by using the appropriate placement of controllers. Wireless Pers. Commun. **110**(2), 1057–1071 (2020)

Phishing Websites, Detection and Analysis: A Survey

Leena I. Sakri, Pushpalatha S. Nikkam, Madhuri Kulkarni, Priyanka Kamath, Shreedevi Subrahmanya Bhat, and Swati Kamat

Abstract Phishing is the despicable utilization of electronic interchanges to trick clients. Phishing assaults resolve to increase delicate data like usernames, passwords, MasterCard information, network qualifications, and the sky is the limit from there. Phishing assaults endeavor to increase touchy, secret data, for example, usernames, passwords, charge card data, network qualifications, and then some. Phishing Websites copy the first sites so clients believe that they are utilizing the first sites. On account of phishing assaults, each individuals and associations are at threat. Phishing assaults might be forestalled by identifying the sites and serving to clients to detect the phishing sites. To distinguish the phishing sites, there have been various strategies applied. Diverse machine learning methods, information mining procedures, neural organization and different calculations have been utilized for anticipating or ordering or distinguishing the phishing sites. This paper aims at surveying on recently proposed phishing detection techniques.

Keywords Phishing · Machine learning · Data mining · Novel neural network · Extreme learning machine · AI meta-learners and extra-trees algorithm · Use case-based reasoning · Forest by penalizing attribute algorithm

1 Introduction

There are over 1.7 billion websites exist, till date. A site is an assortment of website pages, pictures, recordings of related information which gives data. There are various sorts of sites like an individual site, a corporate site for an organization, an administration site, an association site, and so on. Corporate, government and association sites request that the clients enter their own data for adaptation prior to utilizing their sites. As online technology is evolving, there are online advertisements, online shopping, online courses, online, etc. Many of which ask the user for personal

L. I. Sakri · P. S. Nikkam · M. Kulkarni · P. Kamath · S. S. Bhat (✉) · S. Kamat
Department of Information Science and Engineering, SDM College of Engineering and Technology, Dharwad, India

H. K. D. Sarma et al. (eds.), *Contemporary Issues in Communication, Cloud and Big Data Analytics*, Lecture Notes in Networks and Systems 281,
https://doi.org/10.1007/978-981-16-4244-9_2

information. As online financial activities are raising, online fraudulent is also increasing in which phishing is a major concern.

Phishing site is a site which makes the copy of the genuine site with not many changes so client cannot distinguish, in view of which the client falls into the snare of the phishing assailants and part with their own data, for example, financial balance number, online media account subtleties, username, secret key and so forth There are almost 1.4 million phishing sites made each month, as indicated by the Webroot Quarterly Threat Trends Report (covered by Dark Reading). Over 60,000 phishing sites revealed in March 2020 alone.

As the quantity of phishing sites is expanding step by step, there is a need for identifying these sites and spare the clients from these phishing assaults. Phishing is additionally a well-known technique for digital assailants to flexibly malware, through urging victims to download a record or to a hyperlink so it will covertly introduce the malevolent payload in assaults that may be upsetting Trojan malware ransomware of adverse and problematic assaults. The internet business, online installment administrations, and web-based media are the most endured by this assault. The assailants create comparable sort of site page and convey them through messages and different strategies and attempt to contact individuals to pick up their data actually and monetarily.

Creating measures against phishing is a difficult issue since victims give them their information unwittingly and, in this manner, help the assailants. Furthermore, clients normally have an absence of security issues. The wonder of phishing might be obscure. Clients do not consider secrecy to be security as their essential assignment. Numerous methodologies have been proposed by various creators for identification of phishing sites. Our paper indicates the methodologies which has more exactness in forecast and recognition.

2 Methodology

2.1 Novel Neural Network

The application of a novel neural network in the detection of phishing websites" paper published by Feng et al. [1]. The proposed system divided the dataset into training and testing dataset. The training dataset is analyzed for the URL and exact features for which the novel neural network will be applied using design risk minimization and Mante Carlo. From the testing dataset, the exact features will be collected for which the phishing data model will be built. When the user enters the URL, the URL will be tested and it will be predicted as benign or phishing by the system.

Different attributes such as IP Address, length of the URL, presence of @, double slash redirecting "-" symbol in the domain name, HTTPS token, etc. are taken as attributes for which 1, − 1 values are specified. The paper utilizes concealed layer and yield layer for the cycle. A shrouded layer in a manufactured neural organization might be a layer in the middle of information layers and yield layers, any place counterfeit neurons ingest an assortment of weighted data sources partner degreed turn out partner degree yield through an actuation perform. The yield layer in an engineered neural network is the end layer of neurons that produces given yields for the program. Despite the fact that they are made simply like distinctive manufactured neurons with inside the neural network, yield layer neurons can be developed or found in an exceptional manner, for the explanation that they are the end "entertainer" hubs at the network. The calculation will settle nonlinear issues by the functions of activation. This paper has an exactness of 97.71 %.

2.2 *Extreme Learning Machine*

"Effective Classification of Phishing sites supported New Rules by Using Extreme Learning Machines" paper published by Kaytan et al. [2].

There are six steps that are administered during this proposed system.

Step 1 A website is visited.
Step 2 Consistent with the features and their rules check the 30 input attributes.
Step 3 Samples are collected from dataset.
Step 4 Dataset is split as training and testing dataset, i.e. 90% training dataset and 10% testing dataset.
Step 5 Extreme learning machine categorization.
Step 6 Legitimate or phishing website prediction.

Thirty attributes are considered in the input dataset. By considering these 30 attributes, the values and class of the output is calculated. Input dataset can take 0, −1 and 1 as its values. 2 or 3 different values are taken as attributes for the created input. Results of the output dataset obtained during this period may take two different values. An online site is taken under consideration as valid, dubious or spoofing within the obtained rules for input attributes of the dataset. A categorization is finalized and the output is within the type of spoofing or genuine website which is within the dataset. Through the analysis, it is seen that values such as 1, 0, −1 has been utilized for genuine, suspicions and spoofing websites, respectively. From the above work, 95.05% was the classification accuracy, and hence, the absolute best classification exactness was estimated as 95.93%.

2.3 *AI Meta-Learners and Extra-Trees Algorithm*

"AI Meta-Learners and additional-Trees formula for the Detection of Phishing Websites" revealed by Al-Sariera et al. [3] in 2017.

The paper has extended four meta-learners models: AdaBoost-Extra Tree (ABET), Bagging—Extra tree (BET), Rotation Forest—Extra Tree (RoFBET) and

LogitBoost-Extra Tree (LBET) then created using the extra-tree base classifier. These proposed AI-put together meta-learners was fitted with respect to the phishing site datasets and their exhibitions were thus assessed. The general strategy of the examination is part into four modules:

- The initial module involves the preparation of datasets for experimental purposes.
- In the second module, the AI algorithms calculation were introduced with the proper boundaries so as to build up the proposed and other meta learners phishing detection models.
- In the third module, there is a mix of meta-learners and base learners to frame the resultant AI models on the datasets which was passed for the evaluation. With the assistance of the N-overlay cross-approval technique, the AI-based meta-student occasion models was performed. During this exploration work, the N esteem was set to 10. The datasets were isolated into ten equivalent gatherings and afterward prepared on nine of the divided information while testing on the leftover one. This cycle was repeated ten times, and hence, the test information were fluctuated in like manner until all pieces of the information are disjoint and was utilized for instructing Associate in Nursing testing of the model.
- The fourth module involves the evaluation of the developed model. The outcome of the models' were examined using ROC, Accuracy, False Positive (FP) and *F*-measure as these metrics are widely used for examination of AI-based classification models. The proposed models during this exploration work exhibited the strength of AI meta-learners as an insightful algorithm for developing models usable in detecting phishing websites.

BET's model had an accuracy of 97.404%, RoFET's model had an accuracy of 97.449%, ABET's model had an accuracy of 97.485% and LBET's model had an accuracy of 97.576%. All the created models of this examination work accomplished precision above 96% and approximately equal to 98%, and there is a generation of false positive rate which is less than and a false negative rate less than or equal to 0.033.

2.4 Forest by Penalizing Attribute Algorithm

The paper "Phishing Website Detection: Forest by Penalizing Attributes Algorithm and Its Enhanced Variations" published by Alseriera et al. [4]. There is an introduction of 3 Meta-learner models dependent on Forest Penalizing Attribute (Forest PA) algorithm which is used for Phishing Website Detection Model (PWDMs). As there is reduction in the bias and variance in the process of classifying, for the task of classification, the Meta-learner application has been shown to be efficient. This paper uses dataset consisting of 31 attributes, of which only one is class variable with 11,055 instances. Three phishing website detection models utilizing datasets are proposed and executed in this paper. Three proposed phishing discovery models were created by executing these calculations. Forest PA, AdaBoost and Bagging Algorithms.

By implementing these algorithms, the result was we could recognize phishing and genuine websites with the precision of 96.26%. This shows that Forest PA calculation adequately recognizes either site type with high precision.

2.5 Use Case Based Reasoning

The paper "Use Case Based Reasoning for Phishing Detection" published by Hassan Y. A. Abutairand Abdelfettah Belghith in 2017. By this paper, they have launched a Case-Based Reasoning (CBR) Phishing Detection System (CBR-PDS). It is reliant on CBR strategy as a basic or fundamental content. This posed research is hugely compatible as it can easily adapt to recognize current malicious websites with very small datasets, in comparison to different classifiers which need to be steadily educated beforehand.

Case-Base Reasoning Methodology.

CBR Methodology bear a resemblance to personage logic. The principle highlight of CBR is the capacity to examine human intuition for taking care of issues from past encounters or cases as comparable issues having comparative solutions. Case-Based Reasoning consists of four phases.

1. RETRIEVE: It recuperates the most related cases.
2. REUSE: It reuses the findings and data for these circumstances to handle the issues.
3. REVISE: It changes the posed aim.
4. RETAIN: It holds the response for handling similar case later on.

CBR-PDS is inspected in opposition to simple URL capabilities and intra-URL. CBR device starts off evolved the reasoning and retrieves the most related type of cases. Based on those related instances, a significant answer is determined. If the determined answer is malicious internet site, then the present internet site is phishing; otherwise, it is a miles legitimate internet site. In CBR-PDS, not like the traditional system mastering classifiers, the machine can adapt and boldly expect any case with high accuracy. Their mile adaptable to unmistakable records sets sizes. The features present in the website are extracted, clear examination is done, then formulated as instances to be used within the prediction system. CBR-PDS indicates an extravagant accuracy. The rage of accuracy is from 95.62 to 98.03%.

Different methods specified in our paper are presented in the form of table.

S. No.	Methodology	Year of publication	Accuracy
1	Novel neural network	2018	97.71%
2	Extreme learning machine	2017	95.05%
3	AI meta-learners and extra tree algorithm	2017	Above 96% and approx. 98%

(continued)

(continued)

S. No.	Methodology	Year of publication	Accuracy
4	Forest by penalizing attribute algorithm	2020	96.26%
5	Use case-based reasoning	2017	95.62–98.03%

3 Conclusion

For the past few years, phishing has been a serious concern which has got to be tackled. For detection of these phishing websites, different methods are proposed. In our paper, we have selected five proposed papers which have found the ways to detect the phishing websites. We will conclude that the "Use Case Based Reasoning" methodology though has an accuracy ranching from 95.62 to 98.03%, it's not precise. Artificial intelligence Meta-Learners and Extra-Trees Algorithm rather than dividing the dataset into training and testing, it used grouping method before training and testing the dataset. "Novel Neural Network" methodology has the very best and precise accuracy compared to other methodologies; hence, it's the foremost promising method.

References

1. Feng, F., Zhou, Q., Shen, Z., Yang, X., Han, L., Wang, J.Q.: The Application of a Novel Neural Network in the Detection of Phishing Websites (2018)
2. Kaytan, M., Hanbay, D.: Effective Classification of Phishing Web Pages Based on New Rules by Using Extreme Learning Machines 2017.
3. Al-Sariera, Y.A., Adeyemo, V.E., Balogunand, A.O., Alazzawi, A.K.: AI Meta-Learners and Extra-Trees Algorithm for the Detection of Phishing Websites, 2017
4. Alseriera, Y.A., Elijah, A.V., Balogun, A.O.: Phishing Website Detection: Forest by Penalizing Attributes Algorithm and Its Enhanced Variations, 2020
5. Abutair, H.Y.A., Belghith, A.: Use Case Based Reasoning for Phishing Detection, 2017
6. Priya, A., Meenakshi, E.: Detection of Phishing Websites Using C4.5 Data Mining Algorithm, 2017
7. Kumar, M.S., Indrani, B.: Brain Storm Optimization based Association Rule Mining Model for Intelligent Phishing URLs Websites Detection, 2020
8. Riaty, S., Sharieh, A., Bdour, H.A., Jabri, R.: Enhance Detecting Phishing Websites Based on Machine Learning Techniques of Fuzzy Logic with Associative Rules, 2017
9. Babagoli, M., Aghababa, M.P., Solouk, V.: Heuristic Nonlinear Regression Strategy for Detecting Phishing Websites, 2018
10. Adewole, K.S., Akintola1, A.G., Salihu, S.A., Faruk, N., Jimoh, R.G.: Hybrid Rule-Based Model for Phishing URLs Detection, 2019

11. Ubing, A.A., Jasmi, S.K.B., Abdullah, A., Jhanjhi, N.Z., Supramaniam, M.: Phishing Website Detection: An Improved Accuracy through Feature Selection and Ensemble Learning, 2019
12. Sarhan, A.A., Jabri, R., Sharieh, A.: Website Phishing Detection Using Dom-Tree Structure and Cant-MinerPB Algorithm, 2017
13. Patil, V., Thakkar, P., Shah, C., Bhat, T., Godse, S.P.: Detection and Prevention of Phishing Websites using Machine Learning Approach
14. Robic–Butez, P., Win, T.Y.: Detection of Phishing Websites Using Generative Adversarial Network, 2019
15. Parekh, S., Parikh, D.: A New Method for Detection of Phishing Websites: URL Detection, 2018
16. Alnajim, A., Munro, M.: An Anti-Phishing Approach that Uses Training Intervention for Phishing Websites Detection, 2019
17. Dhanalakshmi, R., Prabhu, C., Chellapan, C.: Detection of Phishing Websites and Secure Transactions, 2018
18. Buber, E., Demir, Ö., Sahingoz, O.K.: Feature Selections for the Machine Learning Based Detection of Phishing Websites, 2018
19. Suleman, M.T., Awan, S.M.: Optimization of URL-Based Phishing Websites Detection Through Genetic Algorithms, 2019
20. Somesha, M., Pais, A.R., Rao, R.S., Rathour, V.S.: Efficient Deep Learning Techniques for the Detection of Phishing Websites, 2020
21. James, D.: An Innovative Framework for the Detection and Prediction of Phishing Websites, 2018
22. Kahksha, S.N.: Detection of Phishing Websites using Machine Learning Approach, 2019
23. James, J., Sandhya, L., Thomas, C.: Phishing Website Detection based on Supervised Machine Learning with Wrapper Features Selection, 2019
24. Tan, C.L., Chiew, K.L., Wong, K., Sze, A.: Enhanced Blacklist Method to Detect Phishing Websites, 2016
25. Chiew, K.L., Tan, C.L., Wong, K., Yong, K.S.C., Tiong, W.K.: Particle Swarm Optimization-Based Feature Weighting for Improving Intelligent Phishing Website Detection, 2019
26. Zhang, W., Jiang, Q., Chen, L., Li, C.: A Machine Learning Based Approach for Phishing Detection Using Hyperlinks Information, 2017
27. Lakshmi, V.S., Vijaya, M.S.: Detection of Phishing Websites Based on Probabilistic Neural Networks and K-Medoids Clustering, 2017
28. Zhuang, W., Ye, Y., Li, T., Jiang, Q.: An Intelligent Anti-phishing Strategy Model for Phishing Website Detection, 2017
29. Aburrous, M., Hossain, M.A., Dahal, K., Thabtah, F.: A New Method for Detection of Phishing Websites URL Detection, 2016
30. Chang, E.H., Chiew, K.L., Sze, S.N., Tiong, W.K.: Phishing Detection via Identification of Website Identity, 2018
31. Sharifi, M., Siadati: Phishing Websites Detection through Supervised Learning Networks, 2018
32. Konradt, C., Schilling, A., Werners, B.: Phishing: an economic analysis of cybercrime perpetrators. Comput. Secur. **58**, 39–46 (2016)
33. Jabri, R., Ibrahim, B.: Phishing Websites Detection Using Data Mining Classification. Trans. Mach. Learn. Artif. Intell. **3**(4) (2015)
34. Ali, W., Malebary, S.: Particle Swarm Optimization-Based Feature Weighting for Improving Intelligent Phishing Website Detection, 2020
35. Singh, C., Meenu, S.: Phishing Website Detection Based on Machine Learning, 2020
36. Kelkar, R.A., Vijayalakshmi, A.: ML Based Model for Phishing Website Detection, 2020
37. Jain, A.K., Gupta, B.B.: Phishing Detection: Analysis of Visual Similarity Based Approaches, 2017

38. Abbasi, A., Zhang, Z., Zimbra, D., Chen, H., Nunamaker, J.F.: Detecting fake websites: the contribution of statistical learning theory. Mis Q. 435–461 (2010)
39. Das, R., Hossain, M.M., Islam, S., Siddiki, A.: Learning a Deep Neural Network for Predicting Phishing Website, 2019
40. Sampat, H., Saharkar, M., Pandey, A., Lopes, H.: Detection of Phishing Website Using Machine Learning, 2018

Analysis of Security Attacks in SDN Network: A Comprehensive Survey

Ali Nadim Alhaj and Nitul Dutta

Abstract SDN is a modern model that has attracted many researchers in the networks since its inception. Since this model was mainly based on separating the control level and the data level which set within the same equipment in the old networks. With the suggestion of the feature of separating control from data transmission and the possibility of programming and adding features as needed in a flexible manner, the SDN architecture has a promising future in the world of networks. Despite the many advantages offered by SDN networks, we cannot ignore the security problem associated with SDN since its inception, which has attracted many researchers to work hard to fill these security gaps and reach a secure infrastructure for SDN networks. So, this research will be divided into parts that include the basic properties of the SDN network and its general structure, and the most important problems and attacks that may affect the structure of network and a review of previous research in this field. The best attempts made to find appropriate security solutions for SDN networks will be discussed. Also, the research will mention the most important attempts to design an integrated security design for SDN, and the future needs to reach a completely secure SDN network.

Keywords Network security · Communication security · SDN security · SDN attacks · Vulnerabilities detection

1 Introduction

Network research is extensively progressing in many directions like Internet of things [1], cognitive radio networks [2, 3], dynamic networking [4], and information centric networks [5]. Moreover, currently the (Software-Defined-Networking) is getting attention of ample of researchers as a concept of decoupling routing responsibilities. SDN is an abbreviation for (Software Defined Networking) which is the technology where the (data and control) plan is separated in network devices, so that the role

A. N. Alhaj (✉) · N. Dutta
Department of Computer Engineering, Marwadi University, Rajkot, Gujarat, India
e-mail: ali.alhaj108620@marwadiuniversity.ac.in

H. K. D. Sarma et al. (eds.), *Contemporary Issues in Communication, Cloud and Big Data Analytics*, Lecture Notes in Networks and Systems 281,
https://doi.org/10.1007/978-981-16-4244-9_3

of these devices is limited to passing data, while management and controlling will be in new layer. In traditional networks, data passing and routing are both contained on one device and this made it seem complicated [6]. In SDN technology, all the requests for data routing are fulfilled in a special software (the controller), while the devices simply apply the decisions that the controller sends to them on the data packets, for that, there is no need for the network devices to have awareness of the routing logic, but only to store (caching of controller's decisions about routing).

The main goal of developing SDN networks is to solve the multiple problems associated with traditional networks of stability and complexity in the structure, as the current networks require greater flexibility to solve problems and protect from errors. SDN attempts to make network intelligence central to a single network component by separating network packet forwarding (data level) from routing (control level) operation. The control level in SDN consists of one or more controlling devices, and the controller is the mind of the network and contributes to making decisions. However, central structure has its problems like security problems, flexibility and scalability, so we can say, that is, one of the main issues which found in SDN.

SDN networks have been closely associated with the OpenFlow protocol since its inception, as it has been relied upon in the communication process between the control layer and the data layer since its appearance in 2011 [7]. However, for many companies have ceased to be an exclusive protocol, and other protocols and technologies have been added. With the increased control of the network in a centralized way comes a greater responsibility of the SDN controller to monitor and prevent any attacks on the network. In this research, we will talk about security problem in SDN network, with more details about the attacks and how we can defeat these attacks. Many researches have dealt with SDN networks and their basic structure since their inception and the strengths and weaknesses of these networks such as [6–13].

Other research focused on the security problem in SDN networks, in the absence of an explicit security layer for SDN network, where the different attacks that could be exposed to the network were classified in [14–25]. These attacks are covered in more detail in various researches such as [10, 21, 26–29]. Many researchers tried to work on finding solutions to various security attacks by proposing various security models as in the following research [15, 30–37]. A number of valuable researches was sought in the field of networks, such as [1–3].

This research paper has been divided into sections: In the second section, the definition of the SDN networks, the three layers of the network and the interlayers are mentioned. In the third chapter, the security attacks of the SDN networks are classified into five types, and these types are explained. The fifth and the final chapter are the conclusion.

2 SDN Architecture

SDN architectures generally contain three layers (groups of functions) shown in Fig. 1.

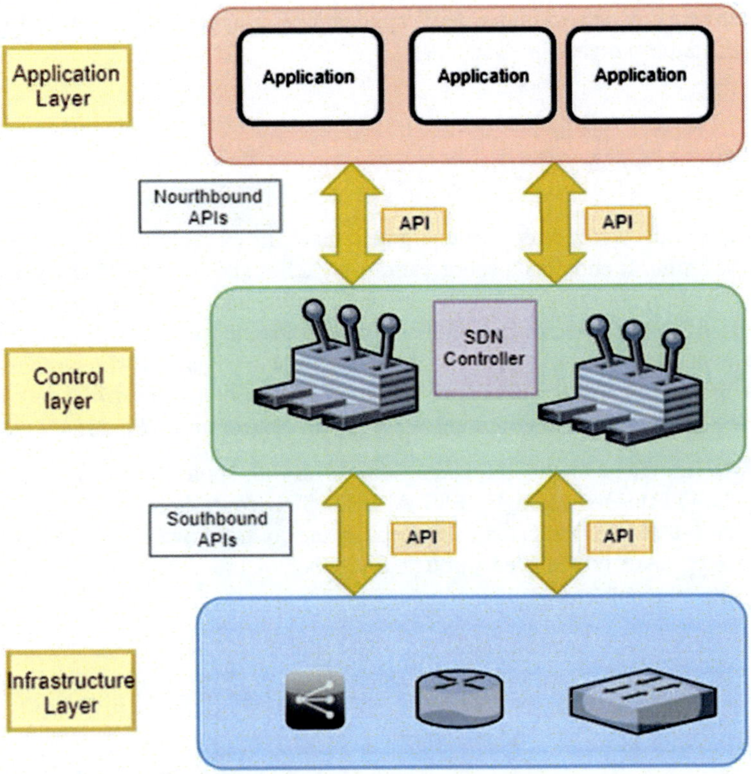

Fig. 1 SDN architecture

2.1 Applications Layer

This layer contains many applications that control the entire network with the help of the controllers and that communicate with it via APIs [11]. This layer can create a complete perception of the network through the data collected from the control layer. It can contain a number of applications specialized in collecting statistical data or assisting in managing the network and creating statistical assessments or some other applications for security purposes [8].

2.2 SDN Controller Layer

SDN controller receives the commands and rules from the application layer and transfers these rules to the lower layer. The network controller collects statistical information about the network from the various devices in the data layer, and it also delivers this data to higher level applications as appropriate. Statistics include network status and various events [9].

2.3 SDN Devices (Infrastructure)

It consists of the various network equipment that forms a backbone network to direct network traffic. It contains a set of various switches and routers. This layer will be the physical layer through which the network virtualization will be placed [8, 9].

APIs in SDN architecture are indicated by north/south interfaces, defining connection way between controllers, applications and network systems. The northern interface helps communication between control units and applications, while the southern interface helps devices to communicate with the control units (Fig. 2).

- **North interface**: intended for communication with applications layer and will be generally achieved through REST APIs for SDN controllers [12, 13].
- **South interface**: Dedicated to connecting the control layer to the data layer, this interface layer is achieved through a number of protocols such as OpenFlow, Ovsdb, Netconf, etc. [8, 13].

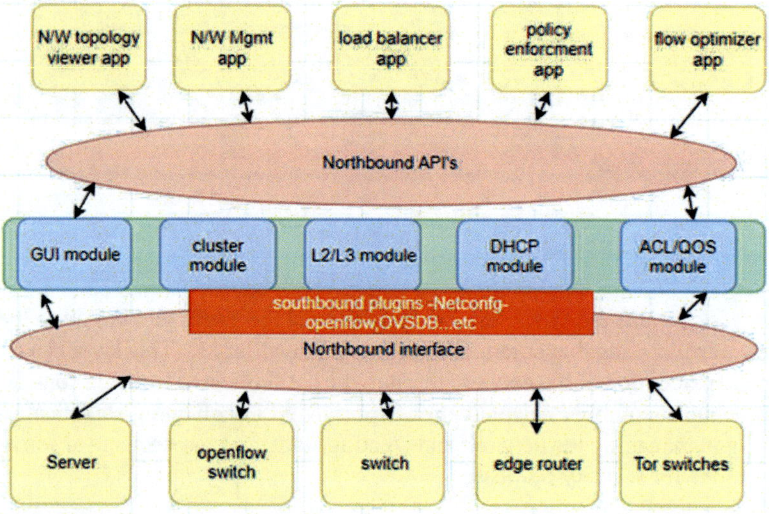

Fig. 2 South and North bound interface

3 Classification of Attacks in SDN Architecture

The security attacks in SDN are classified according to the basic aim of this attack, as the aim of the attack may be unauthorized access to information to access the network for the purpose of monitoring, eavesdropping or stealing information [19].

Elevation of privileges

In these types of attacks, the attacker tries to access the network and log in illegally [18], as a result of security flaws, poor network settings, or the use of prediction and brute force attacks [16, 17].

Disclosure of information

Attackers can target the forwarding behavior of the network by calculating the delay times within the flow table as in [13, 20]. The attackers can eavesdrop on the data passing in a certain path and try to access data stores or impersonate another device in the network to access personal information for a specific user [17, 21].

Tampering

Attackers can exploit the API or use malicious applications to modify the rules in the flow table, which leads to some conflicts in the network [26]. Vulnerabilities and unexamined packet appropriately can be exploited to modify data in the network and access important databases [16, 17].

Denial of service (DoS)

Among the most important causes of service outages are attacks on network resources, for example, the network is overflowing with a large number of packets or attacks that target the rules in the flow table [23]. The attack may target the controller as a central point of failure or it may target any other device in the network [18, 21, 22, 24, 25].

Network data destruction

Various attacks include destroying rules in the flow table and other attacks that include destroying specific application flows through malicious applications in the network, and some attacks targeting service chains, authentication data, and control sessions. Special attacks target controller and some controller mechanism [14, 15, 17, 20].

4 Overview of SDN Attacks

The evolution of networks creates new types of attacks, and specific and unknown risks, which can be exploited at any time. At present, it is difficult for us to identify the vulnerabilities of the SDN networks, as there is no prior information about the logs of attacks on this network. But, a classification of potential attacks can be made

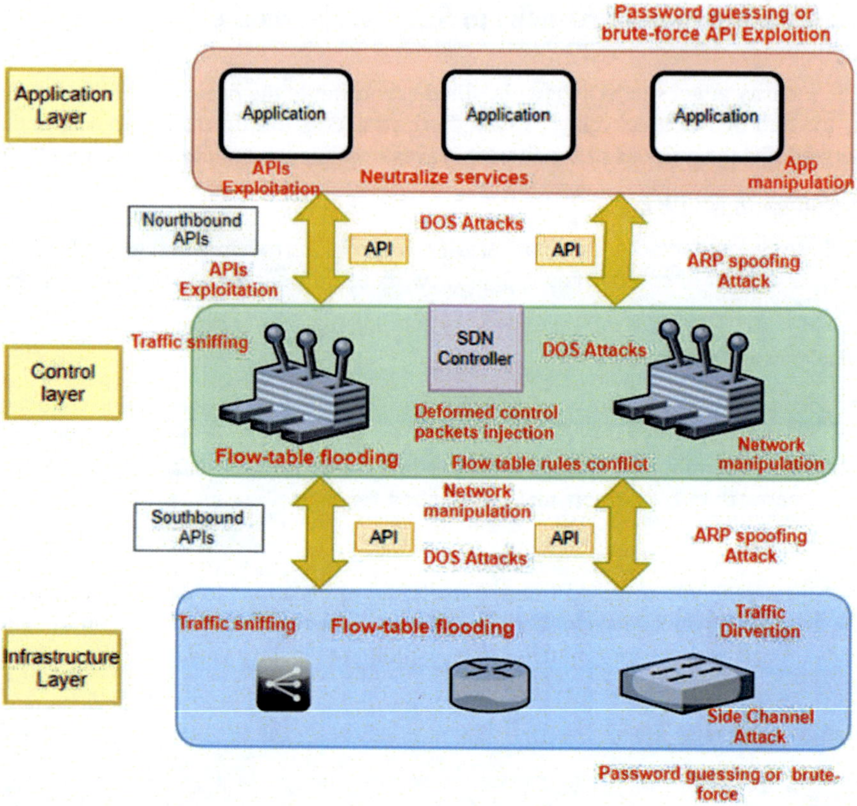

Fig. 3 Attacks in SDN architecture

to be used as a reference and to lay the foundations of security. Figure 3 shows an SDN architecture with potential attacks (in red).

- **App manipulation**: This attack targets the application-level. Exploiting an application vulnerability could lead to malfunctions, interruptions in service, or data eavesdropping. An attacker could access to high privileges to implement an SDN attack and perform illegal operations [10, 21, 27].
- **Neutralize services**: Malicious applications can be used in the application layer to target control packets, where control packets can be prevented from accessing the application intended for processing them. The order of these control packets can also be changed, which affects the order of processing, and control packets can be redirected incorrectly, in addition to the possibility of obtaining sensitive information about the network from this control packets [10, 21].
- **API exploit**: The software component APIs may contain vulnerabilities. This allows the attacker to gain illegal access to the data [27]. An attack can be executed

to expose the information exchange between application layer and control layer and target data for a specific application to stop it from working [10, 21].

- **Network Manipulation**: One of the dangerous attacks in which the attacker targets the entire network and the controllers in particular to intercept data passing through the network and cause-specific damages. [27].
- **Flow table rules conflict**: Attackers may target flow table rules through malicious applications that could create or add new rules to flow table that conflict with existing rules. This could lead to dropping or blocking some data packets in the network and a conflict in network policies, as well [26, 29].
- **Flow-table flooding**: This type of attack is common in SDN networks, the attack is done by sending a large number of packets randomly, but the flow table is limited so the flow table will overflow, and it will affect the controller performance due to the large number of packets that need to be addressed [21, 28].
- **Traffic diversion**: This attack focuses on network devices in the data layer that is forwarding the data, where the attackers can eavesdrop and collect statistical data for the network [27].
- **Side channel attack**: This type of attack targets components at the data level. The most important information that can be useful to the attacker is the time information such as the time required for a data traffic to pass through the network [21].
- **DoS**: It's a common attack which affects all parts of the SDN. Through DoS, an attacker can reduce or completely disable SDN services [10, 21, 27].
- **Deformed control packets injection**: Suspicious control packages can be designed targeting switches in order to induce them to behave strange behavior and to expose weak points in the network and try to exploit these points to carry out various other attacks [21].
- **ARP Spoofing Attack**: It is a common attack under the name MITM, and it can also be called ARP cache poisoning, as the attacker eavesdrops and monitors data movement on the network. The attacker could also try to modify or break traffic. [21, 27, 28].
- **Traffic sniffing**: A hacker or sniffer can use this method to access important data in the network and analyze this data to obtain specific gains, and the attack can be focused mainly at the joints of the network and important elements. Eavesdropping can happen wherever the constant traffic is. The data being intercepted may include information about the network's flow rules [21, 27, 28].
- **Password guessing or brute force**: It is an old attack that could target non-SDN device to gain permission to access the network by predicting a user password or brute force [27] (Table 1).

5 Security Solution for SDN Architecture

SDN appeared with new features that led to the emergence of a large number of security attacks and at the same time helped with a number of security solutions

Table 1 Attacks in SDN architecture

Attack	Layer	Source
App manipulation	Application	[10, 21, 27]
Neutralize services	Application	[10, 21]
API exploit	Application, Northbound interface	[10, 21, 27]
Network Manipulation	Control	[27]
Flow table rules conflict	Control	[26, 29]
Flow-table flooding	Control, data layer	[21, 28]
Traffic diversion	Data layer	[27]
Side channel attack	Data layer	[21]
DoS	Application, control, data, South and North APIs	[10, 21, 27]
Deformed control packets injection	Control, data	[21]
ARP spoofing attack	Control, data	[21, 27, 28]
Traffic sniffing	Application, control, data, South and North APIs	[21, 27, 28]
Password guessing or brute force	Data	[27]

to make the network more secure. The most important of these features: 1—The possibility of central control of the network 2—Programmability 3- Dynamic control of the forwarding process. Much of the security research of SDN relied on the three previous features.

The previous security solutions that are available with traditional networks as in middle-box were initially thought of [15]. In [30], the researchers suggest a central security control unit called slick responsible for verifying data and directing it to specialized security middle-box.

In [31], the researchers proposed a security architecture—Flow Tags—based on middle-box that contains routing information for the network and interacts with the controller through a programmatic interface API which checks the package header and compares it with the specific security policies.

The problem of the two previous solutions is that they ignore the dynamic flow feature, which is a new feature of SDN, and they also propose an amendment to the SDN structure.

In [32], the researchers proposed a logical solution to manage the middle-boxes and to place them in a way that preserves the structure and SDN functions.

This solution lies in programming the network to complete the process of directing data to the middle-boxes. These solutions are not simple and require full knowledge of the expected attacks and routing mechanisms in the network.

Some other researches as in [33, 34] suggested that the switches should be involved in the security operations of checking inspecting and directing by adding security rules to the flow table to allow or prevent the packets from passing through the network.

By taking advantage of the centralization of control and forwarding in SDN and programmability, the middle-boxes were replaced as a security solution with other solutions based on artificial intelligence and machine-learning techniques. In [38], a training system was proposed based on collecting training data from specific points in the network and various protocols and using this data for training to produce a system or application capable of distinguishing normal and deformed packets. Training results have been added as an application that works with the controller.

In [35], work was done to provide a smart IDPS solution for SDN networks by integrating machine-learning mechanisms to help in checking packets. The model was trained based on the data of various attacks regarding the types and characteristics of the attack, where it is checked whether the packets are intact or suspicious. The IDPS model was confirmed and gave acceptable results.

Athena in [36] is an integrated security framework that researchers have developed to help security administrators in networks to deploy security solutions for discover threat and attacks in simple and easy ways. Athena framework is integrated with the controllers deployed in SDN networks.

Athena framework consists of five basic blocks, beginning with the data and statistics collection block, then the second block includes machine-learning algorithms that help to detect security risks, third block is procedures for detecting risks and attacks, fourth block for security managers and the fifth block for the user to view security reports.

Delta in [37] is a framework that has been developed to standardize the vulnerability detection process in SDN. This framework is based on the use of modules and protocol fuzzing techniques to automatically derive new risks and attacks by testing attacks within the SDN environment with different layers and different environments.

6 Conclusion

The new features of SDN networks such as centralized control and programmability have contributed greatly to solving a lot of security problems and some traditional network problems, but these features had a negative impact as it cause to the emergence of a number of new security risks and attacks. If it is possible to know the basic classifications of the attacks, and we tried to find security solutions for these attacks, we can reach a safe network of SDN, and this requires the integration of all efforts and security solutions in this field.

In this research paper, we dealt with the structure of the SDN network and its constituent layers, then the security attacks that could affect the SDN network were classified. After that the security attacks that target the network's structure and the distribution of these attacks on the SDN layers were addressed. In the end, a number of security solutions were mentioned and an emphasis on the possibility of collecting various technologies to support these solutions.

References

1. Sathwara, S., Dutta, N., Pricop, E.: IoT forensic a digital investigation framework for IoT systems. In: 10th IEEE International Conference on Electronics, Computers and Artificial Intelligence (ECAI), pp. 1–5, Romania, 2018
2. Dutta, N., Sarma, H.K.D., Polkowski, Z.: Cluster based routing in cognitive radio adhoc networks: reconnoitering SINR and ETT impact on clustering. Com. Com., (Elsevier), pp. 10–20, vol. 115, 2018
3. Dutta, N., Sarma, H.K.D.: A probability based stable routing for cognitive radio adhoc networks. Wire. Net. **23**(1), 65–78 (2017)
4. Dutta, N., Sarma, H.K.D.: A scheme for dynamic MAP selection in HMIPv6. National Acad. Sci. J. India Sect. A Phys. Sci. (NASA) **90**, 371–382 (2020)
5. Delvadia, K., Dutta, N., Ghinea, G.: An efficient routing strategy for information centric networks. IEEE ANTS, pp. 1–6, Goa, India
6. Abdou, A., van Oorschot, P.C., Wan, T.: Comparative analysis of control plane security of SDN and conventional networks. IEEE Commun. Surv. Tutor. **20**(4), 3542–3559 (2018). https://doi.org/10.1109/COMST.2018.2839348
7. Agborubere, B., Sanchez-Velazquez, E.: OpenFlow communications and TLS security in software-defined networks. In: 2017 IEEE International Conference on Internet of Things (iThings) and IEEE Green Computing and Communications (GreenCom) and IEEE Cyber, Physical and Social Computing (CPSCom) and IEEE Smart Data (SmartData), Exeter, 2017, pp. 560–566. https://doi.org/10.1109/iThings-GreenCom-CPSCom-SmartData.2017.88
8. - Click2Cloud Inc.: Software Defined Networking. 17 Sept 2019. http://blog.click2cloud.net/?p=1286
9. - KNET Solutions (Online Training Centre): RYU SDN Crash Course. https://learning.knetsolutions.in/docs/ryu/
10. Hogg, S.: SDN Security Attack Vectors and SDN Hardening: Securing SDN Deployments Right from the Start (2014)
11. Kreutz, D., Ramos, F., Verissimo, P.: Towards secure and dependable software-defined networks. In: Proceedings of the Second ACM SIGCOMM Workshop on Hot Topics in Software Defined Networking. ACM, pp. 55–60 (2013)
12. Li, C.-S., Liao, W.: Software defined networks. IEEE Commun. Mag. **51**(2), 113113 (2013)
13. Farhady, H., Lee, H., Nakao, A.: Software-defined networking: a survey. Comput. Netw. **81**, 79–95 (2015). https://doi.org/10.1016/j.comnet.2015.02.014
14. Scott-Hayward, S.: Design and deployment of secure, robust, and resilient SDN controllers. In: Proceedings of the 2015 1st IEEE Conference on Network Softwarization (NetSoft), pp. 1–5, 2015. https://doi.org/10.1109/NETSOFT.2015.7258233
15. Scott-Hayward, S., O'Callaghan, G., Sezer, S.: SDN security: a survey. In: 2013 IEEE SDN for Future Networks and Services (SDN4FNS), pp. 1–7, 2013. https://doi.org/10.1109/SDN4FNS.2013.6702553
16. Scott-Hayward, S., Natarajan, S., Sezer, S.: A survey of security in software defined networks. IEEE Commun. Surv. Tutor. **18**(1), 623–654 (2016)
17. Hizver, J.: Taxonomic modeling of security threats in software defined networking. Proc. BlackHat Conf. **2015**, 1–16 (2015)
18. Lindqvist, U., Jonsson, E.: How to systematically classify computer security intrusions. In: Proceedings 1997 IEEE Symposium on Security and Privacy. IEEE, pp. 154–163, 1997. Cat. No. 97CB36097
19. Jouini, M., Rabai, L.B.A., Aissa, A.B.: Classification of security threats in information systems. Proc. Comput. Sci. **32**, 489–496 (2014)
20. Patwardhan, A., Jayarama, D., Limaye, N., Vidhale, S., Parekh, Z., Harfoush, K.: SDN security: information disclosure and flow table overflow attacks. In: 2019 IEEE Global Communications Conference (GLOBECOM), Waikoloa, HI, USA, 2019, pp. 1–6. https://doi.org/10.1109/GLOBECOM38437.2019.9014048

21. Yoon, C., et al.: Flow wars: Systemizing the attack surface and defenses in software-defined networks. IEEE/ACM Trans. Netw. **25**(6), 3514–3530 (2017). https://doi.org/10.1109/TNET.2017.2748159
22. Dayal, N., Srivastava, S.: Analyzing behavior of DDoS attacks to identify DDoS detection features in SDN. In: 2017 9th International Conference on Communication Systems and Networks (COMSNETS), Bangalore, 2017, pp. 274–281. https://doi.org/10.1109/COMSNETS.2017.7945387
23. Nguyen Tri, H.T., Kim, K.: Assessing the impact of resource attack in software defined network. In: 2015 International Conference on Information Networking (ICOIN), Cambodia, 2015, pp. 420–425. https://doi.org/10.1109/ICOIN.2015.7057934
24. Dover, J.M.: A Denial of Service Attack against the Open Floodlight SDN Controller (2013)
25. Dover, J.M.: A Switch Table Vulnerability in the Open Floodlight SDN Controller (2014)
26. Hu, Z., Wang, M., Yan, X., Yin, Y., Luo, Z.: A comprehensive security architecture for SDN. 2015 18th International Conference on Intelligence in Next Generation Networks, Paris, 2015, pp. 30–37. https://doi.org/10.1109/ICIN.2015.7073803
27. Asturias, D.: 9 Types of Software Defined Network Attacks and How to Protect From Them (2017)
28. Benton, K., Camp, L.J., Small, C.: Openflow vulnerability assessment. In: Proceedings of the Second ACM SIGCOMM Workshop on Hot Topics in Software Defined Networking. ACM, pp. 151–152 (2013)
29. Röpke, C.: SDN malware: problems of current protection systems and potential countermeasures. In: Meier, M., Reinhardt, D., Wendzel, S. (Hrsg.), Sicherheit 2016 - Sicherheit, Schutz und Zuverlässigkeit. Bonn: Gesellschaft für Informatik e.V.. (S. 89–100) (2016)
30. Anwer, B., Benson, T., Feamster, N., Levin, D., Rexford, J.: A slick control plane for network middleboxes. Open Networking Summit, 2013 [Online]. Available http://nextstep-esolutions.com/Clients/ONS2.0/pdf/2013/researchtrack/posterpapers/final/ons2013-final51.pdf
31. Fayazbakhsh, S., Sekar, V., Yu, M., Mogul, J.: FlowTags: enforcing network-wide policies in the presence of dynamic middlebox actions. In: Proceedings of the Second Workshop on Hot Topics in Software Defined Networks. ACM, 2013
32. Qazi, Z.A., Tu, C.-C., Chiang, L., Miao, R., Sekar, V., Yu, M.: SIMPLE-fying middlebox policy enforcement using SDN. ACM SIGCOMM, Aug 2013
33. Matias, J., Garay, J., Toledo, N., Unzilla, J., Jacob, E.: Toward an SDN-enabled nfvarchitecture. IEEE Commun. Mag. **53**(4), 187–193 (2015). https://doi.org/10.1109/MCOM.2015.7081093
34. Battula, L.R.: Network security function virtualization(nsfv) towards cloud computing with nfv over openflow infrastructure: challenges and novel approaches. In: 2014 International Conference on Advances in Computing, Communications and Informatics. ICACCI, pp. 1622–1628, 2014. https://doi.org/10.1109/ICACCI.2014.6968453
35. Le, A., Dinh, P., Le, H., Tran, N.C.: Flexible network-based intrusion detection and prevention system on software-defined networks. In: 2015 International Conference on Advanced Computing and Applications. ACOMP, pp. 106–111, 2015. https://doi.org/10.1109/ACOMP.2015.19
36. Lee, S., Kim, J., Shin, S., Porras, P., Yegneswaran, V.: Athena: a framework for scalable anomaly detection in software-defined networks. In: 2017 47th Annual IEEE/IFIP International Conference on Dependable Systems and Networks (DSN), Denver, CO, 2017, pp. 249–260. https://doi.org/10.1109/DSN.2017.42
37. Lee, S., Yoon, C., Lee, C., Shin, S., Yegneswaran, V., Porras, P.A.: DELTA: A Security Assessment Framework for Software-Defined Networks. In: NDSS 2017 Feb 27
38. Tantar, E., Tantar, A.-A., Kantor, M., Engel, T.: On using cognition for anomaly detection in SDN. In: EVOLVE-A Bridge between Probability, Set Oriented Numerics, and Evolutionary Computation VI. Springer, pp. 67–81, 2018

An Overview of 51% Attack Over Bitcoin Network

Raja Siddharth Raju, Sandeep Gurung, and Prativa Rai

Abstract Cryptocurrencies are a new paradigm in terms of digital currency due to various features but mostly because of secure in nature and provide anonymity. Bitcoins are the most prominent example of cryptocurrency based on the proof of work (PoW) consensus mechanism. Bitcoins have been initiated as the exchangeable currency in some parts of the world, but the question arises that what are the major challenges that Bitcoins face even if it is secure in nature. The paper highlights the important attributes of a Bitcoin network, how the mining of the Bitcoin is done, and the consensus protocol on which Bitcoins are based on. It also gives insight into the impact of the "51%" attack on the Bitcoin network and the countermeasures that can be applied as a remedial solution to prevent such an attack.

Keywords Cryptocurrencies · Privacy · Security · 51% attack · Consensus

1 Introduction

Blockchain has been termed as the trust-free economic transaction that maintains anonymity and reliability based on unique technology [1]. Bitcoin or cryptocurrencies are the most known application of blockchain. Cryptocurrencies are slowly taking over the current financial transaction as it provides secure, reliable, and after all, a decentralized communication. Blockchain operates in a decentralized mechanism where it is incorporated with various core technologies such as cryptographic hash, digital signatures, and distributed consensus mechanism. All though, blockchain can be diversified into various other application other than cryptocurrencies [2]. Blockchain architecture is based on cryptographically linked blocks. Blocks are basically the containers that contain the information that is encrypted.

R. S. Raju (✉) · S. Gurung · P. Rai
Sikkim Manipal Institute of Technology, Majitar, Sikkim, India

S. Gurung
e-mail: sandeep.gu@smit.smu.edu.in

P. Rai
e-mail: prativa.r@smit.smu.edu.in

H. K. D. Sarma et al. (eds.), *Contemporary Issues in Communication, Cloud and Big Data Analytics*, Lecture Notes in Networks and Systems 281,
https://doi.org/10.1007/978-981-16-4244-9_4

The first and the foremost block is called Genesis Block [3]. In the Genesis block, the previous hash value is always zero as it is the first and foremost block in the blockchain network.

Features of Blockchain

Some of the features or the key characteristics of blockchain network that makes it unique are as follows [3]:

- Decentralized: Blockchain is based on decentralization where each block is connected to another block which makes a chain. Also, there is no third party involved in this network, and no central server is there.
- Secure: All the data in the network are cryptographically secured using SHA-256 (Secure Hash Algorithm) and make data reliable. It maintains data confidentiality.
- Immutable Ledger: Blockchain network keeps a record of the activities in a network known as a ledger. That ledger is validated by each node in a blockchain network, and every block contains a ledger. That ledger makes sure that it maintains non-repudiation and accountability.
- Transparency: Blockchain is also known to be transparent in nature as we can see that how and where data is sent to.

Versions of Blockchain

There are currently four versions of blockchain that are categorized according to the following implementation given below [4]:

- **Blockchain 1.0**: Blockchain network that is used specifically for the cryptocurrency. It allows financial transactions on the blockchain.
- **Blockchain 2.0**: Blockchain network that is used specifically for smart contracts. Smart contracts are the small piece of computer codes that are used to define conditions that facilitate the two entities to come to a single agreement. Basically, smart contracts lie on the top of the network. Ethereum, one of the cryptocurrencies, uses smart contracts to prevent fraud and hacks.
- **Blockchain 3.0**: Blockchain network that is used for implementing DApps, i.e., decentralized application. It provides decentralized storage to run the code of decentralized applications.
- **Blockchain 4.0**: Blockchain network is specifically designed to make technology compliance with industrial applications. It provides solutions and approaches to meet business demands using blockchain.

Bitcoin

Bitcoins, based on Blockchain 1.0, are the form of digital currency that has been secured with a cryptographic hash function. Bitcoin is a cryptocurrency and the most known application of blockchain technology. Bitcoin was launched in 2009 and intended as the decentralized, trusted, and secure form of digital currency [5]. The general idea of Bitcoin was to reduce digital scarcity and provide a connection between, and the main two entities can come together, transact the money with trust, and eliminate the third party. Thus, Bitcoins can be used to make payments directly to the payee, i.e., P2P (Peer-To-Peer) over the Internet without the need of the bank, i.e., the third party [6]. Some of the key features which attract people to use Bitcoins are trustworthiness, security, anonymity, and non-repudiation [7]: Although, since the creation of Bitcoins, most people have used it for legal purposes like investment and real estate purchases while some have taken the key features of Bitcoins into illegal activities like ransom, drugs, terrorist funding, etc. [8]. Some countries have started Bitcoin trading for purchasing goods such as in Spain where olive oils can be purchased using Bitcoins and in Moscow, vodka's can be bought using Bitcoins. Companies like Microsoft have accepted redeeming Bitcoins as store credits since 2014 and as payment for movies and Xbox games at online store [9].

Security Aspects in Blockchain

As per the analogy of the blockchain, the fundamental concepts of blockchain-like transparency, secure, and anonymity [3] ensures that the blockchain network is less prone to attacks. But in practical terms, blockchain network is prone to attacks like 51% attack [10]. A 51% attack is when a miner block owns more than 51% of the network from the owner of the network. In the latter part of the paper, the authors will give a brief on the mechanism of 51%. An attempt has been made by the authors to conduct a 51% attack on a Bitcoin test network that the authors have created in a virtual machine. Authors have tried to access that Bitcoin network from a remote host to try for 51% attack by mining blocks at a faster rate than the actual miner.

2 Background

As we have seen in the introduction of the Bitcoin, the section generally discusses the insight of the Bitcoin network. The Bitcoin block components are shown in Fig. 1.

Some of the components of Bitcoin block are as follows [2]:

- *Hash Value*: Hash value is a cryptographic value more specifically SHA-256. It is a 64-bit value that is generated by combining the nonce value, transaction data, and block number.
- *Nonce Value*: Nonce value is a 4-byte value which in the initial stage is 0 but increases after every hash operation performed.
- *Previous Block Hash*: Previous block hash is the hash value of the previous block stored in the current block to make a chain of blocks.
- *Next Block Hash*: is the hash value of the previous block stored in the current block in order to make a chain of blocks
- *Merkle Root*: Merkle root is the tree representation of all the hash values of the transaction that are performed in the network. All the transaction values are hashed into one hash value that stands at the top of the Merkle root tree as shown in Fig. 2.
- *Timestamp*: Timestamp keeps the time record at which time the transaction takes place for accountability purpose

Fig. 1 Bitcoin block components in a network [2]

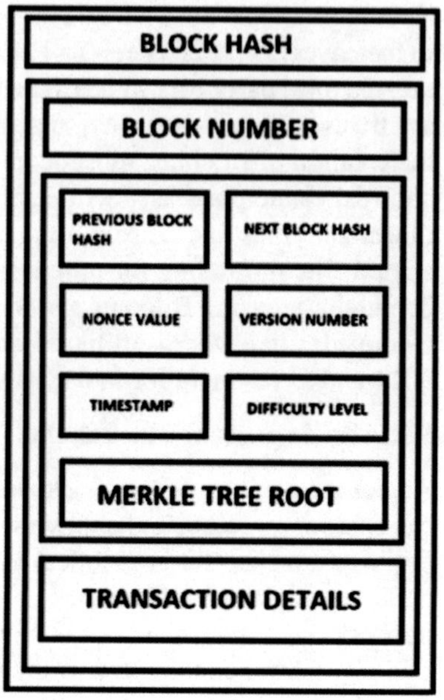

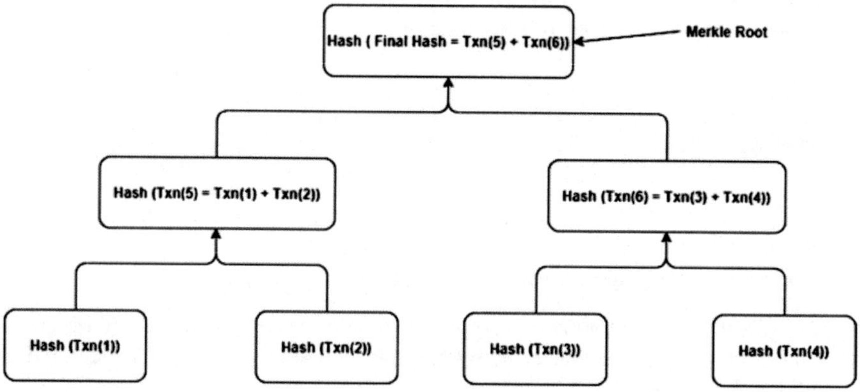

Fig. 2 Merkle root tree

- *Version Number*: Version number refers to the version of the network used by the blocks. The latest version of the Bitcoin network is 0.19.0.1.
- *Difficulty Level*: Difficulty level determines the difficulty of finding out the nonce value of the block. The difficulty is determined by the number of 0's, i.e., the greater number of 0's, the difficulty level is high, and the smaller number of 0's, the difficulty level is low.
- *Transaction Details*: It keeps all the data of transactions that took place in the Bitcoin network.

Consensus mechanism is a protocol in the blockchain which is designed such that all participants can come to a single agreement. Consensus protocol ensures that each and every transaction is a valid transaction, and it is recorded to each and every block in the blockchain network. Since blockchain is decentralized and there is no third party involved to validate the transaction, blockchain allowed every participant in the node to act as a third party and validate the transactions happening in a network. The whole process of the protocol makes the trust with each node, secure payment, and reliable transaction. There are few consensus protocols to validate transactions in the blockchain network [11]. Mostly, two main consensus protocols are used to validate a transaction, that is as follows [12]:

- **PoW (Proof of Work)**: Proof of work is a consensus protocol that is based on the mining process. Proof of work as the name suggests that the miner needs to show the proof of the work that he has done to mine the blocks, and in return, he is rewarded with the cryptocurrencies. Some examples of PoW are Bitcoins, Ethereum, Ripple, etc.
- **PoS (Proof of Stake)**: Proof of stake is a consensus protocol in which how much the stakes the person holds. The more you stake hold in a network, the more chances are that you will become one of the owners of the network. In proof of stake, we do not need to show proof of work.

Bitcoin Mining

Bitcoin mining is the process of generating blocks in a Bitcoin network. The person who mines the network is known as miner. Bitcoin mining works on the basis of a consensus mechanism called proof of work (PoW). Proof of work is a consensus mechanism in which the miner gets the reward in Bitcoins [13]. Bitcoin mining is a process where a miner uses its computational power to find a block. That block is generally identified using a nonce value. A nonce value is a 4-byte value that is initially 0 at starting, and the hash value is generated after every operation is performed. Nonce value of a block is decided when we finally receive the output. Miner finds out the nonce value of the block using computational power [13]. Thus, there is another term associated with Bitcoin mining, i.e., difficulty level. When the miner starts finding the nonce value of the block, the miner uses a brute-force mechanism to find the block. The block hash is led by the number of 0's, so it is said that there is more number of 0's, then the difficulty level of finding that block is also high. So finding the nonce value requires high computational power and even high electricity resource. The miner when he finds out the correct nonce value, the miner is then rewarded (also known as fees) with the Bitcoins. This process is called Bitcoin mining [14]. A digital signature is a mechanism that is used to authenticate data if the data is correct or not [15]. Every block in the blockchain network has a digital signature associated with it. Each block's digital signature is generated on the basis of the Merkle hash and digital signature of the previous block. A digital signature does not validate the blocks instead it validates the transaction. Before sending Bitcoins to a node, the sending node must attach sign the transaction using a digital signature [16]. Hash is a special function that can map large data of arbitrary length onto data of fixed length. Hashes are a special type of encrypted string that could not be reversed back to the original data. Each input data generates a certain unique hash. A little bit of change in the input data generates a totally new hash. Hashes are used to verify data in the blockchain [17].

Bitcoin Transaction

So, we have seen in the previous section that how Bitcoins are mined. Now, in this section, we will discuss how Bitcoins are transferred from one node to another. Basically, what we need before transferring is the sending and receiving address. The sending address is the address from where Bitcoin is transferred, and the receiving address is the address where the Bitcoins have to be transferred. For example, Node A wants to send 1 Bitcoin to Node B. So, Node A needs the address of Node B, so that he can transfer the Bitcoins. Simply Node A enters the address of Node B and sends the address. Some of the features related to the Bitcoin transaction are as follows:

- **Address**: Address is a combination of 26–35 alphanumeric characters that is used to transfer the Bitcoins to that node. There are three types of address that are used to provide the address in Bitcoin network, are as follows [18].
- **Lock time**: Lock time is the time at which the transaction took place. Lock time is present in the transaction ID.
- **Transaction ID**: Transaction ID is a hash value generated using SHA-256 algorithm which contains the transaction details, i.e., sending address, receiving address, the value that has been transferred along with the time stamp
- **Vin**: Vin is containing the sending address that has sent Bitcoins to the receiving address. Vin contains only one address that is sending address.
- **Vout**: Vout contains two addresses. One is the receiving address, and another is the change address.

3 51% Attack

In the Bitcoin network, there is a concept called longest chain rule (LCR) [5]. The author suggested that in the Bitcoin network, a node that has the longest chain of blocks in the Bitcoin network is the node owning the whole Bitcoin network. The rule is simplified as the node having the maximum number of blocks mined is the clear winner in the Bitcoin network as shown in Fig. 3.

A Sybil attack is a kind of attack in which the nodes enter the Bitcoin network, and it tries to perform a transaction to another address. The main idea is that the nodes impersonates as a valid node in the Bitcoin network and tries to transact the Bitcoins. But the Sybil attack cannot be successful as to validate the transaction, the majority of the node must conform to conduct the attack [19]. For example, in a Bitcoin network of 10 nodes, if there are two faulty nodes, the number of nodes that need to confirm the transaction is given by (1).

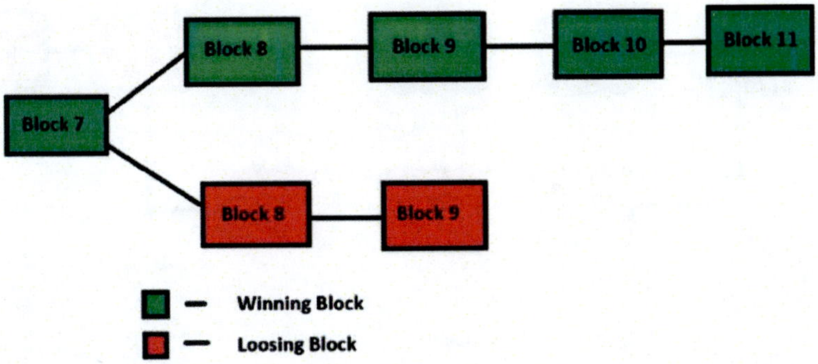

Fig. 3 Longest chain rule

$$(R) = 3(f) + 1 \quad (1)$$

where f is the number of faulty nodes. Let us take an example wherein the network consists of ten nodes out of which two are faulty nodes. According to (1), the confirmation must be that the following two faulty nodes require to validate themselves are:

$$(R) = 3(f) + 1 = (3 \times 2) + 1 = 7$$

Therefore, seven nodes must confirm that two faulty nodes are valid nodes otherwise they will be discarded from the network.

The longest chain rule has given birth to a new kind of attack that is called "51% attack." A 51% attack is a kind of attack conducted over the Bitcoin network. In this attack, the person who has owned the 51% of the Bitcoin network, is the owner of the Bitcoin network. The effect of this network is that all the nodes mining the Bitcoins will go to the attacker who has owned 51% of the network [10].

4 Methodology

Figure 4 represents an approach of a 51% attack in a Bitcoin network, the miner is mining the block, but the attacker gets access to the Bitcoin network. The attacker along with the miner starts mining the blocks, and let us assume that the attacker has high computational power such that he starts mining the blocks at an unprecedented rate. When the attacker mines more blocks than the existing miner and when he owns more than 51% of the nodes in a Bitcoin network, the attacker then is said to be the owner of the network. How Bitcoin longest chain rule has developed the attack called a 51% attack. Gambler Ruin's problem states that where a gambler has unlimited credits and therefore, he plays unlimited trials in order to reach the breakeven point

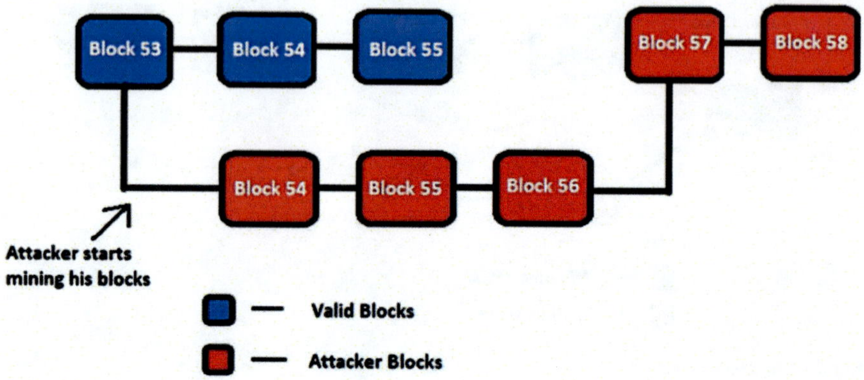

Fig. 4 Block diagram of 51% attack approach

[20]. Similarly, Nakamoto [5] took the same problem in terms of the probability of an attacker to reach the breakeven point.

Equation (2) describes that if the probability of an honest node to find the next block, i.e., '*p*,' is greater than the probability of an attacker to find the next block, i.e., '*q*,' the probability of an attacker to find the next block behind *z* blocks decreases.

$$q_z = \begin{cases} 1, & p < q \\ \left(q/p\right)^z, & p \geq q \end{cases} \tag{2}$$

where $p =$ probability of an honest node finds the next block, $q =$ probability of an attacker finds the next block, and $q_z =$ probability of an attacker to find the next block from *z* blocks behind.

The authors have taken "Bitcoin Client," an open-source API that can be used to connect to the main Bitcoin network or even we can create our test network that is also known as "Bitcoin Regtest." The Bitcoin Regtest can be accessed by the Bitcoin Client API call remotely. To access the Bitcoin Regtest, the system must have a Bitcoin Client installed. The procedure followed by the authors to conduct a 51% attack is as follows:

- **Establish a Bitcoin Regtest**: Initially, the authors have created a "Bitcoin Regtest" in a virtual machine running on Linux Operating System.
- **Miner connects to the network**: One of the authors is acting as a valid miner, connects to the Bitcoin Regtest using Bitcoin Client API call, and starts mining the blocks.
- **The attacker connects to the network**: Another author acting as an attacker connects to the Bitcoin Regtest API call, and then, the attacker starts mining the blocks too.
- **The number of blocks**: Both, attacker and valid miner are mining the blocks. If the number of blocks of an attacker is more than the number of blocks mined by

the valid miner, then the attacker becomes the owner of the Bitcoin Regtest, and if the number of blocks mined by the valid miner is more than the number of blocks mined by the attacker, then valid miner is the owner of the Bitcoin Regtest.

5 Implementation

We have seen what 51% attack is, and now we will see how 51% attack is practically possible. For this, the authors have created a test network of Bitcoin using Bitcoin core. Bitcoin core is an API that allows us to join the real-time Bitcoin network for mining, and we also have an additional feature to create a test network to perform testing. Thus, using Bitcoin core, we have hosted a Bitcoin test network in the local system hosted the Bitcoin test network over the Internet using port forwarding technique for which we can remotely access the Bitcoin network and try to conduct the attack as an attacker.

- We have created a test network, and we have already mined 101 blocks in one address, i.e., node A address. Since 101 blocks are mined and first 100 confirmation exists, node A is rewarded with 50 Bitcoins as shown in Fig. 5.
- Now, an attacker gets connected to the Bitcoin test network which is hosted over the Internet as shown in Fig. 6.
- To confirm the transaction, here is the list of mining of the last 20 blocks in the Bitcoin core client that contains the details of the test network hosted as shown in Fig. 7.

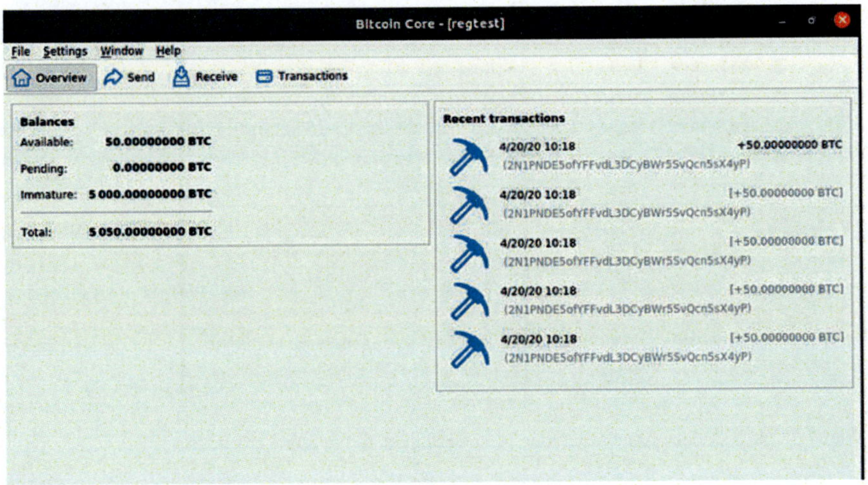

Fig. 5 Bitcoin Regtest network on virtual machine

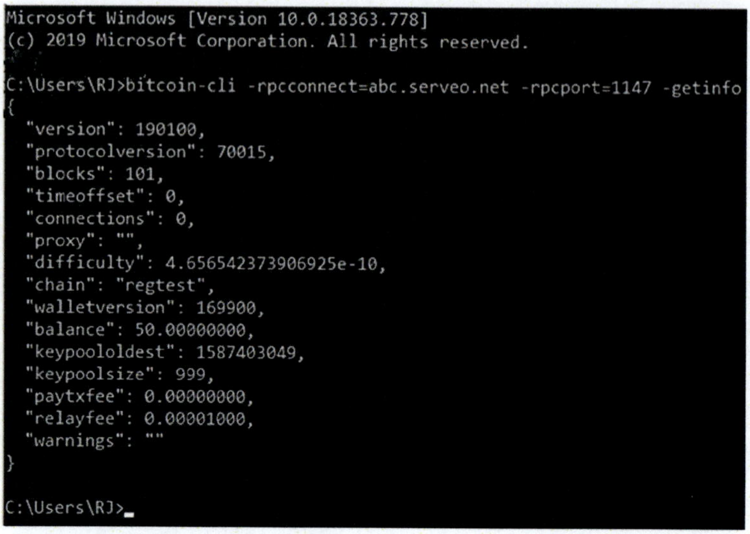

Fig. 6 Attacker has accessed the Bitcoin network

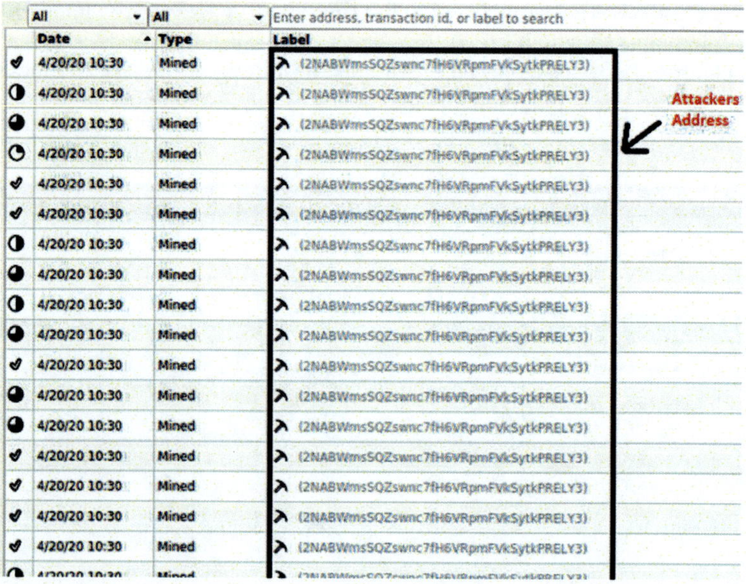

Fig. 7 Last 20 blocks mined by Node A, reward goes to the attacker's address

6 Results and Analysis

As we have seen in the previous section that how the 51% attack is implemented on the Bitcoin network, therefore this section will overview the results of the attack conducted over the Bitcoin test network along with some of the effects that are caused by the 51% attack along with some of the proposed solution to this attack.

Results of the Attack

As we have seen that the attack has been successfully tested, therefore we will see the results of the attacks.

In Fig. 8, we can see that the blue line represents the attacker's address, and the yellow line represents the honest node address. In this graph, we see that for honest node mines the first 100 blocks and get rewarded with 50 Bitcoins. But the attacker gets into the network, and the attacker has high computational power through which he can mine the blocks and the attacker also generates the first 100 blocks. Then, attackers generate the next blocks in an unprecedented such that his number of chain of blocks is higher than the honest node number of chain of blocks. As you can see, the previous owner of the block was the honest node, now the ownership has been taken up by the attacker node.

Attack Case Study

We have seen the results of how this attack affects the Bitcoin network with a 51% attack, now we will see on what calculations the attack is successful. As we have seen from (2), i.e., Gambler Ruin's problem [20] and how author Nakamoto [5] has used that equation in order to identify the probability of an attacker to find the next block and the probability of an honest node to find the next block. So using this data, the authors took data as an assumption, and then, we created a graph to show how the attack success is seen and what all are the factors that can make the attack successful or unsuccessful.

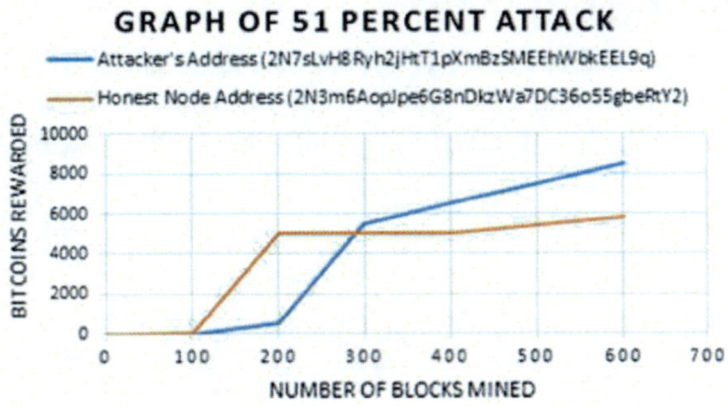

Fig. 8 Graph illustrates the vectors of 51% attacks conducted over the test network

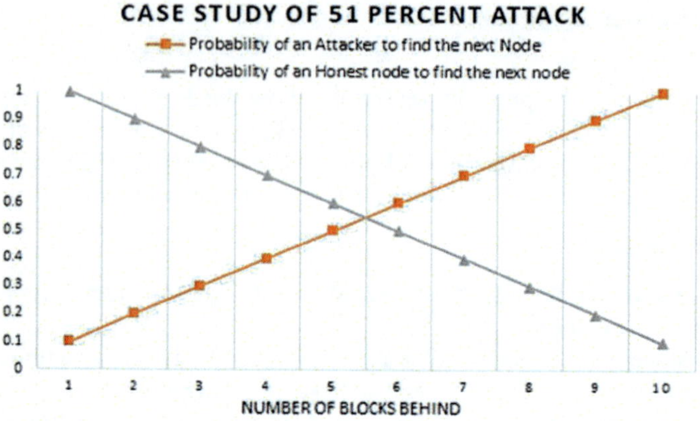

Fig. 9 Graph illustrates the key factors involved in 51% attack

Figure 9 explains that the 51% attack is based on two factors, i.e., number of blocks behind and the probability of an honest node to find the next block. So we can analyze the best case and the worst case of a successful 51% attack.

Best Case

The best case is where the probability of an attacker to find the next node is high. The best case for a successful 51% attacks is:

- When there are a greater number of blocks behind the attacker block
- When the probability of an honest node to find the next block is very low.

Worst Case

The worst case is where the probability of an attacker to find the next block is very low. The factors that might make the 51% attack unsuccessful are as follows:

- When there are a smaller number of blocks or no blocks behind the attacker's block
- When the probability of an honest node to find the next block is high.

Effects of 51% Attack

As you can see how 51% attack can take place in the Bitcoin network and some of the insights about 51% attack in Bitcoin network were also given, now we will discuss some of the effects of 51% attack that tends to harm the Bitcoin network.

Double-Spending Attack

A double-spending attack is a kind of attack in which a node sends some Bitcoins to a different node and each node receives the Bitcoin, but the sending node is debited with only one transaction amount [21]. For example, Node A sends 5 Bitcoins to Node B in a network and again Node A sends 5 Bitcoins but to Node C in the network. They are inside the Bitcoin network unconfirmed transaction and until the miner does not confirm the transaction, it stays in the unconfirmed memory pool. So, in some cases, we can have two miners present in the same network, and if two miners confirm the transaction simultaneously, Node B and Node C will receive their halves, i.e., 5 Bitcoins each but instead of debiting Node A with 10 Bitcoins, Node A is debited only with 5 Bitcoins. This kind of attack is known as a double-spending attack and can be performed using 51% attack.

Mining Effect

In a Bitcoin network, if an attacker performs the 51% attacks and generally owns the network, so he is considered to be the owner of the network. So, if the existing honest nodes perform the mining, all the mining rewards go to the attacker's address as he is the current owner of the Bitcoin network.

Prevention of 51% Attack

Although we can see some of the effects caused by 51% attack, therefore some solutions are proposed for the prevention of 51% attack.

a. **Timestamps**

To prevent the 51% attack, author Aggelos et al. [22] have suggested making strict time constraints to confirm the transaction to make the mining process slower. Timestamps are basically the time at which the blocks are mined, and those timestamps are recorded in the blocks in a network. The reason for implementing strict timestamps is because of how much computational power the attacker has, he cannot achieve the 51% attack as in this model, the timestamp of the previous block and the next block is matched, and if it reaches the reward time, then the reward is given else the consensus rejects the block and iteration are performed again. The minimum valid timestamp is 16 s, so if the timestamps are divisible by 16 s, then it is validated.

b. **Hybrid Consensus (PoW + PoS)**

As we have seen in the earlier section about PoW and PoS consensus mechanism [12], author Rahman [17] has suggested the mechanism of making a hybrid consensus mechanism of PoW and PoS. The main idea of the hybridization was to lower down the probability of an attacker to perform the 51% attack in the Bitcoin network. The author has expressed the model in (3) where he said if a malicious node owns 51% of the network (PoW) and 51% of the network's wealth (PoS), the probability of the malicious node to own the network under hybrid consensus, i.e., PoW + PoS, is defined as:

$$P(\text{Malicious Node}) = (\text{PoW}) + (\text{PoS}) \quad (3)$$

$$P(\text{Malicious node}) = 51\% \times 51\%$$
$$P(\text{Malicious node}) = (51/100) \times (51/100)$$
$$P(\text{Malicious node}) = 26.01\,\%$$

The probability of a malicious node to conduct a 51% attack on the Bitcoin network based on hybrid consensus protocol is low. According to Eq. (3), to own the network:

$$P(\text{Malicious Node}) = 71\% \times 71\%$$
$$P(\text{Malicious Node}) = (71/100) \times (71/100)$$
$$P(\text{Malicious Node}) = 50.41\,\%$$

The malicious node must at least 71% of the network and 71% of the network's wealth to conduct the 51% attack.

c. **Penalty for Delayed Block Submission**

The technique is used to prevent by putting a penalty over the block submission that is delayed. The author Garofollo et al. [23] suggests that penalties are to be imposed over the blocks that remain hidden in the network. The time is calculated as the interval between the block's duration. If any delay is detected, the protection notifies to not to make any fraudulent transactions until the delay is lifted by the security protection. Thus, the delay blocks is calculated using (4).

$$\text{Delayed Blocks} = \sum_{i=1}^{n} n(n+1)/2 \quad (4)$$

where 'n' is the number of blocks between the block height and the recent block. For example, if an honest node mines blocks from 150 to 173 and the attacker node mines blocks from 150 to 174, so the value of n is calculated as:

$$N = \text{Block Height}$$
$$N = 173 - 150$$
$$N = 23$$

So, the computation of delayed blocks is according to (4):

$$\begin{aligned}\text{Delayed Blocks} &= N(N+1)/2\\ \text{Delayed Blocks} &= 23(23+1)/2\\ &= 23(24)/2 = 276\end{aligned}$$

Therefore, 276 is the penalty delay.

d. **Waiting for Confirmations**

To avoid a 51% attack, it is to make sure that the receiver end must wait for the number of confirmations [23]. The greater number of confirmations is there, the probability of an attacker getting success for 51% attack decreases. Hence, it is said that if the Bitcoin transaction is less than 0.14 Bitcoins, one confirmation is enough to validate the transaction. If the transaction value lies between 0.14 Bitcoins to 1.14 Bitcoins, three confirmations are required, and if the transaction value lies between 1.14 Bitcoins to 14.4 Bitcoins or more than 14.4 Bitcoins, six confirmations are required to validate the transactions.

7 Conclusion

In this paper, we have studied blockchain technology and one of its prime application, Bitcoin. Also, we have seen the insights of the Bitcoin network and the functioning of Bitcoin in a Bitcoin network along with the 51% attack, and we have performed a small experiment to show how practically it is possible to conduct a 51% attack over the Bitcoin network. The implementation of 51% attack was done on a test network hosted on the author's system. The Nakamoto [5] longest rule chain vulnerability brought various Bitcoin networks to loss as we have seen how 51% attack has encouraged various people to perform the 51% attack in the real scenarios and has affected various Bitcoin networks. Some of the networks that were victimized by 51% attacks were Bitcoin Gold and ZenCash in 2018 which brought a huge loss of approximately 17.8 million dollars [24]. Also, we have discussed that how much hashing power plays an important role in making the 51% attack possible and we have also seen the case study for a successful attack of a 51% attack, the effects that are caused by 51% attack, and the solutions proposed by various authors to prevent 51% attack. Thus, we have analyzed all the prospects of the Bitcoin network and 51% attack and concluded that more research is needed in this field in terms of maintaining security and anonymity along with transparency.

References

1. Glaser, F.: Pervasive decentralization of digital infrastructures: a framework for blockchain-enabled system and use case analysis. In: 50th Hawaii International Conference on System Sciences (HICSS 2017), Waikoloa, 2017
2. Zheng, Z., Xie, S., Dai, H.N., Chen, X., Wang, H.: Blockchain challenges and opportunities: a survey. Int. J. Web Grid Serv. **14**(4), 352–375 (2018)
3. Yaga, D., Mell, P., Roby, N., Scarfone, K.: Blockchain Technology Overview. arXiv:1906.11078 (2019)
4. Khan, A.G., Zahid, A.H., Hussain, M., Farooq, M., Riaz, U., Alam, T.M.: A journey of WEB and Blockchain towards the Industry 4.0: an overview. In 2019 International Conference on Innovative Computing (ICIC), pp. 1–7. IEEE, Nov 2019
5. Nakamoto, S.: Bitcoin: A Peer-to-Peer Electronic Cash System. Manubot (2019)
6. Abadie, R., Carrington, C.: Disrupting Africa: riding the wave of the digital revolution [Online]. Available https://www.pwc.com/gx/en/issues/high-growth-markets/assets/disrupting-africa-riding-the-wave-of-the-digital-revolution.pdf, 17 Dec 2017
7. De Filippi, P., Haunter, D., Opden Kamp, C. (eds.): A History of Intellectual Property in 50, 2019
8. Kristoufek, L.: What are the main drivers of the Bitcoin price? Evidence from wavelet coherence analysis. PLoS ONE **10**(4), e0123923 (2015). https://doi.org/10.1371/journal.pone.0123923
9. Semova, M., Dimitrova, V., Haralampiev, K.V.: Cryptocurrencies and financing of social and anti-social projects. Paper presented at the conference towards industry 4.0 technology or ideology, St. Kliment Ohridski, Bulgaria [Online]. Available https://www.researchgate.net/publication/324200653_cyptocurrencies_and_financing_of_social_and_anti-social_projects, Nov 2017
10. Kim, S.K., Yeun, C.Y., Damiani, E., Al-Hammadi, Y.: Various perspectives in new blockchain design by using theory of inventive problem solving. In: IEEE International Conference on Blockchain and Cryptocurrency, Seoul, South Korea, May 2019
11. Baliga, A.: Understanding Blockchain Consensus Model 2017. Available online https://pdfs.semanticscholar.org/da8a/37b10bc1521a4d3de925d7ebc44bb606d740.pdf. Accessed on 1 Aug 2018
12. Garay, J., Kiayias, A.: SOK: a consensus taxonomy in the blockchain era. In: Cryptographers' Track at the RSA Conference, pp. 284–318. Springer, Cham, Feb 2020
13. Szalachowski, P., Reijsbergen, D., Homoliak, I., Sun, S.: Strongchain: transparent and collaborative proof-of-work consensus. In: 28th {USENIX} Security Symposium ({USENIX} Security 19, pp. 819–836, 2019
14. Kristoufek, L.: Bitcoin and its mining on the equilibrium path. Energy Econ. **85**, 104588 (2020)
15. Walker, H. How Digital Signatures and Blockchains Can Work Together. 2016. Available online www.cryptomathic.com/news-events/blog/how-digital-signatures-and-blockchains-can-work-together. Accessed on 1 Aug 2018
16. Asolo, B.: 51% Attack Exlpained. 2019. Available online https://www.mycryptopedia.com/52-%attack-explained/. Accessed on 1 Jan 2019
17. Rahman, A.: A Hybrid POW-POS Implementation Against 51% Attack in Cryptocurrency System (Doctoral dissertation, United International University), 2019
18. What is a Bitcoin Address? A 3-Minute Run Through [Online]. Available https://blog.hubspot.com/marketing/Bitcoin-address
19. Zhang, S., Lee, J.H.: Double-spending with a Sybil attack in the Bitcoin decentralized network. IEEE Trans. Industr. Inf. **15**(10), 5715–5722 (2019)
20. Feller, W.: An Introduction to Probability Theory and its Applications, 1957
21. Karame, G.O., Androulaki, E., Capkun, S.: Double-spending fast payments in Bitcoin. In: Proceedings of the 2012 ACM Conference on Computer and Communications Security, pp. 906–917, Oct 2012
22. Kiayias, A., Konstantinou, I., Russell, A., David, B., Oliynykov, R.: A Provably Secure Proof-of-Stake Blockchain Protocol, 2016

23. Rosenfeld, M.: Analysis of Hashrate-Based Double Spending. arXiv:1402.2009, 2014
24. 51% Attacks: The Deathly Scenario for any Blockchain. Online. Available https://www.infosecurity-magazine.com/opinions/51-attacks-blockchain/

An IPS Approach to Secure V-RSU Communication from Blackhole and Wormhole Attacks in VANET

Gaurav Soni, Kamlesh Chandravanshi, Mahendra Ku. Jhariya, and Arjun Rajput

Abstract Vehicles or nodes in Vehicular Ad hoc Network (VANET) are forwarding the traffic information for validation route information. Attacker vehicles are sending false messages of route information and not accepting traffic status packets or data packets. The abnormal behavior of the malicious nodes (Blackhole Attacker) and wormhole attacker is recognized by the reliable security mechanism. This paper proposes an intrusion detection and prevention (IPS) scheme to secure vehicle to RSU (V-RSU) communication from malicious (Blackhole) as well as wormhole attack in VANET. The IPS algorithm is applied to the RSU to recognize the malicious actions of an attacking vehicle by swarm optimization approach. The particle swarm optimization (PSO) confirms the attacker's presence and delivers effective traffic information. In the proposed IPS scheme, vehicles also receive traffic data from the leading vehicles and also forward traffic information to other vehicles. The traffic data exchange is monitored by RSU to identified malicious actions. The main task of the proposed security system is the effective management of vehicles in the presence of an intruder. Simulation results confirm that the proposed IPS scheme with PSO provides better performance in the presence of both attackers in VANET. The performance of previous IDS, attacker and proposed IPS is measure through performance metrics.

Keywords Attacks · IPS · PSO · Routing · RSU · Security · VANET

G. Soni (✉) · A. Rajput
Department of Computer Science and Engineering, Lovely Professional University, Phagwara, Punjab, India

K. Chandravanshi (✉)
Department of Information Technology, LNCT, Bhopal, India

M. Ku. Jhariya
Department of Computer Science and Engineering, MANET, Bhopal, Madhya Pradesh, India

H. K. D. Sarma et al. (eds.), *Contemporary Issues in Communication, Cloud and Big Data Analytics*, Lecture Notes in Networks and Systems 281,
https://doi.org/10.1007/978-981-16-4244-9_5

1 Introduction

In the recent traffic scenario, VANET security is a major issue because many malicious vehicle entrances are reducing network performance [1, 2]. The VANET scenario system is helpful to control traffic smartly. It also requires some advanced resources like the telematics box, OBUs, etc. In VANET, traffic information packets contain important information so that it is necessary to ensure about the packets are accepted or modified by an attacker. In the same way, the responsibility of drivers must be more so that they correctly and timely information about the traffic situation [3, 4]. Particle swarm optimization (PSO) mechanism is based on the social behavior and dynamics of insects, birds and fish [5]. These animals optimize their adaptation to their environment to protect themselves from predators, search for food and companions, etc. If left the random initialized situation, they automatically adapt to optimize their environment. This leads to the stochastic nature of the PSO. The PSO is widely used for secure routing in networks [6]. The swarm intelligence gives information about the attacker's presence and also selecting the secure path for routing in between sender and receiver in the network. The work of route selection is based on higher pheromones value in the network.

2 VANET Mobility Factor

Mobility of vehicles is connected with cars, railways, bicycles, motorcycles and everything that moves on wheels on the roads [7–9]. As for cars in VANET, many factors influence their mobility, for example in street construction the traffic is determined by streets, their directions, traffic lights and road signs. The small areas surrounded by streets can be called as block size [10]. The size will help determine intersections and therefore how often the vehicle decelerates and stops. In motion control mechanism, traffic lights and stop signs are the main signs that have a predetermined location and help make any proposed mobility model more realistic. The interdependent vehicle traffic explains as each and every vehicle is pretentious by the motion of the nearby vehicles. The average speed is maintained and if it faster it will control then change position or location. Besides, speed limits affect the average vehicle speed [11].

3 Literature Survey

The swam technique is also possible to use in a VANET routing algorithm [12]. The algorithm is based on IWD (Intelligent Water Droplets) and uses a fidelity model that provides the high-operating speed that is essential in dynamically moving networks where nodes move at high speeds, and the network topology is unstable. But the trust-based scheme is based on bandwidth and latency, and malware detection is not mentioned. Ahmad et al. [13] propose a robust MiTM (Man-in-The-Middle) trust structure to prevent malicious information from being exchanged by unauthorized nodes. Confidence calculations do not provide a new approach to attack detection. Nandy et al. [14] were combine the idea of a rating table of vehicles according to their behavior when using the network. Also, develop a collaborative learning and distribution mechanism to update the scoreboard and finally identify intruders on the network.

4 Problem Statement

The safety of the special vehicle network (VANET) is a serious problem because their survival is badly connected with crucial fatal situations. A self-organizing network is easily affected by blackhole and wormhole attackers who perform malicious actions to reduce network performance by suppressing traffic information, consuming channel bandwidth, and proving poor traffic conditions. Important traffic information must not be altered by a malicious vehicle.

5 Proposed IPS Algorithm

The attack behavior is detected and prevented by the pheromone detection-based method which values in between 0 and 1 in PSO. If any terminal or RSU detect PSO fitness (PSf) value is low that means data drop is higher than 30%. It means the node is treated like an attacker and block by RSU and sends the alert message to all other vehicles. So that in future no bad activity happens in the network. In the given algorithm step by step procedure are mentioned.

Input:

φ: Network Area 1652*1652m^2

V_n: $\forall V_n \in \varphi$ // Vehicle in network, $\sum_{i=1}^{n} V_i$: V_{S1}-----V_{Sn} // Route Requester Vehicle , $\sum_{j=1}^{k} V_k$: V_{R1}------V_{Rm} // Route Replier Vehicle

T: Receiver Vehicle, R_{req}: route request packet , Ψ: Vehicle Radio Range

R_{prot}: AODV, $U_n = u_1$----$u_n \in \varphi$ // terminal, RSU_k: Road side unit $\in \varphi$

A_t: Worm (A_1) & Blackhole (A_2) attack, Q_n: q_1, q_2, q_3---q_n set of wormhole, Prop: ground way

Output: Packet delivery ratio, Throughput, Delay

Step 1 Procedure to route request and reply

V_i call $R_{pro}(V_i, V_p, \Psi)$ use U_n
V_i broadcast $R_{pro}(V_i, V_p, \Psi)$

Step 2 If V_m = = Available & V_m != V_P **Then**
Successor of V_i available
$V_m \leftarrow$ send $R_{pro}(V_i, V_p, \Psi)$ to next V_j
$V_m \leftarrow V_{m+1}$ in route table
Else if V_m = = Available & $V_m == V_P$ **Then**
V_P receive $R_{pro}(V_i, V_p, \Psi)$
Send acknowledgment to V_i by reverse path
V_i send traffic status & V_p receive traffic status
Else
V_p not in Ψ or φ
V_i expire TTL
End if

Step 3 If V_l in Ψ && V_l receive r_{req} **Then** ***// Attacker activity type: Blackhole***
$V_l \leftarrow$ generate h_{seq} capture V_p address
V_l update route packet
Bind_reply(V_p, V_i, h_{seq})
V_i receiver fresh reply & spoofed by V_l
V_i Send data to V_l
If packets == "UDP/TCP" **Then**
Capture packet by V_l
Not forward to V_p
End if
Else
V_i watch network activity for attack spread
End if

Step 4 V_i sends data (V_i, V_p, data, Ψ) ***// Attack activity type: wormhole***
If q_1 in route of V_i,V_p **Then**
q_1 forward to diverted node q_2
q_2 capture data & not forward to next hop or V_p
End if

//Detection by RSU the Confirm the Attacker existence

Step 5 f_z: {0.0,0.1,0.2,……..1.0} // pheromone value

$PH: \prod_{p=0}^{1} h_p$ //Pheromone value

$L: \prod_{k=1}^{n} V_k$ // Intermediate vehicles

λ: pause time in second, Y: total route packets by links, fn: node fitness

ACK: Acknowledgement, R_d: reliable data

$$pdr = \frac{\text{data receives}}{\text{data send}} * 100$$

While route-execute(V_i, V_p, Ψ) **do // normal route by PSO technique**

V_i broadcast R_{pro}(V_i, V_p, Ψ)

V_p receives route packets && send ACK to V_i

End While

Step 6 While V_k in V_i to V_p **do**

Calculate $ph = \frac{V_{k-a}}{A}$

If ph(V_k) > 0.5(f_n) **Then**

V_k as reliable and send R_d to V_p

Calculate Y for f_n calculation

Else If ph(V_k) ≤ 0.5(f_n) **Then**

V_k set suspicious & monitored by RSU_i

RSU_i classify attack symptoms of V_k

If V_k generate h_{seq} & reason == "loop" **Then**

Attack type: blackhole (A_2)

Block attacker node

Else if V_k update route & V_k create new link (q_1,q_2) **Then**

V_k← set as q_1

V_{k+1}← q_2

If q_1 forward data to q_2 **Then**

q_2 ←drop data

q_1, q_2 set as wormhole (A_1) by RSU_i

Blocked q_1, q_2 by RSU_i

New path search by route call

Else

Unknown attack watch and block by RSU_i

End if

End if

End if

End do

6 Simulator Tool Overview and Performance Parameters

Modeling of intruders and security modules is performed using network simulator (NS-2) version 2.31 [15]. The simulation parameters such as the simulation area are 1652 m * 1652 m. The number of vehicles is 69, and the antenna type is omni-directional. The maximum speed of the vehicle is 50 m/s. Simulation time is considered 290 s and the size of packets are 512 bytes.

7 Results

In this section proposed IPS, attackers and previous IDS performance are measured, and the performance of the proposed scheme is better.

7.1 Throughput Analysis

The proposed IPS performance in the presence of a wormhole is mentioned in Fig. 1a. The previous wormhole-IDS is a secure network and also improve the performance in the presence of three wormholes nodes and 10 Malicious (Blackhole) nodes but it is not efficient in term of improving performance and reducing detection overhead. The proposed wormhole-IPS, performance, is better than the existing IDS scheme. During a Malicious attack, the number of packets dropping is high. The proposed malicious (Blackhole) IPS provide a better performance as compared to the existing IDS scheme. The efficient ACO approach in routing provides a better way of communication between the vehicles. The performance of the proposed IPS in the presence of IPS is mentioned in Fig. 1b.

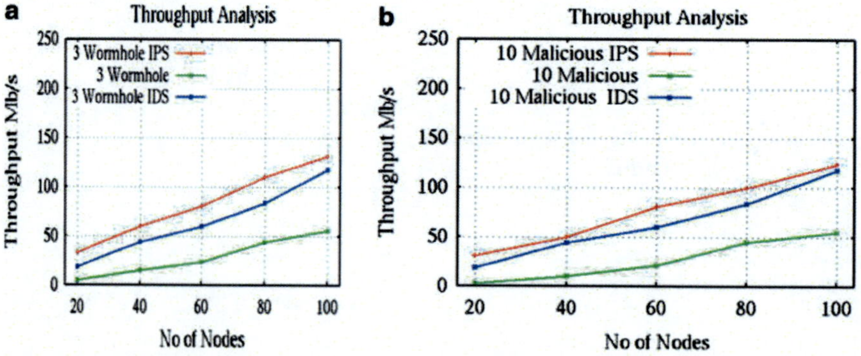

Fig. 1 **a** Throughput analysis (wormhole). **b** Throughput analysis (malicious)

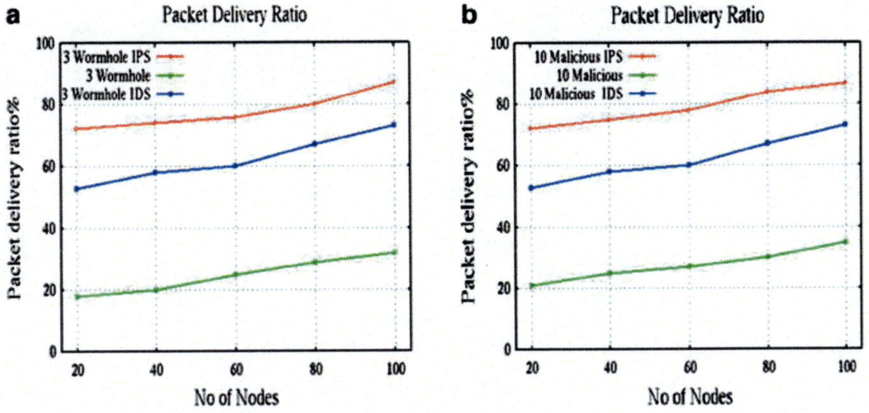

Fig. 2 **a** PDR analysis (wormhole). **b** PDR analysis (malicious)

7.2 *PDR Analysis*

PDR analysis in proposed wormhole-IPS, wormhole, and wormhole-IDS in the presence of 3 wormhole nodes are measured in this graph. Successful data reception is about 72–88% in the presence of proposed IPS. But in only wormhole attack, the percentage of data reception is 18–30% in all node density scenarios. The previous IDS scheme results showing results of about 55–75% in the same scenarios. On the other hand, the IPS provides a 13% improvement in performance mentioned in Fig. 2a.

Successful data receiving in the presence of 10 blackhole nodes are measure in between 20 and 38% vehicles scenario. PDR performance of the previous IDS scheme is measure in between 52 and 75% vehicle density and the proposed IPS or RSU-based communication is measure in between 72 and 89%. The performance is mention in Fig. 2b.

7.3 *Average Delay Analysis*

In traffic information packets, forwarding and receiving are completely based on the vehicles in VANET. The delay in the presence of an attacker in the network is high, i.e., about 3 times more as equal to the existing IDS and proposed IPS approach. The delay analysis in the presence of a wormhole attacker is mentioned in Fig. 3a. The presence of malicious attackers dumps a large amount of data in the network. The security against malicious IPS is also better than the previous IDS scheme and also effective to protect the network and reduce delay. The delay analysis in the presence of blackhole (Malicious) nodes attack is mentioned in Fig. 3b.

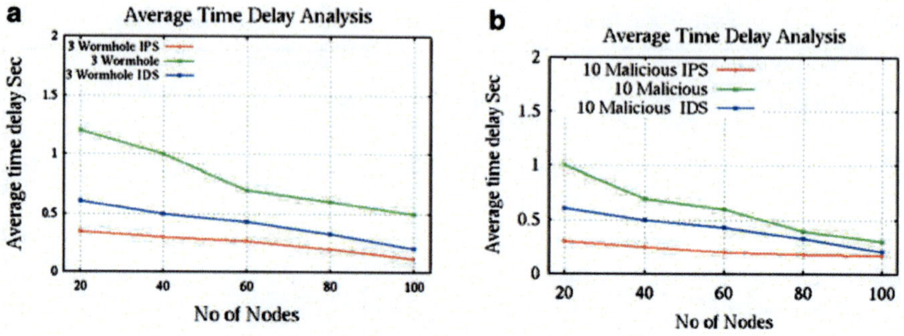

Fig. 3 **a** Delay analysis (malicious). **b** Delay analysis (wormhole)

8 Conclusion and Future Work

Safety criteria and proper traffic transferring are being addressed to improve performance in VANET. This paper proposed a new secure IPS algorithm to detect and prevent wormhole as well as malicious (Blackhole) vehicles with the help of PSO to disable their communication capabilities for further communication over the network. The proposed RSU-based IPS has not only generated a warning but also stops the malicious activity of the attacker. The route selection is based on higher pheromones value in the network. V to RSU communication ensures security and transmit information about the attacker to all RSU at rest and the vehicles surrounding them. After all, this information is transmitted for blocking malicious vehicles. The IPS security with PSO is detected intruders infection and minimizes the loss of useful traffic packets. Minimizing displacement represents a better vehicle movement. The proposed security mechanism recovers performance by about 90% over the typical VANET scenario. In future, proposes the scheme for high-speed vehicle measurement. In this approach, the RSUs are trying to provide a clear path for high-speed vehicles and identify a vehicle that has changed or does not run on the road according to vehicle communication policy.

References

1. Wang, Y., Li, F.: Vehicular ad hoc networks. In: Guide to Wireless Ad Hoc Networks. Book on Computer Communication and Networks. Springer (2009)
2. Toor, Y., Muhlethaler, P., Laouiti, A.: Vehicle ad hoc networks: applications and related technical issues. IEEE Commun. Surv. Tutor. 74–88, 3rd quarter (2008)
3. Mokhtar, B., Azab, M.: Survey on security issues in vehicular ad hoc networks. Alex. Eng. J. 1–11 (2015)
4. Muhammad, A., Wang, G., Bhuiyan, Z.A., Wang, T., Chen, C.: A survey on security attacks in VANETs: communication, applications and challenges. Veh. Commun. **19** (2019)
5. Yang, X.-S., Mehmet, K.: Swarm Intelligence and Bio-Inspired Computation (2013)

6. Li, G., Boukhatem, L., Wu, J.: Adaptive quality-of-service based routing for vehicular ad hoc networks with ant colony optimization. IEEE Trans. Veh. Technol. **66**(4), 3249–3264 (2017)
7. Hao, Y., Cheng, Y., Zhou, C., Song, W.: A distributed key management framework with cooperative message authentication in VANETs. IEEE J. Sel. Areas Commun. **29**(3), 616–629 (2011)
8. Mahajan, A., Potnis, N., Gopalan, K., Wang, A.: Urban mobility models for VANETs. In: Proceeding of 2nd Workshop on Next Generation Wireless Networks (2006)
9. Gaikwad, D.S., Zaveri, M.: A novel mobility model for realistic behavior in vehicular ad hoc networks. In: 11th IEEE International Conference on Computer and Information Technology, Cyprus (2011)
10. Ghafoor, K.Z., Mohammed, M.A.: Routing protocols in vehicular ad hoc networks: survey and research challenges. Netw. Protoc. Algor. **5**(4), 39–83 (2013)
11. Taleb, T., Sakhaee, E., Jamalipour, A., Hashimoto, K., Kato, N., Nemoto, Y.: A stable routing protocol to support ITS services in VANET networks. IEEE Trans. Veh. Technol. **56**(6), 3337–3347 (2007)
12. Krundyshev, V., Kalinin, M., Zegzhda, P.: Artificial swarm algorithm for VANET protection against routing attacks. In: Industrial Cyber-Physical Systems (ICPS) (2018)
13. Ahmad, F., Kurugollu, F., Adnane, A., Hussain, R., Hussain, F.: MARINE: Man-in-the-middle Attack Resistant trust model IN connEcted vehicles. IEEE Internet Things J. **7**(4), 3310–3322 (2020)
14. Nandy, T., Noor, M., Bhattacharyya, S., Idris, M.Y.I.B.: T-BCIDS: trust-based collaborative ıntrusion detection system for VANET. In: National Conference on Emerging Trends on Sustainable Technology and Engineering Applications (NCETSTEA) (2020)
15. Network Simulator-ns-2. https://www.isi.edu/nsnam/ns/tutorial/index.html

BER Analysis of FBMC for 5G Communication

Balwant Singh, Malay Ranjan Tripathy, and Rishi Asthana

Abstract The 4G up-gradation was essential because of the exponential increment of the wireless applications. So, 5G was introduced with large capacity, small latency, improved reliability, high data rate, and better quality of services (QoS). 4G uses orthogonal frequency-division multiplexing (OFDM) which has a decent quality of long-distance communication, by eliminating intersymbol interference (ISI) and improving the signal-to-noise ratio (SNR), but it has a spectrum wastage problem. However, the improved data rate is possible by spectrum utilization. 5G adopted various techniques for the better utilization of spectrum in current scenarios. The new generation waveforms, i.e., filtered orthogonal frequency-division multiplexing (F-OFDM), generalized frequency-division multiplexing (GFDM), universal filtered multicarrier (UFMC), and filter bank multicarrier (FBMC), are one of them. This paper will discuss FBMC which is the alternative technology of OFDM with better features by building subcarrier waveform based on prototype filters. Here, FBMC is investigated in terms of BER (bit error rate) by varying several parameters such as the size of inverse fast Fourier transforms/fast Fourier transform (IFFT/FFT), length of guard band, and length of symbols. Results of simulation will show that the bit error rate (BER) of FBMC is not affected by the variation of symbol length, while advancement in the size of IFFT/FFT increases BER, on the given parameter while the BER reduces for large size of the guard band of each subcarrier.

Keywords Filter Bank Multicarrier · IFFT/FFT · Offset QAM · Orthogonal Frequency-Division Multiplexing · Bit Error Rate

---

B. Singh (✉) · M. R. Tripathy
Amity School of Engineering and Technology (ASET), Noida, Uttar Pradesh 201303, India

M. R. Tripathy
e-mail: mrtripathy@amity.edu

R. Asthana
Goel Institute of Technology and Management, Lucknow, Uttar Pradesh 201306, India

H. K. D. Sarma et al. (eds.), *Contemporary Issues in Communication, Cloud and Big Data Analytics*, Lecture Notes in Networks and Systems 281,
https://doi.org/10.1007/978-981-16-4244-9_6

1 Introduction

5G provides data rate up to 20 Gbps, the latency of one-millisecond, bandwidth per unit area 1000×, the number of connected devices per unit area 100× (compared with 4G), availability of 99.99%, complete coverage (100%), reduction in the network energy usage up to 90%, and 10-year battery life. 5G accepted some new technologies and concepts, such as D2D, new radio frequencies, massive MIMO, edge computing, small cell, beam forming, the convergence of Wi-Fi and cellular, non-orthogonal frequency-division multiple access (NOMA), and software-defined networking/network function virtualization (SDN/NFV) [1].

OFDM is the greatest choice for point-to-point communication in 4G [2] with some drawbacks:

i. The OFDM requires perfect synchronization between a user node and BS [3], but it is difficult to establish because of the mobile environment, i.e., Doppler shift of different users [4]. Multiuser cancelation methods [5–8] can solve this problem, but it is complex to implement. So that one of the principal properties, i.e., simplicity, of OFDM will drop.
ii. The OFDM has carrier aggregation [9], which is the pitiable response of subcarrier filters for the IFFT/FFT filter bank. Due to the pitiable response of subcarrier filters, egress noise will be generated on the users. This effect can overcome by sidelobe suppression techniques [10–12]. But again, these techniques add complexity to the transmitter and provide loss in the bandwidth efficiency.

FFT and IFFT must be perfectly aligned to get the proper result. However, multipath propagation causes multicarrier symbol overlaps at the receiver side. So, there will be a problem in the demodulation of the signal because of the inter-symbol interference (ISI). This problem can be solved:

i. By the rising of the guard band length and symbol duration [13], this scheme is used in OFDM.
ii. By adding some additional processing (requires a bank of filter) to the FFT and retaining the symbol duration and timing as they are [14], this concept is used in FBMC.

FBMC is an alternative to OFDM, which can reduce ingress noise and egress noise more effectively by using high-quality filters. It is also able to reduce the synchronization problem [15]. Nowadays, three types of FBMC system are defined:

i. Cosine modulated multi-tone (CMT) [16].
ii. Staggered multi-tone (SMT) [17].
iii. Adopted method of frequency division multiplexing [18].

Here, CMT is based on the changing concept and SMT is introduced by Saltzberg. Recently, LTE and WLAN adopted the FBMC to improve the performance of D2D communication. The earlier communication system uses CP-OFDM, but it has loss of

Table 1 FBMC versus OFDM

FBMC	OFDM
Low sidelobes	Large and interfering sidelobes
A cyclic prefix (CP) not present	CP present
Sensitive to carrier frequency offset is low	Highly sensitive
Limited flexibility while adopting the MIMO technique	High flexibility
High spectrum sensing resolution	Low spectrum sensing resolution
High complexity	Low complexity
Multiple access interface (MAI) repressed	MAI present

spectral efficiency (SE) due to cyclic prefix (CP) insertion, high out of band radiation, and high sensitivity to narrowband interferences. To achieve the same, sub-channels of FBMC were designed which controls spectrum very effectively. This design will also remove the CP and increase SE. According to the requirement of the reception and selectivity, the filter bank of FBMC provides necessary frequency isolation (sufficient out of band attenuation of the sub-filters). If a vacant sub-channel is present between allocated sub-channels, then FBMC spectrally separates all allocated sub-channels. Accordingly, users do not require synchronization before the transmission of the subcarrier. This is a very energetic concept for future networks. FBMC also provides the option to simultaneously carry out spectrum sensing and transmission function to the same device [19]. Table 1 presents the basic comparison between FBMC and OFDM [20].

In this paper, the analysis of FBMC is done by varying different parameters such as the size of IFFT/FFT, length of the guard band on both sides of the subcarrier symbol, and length of the symbol. Section 2 of this paper describes design facets of the FBMC transmitter and receiver. Section 3 demonstrate results where BER of FBMC is not affected by the variation of symbol length while advancement in the size of IFFT/FFT increases BER, on the given parameter while the BER reduces for large size of the guard band of each subcarrier and finally Section 4 presents conclusions of this paper.

2 Design Aspects

FBMC is a subclass of multicarrier systems where IFFT and FFT can serve as multicarrier modulators and demodulators, respectively. The output of the IFFT is defined as

$$x(n) = \sum_{j=0}^{M-1} X_j(mM)e^{i2\pi \frac{j(n-mN)}{M}} \quad (1)$$

for $mM \leq n \leq (m+1)M$ where M is the size of IFFT and set of M data samples, i.e., $X_j(nM)$ with $0 \leq j \leq M-1$ is the input of IFFT. The set of 'M' samples is the multicarrier symbol, while 'm' is the symbol index. Note that the successive multicarrier symbols will not overlap in the time domain. FBMC perform FFT operation at the receiver and output of the FFT is defined as

$$X_j(nM) = \frac{1}{M} \sum_{n=mM}^{mM+M-1} x(n)e^{-i2\pi \frac{j(n-mM)}{M}} \quad (2)$$

FBMC uses the set of synthesis and analysis filters which is offset quadrature amplitude modulation filter bank multicarrier (OQAM/FBMC). This prototype filters can determine stopband attenuation, inter-symbol interference (ISI), and inter-channel interference (ICI). Oversampling factor (K) defines the characteristics of prototype filters. Note that the characteristics of prototype filter are the ratio of the impulse response duration of the filter and the symbol period of multicarrier. Additionally, the oversampling factor is the number of the frequency coefficient which are introduced between the FFT filter coefficients. Design of the prototype filters uses the Nyquist criterion. Furthermore, the 'background noise' power is an important parameter for design of the prototype filter, and it is described as power of interference power due to the non-orthogonality of the carriers, beyond the neighboring sub-channels [14].

Implementation of the filter bank requires an extension in the IFFT/FFT. This operation will generate complex computations that can be reduced by the polyphase network (PPN) [21]. This scheme is known as PPN-FFT [22]. FBMC does not uses CP so that they do not have much effect on the interferences. This paper will consider, the number of multicarrier symbols is 'k' and the prototype filter order '$2k - 1$' where 'k' maybe 2, 3, and 4 [23]. At this point, the length of IFFT is $N * k$. Each symbol of subcarriers overlapped due to delay of $N/2$ (N is the number of subcarriers). The capacity of the FBMC can be increased by applying OQAM [24]. Figure 1 shows the transmitter, where input data bits, mapped to the designed data symbol, before OQAM (offset QAM) processing. Serial to the parallel conversion will be performed earlier to the extended IFFT operation. The filter bank approach is used in FBMC instead of the FFT approach (used in OFDM). Every block of data extends the time window so that the multicarrier symbol overlaps. This interference between symbols can be avoided if the filter of the channel satisfies the Nyquist criterion. Remember

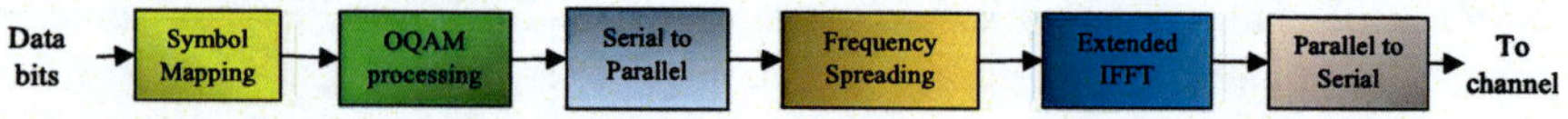

Fig. 1 FBMC transmitter

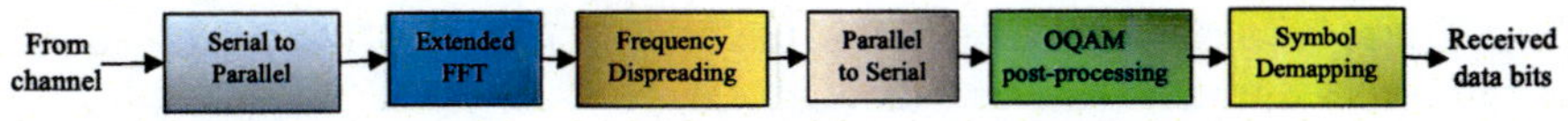

Fig. 2 FBMC receiver

that spreading must be completed before the extended IFFT operation. The data element is spread over several IFFT inputs, and this operation is known as weighted frequency spreading. Afterward, parallel to serial conversion completes in the FBMC transmitter, earlier to communicating the data symbol to the receiver, by channel. Amplitude and phase distortion, timing offset, frequency offset, and noise effect the transmitted information in the channel.

One point must be noted that the independent group of sub-channels can be allocated to different users and user synchronization is not mandatory. Amplitude and phase distortion, time offset, and frequency offset will be compensated by a sub-channel equalizer. The time-domain or the frequency-domain channels can be implemented and depends upon filter bank implementation of the receiver.

Figure 2 presents the FBMC receiver; here and now, the receiver performs serial to parallel conversion before FFT operation. Data elements are recovered with the help of a weighted dispreading operation. All over again, parallel to serial conversion was performed before proceeding with OQAM processing. Subsequently, received symbol is damped and gets the transmitted symbol.

3 Results and Discussion

The transmitter is mapping, the data symbols then the OQAM modulator produces even and odd symbols. These symbols are up-sampled by factor K. Afterward, these symbols will be padded with guard bits and filtered by a prototype filter. The transmitter will remove 1/2 filter delay, then compute the IFFT length of the transmitted symbol. The resulting transmitted symbol is the sum of the delayed real and imaginary symbols.

Additive white Gaussian noise (AWGN) is added on the channels which represent noisy conditions. Now, receiver will perform an FFT operation along with matched filtering and prototype filtering. This operation will overcome the effect of interference. Subsequently, the receiver will remove $K - 1$ elements and guard band. Next, OQAM post-processing and down-sampling (by a factor of $2K$) operations are performed. Finlay, we extract the imaginary part (which is the Kth sample after the real one) and the real part (which is the Kth samples after the imaginary one) of the data signals. This data will be normalized through the up-sampling factor before the de-mapping operation.

This paper works for 512-point FFT, 1024-point FFT, and 2048-point FFT. Length of the guard band varies from 20 to 1000 units. Here, matched filtering is implemented, which is characterized by oversampling factor K. At this juncture,

FBMC uses frequency spreading. Here, the value of K is 4, SNR is 15 dB, and the number of bits per subcarrier is 8 (256QAM).

Let us consider the number of complex symbols per OFDM symbol is L_N, the number of FFT points is F_N, guard bands on each side is B_G, the number of transmitted bits is TB_N, and the number of bits per subcarrier is B_N. The relation between these parameters is

$$L_N = F_N - 2 * B_G \tag{3}$$

$$TB_N = B_N * L_N/2 \tag{4}$$

According to Eq. (3), if B_G increases beyond 250, 500, and 1000 for 512-point FFT, 1024-point FFT, and 2018-point FFT, respectively, then L_N will be negative. So, the simulation will vary the above parameter accordingly. B_N is 8 (256QAM). Now, according to Eq. (4), TB_N will be 4-times of the L_N.

Now, this paper is going to analyze the effect of the variation in the symbol length (SL_S) for 512-point FFT, 1024-point FFT, and 2048-point FFT. Variation of SL_S, not affecting so much the BER of FBMC (Fig. 3). Note that the BER is low for 512-point FFT and high for 2048-point FFT. Subsequently, the increment in length of the guard band on both sides of subcarriers reduces BER up to zero (Fig. 4). Results also show that the 512-point FFT requires 250 units of guard band, 1024-point FFT requires 500 units of guard band, and 2048-point FFT requires 1000 units of guard band to get the zero value of the BER.

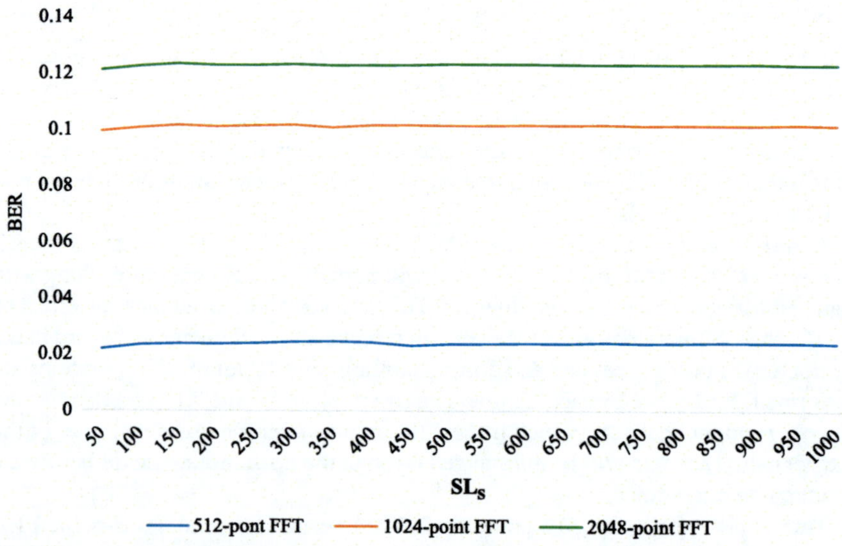

Fig. 3 Length of symbol versus BER

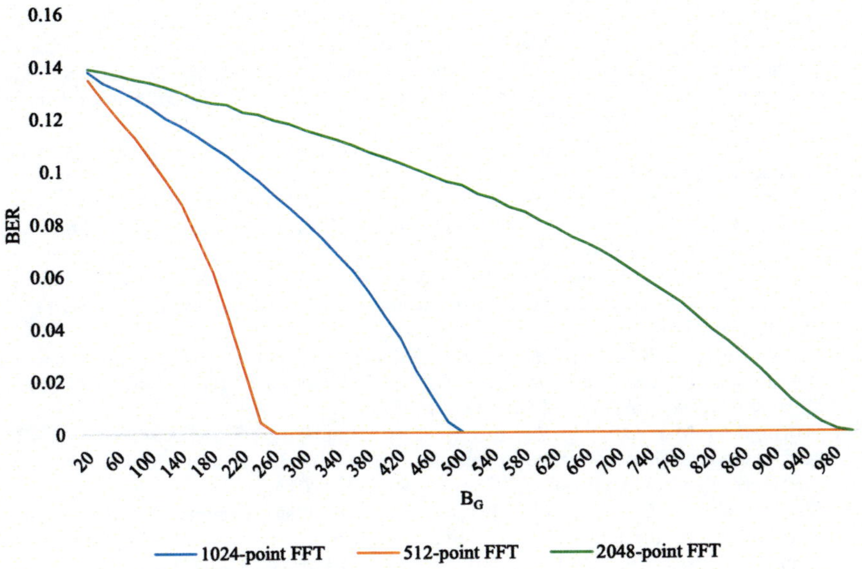

Fig. 4 Size of guard band versus BER

4 Conclusions

FBMC offers higher spectral efficiency in comparison to OFDM, but FBMC has a larger filter delay due to per subcarrier filtering. FBMC also requires modification for MIMO processing due to the OQAM processing. Here, simulation results for given parameters demonstrate that the variation of the symbol length in FBMC not affecting BER so much but the size of IFFT/FFT affects obvious, i.e., increment in the size of IFFT/FFT grows the BER. Over again, BER of FBMC can be reduced up to zero by the increasing size of the guard band on each side of the subcarrier so that the large size of FFT requires a hefty guard band for the reduction BER. Among the various technologies used for 5G, the waveform with better spectral control is important for the growth of spectral efficiency. FBMC is the alternative technology that has better features compared to OFDM by building a subcarrier waveform based on prototype filters.

References

1. Shafi, M., et al.: 5G: a tutorial overview of standards, trials, challenges, deployment, and practice. IEEE J. Sel. Areas Commun. **35**(6), 1201–1221 (2017). https://doi.org/10.1109/JSAC.2017.2692307
2. Haque, M.M., Rahman, M.S., Kim, K.D.: Performance analysis of MIMO-OFDM for 4G wireless systems under Rayleigh fading channel. Int. J. Multimed. Ubiquitous Eng. **8**(1), 29–40

(2013)
3. Zhang, W., Gao, F., Yao, B.: Blind CFO estimation for multiuser OFDM uplink with large number of receive antennas. In: International Conference on Acoustics, Speech, and Signal Processing ICASSP, Proceedings, May 2016, vol. 2016, no. 9, pp. 3721–3725. https://doi.org/10.1109/ICASSP.2016.7472372
4. Rotta, P.R., Tillotson, B.J.: Apparatus and method for correcting doppler shift in mobile communication systems. Google Patents, 20 Nov 2007
5. Huang, D., Ben Letaief, K.: An interference-cancellation scheme for carrier frequency offsets correction in OFDMA systems. IEEE Trans. Commun. **53**(7), 1155–1165 (2005). https://doi.org/10.1109/TCOMM.2005.851558
6. Lee, K., Lee, I.: CFO compensation for uplink OFDMA systems with conjugated gradient. In: IEEE International Conference on Communications, pp. 1–5 (2011). https://doi.org/10.1109/icc.2011.5962693
7. Lee, K., Lee, S.R., Moon, S.H., Lee, I.: MMSE-based CFO compensation for uplink OFDMA systems with conjugate gradient. IEEE Trans. Wirel. Commun. **11**(8), 2767–2775 (2012). https://doi.org/10.1109/TWC.2012.052512.110811
8. Farhang, A., Marchetti, N., Doyle, L.E.: Low complexity LS and MMSE based CFO compensation techniques for the uplink of OFDMA systems. In: 2013 IEEE International Conference on Communications (ICC), pp. 5748–5753 (2013)
9. Lin, S.P., Chen, Y.F., Tseng, S.M.: Iterative smoothing filtering schemes by using clipping noise-assisted signals for PAPR reduction in OFDM-based carrier aggregation systems. IET Commun. **13**(6), 802–808 (2019). https://doi.org/10.1049/iet-com.2018.5421
10. Brandes, S., Cosovic, I., Schnell, M.: Sidelobe suppression in OFDM systems by insertion of cancellation carriers. In: IEEE Vehicular Technology Conference, vol. 1, pp. 152–156 (2005). https://doi.org/10.1109/VETECF.2005.1557490
11. Yuan, Z., Pagadarai, S., Wyglinski, A.M.: Cancellation carrier technique using genetic algorithm for OFDM sidelobe suppression. In: MILCOM 2008–2008 IEEE Military Communications Conference, pp. 1–5 (2008)
12. Selim, A., Macaluso, I., Doyle, L.: Efficient sidelobe suppression for OFDM systems using advanced cancellation carriers. In: 2013 IEEE International Conference on Communications (ICC), pp. 4687–4692 (2013)
13. Vlachos, K., Ferreira, F., Sygletos, S.: Performance evaluation of a reconfigurable optical add drop multiplexer design for high-order regular and offset-QAM signals. In: International Conference on Transparent Optical Networks, July 2018, vol. 2018, pp. 1–4 (2018). https://doi.org/10.1109/ICTON.2018.8473580
14. Bellanger, M., et al.: FBMC physical layer: a primer. PHYDYAS **25**(4), 1–31 (2010). [Online]. Available: http://www.ict-phydyas.org/teamspace/internal-folder/FBMC-Primer_06-2010.pdf
15. Premnath, S.N., Wasden, D., Kasera, S.K., Patwari, N., Farhang-Boroujeny, B.: Beyond OFDM: best-effort dynamic spectrum access using filterbank multicarrier. IEEE/ACM Trans. Netw. **21**(3), 869–882 (2012)
16. Chang, R.: High-speed multichannel data transmission with bandlimited orthogonal signals. Bell Syst. Tech. J. **45**(10), 1775–1796 (1966)
17. Saltzberg, B.R.: Performance of an efficient parallel data transmission system. IEEE Trans. Commun. Technol. **COM-15**(6):805–811 (1967). https://doi.org/10.1109/TCOM.1967.1089674
18. Cherubini, G., Eleftheriou, E., Ölçer, S.: Filtered multitone modulation for very high-speed digital subscriber lines. IEEE J. Sel. Areas Commun. **20**(5), 1016–1028 (2002). https://doi.org/10.1109/JSAC.2002.1007382
19. Stitz, T.H.: Filter bank techniques for the physical layer in wireless communications. Tek. yliopisto. Julk. Univ. **919**(2010) (2010). https://doi.org/10.2459/JCM.0b013e328343e9e0
20. Kansal, P., Shankhwar, A.K.: FBMC vs OFDM waveform contenders for 5G wireless communication system. Wirel. Eng. Technol. **08**(04), 59–70 (2017). https://doi.org/10.4236/wet.2017.84005

21. Bellanger, M.G., Bonnerot, G., Coudreuse, M.: Digital filtering by polyphase network: application to sample-rate alteration and filter banks. IEEE Trans. Acoust. **24**(2), 109–114 (1976). https://doi.org/10.1109/TASSP.1976.1162788
22. Heute, U., Vary, P.: A digital filter bank with polyphase network and FFT hardware: measurements and applications. Signal Process. **3**(4), 307–319 (1981). https://doi.org/10.1016/0165-1684(81)90001-3
23. Viholainen, A., Bellanger, M., Huchard, M.: PHYDYAS project deliverable 5. 1: prototype filter and structure optimization. FP7-ICT. Tech. Rep. (2010)
24. Dashti, S., Fakhraie, S.M.: Analysis and design of OFDM/OQAM system with hexagonal lattice based on filterbank theory. In: 2014 7th International Symposium on Telecommunications (IST 2014), vol. 50, no. 5, pp. 383–387 (2014). https://doi.org/10.1109/ISTEL.2014.7000734

Impact of TCP-SYN Flood Attack in Cloud

Anurag Sharma, Md. Ruhul Islam, and Dhruba Ningombam

Abstract The given paper is focused on experimental study of one particular category of DoS attack (Denial of Service) in cloud computing network known as TCP SYN flood attack, and its effect on the resource availability and the cloud service factors. The attack typically takes up the general services of the cloud and the resources of the cloud, thus denying of proper services to the genuine users. The resources of the cloud server are drained and hence any other incoming requests would not be responded, thereby denying the access to the legitimate users. The assumption here is that the data being transmitted between client, and server is a multimedia data. In this paper, we have tried an experimental study on TCP SYN Flood attack and tried to see out the various parameters affected by the attack.

Keywords Cloud computing · TCP · SYN flood attack · Availability · Denial of service

1 Introduction

The basic concept of cloud computing is providing various computational facilities over the internet, where the user might want some storage services, computing power, and the ability to access it anywhere only with the help of an internet connection. With the advancement of cloud computing, the process of on-demand infrastructure services has become simpler and has benefited various organizations since the cloud works on pay-per-use model and it also has the ability to increase or decrease the need of resources [1]. The most widely used services in cloud includes Gmail, Facebook,

A. Sharma (✉) · Md. R. Islam · D. Ningombam
Department of Computer Science and Engineering, Sikkim Manipal Institute of Technology, Sikkim Manipal University, Majitar, India

Md. R. Islam
e-mail: md.i@smit.smu.edu.in

D. Ningombam
e-mail: dhruba.n@smit.smu.edu.in

H. K. D. Sarma et al. (eds.), *Contemporary Issues in Communication, Cloud and Big Data Analytics*, Lecture Notes in Networks and Systems 281,
https://doi.org/10.1007/978-981-16-4244-9_7

and OneDrive, etc., these services can be accessed anywhere in any devices like mobile phones, laptops, tablets provided an internet connection.

Security is and always has been a major concern for a technology like cloud computing. The cloud services can be threatened from various types of risks which might cause a great loss to the user or the cloud service provider. The cloud service availability in the modern day has been affected by a huge barrier which has proven to be fatal which is known as denial of service (DoS), where the service of the cloud to the end users are taken up by the intruder and so that no genuine users can use it. This might be disastrous financially as the computational cost might be very high. Such attack can have a great long term effect on the cloud in terms of resources and or financially, short term for customers as their work might not be done and long term for cloud service providers as they might lose their customers [2].

Classification of Denial of Service attacks [2]:

- Volume Based Attacks: This type of attack covers UDP flooding, ICMP flooding, and spoofed-packet flooding. The aim of the attacker is to take up the resources of the site being attacked, and the measurement is done in bits per second (Bps).
- Protocol Attacks: The mentioned attack comprises of SYN flood attack, fragmented packet, Ping of Death, reflection-based DoS, amplification-based DoS attack etc. The mentioned attack takes up the server resources, or of those in-between communication mediums, the measurement is done in packets per second (Pps).
- Application Layer Attacks: The mentioned attack comprises of Slowloris attacks Slow Post attack, attack that aims Apache, Windows. This consists of genuine and relevant requests; the aim here is to clatter the web server. This attack is very challenging to distinguish from legitimate traffic. The measurement is done in requests per second (Rps) (Fig. 1).

Information Security Principles

In order to have a secure cloud communication which is secure, there are certain principles which we need to follow; those principles are known as information security principles. One of the known security models which deal with the aspect of information security is CIA. The C stands for Confidentiality; I stand for Integrity and A stands for Availability.

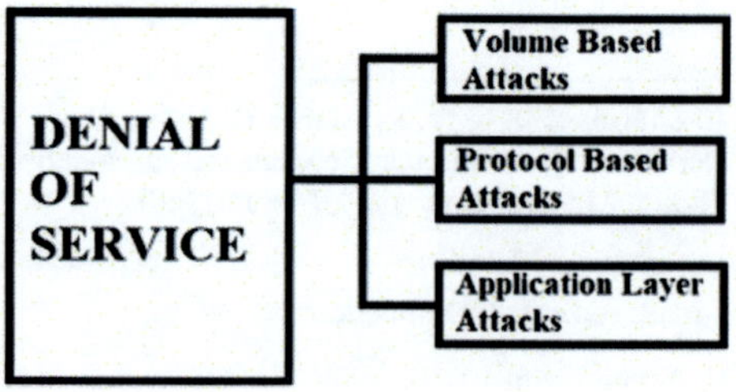

Fig. 1 Classification of Denial of Service attacks [2]

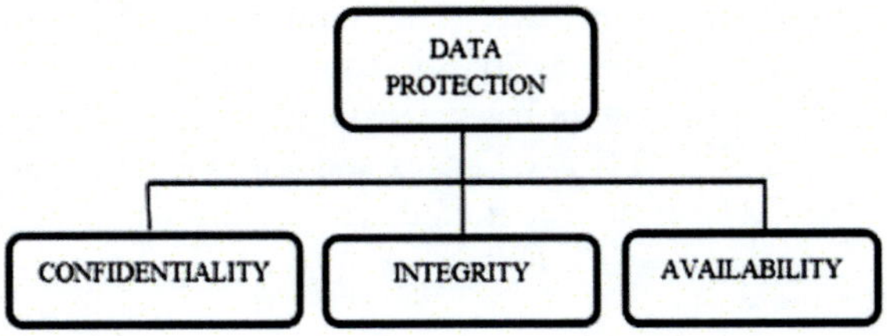

Fig. 2 Division of data protection into CIA

- Confidentiality: Confidentiality refers to the protection of information for the users who are not authorized. The main aim is to hide the information from others so that they may not access it. The increase in users in cloud has led to increase in threats in clouds so the confidential data needs to be saved.
- Integrity: Integrity refers to the data being consistent and seeing how accurate the data is. The data should not be manipulated by any unauthorized users. The data should not be altered, while being transmitted to other system.
- Availability: Availability refers that the data should be available every time the user wants to access it, however and whenever the user needs it. It basically refers to that the data should be always available for the authorized users to access [3]. The SYN flood attack basically attacks the availability of the CIA (Fig. 2).

Our contributions are summarized as follows:

- To the finest of our awareness, the offered paper has been advanced from the various former research papers that are referred by incorporating and refining all the works onto one project.
- We had hosted a real time website in the cloud where the penetration testing or attacking was carried out.

2 Related Work

In [4], an analysis has been done on a type of denial of service attack in mobile Ad hoc networks termed as SYN Flooding attack. The attacker sends multiple requests with spoofed addresses, so thus all the resources are occupied. An analysis has been done on how the attack can affect the normal working.

In [5], the authors have designed a system for TCP SYN flood attack and then, they have tried to develop a mitigation and detection technique which could trace if an attack is being done in the system using dendritic cell algorithm (DCA). This technique is being called as detection of DDoS using artificial immune systems.

In [6], the authors in their paper describe various types of DDOS attacks strategies and the amount of increase in attack in the recent years. The authors have described various types of defensive mechanisms and various mitigation techniques. Various types of recent attacks and various recent researches have been presented as well.

The challenges involved, and future scope of the project has been discussed. The detection includes use of Snort and Dward.

In [7], the author has created a virtualized environment, and then the author has done a TCP SYN flood attack and after that the author has tried to detect the attack in cloud environment grounded on header's statistical features of TCP/IP.

3 Methodology

In the presented paper, an experimental analysis is done and the denial of service attack is performed, and it is then analyzed how the system reacts normally comparing to the system when it is flooded with spoofed requests:

TCP SYN Flood Attack: In this experiment, we perform a TCP SYN flooding attack in a real world scenario in a website. We send a succession of SYN requests, with spoofed IP addresses, to the target cloud server to consume server resources in such a manner that the system would not be able to respond to the legitimate requests. It overwhelms the server in such a way that for each of the incoming packets requests the server has to establish a connection. The server then sends a SYN-ACK packet, but since the series of the requests from the user side were not legitimate so the three way handshaking won't be completed. The half-open connections therefore saturate the server and no legitimate request can now establish a connection until the server resources are reset in a timeout [5] (Fig. 3).

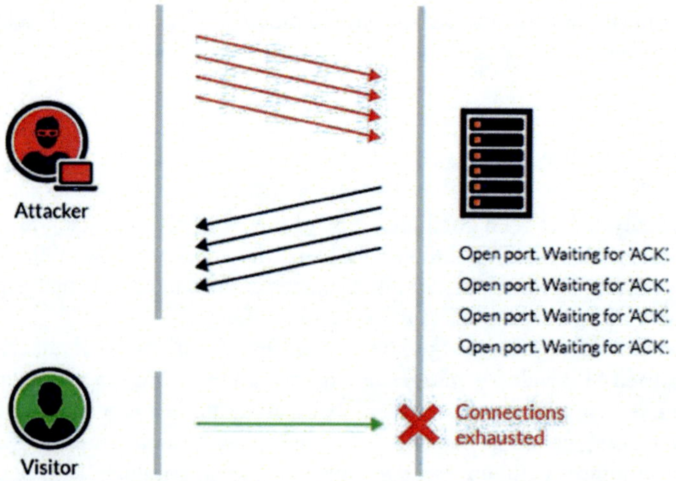

Fig. 3 TCP- SYN flood exploitation

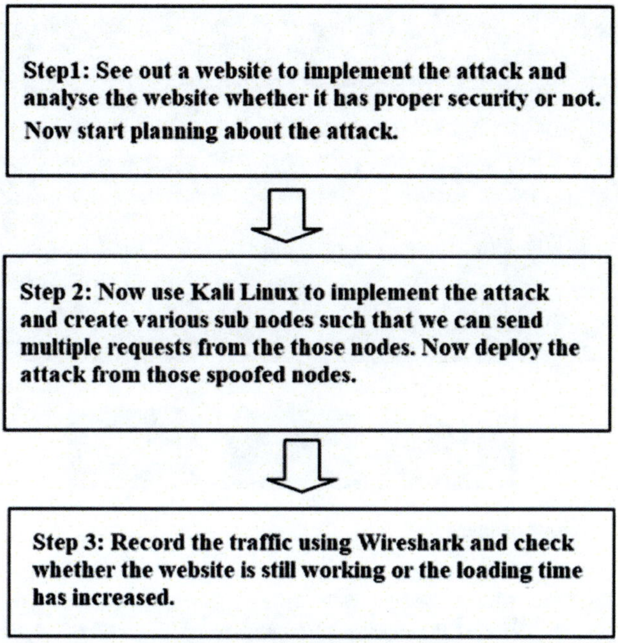

Flowchart for TCP SYN-Flood Attack

4 Experimental Results

The first part of implementation of the attack was to host a website. A website was hosted, and was made available in the internet, as we can clearly see in Fig. 4. The penetration test or the attack is going to be carried out in the Kali Linux which is a Linux based OS known for ethical hacking and penetration testing.

The next step was to identify the IP address of the victim. Using the command "tracert www.websitename.com" in the command line, we get the IP address of the victim. Since various types of penetration testing tools are readily available, and the next step was to shortlist various tools like Metasploit Project (msf5) which is a penetration testing tool and using the tool we carried out the attack as seen in Fig. 5 where we flood the victim with numerous spoofed packets.

The use of Wireshark came in the next step where we were seeing and recording the traffic being sent out to the victim or penetrating the victim with huge number of packets seen in Fig. 6. Wireshark is basically used to analyse the traffic in a network.

The attack on being continuously carried out finally made the website to go on in passive mode (unresponsive) and thus was not able to respond to any new requests as seen in Fig. 7.

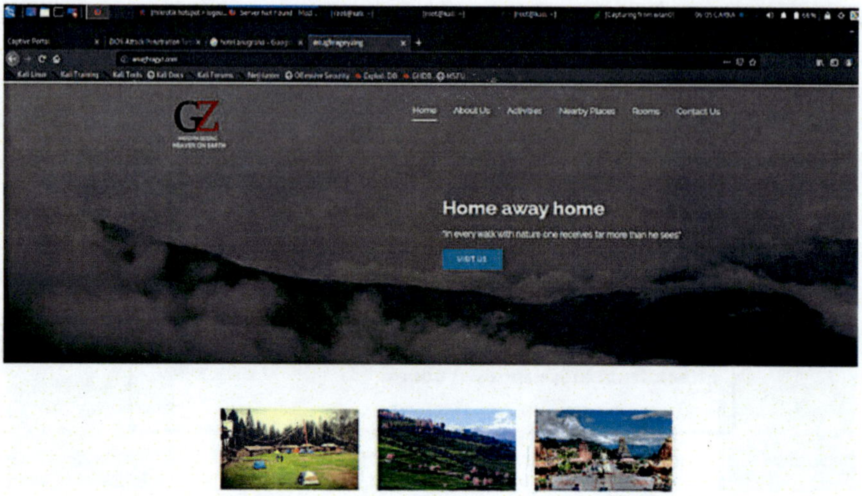

Fig. 4 Host the website to attack

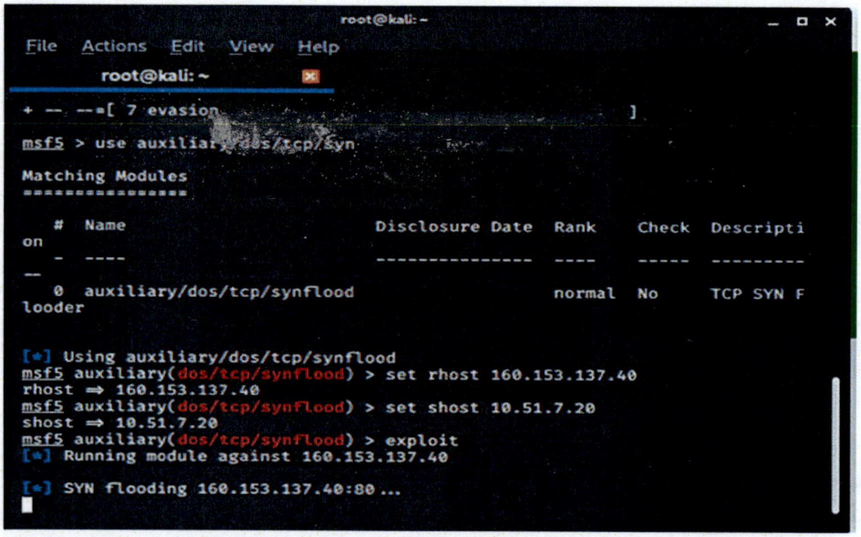

Fig. 5 Start implementing the attack in Kali Linux using Metasploit Project (msf5)

The resulting graph in Fig. 8 shows the traffic under the normal condition and when it is under a TCP SYN flood attack.

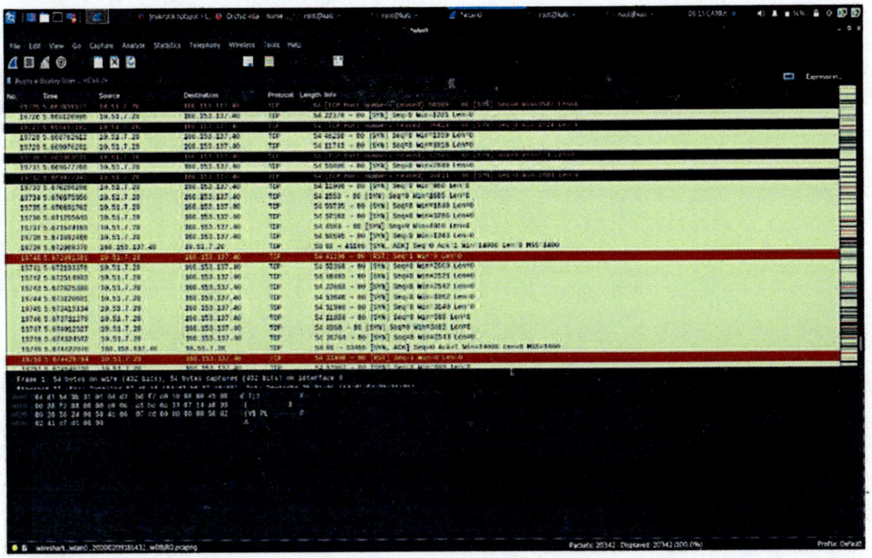

Fig. 6 Recording the traffic using WireShark

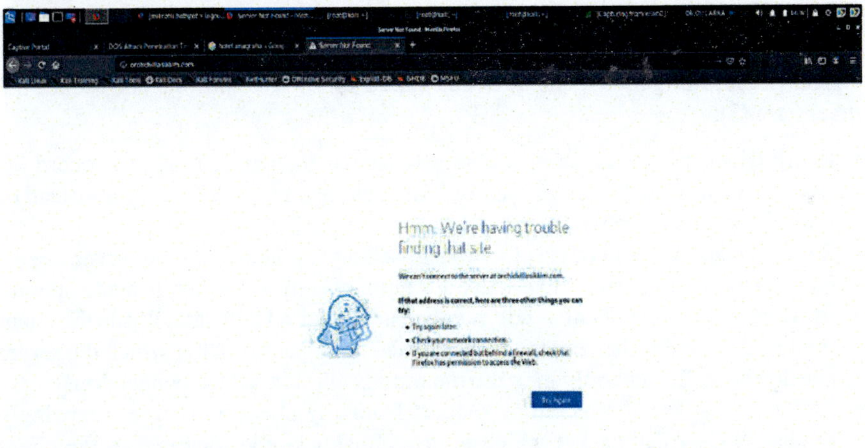

Fig. 7 After the attack, the website completely goes down

5 Analysis

The resulting graph in Fig 8 clearly shows the increase in traffic and how it can be dangerous in causing harm to the system. The result of the attack is such that the website has failed to respond for a long period of time as seen in Fig 7 so thus proving

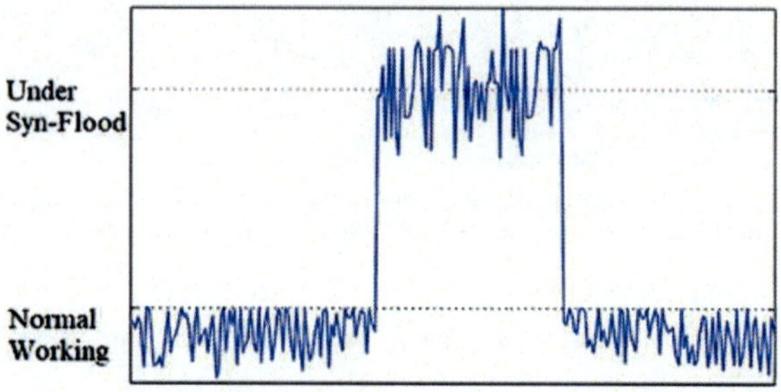

Fig. 8 Graph showing normal traffic to when the traffic when the attack is initiated

how harmful the attack can be if this is implement in a system where the legitimate users are in huge number. The SYN-flood attack as demonstrated in Figs. 4, 5, 6, 7 and 8 has clearly exploited the three-way handshaking protocol of TCP as here we send the SYN packet from spoofed IP addresses and the server responds with a SYN-Ack but doesn't get back Ack from the user thus creating a half open connection and thus a huge number of requests is sent to the server to put down the website for a short span of time. The resulting graph in Fig. 16 shows the result consumption normally and in under the attack.

The technological advancement has enabled various security measures, nevertheless there are ways in which the intruders are able to attack and the servers are still vulnerable to SYN flood attacks. So prevention of these types of attacks still remains the priority for the cloud service providers.

The various mitigation techniques that can be followed to prevent and reduce the attacks are [8]:

- Micro Blocks: In this case, the administrator has the ability to allot a record (as low as 16 bytes) in the server memory for each incoming SYN request instead of a whole connection object.
- SYN cookies: The use of cryptographic hashing is done where the server directs its SYN-ACK response with a sequence number (seq-no) which is made up with the IP address of the client + port number, and other related data. While the client replies, this hash is then contained within the ACK packet. After which the server checks the ACK, and only then the memory is allotted for the connection.
- RST cookies: For the incoming request from any client, the server knowingly sends a SYN-ACK packet which is invalid. This would then end in the client creating an RST packet, thus initiating the server that something is incorrect. When the server receives this, it knows that the request is genuine, the client is then logged in, and the succeeding inbound connection from the client is accepted.
- Stack Tweaking: Here the admin can twist TCP stacks to considerably reduce the outcome of SYN flood. It might involve either of decreasing the duration of timeout until the stack releases memory which is allotted to a connection, or choosing to release any incoming connections [9].

6 Conclusion and Future Work

There is no any doubt that the current technological development in the field of cloud computing has proven to be of huge benefit to most of the industries, start-up and consumers in terms of availability of different types of services with the help of internet across the world but the threat in cloud security still is a huge problem which might be a roadblock for organizations which are trying to fully migrate on to cloud.

Denial of service is hugely problematic by looking at the outcome that might cause a big loss in terms of resources or finance. The cost and time to recover from these issues might not be easy, which is sufficient for organizations to back out. Collective measures of protection need to be done which would reduce the influence and reduce the scale of such attacks. The administrator should detect the issues in networks in an active manner to reduce the threats of those attacks and protect the devices and overall protect themselves from such attacks.

Denial of service has been a major problem since the inception of cloud computing so in future we want to see out ways in which DoS can be prevented or we can decrease the frequency of this attack by creating various counter measures.

References

1. Byrne, D., Corrado, C., Sichel, D.E.: The Rise of Cloud Computing: Minding Your P'S, Q'S and K'S. National Bureau of Economic Research 2018. http://www.nber.org/papers/w25188. Working Paper 25188
2. Rong, C., Nguyen, S.T., Jaatun, M.G.: A survey on security challenges in Cloud Computing. SciVerse ScienceDirect
3. Bollinadi, M., Damera, V.K.: Cloud computing: security issues and research challenges. J. Netw. Commun. Emerging Technol. (JNCET) **7**(11) (2017)
4. Geetha, K., Sreenath, N.: SYN flooding attack—identification and analysis. In: International Conference on Information Communication and Embedded Systems (ICICES) 2014 ISBN No. 978-1-4799-3834-6/14/$31.00©2014 IEEE—S.A. Engineering College, Chennai, Tamil Nadu, India
5. Ramadhan, G., Kurniawan, Y., Kim, C.-S.: Design of TCP SYN flood DDoS attack detection using artificial immune systems. In: 2016 IEEE 6th International Conference on System Engineering and Technology (ICSET) October 3–4, 2016 Bandung–Indonesia
6. Mahjabin, T., Xiao, Y., Sun, G., Jiang, W.: A survey of distributed denial-of-service attack, prevention, and mitigation techniques. Int. J. Distrib. Sensor Netw. **13**(12) (2017)
7. AL-Hawawreh, M.S.: SYN flood attack detection in cloud environment based on TCP/IP header statistical features. In: 2017 8th International Conference on Information Technology (ICIT)
8. Srinivasan, K., Mubarakali, A., Alqahtani A.S., Dinesh Kumar, A.: A Survey on the Impact of DDoS Attacks in Cloud Computing: Prevention, Detection and Mitigation Techniques. In: Balaji, S., et al. (eds.) ICICV 2019, LNDECT, vol. 33, pp. 252–270. Springer Nature Switzerland AG 2020
9. Jamaluddin, M., Touqir Anwar, M., Saira, K., Wani, M.Y.: DDoS SYN flooding; mitigation and prevention. Int. J. Sci. Eng. Res. **5**(12), 484 (2014). ISSN 2229-5518

An Efficient Cooperative Caching with Request Forwarding Strategy in Information-Centric Networking

Krishna Delvadia and Nitul Dutta

Abstract Information-centric network emphasizes on translating the current Internet paradigm from host driven to content driven. A numerous data-centric features like content driven routing, random network topology, and pervasive caching make ICN caching distinct from state of the art of work. This paper emphasizes on the cooperative caching in ICN and contributes a dominating-set-oriented cooperative caching strategy. It considers placement of data and request message forwarding in a strong co-relation. Here, the concept of connected dominating set has been used to create a virtual backbone that includes core routers where most popular content can be cached. The betweenness centrality parameter of that core router helps the routing strategy to forward request packet to the node where content is most likely to be stored which leads to faster retrieval of data. The proposed strategy will improve the performance parameters like content retrieval latency, cache hit ratio, mean hop distance, total transportation cost, and network overhead. The paper also discusses the working of proposed strategy with related algorithms, illustrations, and its benefits over state-of-the-art research in context of network performance.

Keywords Connected dominating set · Betweenness centrality · Cooperative caching · Request forwarding · Information centric networks

1 Introduction

The field of communication has witnessed a remarkable transformation in past couple of decades. This evolution covers diversified research directions like CCN [1], CRN [2, 3], Internet of things [4], etc. Information-centric network is evolving as a new Internet paradigm, in which the focal point of communication is named-content instead of node address [1]. The content chunk is recognized by unique name, routed,

K. Delvadia (✉)
Department of Information Technology, Uka Tarsadia University, Surat, Gujarat, India

K. Delvadia · N. Dutta
Department of Computer Engineering, Marwadi University, Rajkot, Gujarat, India
e-mail: nituldutta@ieee.org

H. K. D. Sarma et al. (eds.), *Contemporary Issues in Communication, Cloud and Big Data Analytics*, Lecture Notes in Networks and Systems 281,
https://doi.org/10.1007/978-981-16-4244-9_8

and requested with the help of names in place of their actual location addresses [5]. The content chunk can get cached inside intermediate content routers when it travels from content source to requestor. This cached copy will be utilized to satisfy further identical content requests.

In-network caching is one of the important characteristics in ICN and gets differentiated from existing overlay caching solutions like CDN, P2P, and Web caching. In content delivery networks (CDN), the caching nodes are deployed in a hierarchical structure at specified locations and take cooperative storage decisions within the hierarchy of caches. Though the caching nodes of ICN cover the entire Internet and are organized as a random network topology instead of hierarchical organization, the important point is temporary data copies may exist in any content router with highly volatile nature. The challenging task is to forward data requests to the nearest content holding cache [6].

In traditional scenario as depicted in Fig. 1, when content router R_1 gets request for content chunk c1, it first verifies its own cache. If it does not have a copy of c1, it sends request to origin server. In between, the request can get fulfilled by the router R_6, along the way to origin server. The ideal route is to send the request toward router R_4 which is just one hop away from R_1 and holding content c1. The reason behind this is that the decision related to content placement and request forwarding has been taken by router in a non-cooperative fashion. An efficient cooperative caching strategy is the solution to the mentioned problem.

The researchers have investigated cooperative caching strategy in order to improve network level performance [7]. The cooperative Web caching makes use of priori information in order to have an optimal placement of data. In opposite to that, ICN requires a simple cooperative caching strategy to support new functionalities like ubiquitous caching nodes and line speed need. The line speed need limits the caching complexity in content routers, and ubiquitous caching raises the data availability but

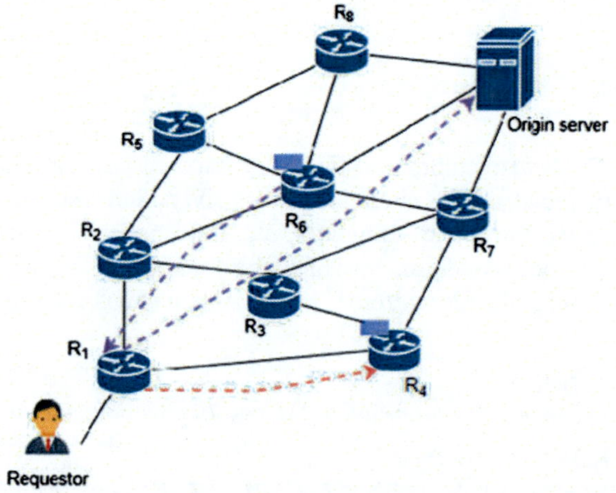

Fig. 1 State-of-the-art caching and forwarding strategy in ICN

creates the issue of how to efficiently forward content requests [8]. The information related to caching location of content chunks should be utilized in the request routing procedure to locate a nearest replica. Though, majority of state-of-the-art routing solutions forward request packets independently without taking caching related information into consideration.

In this paper, we contribute a cooperative caching strategy tightly coupled with request forwarding. The dominating set conception has been used to create a virtual backbone network on which routing and caching strategies are deployed. The dominating set theory has been introduced in proposed ICN scenario to divide random network topology of ICN into two hierarchical levels. Then, the cooperative caching mechanisms among two levels can be designed.

The major contribution of the research work is as follows:

- Use the concept of dominating set for exploring information about network topology and create an ICN enabled virtual backbone network.
- Contribute a cooperative caching strategy by taking placement of content and routing into consideration.
- Discussion on the improvement of the network performance parameters with adaptation of proposed approach. The comparative analysis of proposed approach over state-of-the-art solutions has been presented.

The organization of rest of paper is as follows. Section 2 depicts a survey of existing related research work. Section 3 presents the proposed approach of cooperative caching mechanism coupled with request routing followed by algorithms for the same. Section 4 discusses strength of proposed approach over state-of-the-art techniques through comparative analysis. Finally, conclusion and future work are mentioned in Sect. 5.

2 Literature Survey

The concept of dominating set has been widely adapted as an efficient solution for improvement in network-level performance of wireless networks [9]. The wireless network does not have any physical network as a backbone and adapt dominating set to identify core routers as gateway nodes. In the proposed work, DS concept is used to differentiate caching nodes within ICN by making random topology a hierarchical one. This also simplifies the cooperative caching mechanism. The operation of cooperative caching can be achieved in two ways: implicit and explicit. In case of explicit cooperation [7, 10], exchange and communication of information among caches will take place to realize optimal placement of content. This incurs high overhead and complex cooperation algorithm. In case of implicit cooperation [11], no such exchange and communication of information is needed. It depends on local caching strategies to attain better performance. Implicit cooperation is suitable for ICN because of its simplicity and less network overhead. The proposed work incorporates the method of implicit cooperation.

The authors in [12] provide an implicit cooperation-based cache networks named breadcrumbs, in which the intermediate content routers cache the needed content along the path from hitting source to requestor and maintains state information for further identical requests. The authors in [12] adapts a whole course caching method that cache content and preserve state at all in-between content routers along the path from hitting source to requestor. This method promotes duplication of content to minimize content retrieval delay, still, it leads to redundant caching [13]. In oppose to that, proposed work uses an index variable to store cache trail but minimizes the duplication in caching. The authors in [14] contribute an implicit cooperation but are restricted to hierarchical paradigm. The authors in [15] provide a probability-based implicit cooperation for placement of content, named ProbCache. They have emphasized on approximation of path's cache capacity and cache data using probabilistic approach to minimize caching duplication. Though, ProbCache does not involve joint consideration of content placement as well as request packet routing. The proposed approach aims to be superior performer over ProbCache as it performs well compare to whole course caching.

The authors in [16] propose an integrated caching as well as interest routing method for ICN. The scheme has been designed to cache digital data efficiently by exploiting the benefits of collaborative and non-collaborative caching. The scheme directly forwards the content request to the origin server without doing request flooding. The authors in [17] provide a novel local cooperation-based approach that attains better performance without incurring higher signaling congestion. It is also incrementally implementable, fault-tolerant, and lightweight in nature.

The above-mentioned cooperative caching and request routing solution emphasize more on request forwarding without exploiting built-in ICN features. The proposed routing solution forwards interest packet to the nearest content holder with the knowledge of content placement strategy. It aims to deliver requested content to user through fastest possible route, so that content retrieval delay, mean hop distance, and total transportation cost can be minimized. It considers betweenness centrality value of a router while forwarding interest packet, to accelerate the probability of cache hit. This creates base for proposed solution to prove its superiority over state-of-the-art research solutions. Comparative analysis for state-of-the-art mechanisms related to cooperative caching and routing in ICN is presented inside Table 1.

3 Cooperative Caching Strategy

This section demonstrates the procedure to build a virtual backbone network in ICN and the way it decomposes the entire network topology into distinct subnetworks. Then, we construct cooperative caching joint request forwarding strategy based on the decomposed network.

Table 1 Comparative analysis for the existing cooperative caching and routing mechanisms

Mechanism	Category	Content replication	Cache retrieval	Reducing caching redundancy	Improving cache availability	Communication overhead	Aims to reduce content retrieval latency	Mean hop distance
[7]	Explicit cooperation	Caching everything	Available along the path	No	Yes	High	No	High
[10]	Explicit cooperation	Caching everything	Available along the path	No	Yes	High	No	High
[11]	implicit cooperation	Caching everything	Available along the path	No	Yes	Low	No	High
[12]	On-path caching, implicit cooperation	Caching everything	Available along the path	No	Yes	Low	No	High
[13]	Explicit cooperation	Caching everything	Available along the path	No	Yes	High	No	High
[14]	Implicit cooperation	optimal object placement algorithm	Available at peer nodes	No	No	Low	No	High
[15]	Probabilistic implicit cooperation	Probability caching + implicit coordination	Cache content probabilistically	Yes	No	Low	No	Less
[16]	Heterogeneous caching	Caching at "core" router	Available at core and ordinary nodes based on content popularity	Yes	No	Low	No	Less

(continued)

Table 1 (continued)

Mechanism	Category	Content replication	Cache retrieval	Reducing caching redundancy	Improving cache availability	Communication overhead	Aims to reduce content retrieval latency	Mean hop distance
Proposed strategy	Heterogeneous caching	Caching at "core" router	Available at core and ordinary nodes based on content popularity	Yes	No	Low	Yes	Minimum

3.1 Dominating Set Construction

A random topology of ICN can be visualized by an undirected graph $G = (V, E)$, where V contains n nodes and E contains e edges. In graph theory, A *dominating set for* graph G comprises of a subset $V' \in V$, so that every node in $V - V'$ is neighbor to some node present in set V'. A connected dominating set for given graph G is a set C of vertices that satisfy below-mentioned properties:

Any node in C can reach out any other node in C by a route that remains entirely in C. This means that C induces a connected subgraph for G.
Each node in G either belongs to C or is neighbor to a node in C. This means that C is a dominating set of G.

The proposed approach selects to build connected dominating set (CDS) for ICN topology because the nodes in C can do communication among each other without use of nodes in $V–C$.

The CDS construction mechanisms have been referred well in the existing network literature. Proposed approach adopts one traditional procedure presented in [6]. Figure 3 shows the in-detail procedure to formulate a connected dominating set for any graph G. The illustration for the same in context of ICN is depicted in Fig. 2. Initially, we highlight node R6 with blue and its adjacent vertices R2, R3, R4, R7, R8, R9, R10 with sky-blue. Then, we highlight R2, R7, R10 as blue and its adjacent vertices R1, R5, R11 as sky-blue. As per the CDS construction mechanism, all vertices are now highlighted with blue or sky-blue. Following this, all the blue nodes in the output-set of algorithm 1 will construct a CDS, and the sky-blue nodes play the role of ordinary nodes. The CDS construction mechanism converts the random ICN topology into a two-level hierarchical organization. The nodes present inside CDS called as core nodes are present at the top hierarchy. The rest of the nodes called as ordinary nodes are present at the bottom hierarchy.

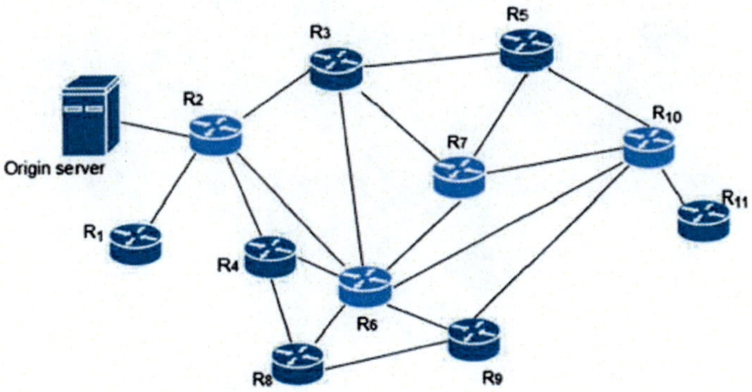

Fig. 2 Construction of CDS in ICN scenario

Algorithm for CDS construction mechanism:
Step 1: Initially highlight each vertex v with yellow.
Step 2: Choose the vertex v that has highest degree and highlight it with blue. Highlight each neighbor of v with sky-blue.
Step-3: Choose a sky-blue vertex v that has maximum yellow neighboring nodes. Highlight it with blue and each of its adjacent yellow nodes with sky-blue.
Step-4: Go to step no. 3 until each of the vertex v is highlighted with sky-blue or blue.

Fig. 3 CDS construction mechanism

3.2 Network Decomposition

The ICN network topology is decomposed based on the role like core nodes and ordinary nodes. The core nodes present inside CDS will form a virtual backbone network for ICN. The cooperative caching mechanism is focused around core nodes. The core nodes are used to break down the complex ICN topology into distinct subnetworks. The key point is that for any ICN network topology and related CDS, the decomposition of network into groups is implemented artificially while keeping in mind topological information and real-time needs [6].

Each node present inside CDS is termed as a hub node that acts as a central controller for several ordinary nodes. Following this, the network is decomposed into distinct subgroups. For example consider the network topology in Fig. 2, the decomposition of network for the same is depicted in Fig. 4. A dotted geometry shows a subnetwork. Each of subnetworks contains one hub (core router) and several ordinary routers. Core routers and ordinary routers can be differentiated based on their assigned roles and resources for example heterogeneous caching capacity. Core routers act as a gateway nodes and have larger cache capacity.

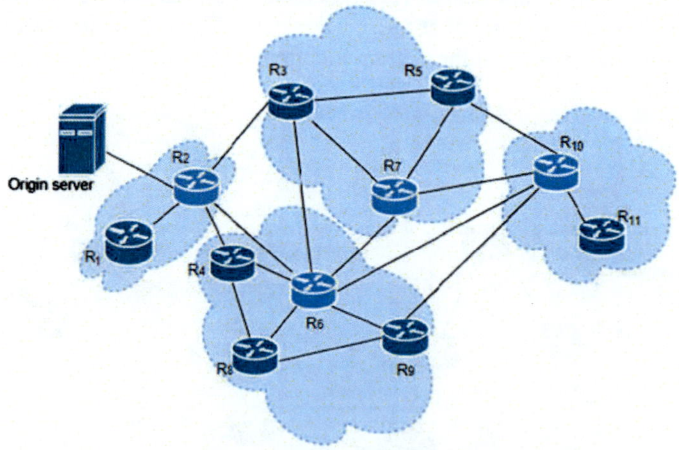

Fig. 4 Illustration for network decomposition event in ICN scenario

3.3 Cooperative Caching with Request Forwarding Strategy

In this section, we discuss the cooperative caching with request forwarding strategy. The proposed cooperative strategy is designed from the point of view of caching method and request forwarding. The majority of existing literature in ICN treats routing and caching as a two independent procedures. Traditionally, caching method is chosen as per policy and request forwarding refers forwarding information base (FIB) table. Proposed approach considers caching and routing as strongly co-related processes. The caching mechanism takes content placement related decision with consideration of available routing information. In oppose to that, the forwarding table is built depending on the caching mechanism as well as routing procedure.

(1) Caching mechanism

The key objective of proposed caching mechanism is to avoid excessive duplication of content inside network. In order to achieve the same, core routers will cache the content that is most popular. So, in general, cache variety can be achieved. The proposed strategy is formulated as follows:

- In contradiction to traditional on-path caching mechanism, whenever data packet travel from hitting source to requestor, only core routers along the path cache the data packet. The ordinary routers along the path do not cache the data packet.
- The hub router will use the least frequently used (LFU) caching mechanism to replace content whenever the cache capacity exceeds.
- A deleted content from hub router is sent randomly toward anyone of its ordinary routers located at one-hop distance. A caching index is maintained at hub router whenever they remove any content. Index entry contains two fields (content ID, egress router), namely unique content identifier, and the egress router to which the content is forwarded.
- The ordinary routers adapt LFU caching mechanism to replace content and discard the eliminated contents directly.
- The most popular chunks of content will be pushed toward hub routers and less popular contents toward ordinary routers. In order to reduce the caching duplication, content chunk will not be present inside core and ordinary routers at the same time. The significance of index entry in core router is that a replica of that content has been cached at one hop away distance. This can be used to direct the subsequent requests. Router will eliminate an index entry as per defined time to live attribute (TTL). The TTL attribute can be calculated by the average content survival duration in ordinary routers. It can also be modeled as a statistically experimented attribute.

(2) Request forwarding

The traditional routing solution takes request forwarding decision independently as per the FIB table. The proposed routing strategy refers FIB table and also exploits the caching information. It also use the concept of betweenness centrality (BC) parameter related to router. BC is a centrality measure in network based on shortest routes. For every pair of routers inside network, there exists minimum one shortest path between the routers, so that any of the following case becomes minimal. (1) Total links that the path travels along (Topology with no weights) (2) summation of link weights (Topology with weights). The BC value for router is the count of these shortest paths that travel via this router. BC shows the degree of which routers stand among each other. The CR with higher BC value leads to the fact that it has more control over entire network as majority of content chunks might have selected that router to forward interest/data packets. This fact will increase the likelihood for cache hit at that node for needed content. The BC value for CR will be calculated using following Eq. 1.

$$\mathrm{BC}(R) = \frac{\sum_{A\#R\#B} \sigma_{AB}(R)}{\sigma_{AB}} \tag{1}$$

where σ_{AB} denotes the number of shortest routes from CR called A to CR called B. $\sigma_{AB}(R)$ denotes the number of those paths that goes through CR called R. The BC value of a CR scales down with the number of CR pairs as directed by the summation indices. So, the calculation may be rescaled with division of it by the number of CR pairs, excluding R so that BC(R) $\in [0, 1]$. The division is performed by $(N - 1)$ $(N - 2)/2$ in case of undirected network topology where N means the total CRs within topology.

The in-detail routing strategy is depicted in Fig. 5. The primary idea of algorithm is to forward the unsatisfied request packets firstly toward hub routers of that subnetwork. The hub routers store the content with highest popularity. It also uses an index structure to store the storage details related to their ordinary routers. Because of this fact, request packets have higher likelihood for cache hit at hub routers. When the content search at hub router becomes unsuccessful, then the ordinary content routers will be examined based on the index entry maintained at hub router. Proposed strategy forwards the interest packet toward CR with highest BC value, when it does not find any index entry for requested chunk in hub router. This mechanism aims to fetch content from fastest possible route in order to reduce content retrieval delay and increase user level performance parameters.

4 Implementation Plan

The implementation of the proposed protocol will be carried out inside ndnSIM, NS-3 driven simulator. The performance evaluation will be done with respect to realistic Internet-based network topologies like US26, Euro28, and BRITE. The LFU caching mechanism will be adapted at hub as well as ordinary content routers to preserve popular content inside network. Using CDS algorithm, the entire network topology will be decomposed into disjoint subnetworks with single hub router and multiple ordinary routers within it. The most popular and highly requested content will be

```
Algorithm 2: Request Forwarding Strategy
INPUT: Content router CRi gets an interest packet
BEGIN
Verify inside its own CS (content store);
if CS contains data chunk then
    Return Data packet to requestor;
else
  if content router CRi is a hub router then
      Scan the index data structure maintained at CRi;
      if CRi has an index entry (Content-name, ordinary CR
          name) related to requested chunk details
      then
          Forward the request packet towards related ordinary
          content router;
      else
          Forward the request packet towards nearest 1-hop CRi
          with max BC value;
      end if
  else
    Forward content request to its related hub router in that
    subnetwork;
  end if
end if
```

Fig. 5 Request packet forwarding strategy

stored at hub sites which will eventually increase cache hit ratio and decrease content retrieval latency as per the directions imparted by proposed cooperative caching with request forwarding strategy in information-centric networking. The scalability aspect of proposed protocol will also be examined in order to validate protocol performance in real-time network scenario.

5 Conclusion

This paper utilizes the concept of connected dominating set to build backbone network for ICN. This actually decomposes the entire ICN topology into two-level hierarchical paradigm. The proposed cooperative strategy considers placement of content along with request packet forwarding. The proposed approach aims to forward content requests toward caches that have higher probability of having desired content. The proposed approach uses the concept of CDS and BC value of node to accomplish its objective. This lead to reduction in user level performance parameter like content retrieval delay. In future, we will investigate the performance of proposed strategy for realistic Internet topologies in ndnSIM-2.0, NS-3 based NDN simulator.

References

1. Jacobson, V., Smetters, D.K., Thornton, J.D., Plass, M.F., Briggs, N.H., Braynard, R.L.: Networking named content. In: ACM International Conference on Emerging Networking Experiments and Technologies, New York, United States, December, 2009
2. Dutta, N., Sarma, H.K.D., Polkowski, Z.: Cluster based routing in cognitive radio adhoc networks: reconnoitering SINR and ETT impact on clustering. Comput. Commun. **115**, 10–20 (2018)
3. Dutta, N., Sarma, H.K.D.: A probability based stable routing for cognitive radio adhoc networks. Wirel. Netw. **23**(1), 65–78, 2017
4. Sathwara, S., Dutta, N., Pricop, E.: IoT forensic a digital investigation framework for IoT systems. In: 10th IEEE International Conference on Electronics, Computers and Artificial Intelligence (ECAI), pp. 1–5, Romania, 2018
5. Delvadia, K., Dutta, N., Ghinea, G.: An efficient routing strategy for information centric networks. In: IEEE International Conference on Advanced Networks and Telecommunications Systems, Goa, India, December, 2019
6. Xu, Y., Li, Y., Lin, T., Zhang, G., Wang, Z., Ci, S.: A dominating set-based collaborative caching with request routing in content centric networking. In: IEEE International Conference on Communications (ICC), Budapest, Hungary, June, 2013
7. Dai, J., Hu, Z., Li, B., Liu, J., Li, B.: Collaborative Hierarchical Caching with Dynamic Request Routing for Massive Content Distribution. IEEE INFOCOM, Florida, USA, March 2012
8. Rossi, D., Rossini, G.: Caching Performance of Content Centric Networks Under Multi-Path Routing (and more). Technical Report, Telecom ParisTech, 2011
9. Cartigny, J., Simplot, D., Stojmenovic, I.: Localized Minimum-Energy Broadcasting in Ad-hoc Networks. IEEE INFOCOM San Francisco, USA, March, 2003
10. Bektas, T., Cordeau, J.F., Erkut, E., Laporte, G.: Exact algorithms for the joint object placement and request routing problem in content distribution networks. Comput. Oper. Res. **35**(12), 3860–3884 (2008)
11. Laoutaris, N., Zervas, G., Bestavros, A., Kollios, G.: The Cache Inference Problem and its Application to Content and Request Routing. IEEE INFOCOM, Alaska, USA, May, 2007
12. Rosensweig, E.J., Kurose, J.: Breadcrumbs, Efficient, Best-Effort Content Location in Cache Networks. IEEE INFOCOM, Rio de Janeiro, Brazil, April, 2009
13. Jiang, A.X., Bruck, J.: Optimal content placement for en-route web caching. In: IEEE International Symposium on Network Computing and Applications, pp. 9–16, 2003
14. Li, Y., Lin, T., Tang, H., Sun, P.: A chunk caching location and searching scheme in content centric networking. In: IEEE International Conference on Communications, Ottawa, Canada, June, 2012
15. Psaras, I., Chai, W.K., Pavlou, G.: Probabilistic in-network caching for information-centric networks. In: ACM Workshop on Information Centric Networking (ICN), Helsinki, Finland, August, 2012
16. Thar, K., Tran, N.H., Ullah, S., Oo, T.Z., Hong, C.S.: Online caching and cooperative forwarding in information centric networking. IEEE Access **6**, 59679–59694 (2018)
17. Saha, S., Lukyanenko, A., Ylä-Jääski, A.: Cooperative Caching Through Routing Control in Information-Centric Networks. IEEE INFOCOM, Turin, Italy, April, 2013

Instabilities of Consensus

Priya Ranjan

Abstract Consensus protocol analysis suffers from unrealistic instantaneous information communication assumption. To address this, we shall consider arbitrarily large delays in consensus protocols and explore its impact on dynamical behavior of convergence. We formulate a conjecture about emergence of slowly oscillating periodic (SOP) orbits as delay becomes significant from sine-like waves to square-waves. Further, dynamical stability behavior of (i) minimally connected multi-agent configuration, i.e., a chain, (ii) fully connected graph, and finally the intermediate configuration, (iii) multiple agents connected in a circular fashion with exactly one full length closed path will be reported. It will be shown that for chain configuration, period doubling behavior will be observed irrespective of number of agents. For a fully connected network configuration guranteed and stable convergence to consensus point with equal contributions from all agents, i.e., peer-setup will be proved irrespective of delay and number of agents. Surprisingly, it will be demonstrated that multiple agents connected in a full length closed path will converge to consensus with equal contribution from all nodes irrespective of delay if number of agents is odd, while they will oscillate in the period doubling fashion if number of agents is even. A bifurcation diagram with link strength parameter which transitions a chain into fully connected closed path will be presented to illustrate interesting branching processing from period doubling to single equilibrium consensus point. Connection with theory of circulant matrix [11, 12] will be pointed out for future deeper investigations.

Keywords Onsensus · Delay · Slowly oscillating periodic orbit · Hopf

1 Introduction

Accessibility to economic, reliable, and high-bandwidth network services have enabled many new technological phenomena like consensus among remote agents. While most of the stress in the scientific research community is on proving the existence of consensus with instantaneous information availability [2], there are papers

P. Ranjan (✉)
Department of ECE, SRM-AP University, Mangalgiri, Andhra Pradesh, India
e-mail: ranjan.p@srmap.edu.in

H. K. D. Sarma et al. (eds.), *Contemporary Issues in Communication, Cloud and Big Data Analytics*, Lecture Notes in Networks and Systems 281,
https://doi.org/10.1007/978-981-16-4244-9_9

modeling different delays in the information exchanges among nods x_i and x_j for $i, j = 1, 2, \ldots, n$ in the community of n-nodes [2] where a local linearization-based approach has been taken for finite delays. In contrast with earlier models studied [17], we take an approach where communication delay problem has been looked from non-linear analysis perspective, and we take the model with communication delay $T \to \infty$ and discover a rich set of dynamical behavior which are amenable to difference equations as compared to differential equations in the limit of $T \to 0$. We seek stability results for all values of $T \gg 0$ [3–7]. It seems like nature first made difference equations in the backend and then wrapped the frontend of delay-differential equation. In particular, the objective of this work is to explore into the phenomena of delay-induced instability in consensus over an arbitrary network configuration of multi-agent system. In general, this work will contribute toward answering these questions:

1. What is the impact of introducing large delay $T \gg 0$? It turns out that introduction of large delay leads to a difference equation as compared to delay-differential equation whose dynamical behavior in turn determines the same of underlying delay-differential Eq. [7]. In this era of digital computation, clearly, difference equations (even in matrix format) are far easier and quicker to compute/simulate as compared to their underlying delay-differential equations which take large amount of memory as amount of delay increases, slowing down the simulation performance drastically.
2. All the nonlinearity in the system is captured in function $f(\cdot)$ both as direct application and in its inverse. Future values depend on $f(\cdot)$ but not as deeply and significantly as we would have imagined which is evident from Eq. 10 where it seems like network structure has more influence in future evolution of states of multi-agents, participating in the consensus.
3. The most important observation is about impact of underlying graph or connectivity structure which is solely responsible for multiple dynamical phenomena even when there is no major nonlinearity present. As we see that future iterates of $[D^{-1}A]^m \to C, 0 < m \to \infty$ where D is degree matrix and A is an adjacency matrix of underlying graph of multi-agents participating in the consensus. C is a consensus matrix with all equal entries $\frac{1}{n}$. We shall demonstrate that impact of $D^{-1}A$ matrix which in turn will decide that which multi-agent configurations are going to oscillate as a difference equation which in turn will result in a slowly oscillating periodic orbit(SOP) and which configurations of multi-agents as coded in D and A will converge the peer-structure consensus with equal contributions from each agent or a rank-based power structure where one node has more contribution than the others.
4. In particular, it will be shown that the dynamical behavior from diverging branches of period doubling will merge into once solution branch toward consensus point with equal participation with link gain as a bifurcation parameter. We believe that this is a first period doubling bifurcation study for multi-agent consensus. It should be remembered that period doubling mode is lack of consensus in time; i.e., nodes will have oscillating opinions without reaching actual

consensus points. This instability is serious in the sense that its not real difference of opinion or stands among agents which is hampering consensus, rather its delayed information communication structure or stale states which are causing this oscillation. Clearly, this kind of instabilities can be controlled by introducing more low-latency links which may be costly in real life. Lack of hi-bandwidth and low-latency communication resources is responsible for this kind of instability.

This work is organized as follows. Model development and framework is provided in Sect. 2. Section 3 described the stable behavior of fully connected graphs. Section 4 discusses instability of linear chain configuration. Section 5 talks about ring configuration stability behavior for even n and Sect. 7 discusses the same for odd n. Section 6 talks about the stability of intermediate configuration between a linear chain and ring. Section 8 discusses bifurcation diagrams as linear chain approaches a ring with connectivity strength as a bifurcation parameter. Section 9 illustrates the competitive behavior and finally we collect conclusions in Sect. 10.

2 Model Development and Framework

We start with simple model of generalized absolute nonlinear flow in consensus over an undirected graph $G(V, E)$ with degree matrix D and adjacency matrix A.

$$\dot{x_i}(t) = \sum_{j=1, j \in Nbd_i} k_{ij}(f_j(x_j(t)) - f_i(x_i(t))) \tag{1}$$

Equation 1 is described as absolute nonlinear flow in [2] for i_{th} agent's state evolution equation without feedback delays which can be incurred due to propagation, queueing, or processing and can be significant in the case of unavailability of underlying physical layer or during congestion [7]. In this work, a significantly delayed version of Eq. 1 is proposed where for ease of presentation and delay symbol book-keeping, information delay has been assumed to $0 \ll T$ over all links which can be thought of maximum of delays. Nbd_i is a representation of neighborhood of node i, i.e., collection of nodes directly connected to node i.

Agent i is supposed to have ready access to $f_i(x_i)$ as compared to $f_j(x_j)$, $j \neq i$ which is supposed to travel over the presumably congested communication channel, and hence, only a stale version of $f_j(x_j(t - T))$ is going to be available at node i which is a T-delayed version of $f_j(x_j)$ with $T \gg 0$. In fact, we can admit arbitrarily large T and provide the stability results in the limit of $T \to \infty$. Rewriting a more realistic version of Eq. 1 leads to,

$$\dot{x_i}(t) = \sum_{j=1, j \in Nbd_i} k_{ij}(f_j(x_j(t - T)) - f_i(x_i(t))) \tag{2}$$

Doing a traditional time scaling $T\tau = t$ leads to $T d\tau = dt$. Rewriting Eq. 2 with new scaled time τ will lead to

$$\begin{aligned}
\frac{\mathrm{d}x_i}{\mathrm{d}t}(t) &= \sum_{j=1, j\in Nbd_i} k_{ij}(f_j(x_j(t-T)) - f_i(x_i(t))) \\
\frac{\mathrm{d}x_i}{T\mathrm{d}\tau}(T\tau) &= \sum_{j=1, j\in Nbd_i} k_{ij}(f_j(x_j(T\tau-T)) - f_i(x_i(T\tau))) \\
\frac{\mathrm{d}x_i}{T\mathrm{d}\tau}(\tau) &= \sum_{j=1, j\in Nbd_i} k_{ij}(f_j(x_j(\tau-1)) - f_i(x_i(\tau))) \\
\text{let } \nu &= \frac{1}{T} \\
\nu\frac{\mathrm{d}x_i}{\mathrm{d}\tau}(\tau) &= \sum_{j=1, j\in Nbd_i} k_{ij}(f_j(x_j(\tau-1)) - f_i(x_i(\tau)))
\end{aligned} \tag{3}$$

where Eq. 3 can be seen as a singular perturbation of $\sum_{j=1, j\in Nbd_i} k_{ij}(f_j(x_j(\tau - 1)) - f_i(x_i(\tau)))$ as $\nu \to 0$ with $T \to \infty$ has been researched heavily in [1, 3–7]. It turns out that this kind of delay-differential equations behaves like a difference equation in the limit of $0 \ll T \to \infty$. Stability and dynamical properties of underlying discrete time maps will determine the dynamical behavior of underlying governing differential Eq. 3. Further, their oscillatory behavior is known as slowly oscillating periodic (SOP) orbit which becomes more like square wave as compared to sine wave when $t \to \infty$ [1, 3, 6]. Some simple cases are presented to demonstrate the main role played by underying graph structure modeling or presenting the multi-agent configuration.

2.1 When $k_{ij} = 1, \forall i = 1, \ldots, n$ and $j = 1, \ldots, n$

Coming back to Eq. 1 is written as

$$\dot{x}_i(t) = \sum_{j=1, j\in Nbd_i} (f_j(x_j(t)) - f_i(x_i(t))) \equiv -Ł(G)f(x) \tag{4}$$

where $Ł(G) := D(G) - A(G)$ is Laplacian of underlying connecting graph G, $D(G)$ is degree matrix, and $A(G)$ is adjacency matrix. Taking delay $T \gg 0$ into account and using graph-theoretic language, Eq. 4 can be rewritten as,

$$\nu\dot{x}(t) = A(G)f(x(t-1)) - D(G)f(x(t)) \tag{5}$$

Now the difference equation which is going to determine the dynamical behavior of Eq. 5 can be written after time scaling and letting $T \to \infty$ as

$$Df(x(\tau)) = Af(x(\tau - 1)) \tag{6}$$

Underlying difference equation for Eq. 6 can be written as:

$$Df(x_{n+1}) = Af(x_n)$$

$$f(x_{n+1}) = D^{-1}Af(x_n)$$

D is diagonal degree matrix with $rank(D) = n$, hence it is invertible. $D^{-1}A$ is also known as transition matrix in Markov Chain literature[8].

$$x_{n+1} = f^{-1}D^{-1}Af(x_n) \tag{7}$$

Assuming that f_i are monotone, and hence invertible.

What is even more interesting is second iterate (Eq. 9) of difference equation (Eq. 7) and $m - th$ iterate in general as given by Eq. 10:

$$\begin{aligned} x_{n+2} &= f^{-1}D^{-1}Af(f^{-1}D^{-1}Af(x_n)) \\ x_{n+2} &= f^{-1}D^{-1}AD^{-1}Af(x_n) \end{aligned} \tag{8}$$

$$x_{n+2} = f^{-1}[D^{-1}A]^2 f(x_n) \tag{9}$$

This gives second iterate!.

$$x_{n+m} = f^{-1}[D^{-1}A]^m f(x_n) \tag{10}$$

which is a generalization to $m - th$ iterate!.

2.2 *Observations*

Equation 10 shows the impact of three different factors in a very telling fashion.

1. Impact of introducing delay leads to a structure as difference equation as compared to differential equation. Difference equations are far easier and quicker to compute/simulate as compared to their differential equation envelope.
2. All the nonlinearity in the system is captured in function $f(\cdot)$ both as direct application and in its inverse. Future values depend on $f(\cdot)$ but not as deeply and significantly as we would have imagined which is evident from Eq. 10
3. Final and the most important observation is about impact of underlying graph or connectivity. As we see that future iterates have significant influence of $[D^{-1}A]^m$ which only become even more prominent asymptotically as $m \to \infty$. This is an

innovative and significant observation, and it can be precomputed to speed up values of future iterates once D and A are known from the graph.

2.3 Impact of Large Delay $T \gg 0$

To motivate the difficulties arising due to introduction of significant delay in consensus equation with $f_i = 1$ for $i = 1, 2$, a simple simulation of two node has been proposed. Its behavior has been documented for various values of delays which provides deep insights into its behavior as $T \to \infty$ which is more realistic and robust in the event of large delays.

$$\dot{x_1}(t) = x_2(t - T) - x_1(t) \tag{11}$$

$$\dot{x_2}(t) = x_1(t - T) - x_2(t) \tag{12}$$

This simple simulink simulation shows the serious influence of large delay as it goes from $T = 1$ in Fig. 1 to $T = 500$ in Fig. 7. While influence of delay can be easily ignored for all practical purposes when $T = 1$, it has significant oscillations for simulation time of 500,000 for $T = 500$ and still refuses to settle. What looked rock stable for $T = 0$ is shaking and oscillating wildly for $T = 500$, and this work is precisely exploring these instabilities in the limit of $T \to \infty$. One should note the different x-scales in Figs. 1, 2, 3, 4, 5, 6 and 7 which show the increasingly large time taken for the consensus to stabilize.

Rewriting Eqs. 11 and 12 in graph-theoretic notations:

$$\dot{x}(t) = -Dx(t) + Ax(t - T) \quad \text{where } x = [x_1 \;\; x_2]^T \tag{13}$$

Fig. 1 $T = 1$

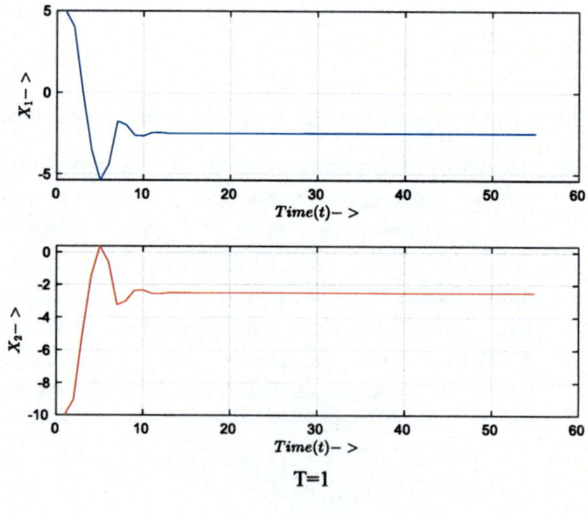

Fig. 2 $T = 5$

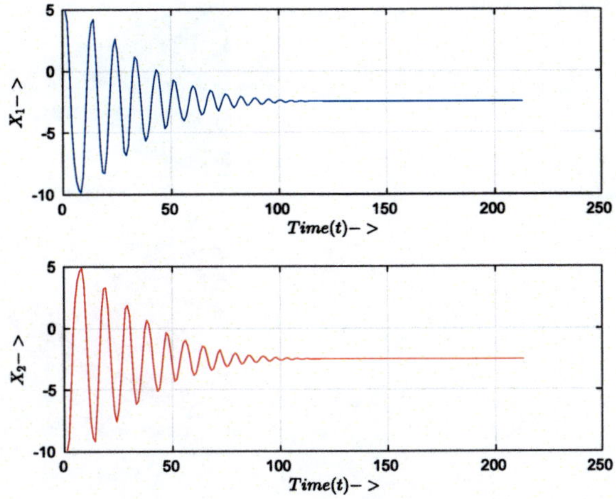

T=5

Fig. 3 $T = 10$

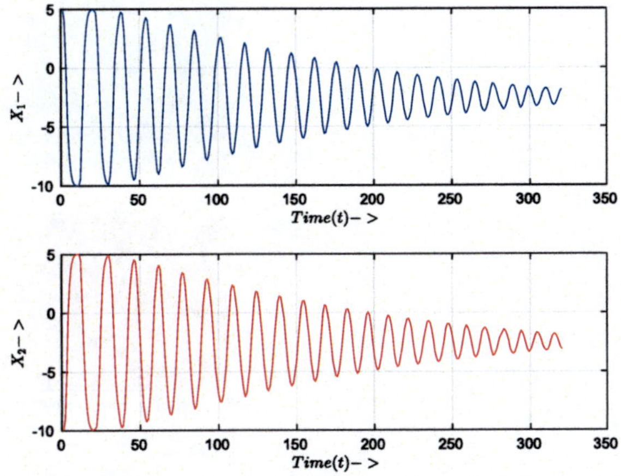

T=10

where $D = \begin{bmatrix} 1 & 0 \\ 0 & 1 \end{bmatrix}$ as degree matrix for two nodes and $A = \begin{bmatrix} 0 & 1 \\ 1 & 0 \end{bmatrix}$ as adjacency matrix for two nodes. Computing difference equation for these two nodes who are trying to reach a consensus is given by

Fig. 4 $T = 50$

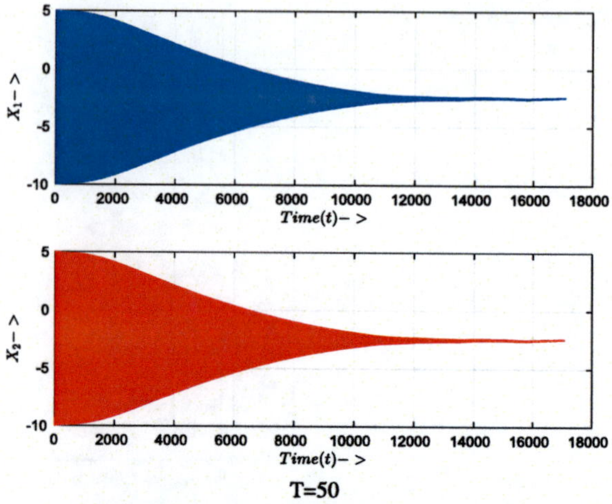

Fig. 5 $T = 100$

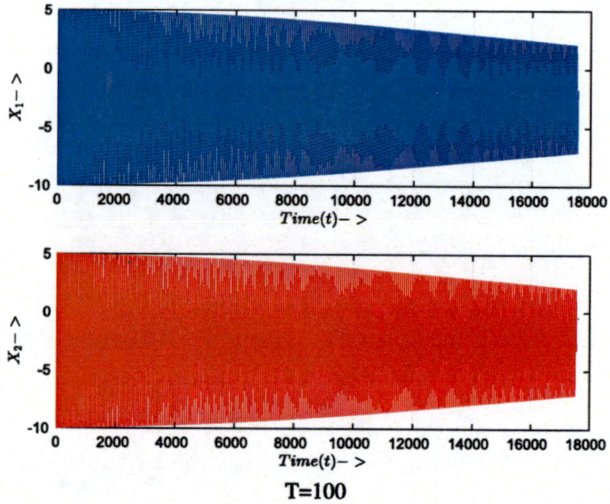

$$x_{n+1} = D^{-1}Ax_n$$

$$x_{n+1} = \begin{bmatrix} 0 & 1 \\ 1 & 0 \end{bmatrix} x_n$$

$$x_{n+2} = \begin{bmatrix} 0 & 1 \\ 1 & 0 \end{bmatrix} * \begin{bmatrix} 0 & 1 \\ 1 & 0 \end{bmatrix} x_n = \begin{bmatrix} 1 & 0 \\ 0 & 1 \end{bmatrix} x_n \tag{14}$$

$$x_{n+2} = x_n \tag{15}$$

This essentially shows that there is a period doubling mode of consensus system as $T \to \infty$. System is NOT converging and just swapping the initial condition of $x_1(0)$ to $x_2(0)$ and vice-versa without any convergence toward the actual consensus value. Period doubling instability is not limited to consensus situations, and they manifest

Fig. 6 $T = 250$

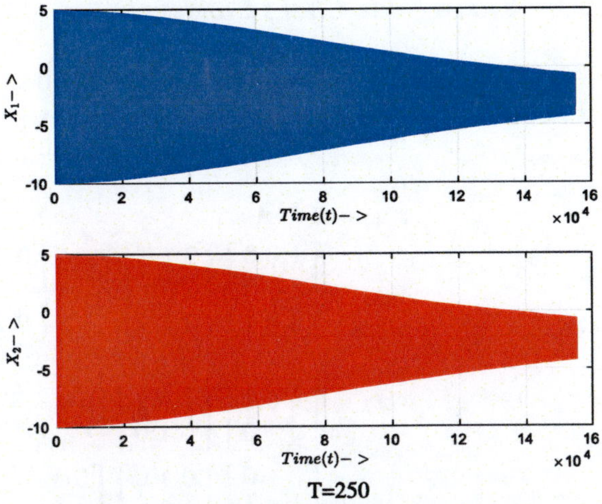

Fig. 7 $T = 500$

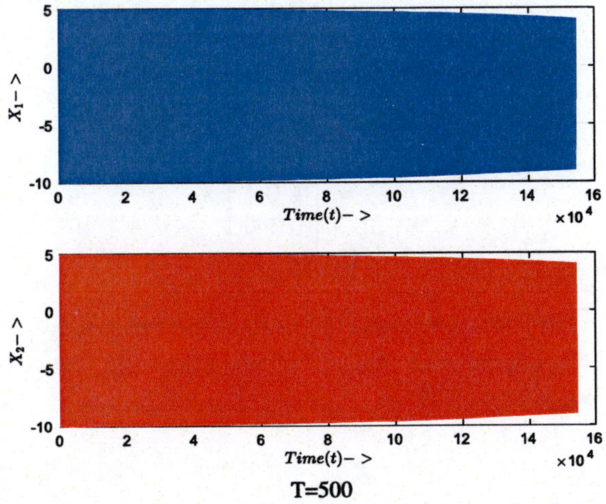

themselves in the form of count-to-infinity problem in distance vector routing [9]. The precise reason for such instabilities is limited local informatic view among nodes which is often counted as a blessing for engineering systems. This precise blessing turns into a curse when it manifests as period doubling and count-to-infinity problems.

2.4 Three Node Chain Consensus

For this scenario as shown in Fig. 10, $A = \begin{bmatrix} 0 & 1 & 0 \\ 1 & 0 & 1 \\ 0 & 1 & 0 \end{bmatrix}$ and $D = \begin{bmatrix} 1 & 0 & 0 \\ 0 & 2 & 0 \\ 0 & 0 & 1 \end{bmatrix}$.

$$
\begin{aligned}
x_{n+1} &= D^{-1} A x_n \\
x_{n+1} &= \begin{bmatrix} 1 & 0 & 0 \\ 0 & \frac{1}{2} & 0 \\ 0 & 0 & 1 \end{bmatrix} \begin{bmatrix} 0 & 1 & 0 \\ 1 & 0 & 1 \\ 0 & 1 & 0 \end{bmatrix} x_n = \begin{bmatrix} 0 & 1 & 0 \\ \frac{1}{2} & 0 & \frac{1}{2} \\ 0 & 1 & 0 \end{bmatrix} x_n \\
x_{n+2} &= \begin{bmatrix} 0 & 1 & 0 \\ \frac{1}{2} & 0 & \frac{1}{2} \\ 0 & 1 & 0 \end{bmatrix} \begin{bmatrix} 0 & 1 & 0 \\ \frac{1}{2} & 0 & \frac{1}{2} \\ 0 & 1 & 0 \end{bmatrix} x_n = \begin{bmatrix} \frac{1}{2} & 0 & \frac{1}{2} \\ 0 & 1 & 0 \\ \frac{1}{2} & 0 & \frac{1}{2} \end{bmatrix} x_n
\end{aligned}
\tag{16}
$$

$$
\begin{aligned}
x_{n+3} &= \begin{bmatrix} 0 & 1 & 0 \\ \frac{1}{2} & 0 & \frac{1}{2} \\ 0 & 1 & 0 \end{bmatrix} \begin{bmatrix} 0 & 1 & 0 \\ \frac{1}{2} & 0 & \frac{1}{2} \\ 0 & 1 & 0 \end{bmatrix} \begin{bmatrix} 0 & 1 & 0 \\ \frac{1}{2} & 0 & \frac{1}{2} \\ 0 & 1 & 0 \end{bmatrix} x_n \\
&= \begin{bmatrix} 0 & 1 & 0 \\ \frac{1}{2} & 0 & \frac{1}{2} \\ 0 & 1 & 0 \end{bmatrix} \begin{bmatrix} \frac{1}{2} & 0 & \frac{1}{2} \\ 0 & 1 & 0 \\ \frac{1}{2} & 0 & \frac{1}{2} \end{bmatrix} x_n \\
&= \begin{bmatrix} 0 & 1 & 0 \\ \frac{1}{2} & 0 & \frac{1}{2} \\ 0 & 1 & 0 \end{bmatrix} x_n \\
&= x_{n+1}
\end{aligned}
\tag{17}
$$

It is very interesting to note that Eq. 17 again demonstrated period doubling mode where system returns to it previous state after two iterations rather than converging to consensus values.

2.5 Three Node Ring Consensus

For this scenario as shown in Figs. 8 and 9, $A_{3\times 3} = \begin{bmatrix} 0 & 1 & 1 \\ 1 & 0 & 1 \\ 1 & 1 & 0 \end{bmatrix}$ and $D_{3\times 3} = \begin{bmatrix} 2 & 0 & 0 \\ 0 & 2 & 0 \\ 0 & 0 & 2 \end{bmatrix}$.

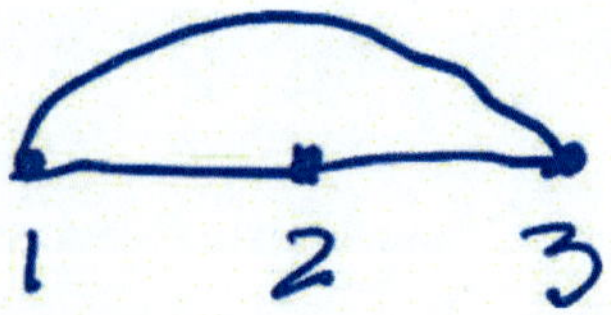

Fig. 8 Three nodes in a ring configuration with 1st node connected to 3rd node

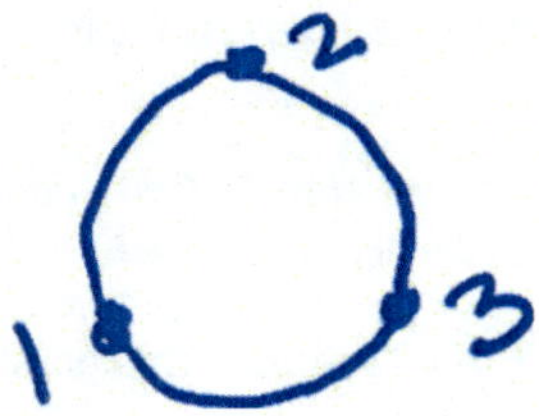

Fig. 9 Three nodes in an equivalent circular/ring configuration with 1st node connected to 3rd node

$$x_{n+1} = D_{3\times3}^{-1} A_{3\times3} x_n \tag{18}$$

$$x_{n+1} = \begin{bmatrix} \frac{1}{2} & 0 & 0 \\ 0 & \frac{1}{2} & 0 \\ 0 & 0 & \frac{1}{2} \end{bmatrix} * \begin{bmatrix} 0 & 1 & 1 \\ 1 & 0 & 1 \\ 1 & 1 & 0 \end{bmatrix} x_n = \begin{bmatrix} 0 & \frac{1}{2} & \frac{1}{2} \\ \frac{1}{2} & 0 & \frac{1}{2} \\ \frac{1}{2} & \frac{1}{2} & 0 \end{bmatrix} x_n$$

$$x_{n+2} = \begin{bmatrix} 0 & 1 & 0 \\ \frac{1}{2} & 0 & \frac{1}{2} \\ 0 & 1 & 0 \end{bmatrix} * \begin{bmatrix} 0 & 1 & 0 \\ \frac{1}{2} & 0 & \frac{1}{2} \\ 0 & 1 & 0 \end{bmatrix} x_n = \begin{bmatrix} \frac{1}{2} & \frac{1}{4} & \frac{1}{4} \\ \frac{1}{4} & \frac{1}{2} & \frac{1}{4} \\ \frac{1}{4} & \frac{1}{4} & \frac{1}{2} \end{bmatrix} x_n$$

$$(D^{-1}A)^{15} \rightarrow \begin{bmatrix} \frac{1}{3} & \frac{1}{3} & \frac{1}{3} \\ \frac{1}{3} & \frac{1}{3} & \frac{1}{3} \\ \frac{1}{3} & \frac{1}{3} & \frac{1}{3} \end{bmatrix} \tag{19}$$

Further iterations do not seem to change it!

$$x_{n+1}^i = \frac{1}{3} \sum_{i=1}^{i=3} x_0^i \text{for all } n > 15, \forall i \tag{20}$$

In three node ring case, consensus is stable and robust for all values of $T \gg 0$ and is guranteed. It is interesting to note that instability induced by delay is somehow being compensated by full connectivity of 3-node ring. There is a clear trade-off between impact of destabilization by significantly large $T \gg 0$ and stabilizing effect of full connectivity leading to stable consensus behavior for all values of T, irrespective of its size. **Lesson learnt is high connectivity for fast and robust consensus.** Now, an optimal control problem can be formulated if we account for the cost of connectivity and stability it induces on consensus!.

3 Fully Connected Graph Will Achieve Consensus for All *n*

Next we prove that strongly connected or fully connected graph will always converge to its consensus value irrespective of $0 \ll T \rightarrow \infty$ which is in consonance with similar result shown proved in [10]. Our result clearly is dynamically stronger because we show full convergence for arbitrary large communication delays, whereas results in [10] are proved for instantaneous interactions which can be thought of a limiting case of our model as $\lim_{T \rightarrow 0}$. Our case stands as more robust even in the event of delayed interactions and will always perform better than this worst case if delay is not very large.

Proposition 1 *Fully connected graph will always converge to its consensus value irrespective of number of nodes (n can be even or odd) and $T \gg 0$.*

Proof To prove this point, we need to show that $(D^{-1}A)^k, k \to \infty$ is going to converge to consensus matrix $C = \begin{bmatrix} \frac{1}{n} & \cdot & \cdot & \frac{1}{n} \\ \frac{1}{n} & \cdot & \cdot & \frac{1}{n} \\ \cdot & \cdot & \cdot & \cdot \\ \frac{1}{n} & \cdot & \cdot & \frac{1}{n} \end{bmatrix}$, i.e., to an nxn matrix with all entries $c_{ij}, \forall 1 \le i \le n, 1 \le j \le n$ being $\frac{1}{n}$.

Let us do this calculation for a fully connected graph with n nodes:

$$D = \begin{bmatrix} n-1 & 0 & 0 \ldots & 0 \\ 0 & n-1 & 0 \ldots & 0 \\ \cdot & \cdot & \cdot\cdot\cdot\cdot & \cdot \\ 0 & 0 & 0 \ldots & n-1 \end{bmatrix}$$

$$D^{-1} = \begin{bmatrix} \frac{1}{n-1} & 0 & 0 \ldots & 0 \\ 0 & \frac{1}{n-1} & 0 \ldots & 0 \\ \cdot & \cdot & \cdot\cdot\cdot\cdot & \cdot \\ 0 & 0 & 0 \ldots & \frac{1}{n-1} \end{bmatrix}$$

$$A = \begin{bmatrix} 0 & 1 & 1 & \ldots & 1 \\ 1 & 0 & 1 & \ldots & 1 \\ \cdot & \cdot & \cdot & \cdot\cdot\cdot & \cdot \\ 1 & 1 & 1 & \ldots & 0 \end{bmatrix}$$

$$D^{-1}A = \begin{bmatrix} 0 & \frac{1}{n-1} & \frac{1}{n-1} & \cdots & \frac{1}{n-1} \\ \frac{1}{n-1} & 0 & \frac{1}{n-1} & \cdots & \frac{1}{n-1} \\ \cdot & \cdot & \cdot & \cdots & \cdot \\ \frac{1}{n-1} & \frac{1}{n-1} & \frac{1}{n-1} & \cdots & 0 \end{bmatrix}$$

$$(D^{-1}A)C = \begin{bmatrix} 0 & \frac{1}{n-1} & \frac{1}{n-1} & \cdots & \frac{1}{n-1} \\ \frac{1}{n-1} & 0 & \frac{1}{n-1} & \cdots & \frac{1}{n-1} \\ \cdot & \cdot & \cdot & \cdots & \cdot \\ \frac{1}{n-1} & \frac{1}{n-1} & \frac{1}{n-1} & \cdots & 0 \end{bmatrix} \begin{bmatrix} \frac{1}{n} & \cdot & \cdot & \frac{1}{n} \\ \frac{1}{n} & \cdot & \cdot & \frac{1}{n} \\ \cdot & \cdot & \cdot & \cdot \\ \frac{1}{n} & \cdot & \cdot & \frac{1}{n} \end{bmatrix}$$

$$= \begin{bmatrix} \frac{n-1}{n(n-1)} & \cdot & \cdot & \frac{n-1}{n(n-1)} \\ \frac{n-1}{n(n-1)} & \cdot & \cdot & \frac{n-1}{n(n-1)} \\ \cdot & \cdot & \cdot & \cdot \\ \frac{n-1}{n(n-1)} & \cdot & \cdot & \frac{n-1}{n(n-1)} \end{bmatrix} \equiv C \tag{21}$$

What this calculation shows that consensus matrix C is invariant under the action of $D^{-1}A$ for fully connected graph irrespective of even or odd n and $T \gg 0$; i.e., once system reaches to $x_n = Cx_0$, then it will stay there which is an extremely desireable property from stable, robust, and delay-independent consensus point of view for all values of $T \gg 0$ and $\forall n$. Since all the nodes are contributing equally ($\frac{1}{n}$) in consensus point, it is a peer-system rather than leader and follower configuration.

We also note that $D^{-1}A$ for fully connected graph is a circulant matrix [11, 12]. Next we demonstrate the path of $[D^{-1}A]^k$ as $k \to \infty$ as it will converge to C and will lead to robust, delay-independent, and stable consensus.

$$(D^{-1}A)^2 = \begin{bmatrix} \frac{1}{n-1} & (\frac{1}{n-1} - \frac{1}{(n-1)^2}) & \dots & (\frac{1}{n-1} - \frac{1}{(n-1)^2}) \\ (\frac{1}{n-1} - \frac{1}{(n-1)^2}) & \frac{1}{n-1} & \dots & (\frac{1}{n-1} - \frac{1}{(n-1)^2}) \\ . & . & . & \dots & . \\ . & . & . & \dots & . \\ (\frac{1}{n-1} - \frac{1}{(n-1)^2}) & (\frac{1}{n-1} - \frac{1}{(n-1)^2}) & \dots & \frac{1}{n-1} \end{bmatrix} \tag{22}$$

Clearly, $[D^{-1}A]^2$ has two kind of elements, one which is diagonal $d_{ii} = \frac{1}{n-1}, \forall i = 1, 2, \dots, n$ and then we have off-diagonal elements $d_{ij} = (\frac{1}{n-1} - \frac{1}{(n-1)^2}), \forall i \neq j$ and $i, j = 1, 2, ..n$. Next as we multiply Eq. 22 with itself to get expression for $[D^{-1}A]^4$. For this fourth power, we can easily show that $d_{ii} = (\frac{1}{n-1})^2 + (n-1)(\frac{1}{n-1} - \frac{1}{(n-1)^2})^2 = \left[\frac{1}{n-1} - \frac{1}{(n-1)^2} + \frac{1}{(n-1)^3}\right], \forall i = 1, 2, \dots, n$ and off-diagonal elements

$$\begin{aligned} d_{ij} &= \left[2\frac{(n-1)(n-2)}{(n-1)^4} + (n-2)\left[\frac{1}{n-1} - \frac{1}{(n-1)^2}\right]^2\right] \\ &= (n-2)\left[\frac{1}{(n-1)^2} + \frac{1}{(n-1)^4}\right] \\ &= \left[\frac{1}{n-1} - \frac{1}{(n-1)^2} + \frac{1}{(n-1)^3} - \frac{1}{(n-1)^4}\right], \forall i \neq j. \end{aligned} \tag{23}$$

Further it can be shown by direction multiplication that $[D^{-1}A]^8$

$$\begin{aligned} d_{ij} &= \frac{1}{n-1} - \frac{1}{(n-1)^2} + \frac{1}{(n-1)^3} - \frac{1}{(n-1)^4} \\ &+ \frac{1}{(n-1)^5} - \frac{1}{(n-1)^6} + \frac{1}{(n-1)^7} - \frac{1}{(n-1)^8}, \forall i \neq j. \\ &= \sum_{k=1}^{k=8}\left[(-)^{k+1}\frac{1}{(n-1)^k}\right] \end{aligned} \tag{24}$$

$$d_{ii} = \frac{1}{n-1} - \frac{1}{(n-1)^2} + \frac{1}{(n-1)^3} - \frac{1}{(n-1)^4} + \frac{1}{(n-1)^5} - \frac{1}{(n-1)^6} + \frac{1}{(n-1)^7}, \forall i = 1, 2, \ldots, n.$$
$$= \sum_{k=1}^{k=7} \left[(-)^{k+1} \frac{1}{(n-1)^k} \right] \quad (25)$$

Clearly, there is a pattern here, and we can compute the asymptotic values of matrix elements for matrix $[D^{-1}A]^k$ as $k \to \infty$ solving for geometric progressions given in Eq. 24 and Eq. 25. A simple geometric progression formula will show that both d_{ii} and d_{ij} will converge to the same value $\frac{\frac{1}{n-1}}{1+\frac{1}{n-1}} = \frac{1}{n}$ as $k \to \infty$ in powers of 2, i.e., 2, 4, 8, 16, 32, ... This proves that $[D^{-1}A]^k \to C$ as $k \to \infty$. We have already shown that once system reaches C, it will stay there as C is invariant under the action of $D^{-1}A$. ■

While it is comforting to know about convergence in a fully connected network, it may be an overkill and luxury of a fully connected network may not be available in many resource constrained situations like natural calamity or war-related scenarios. This motivates the exploration of dynamical behavior or convergence in sparsely connected multi-agent systems.

4 Non-convergence and Period Two Oscillation of $[D^{-1}A]^k$, $k \to \infty$ for All n in Linear Chain Configuration

Linear chain configurations as compared to the ring configurations are shown in Figs. 10 and 11.

Proposition 2 *$[D^{-1}A]^k, k \to \infty$ will NOT converge to consensus matrix C, $\forall$ n in linear chain configuration.*

Fig. 10 Three nodes in a linear chain configuration

Fig. 11 Four nodes in a linear chain configuration

Proof This is equivalent to calculation of eigenvalues of matrix $D^{-1}A$ for all n irrespective of its even or odd. Further, it needs to be shown that $\lambda = -1$ which is responsible for period doubling nature of $[D^{-1}A]^k, k \to \infty$.

$$D^{-1}A = \begin{bmatrix} 0 & 1 & . & . & . & 0 \\ \frac{1}{2} & 0 & \frac{1}{2} & . & . & 0 \\ 0 & \frac{1}{2} & 0 & \frac{1}{2} & . & 0 \\ . & . & . & . & . & . \\ 0 & . & . & \frac{1}{2} & 0 & \frac{1}{2} \\ 0 & 0 & 0 & 0 & 1 & 0 \end{bmatrix} \tag{26}$$

It can be easily observed that matrix $[D^{-1}A]$ is unsymmetric tridiagonal matrix, also known as Jacobi matrix [13] which can be made symmetric by similarity transformation [14]. Post application of similarity transformation [14], a symmetric version of $D^{-1}A$, $(D^{-1}A)_{sym}$ obtained as below. Eigenvalue of $D^{-1}A$, $(D^{-1}A)_{sym}$ are same, and we just need to show that -1 is an eigenvalue of $(D^{-1}A)_{sym}$ with corresponding eigenvector

$$v = \begin{bmatrix} 1 \\ (-1)^1\sqrt{2} \\ (-1)^2\sqrt{2} \\ . \\ . \\ (-1)^{(n-2)}\sqrt{2} \\ (-1)^{(n-1)} \end{bmatrix} \tag{27}$$

$$(D^{-1}A)_{sym} = \begin{bmatrix} 0 & \frac{1}{\sqrt{2}} & . & . & . & 0 \\ \frac{1}{\sqrt{2}} & 0 & \frac{1}{2} & . & . & 0 \\ 0 & \frac{1}{2} & 0 & \frac{1}{2} & . & 0 \\ . & . & . & . & . & . \\ 0 & . & . & \frac{1}{2} & 0 & \frac{1}{\sqrt{2}} \\ 0 & 0 & 0 & 0 & \frac{1}{\sqrt{2}} & 0 \end{bmatrix}$$

$$\begin{aligned} (D^{-1}A)_{sym}v &= \begin{bmatrix} 0 & \frac{1}{\sqrt{2}} & . & . & . & 0 \\ \frac{1}{\sqrt{2}} & 0 & \frac{1}{2} & . & . & 0 \\ 0 & \frac{1}{2} & 0 & \frac{1}{2} & . & 0 \\ . & . & . & . & . & . \\ 0 & . & . & \frac{1}{2} & 0 & \frac{1}{\sqrt{2}} \\ 0 & 0 & 0 & 0 & \frac{1}{\sqrt{2}} & 0 \end{bmatrix} \begin{bmatrix} 1 \\ (-1)^1\sqrt{2} \\ (-1)^2\sqrt{2} \\ . \\ . \\ (-1)^{(n-2)}\sqrt{2} \\ (-1)^{(n-1)} \end{bmatrix} \\ &= (-1)v. \end{aligned} \tag{28}$$

It shows that -1 is an eigenvalue of $(D^{-1}A)_{sym}$ and also of $D^{-1}A$ from [14]. This eigenvalue -1 is essentially responsible for period doubling behavior of $D^{-1}A$ irrespective of even or odd n. Hence proved. ■

This period doubling behavior will manifest as a slowly oscillating periodic orbit of period approximately $2T$ in the delay-differential equation of consensus especially becoming prominent when delay T is large.

Example 1 Let us look at linear chain connected multi-agent configuration of four nodes as shown in Fig. 11.

$$D_{4\times4} = \begin{bmatrix} 1\,0\,0\,0 \\ 0\,2\,0\,0 \\ 0\,0\,2\,0 \\ 0\,0\,0\,1 \end{bmatrix}$$

$$D_{4\times4}^{-1} = \begin{bmatrix} 1 & 0 & 0 & 0 \\ 0 & \frac{1}{2} & 0 & 0 \\ 0 & 0 & \frac{1}{2} & 0 \\ 0 & 0 & 0 & 1 \end{bmatrix}$$

$$A_{4\times4} = \begin{bmatrix} 0\,1\,0\,0 \\ 1\,0\,1\,0 \\ 0\,1\,0\,1 \\ 0\,0\,1\,0 \end{bmatrix}$$

$$D_{4\times4}^{-1}A_{4\times4} = \begin{bmatrix} 0 & 1 & 0 & 0 \\ \frac{1}{2} & 0 & \frac{1}{2} & 0 \\ 0 & \frac{1}{2} & 0 & \frac{1}{2} \\ 0 & 0 & 1 & 0 \end{bmatrix} \equiv b_{4\times4} \tag{29}$$

$$b_{4\times4} = \begin{bmatrix} -1.0000 & -0.0000 & 0.0000 & 1.0000 \end{bmatrix} \tag{30}$$

Eigenvalues of $b_{4\times4}$

$$[b_{4\times4}]^{500} = [b_{4\times4}]^{502} = \begin{bmatrix} \frac{1}{3} & 0 & \frac{2}{3} & 0 \\ 0 & \frac{2}{3} & 0 & \frac{1}{3} \\ \frac{1}{3} & 0 & \frac{2}{3} & 0 \\ 0 & \frac{2}{3} & 0 & \frac{1}{3} \end{bmatrix} \tag{31}$$

$$[b_{4\times4}]^{501} = [b_{4\times4}]^{503} = \begin{bmatrix} 0 & \frac{2}{3} & 0 & \frac{1}{3} \\ \frac{1}{3} & 0 & \frac{2}{3} & 0 \\ 0 & \frac{2}{3} & 0 & \frac{1}{3} \\ \frac{1}{3} & 0 & \frac{2}{3} & 0 \end{bmatrix} \tag{32}$$

Clearly, an eigenvalue of -1 is observed in Eq. 31, and asymptotic period doubling behavior is clear as $k \rightarrow 500$ from Eq. 31 and from Eq. 32.

5 Non-convergence and Period Two Oscillation of $[D^{-1}A]^k$, $k \rightarrow \infty$ for All Even n in Longest Path Ring/circular Configuration

Proposition 3 $[D^{-1}A]^k, k \rightarrow \infty$ *will NOT converge to consensus matrix C for all even n in longest path ring/circular configuration.*

Proof This is equivalent to calculation of eigenvalues of matrix $D^{-1}A$ for all even n. Further, it needs to be shown that $\lambda_{\frac{n}{2}} = -1$ which is responsible for period doubling nature of $D^{-1}A$ for even n. As it has been pointed out earlier that $D^{-1}A$ is a circulant matrix and due to theory of circulant matrix and discrete Fourier transform theory, closed form expressions for eigenvalues of $D^{-1}A$ are available [11, 12].

$$\lambda_{\frac{n}{2}} = \left[1\ w^{\frac{n}{2}}\ (w^{\frac{n}{2}})^2\ .\ (w^{\frac{n}{2}})^{(n-1)}\right] \cdot \begin{bmatrix} 0 & \frac{1}{2} & 0 & \dots & \frac{1}{2} \\ \frac{1}{2} & 0 & \frac{1}{2} & \dots & 0 \\ . & . & . & . & . \\ \frac{1}{2} & 0 & 0 & \dots & \frac{1}{2} \end{bmatrix} \frac{1}{n} \begin{bmatrix} 1 \\ (w^{\frac{n}{2}})^* \\ ((w^{\frac{n}{2}})^2)^* \\ . \\ ((w^{\frac{n}{2}})^{(n-1)})^* \end{bmatrix} \tag{33}$$

where $w = \exp^{-j\frac{2\pi}{n}} \implies w^{\frac{n}{2}} = \exp^{-j\pi} = -1$ and putting $w^{\frac{n}{2}} = -1$ in Eq. 33 gives rise to,

$$\lambda_{\frac{n}{2}} = \left[1\ -1\ 1\ .\ -1\right].$$

$$\begin{bmatrix} 0 & \frac{1}{2} & 0 & \dots & \frac{1}{2} \\ \frac{1}{2} & 0 & \frac{1}{2} & \dots & 0 \\ . & . & . & . & . \\ \frac{1}{2} & 0 & 0 & \dots & \frac{1}{2} \end{bmatrix} \frac{1}{n} \begin{bmatrix} 1 \\ -1 \\ 1 \\ . \\ -1 \end{bmatrix} \tag{34}$$

$$= \left[1\ -1\ 1 \cdot -1\right] \frac{1}{n} \begin{bmatrix} -1 \\ 1 \\ -1 \\ . \\ 1 \end{bmatrix} \tag{35}$$

$$= \frac{\sum_{i=1}{}^{n} -1}{n} \equiv -1. \tag{36}$$

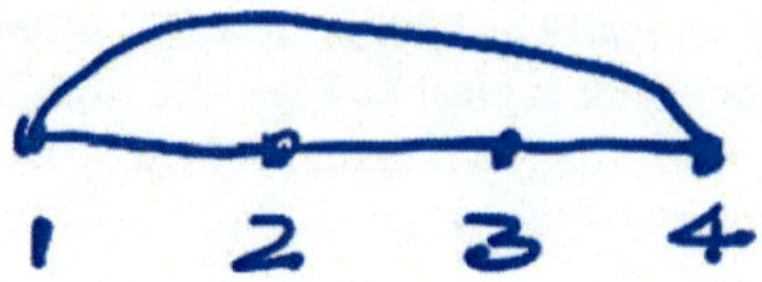

Fig. 12 Four nodes in a linear chain configuration with 1st node connected to 4th node

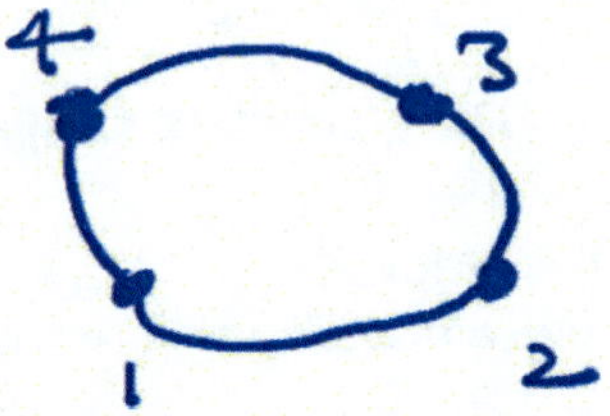

Fig. 13 Four nodes in an equivalent circular/ring configuration with 1st node connected to 4th node

Hence, proved. Alternatively, this can be easily proved by evaluating characteristic polynomial $P(t)$ of circulant matrix $D^{-1}A$ at $t = -1$ as -1 is a valid root of unity for all even values of $n > 0$. ∎

Example 2 Let us look at circularly connected multi-agent configuration of four nodes as shown in Figs. 12 and 13.

Degree matrix $D_{4\times4}$ and Adjacency Matrix $A_{4\times4}$ for the configuration given in Fig. 16 are as follows:

$$D_{4\times4} = \begin{bmatrix} 2 & 0 & 0 & 0 \\ 0 & 2 & 0 & 0 \\ 0 & 0 & 2 & 0 \\ 0 & 0 & 0 & 2 \end{bmatrix} \tag{37}$$

$$D_{4\times4}^{-1} = \begin{bmatrix} \frac{1}{2} & 0 & 0 & 0 \\ 0 & \frac{1}{2} & 0 & 0 \\ 0 & 0 & \frac{1}{2} & 0 \\ 0 & 0 & 0 & \frac{1}{2} \end{bmatrix} \tag{38}$$

$$A_{4\times4} = \begin{bmatrix} 0 & 1 & 0 & 1 \\ 1 & 0 & 1 & 0 \\ 0 & 1 & 0 & 1 \\ 1 & 0 & 1 & 0 \end{bmatrix} \tag{39}$$

$$D_{4\times4}^{-1}A_{4\times4} = \begin{bmatrix} 0 & \frac{1}{2} & 0 & \frac{1}{2} \\ \frac{1}{2} & 0 & \frac{1}{2} & 0 \\ 0 & \frac{1}{2} & 0 & \frac{1}{2} \\ \frac{1}{2} & 0 & \frac{1}{2} & 0 \end{bmatrix} \tag{40}$$

$$W_{4\times 4} = \begin{bmatrix} 1 & 1 & 1 & 1 \\ 1 & -i & -1 & i \\ 1 & -1 & i & -1 \\ 1 & i & -1 & -i \end{bmatrix}$$

where $W_{4\times 4}$ is DFT matrix for dimension 4

$$W_{4\times 4}[D_{4\times 4}^{-1} A_{4\times 4}] W_{4\times 4}^{-1} = \begin{bmatrix} 1 & 0 & 0 & 0 \\ 0 & 0 & 0 & 0 \\ 0 & 0 & -1 & 0 \\ 0 & 0 & 0 & 0 \end{bmatrix} \equiv D_1 \tag{41}$$

Clearly, presence of -1 as eigenvalue at $D_1(3, 3)$ shows the presence of period doubling bifurcation which is going to be a hallmark of all such systems with even dimension containing longest path ring/circular configuration. ■

Appearance of period doubling in difference equation translates to appearance of slowly oscillating periodic orbit with period approximately $2T$ in delay-differential equation space where agents are just swapping different values without actually converging. Next, we describe the behavior of the same system with dimension n as an odd number with longest path ring/circular configuration.

6 Convergence of $[D^{-1}A]^k$, $k \to \infty$ to a Different Consensus Matrix with Ranking of Nodes for $n = 4$ and Node1-Node3 Link in Linear Chain Configuration

Next configuration is a linear chain of four nodes with a node1-node3 link as shown in Fig. 14 which leads to dynamical stability, but consensus point does not have equal contributions from all nodes, and hence, there is a leader-follower ordering among the nodes with node3 being the leader, and node4 is follower of lowest rank.

$$D_{4\times 4} = \begin{bmatrix} 2 & 0 & 0 & 0 \\ 0 & 2 & 0 & 0 \\ 0 & 0 & 3 & 0 \\ 0 & 0 & 0 & 1 \end{bmatrix} \tag{42}$$

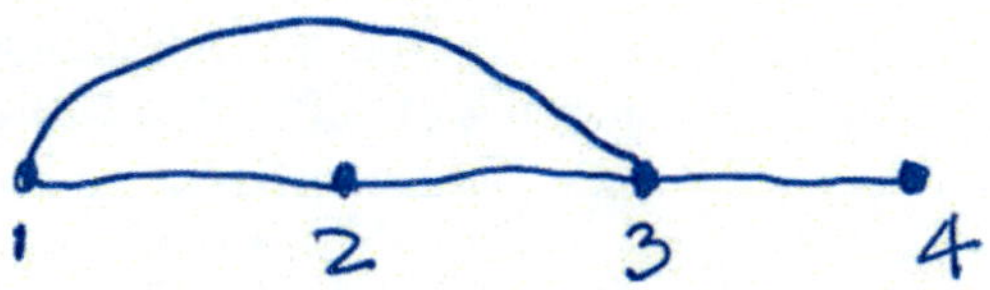

Fig. 14 Four nodes in a linear chain configuration with 1st node connected to 3rd node

$$D_{4\times 4}^{-1} = \begin{bmatrix} \frac{1}{2} & 0 & 0 & 0 \\ 0 & \frac{1}{2} & 0 & 0 \\ 0 & 0 & \frac{1}{3} & 0 \\ 0 & 0 & 0 & 1 \end{bmatrix} \tag{43}$$

$$A_{4\times 4} = \begin{bmatrix} 0 & 1 & 1 & 0 \\ 1 & 0 & 1 & 0 \\ 1 & 1 & 0 & 1 \\ 0 & 0 & 1 & 0 \end{bmatrix} \tag{44}$$

$$D_{4\times 4}^{-1} A_{4\times 4} = \begin{bmatrix} 0 & \frac{1}{2} & 0 & \frac{1}{2} \\ \frac{1}{2} & 0 & \frac{1}{2} & \frac{1}{3} \\ \frac{1}{3} & \frac{1}{3} & 0 & \frac{1}{2} \\ 0 & 0 & 1 & 0 \end{bmatrix} \tag{45}$$

$$[D_{4\times 4}^{-1} A_{4\times 4}]^{501} = [D_{4\times 4}^{-1} A_{4\times 4}]^{500} = \begin{bmatrix} \frac{1}{4} & \frac{1}{4} & \frac{3}{8} & \frac{1}{8} \\ \frac{1}{4} & \frac{1}{4} & \frac{3}{8} & \frac{1}{8} \\ \frac{1}{4} & \frac{1}{4} & \frac{3}{8} & \frac{1}{8} \\ \frac{1}{4} & \frac{1}{4} & \frac{3}{8} & \frac{1}{8} \end{bmatrix} \tag{46}$$

What is being observed from Eq. 46 that while dynamical convergence to an equilibrium point which is different than consensus point with equal contribution from each node is there. This equilibrium point has heterogenous contributions from different nodes, i.e., node1 and node2 contribute 25%, while node3 contributes 37.5%, and finally, node four contributes a minimal of 12.5% leading to node3 emerging as a leader. Dynamically, instability induced by large delay $T \gg 0$ is being compensated by extra link which has been added between node1 and node3. Lesson is clearly increasing connectivity and communication for achieving dynamical stability of consensus process in the event of arbitrarily large communication delays.

7 Convergence of $[D^{-1}A]^k$, $k \to \infty$ to Consensus Matrix C for All Odd n in Longest Path Ring/circular Configuration

Next we explore the convergence behavior of n multi-agents in a ring or circular configuration.

Proposition 4 *$[D^{-1}A]^k, k \to \infty$ will converge to consensus matrix C for all odd n in longest path ring/circular configuration.*

Proof Theory of circulant matrix and discrete Fourier transform (DFT) will be deployed to achieve this [11, 12]. First let us diagonalize $[D^{-1}A]$ using the theory of DFT for circulant matrix as $[D^{-1}A]$ is circulant for all n. We can diagonalize to $D_1 = [DFT]^{-1} * [D^{-1}A] * DFT$ using theory of circulant matrix and DFT. For any odd n, diagonal matrix D_1 will one eigenvalue 1, and rest will be smaller than 1 in magnitude [11, 12]. This means that as we do $D_1^n, k \to \infty$, all diagonal entries will tend to zero except the one whose value is 1, and it will be only non-zero entry $D_1(1, 1) = 1$ left in the matrix. Since, what we are interested in computing $[D^{-1}A]^n$, $k \to \infty$. This can be achieved by computing D_1^k, $k \to \infty$ and then using the formula that $[D^{-1}A]^k = [DFT] * D_1^k * [DFT]^{-1}$. This essentially means $[DFT] * Y * [DFT]^{-1}$ where Y is a matrix with only $Y(1, 1) = 1$, and all other entries are 0. Performing this calculation will show convergence to consensus matrix C with all the entries being $1/n$. ■

Example 3 Let us look at circularly connected multi-agent configuration of five nodes as shown in Figs. 15 and 16.

Degree matrix $D_{5\times5}$ and Adjacency Matrix $A_{5\times5}$ for the configuration given in Fig. 16 are as follows:

$$D_{5\times5} = \begin{bmatrix} 2\,0\,0 & 0 & 0 \\ 0\,2\,0 & \ldots & 0 \\ 0\,0\,2 & 0 & 0 \\ 0\,0\,0 & 2 & 0 \\ 0\,0\,0 & 0 & 2 \end{bmatrix} \tag{47}$$

$$D_{5\times5}^{-1} = \begin{bmatrix} \frac{1}{2} & 0 & 0 & 0 & 0 \\ 0 & \frac{1}{2} & 0 & 0 & 0 \\ 0 & 0 & \frac{1}{2} & 0 & 0 \\ 0 & 0 & 0 & \frac{1}{2} & 0 \\ 0 & 0 & 0 & 0 & \frac{1}{2} \end{bmatrix} \tag{48}$$

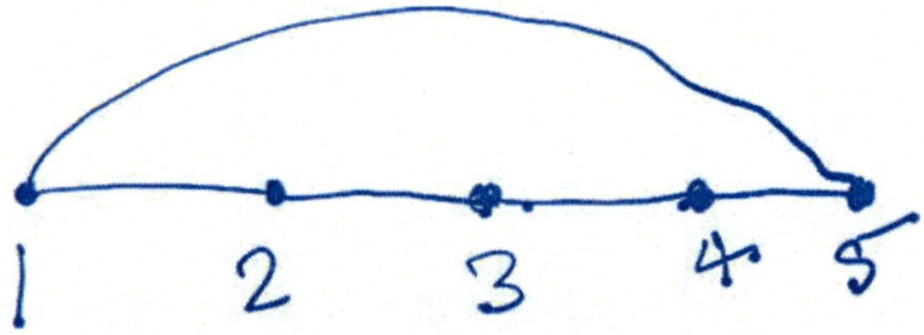

Fig. 15 Five nodes in a linear chain configuration with 1st node connected to 5th node

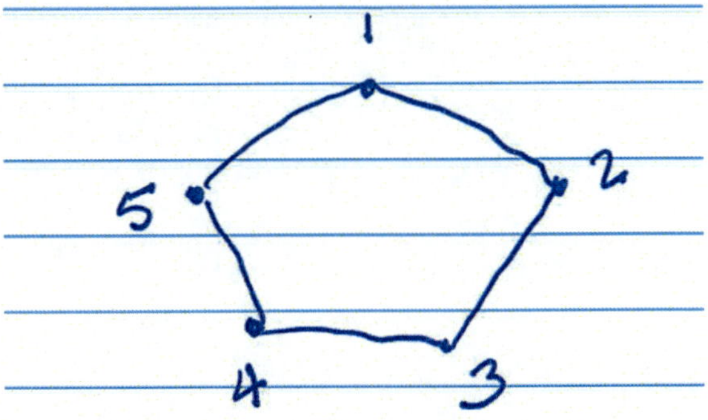

Fig. 16 Five nodes in an equivalent circular/ring configuration with 1st node connected to 5th node

$$A_{5\times5} = \begin{bmatrix} 0 & 1 & 0 & 0 & 1 \\ 1 & 0 & 1 & 0 & 0 \\ 0 & 1 & 0 & 1 & 0 \\ 0 & 0 & 1 & 0 & 1 \\ 1 & 0 & 0 & 1 & 0 \end{bmatrix} \tag{49}$$

$$D_{5\times5}^{-1}A_{5\times5} = \begin{bmatrix} 0 & \frac{1}{2} & 0 & 0 & \frac{1}{2} \\ \frac{1}{2} & 0 & \frac{1}{2} & 0 & 0 \\ 0 & \frac{1}{2} & 0 & \frac{1}{2} & 0 \\ 0 & 0 & \frac{1}{2} & 0 & \frac{1}{2} \\ \frac{1}{2} & 0 & 0 & \frac{1}{2} & 0 \end{bmatrix} \tag{50}$$

As outlined earlier, $[D^{-1}A]^k, k \to \infty$ will determine the asymptotic dynamical stability behavior of this consensus system. It is important to note that $[D^{-1}A]$ as given by Eq. 50 is a *Circulant* matrix, and any circular matrix C_1 can be factored as $C_1 = \frac{1}{N}WD_1W^*$ where elements of matrix W is given by $w_{k,l} = \exp^{j2\pi k\frac{l}{N}}$, D_1 is a diagonal matrix [11, 12], and N is dimension of the matrix. Further, it can be easily shown that $\frac{1}{N}W.W^* = I_{N\times N}$ with $I_{N\times N}$ being identity matrix of dimension N or $W^{-1} = \frac{1}{N}W^*$ which is very useful in computing $[D^{-1}A]^k, k \to \infty$ because $[D^{-1}A]^2 = \frac{1}{N^2}WD_1W^*.W.D_1.W^* = \frac{1}{N}W.{D_1}^2W^*$ or in general $[D^{-1}A]^k = \frac{1}{N^k}WD_1W^*.W.D_1.i, \ldots, W^* = \frac{1}{N}W.{D_1}^kW^*$. In summary, eigenvalues of circulant matrix C_1 given by diagonal matrix D_1 will determine the asymptotic dynamical behavior of Eq. 50 [11, 12]. This resolves the problem of essentially computing the diagonal representation of $[D^{-1}A]$ and evaluating the $k-th$ power of its eigenvalues for which thanks to the theory of circulant matrix and DFT, closed form expression exists [11, 12] helping us prove the consensus for odd number of multi-agents for all values of delay $T \gg 0$.

$$W_{5\times5} = \begin{bmatrix} 1 & 1 & 1 & 1 & \\ 1 & 0.30-0.95i & -0.80-0.58i & -0.80+0.58i & 0.30+0.95i \\ 1 & -0.80-0.58i & 0.30+0.95i & 0.30-0.95i & -0.80+0.58i \\ 1 & -0.80+0.58i & 0.30-0.95i & 0.30+0.95i & -0.80-0.58i \\ 1 & 0.30+0.95i & -0.80+0.58i & -0.80-0.58i & 0.30-0.95i \end{bmatrix}$$

where $W_{5\times5}$ is DFT matrix for dimension 5

$$W_{5\times5}[D_{5\times5}^{-1}A_{5\times5}]W_{5\times5}^{-1} = \begin{bmatrix} 1 & 0 & 0 & 0 & 0 & 0 \\ 0 & 0.30 & 0 & 0 & 0 & \\ 0 & 0 & -0.80 & 0 & 0 & \\ 0 & 0 & 0 & -0.80 & 0 & \\ 0 & 0 & 0 & 0 & 0 & 0.30 \end{bmatrix} \equiv D_1 \tag{51}$$

From expression of D_1, we can see that except one eigenvalue being 1, rest are all less than one in magnitude. This can also be proven for circulant matrix $[D^{-1}A]$ using theory of circulant matrix which provides explicit expressions for eigenvalues [11, 12]. Clearly because of the magnitudes being less than zero, all other eigenvalues except 1 will tend to 0 as powers of

$$\lim_{k\to\infty} D_1^k = \begin{bmatrix} 1&0&0&0&0 \\ 0&0&0&0&0 \\ 0&0&0&0&0 \\ 0&0&0&0&0 \\ 0&0&0&0&0 \end{bmatrix}$$

is computed. Once we know $D_1^k, k \to \infty$, we can compute

$$\lim_{k\to\infty} [D_{5\times5}^{-1} A_{5\times5}]^k =$$

$$W_{5\times5}(\lim_{k\to\infty} D_1^k) W_{5\times5}^{-1} = \begin{bmatrix} \frac{1}{5}&\frac{1}{5}&\frac{1}{5}&\frac{1}{5}&\frac{1}{5} \\ \frac{1}{5}&\frac{1}{5}&\frac{1}{5}&\frac{1}{5}&\frac{1}{5} \\ \frac{1}{5}&\frac{1}{5}&\frac{1}{5}&\frac{1}{5}&\frac{1}{5} \\ \frac{1}{5}&\frac{1}{5}&\frac{1}{5}&\frac{1}{5}&\frac{1}{5} \\ \frac{1}{5}&\frac{1}{5}&\frac{1}{5}&\frac{1}{5}&\frac{1}{5} \end{bmatrix}$$

.

This in turn proves convergence to consensus value for five agents in a maximum length ring configuration. It is clear that there is no leader-follower structure, rather its a peer-group!. ■

8 Bifurcation Diagram and Merger of Two Solution Branches into One as Weight for Node1–Node3 Link Increases in Cooperation Mode!

To understand the emergence of consensus as strength δ of node1-node3 link increases, a three node chain network is considered as shown in Fig. 17. Dynamical behavior of this configuration with slowly increasing δ is plotted as a bifurcation diagram on x-axis and y-axis carries the value of first three entries of first row of $[D_{3\times3}^{-1} A_{3\times3}]^k$, $k \to \infty$ as shown in Fig. 18, where it is observed that period-two solution branches merge together to give a single consensus value which in turn converges to ideal convergence weight in the consensus matrix 0.33 (shown in black line) when $\delta = 1$ and later node1's and node3's contributions dominate when $\delta > 1$.

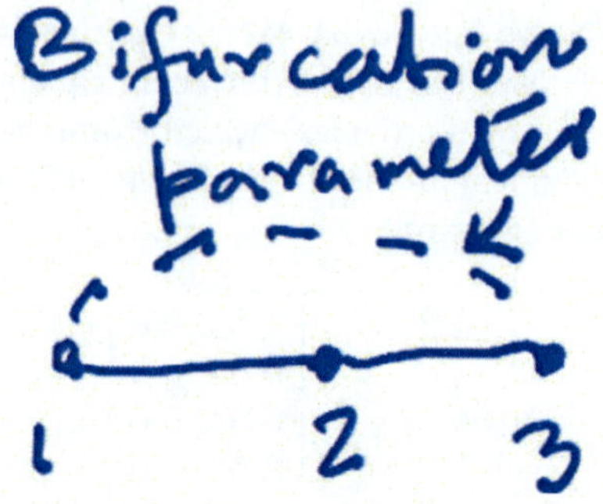

Fig. 17 Three node chain configuration as the weight for node1–node3 link is increased slowly from 10^{-9} to 1+ to look for dynamical behavior of consensus problem and plot the solutions as bifurcation diagram

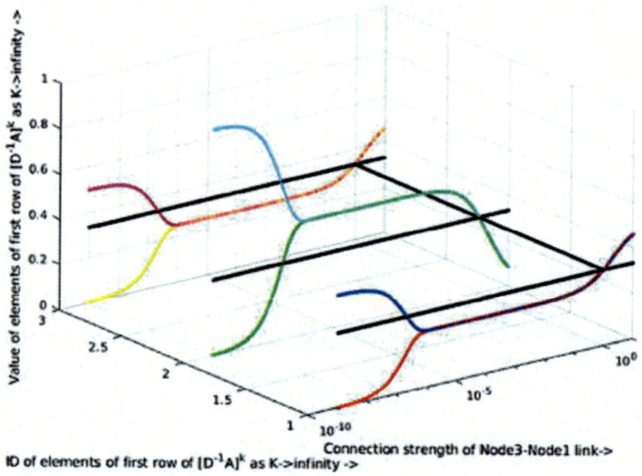

Fig. 18 Bifurcation diagram as weight of node1–node3 link increases. Three plot demonstrate the dynamical behavior of first three entries of $[D^{-1}A]^k$ in first row as $k \to \infty$. It is clear that as the weight of node1–node3 link increases, consensus is stabilized with different contributions from initial conditions of different nodes until it reaches 1 where each node contributes equally. If node1–node3 weight increases more than 1, we see node1 and node3 dominating over node2 as their contributions increase. Black line shows the value of 0.33 for peer-level consensus where all these weights should converge for ideal consensus with equal contribution from all nodes

9 Bifurcation Diagram as Weight for Node1–Node3 Link Increases in Competition Mode with Extremely Small $\delta < 0$!

To understand the emergence of consensus as strength δ of node1-node3 link decreases driving the system in competition, a three node chain network is considered as shown in Fig. 17. Bifurcation diagram shown in Fig. 19 demonstrates that as δ decreases from $-10^{-9} \to -10^{-7}$, two branches of solutions are maintained, and

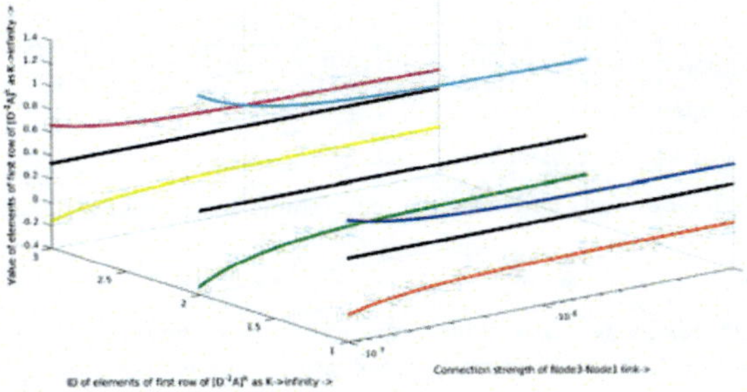

Fig. 19 Bifurcation diagram as weight as δ decreases from $-10^{-9} \rightarrow -10^{-7}$, two branches of solution diverge slowly with competitive mode of system trying to dominate the cooperative mode

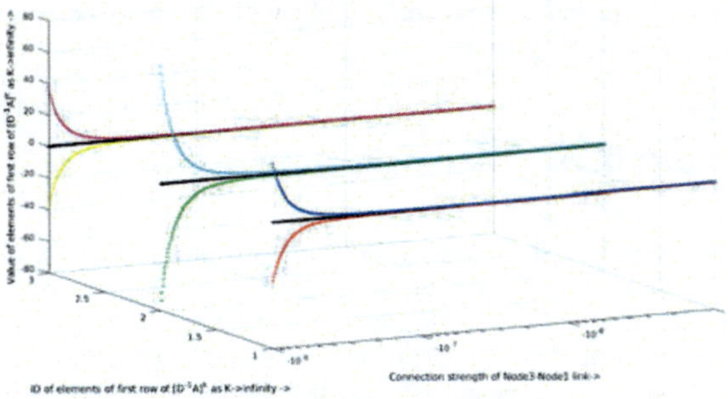

Fig. 20 Bifurcation diagram as weight as δ decreases from $-10^{-9} \rightarrow -10^{-6}$, two branches of solution diverge drastically with competitive mode of system dominating the cooperative mode

they diverge with decreasing δ in relatively small amount. This becomes even more prominent in Fig. 20 as δ decreases from $-10^{-9} \rightarrow -10^{-6}$, and two branches of solution diverge drastically with competitive mode of system dominating the cooperative mode.

10 Conclusion

Delay induced instabilities in consensus protocols are deeply interesting and very practical in this era of serious remote operation. Some of the major issues are number of agents participating and their organization templates, i.e., chain or ring or some other configuration. These dynamical behaviors can be parametrized by number of agents participating, and their configurations are shown in Table 1.

Looking at its relevance and difficulties for next generation of consensus engineers, it needs more attention from nonlinear dynamical systems research community.

Table 1 Dynamical behavior of consensus

Configurations	Even number of agents	Odd number of agents
Chain	Period doubling	Period doubling
Ring/circular	Period doubling	Stable fixed point
Fully connected/star	Stable fixed point	Stable fixed point

Period doubling and convergence to stable fixed point in Consensus

Acknowledgements Author would like to acknowledge deep gratitude toward Prof. A. F. Ivanov and Prof. A. N. Sharkovsky from whom he has learnt a lot. He is also grateful to researchers in circulant matrix [11, 12], FFT-based methods for diagonalizing circulant matrices and tridigonal matrix research communities whose work is directly relevant in solving consensus problems in circular/ring and in linear chain configurations due to their permutation structure.

References

1. Chow, S., Hale, J., Huang, W.: From sine waves to square waves in delay equations. Proc. R. Soc. Edinb.: Sect. Math. **120**(3–4), 223–229 (1992). https://doi.org/10.1017/S0308210500032108
2. Srivastava, V., Moehlis, J., Bullo, F.: On bifurcations in nonlinear consensus networks. J. Nonlinear Sci. **21** (2011). https://doi.org/10.1007/s00332-011-9103-4
3. Ivanov, A.F., Sharkovsky, A.N.: Oscillations in singularly perturbed delay equations. In: Dynamics Reported, pp. 164–224. Springer, Berlin (1992)
4. Ivanov, A.F., Pinto, M.A., Trofimchuk, S.I.: Global behavior in nonlinear systems with delayed feedback. In: Proceedings of the 39th IEEE Conference on Decision and Control (Cat. No. 00CH37187), vol. 5, pp. 4420–4421. IEEE (2000)
5. Verliest, Erik I., Ivanov, Anatoli F.: Robust stability of systems with delayed feedback. Circuits, Syst. Sig. Processing **13**(2–3), 213–222 (1994)
6. Liz, Eduardo, Tkachenko, Victor, Trofimchuk, Sergei: A global stability criterion for scalar functional differential equations. SIAM J. Math. Anal. **35**(3), 596–622 (2003)
7. Ranjan, Priya, La, Richard J., Abed, Eyad H.: Global stability conditions for rate control with arbitrary communication delays. IEEE/ACM Trans. Netw. **14**(1), 94–107 (2006)
8. https://csustan.csustan.edu/~tom/Clustering/GraphLaplacian-tutorial.pdf
9. https://www.geeksforgeeks.org/route-poisoning-and-count-to-infinity-problem-in-routing/
10. Alvim, M.S., Amorim, B., Knight, S., Quintero, S., Valencia, F.: Polarization and Belief Convergence of Agents in Strongly-Connected Influence Graphs, https://arxiv.org/abs/2012.02703 (2020)
11. https://www.cs.unm.edu/~williams/cs530/dft.pdf
12. Davis, P.J., Circulant matrices. American Mathematical Soc. (2013)
13. https://encyclopediaofmath.org/index.php?title=Jacobi_matrix
14. https://en.wikipedia.org/wiki/Tridiagonal_matrix
15. Devriendt, Karel, Lambiotte, Renaud: Nonlinear network dynamics with Consensus–Dissensus bifurcation. J. Nonlinear Sci. **31**(1), 1–34 (2021)
16. Homs-Dones, M., Devriendt, K., Lambiotte, R.: Nonlinear consensus on networks: equilibria, effective resistance and trees of motifs. arXiv preprint arXiv:2008.12022 (2020)
17. Seuret, A., Dimarogonas, D.V., Johansson, K.H.: Consensus under communication delays. In: 2008 47th IEEE Conference on Decision and Control, Cancun, 2008, pp. 4922–4927. https://doi.org/10.1109/CDC.2008.4739278

Delay-Based Approach for Prevention of Rushing Attack in MANETs

Ashwin Adarsh, Tshering Lhamu Tamang, Payash Pradhan, Vikash Kumar Singh, Biswaraj Sen, and Kalpana Sharma

Abstract Mobile ad hoc networks (MANETs) are self-organized, self-configuring collection of autonomous mobile nodes that temporarily forms network and does not require any infrastructure or centralized administration for data communication. The network formed by MANET is vulnerable to threats from malicious nodes which can lead to disruption of the network. The routing protocols available for MANETs also do not provide any mechanism to counter threats or identify malicious nodes or behaviour in the network. In this context, this paper is aimed to find and evaluate the performance of routing protocols when exposed to threats. This paper considers rushing attack and presents a detailed study on their impact in MANET routing. A simulation-based performance analysis on one reactive routing protocols viz ad hoc on-demand routing protocol (AODV) has been performed, and a defence mechanism based on processing delay has been devised to prevent rushing attack. The paper contains various sections like Sect. 1 contains introduction followed by few existing approaches for countering rushing attack in Sect. 2. Section 3 provides the delay-based approach for preventing rushing attack. Sections 4 and 5 present the algorithm for countering rushing attack and simulation environment, respectively. Results and discussion are given in Sect. 6 followed by comparison with other model in Sect. 7. The paper ends with the detailed conclusion given in Sect. 8.

A. Adarsh · P. Pradhan · V. K. Singh (✉) · B. Sen · K. Sharma
Department of Computer Science & Engineering, Sikkim Manipal Institute of Technology, Sikkim Manipal University, Bangalore, India
e-mail: vikash.s@smit.smu.edu.in

A. Adarsh
e-mail: ashwin_201700294@smit.smu.edu.in

B. Sen
e-mail: biswaraj.s@smit.smu.edu.in

K. Sharma
e-mail: kalpana.s@smit.smu.edu.in

T. L. Tamang
Department of Computer Science & Engineering, VMWARE Technologies Bangalore, Bangalore, India

H. K. D. Sarma et al. (eds.), *Contemporary Issues in Communication, Cloud and Big Data Analytics*, Lecture Notes in Networks and Systems 281,
https://doi.org/10.1007/978-981-16-4244-9_10

Keywords MANET · AODV · Rushing attack · RREQ · RREP · Delay

1 Introduction

The on-demand routing protocols in MANETs rely on route discovery process by initiating and exchange of route request (RREQ) and route reply (RREP) packets in the network [11]. According to Mohapatra and Krishnamurthy [15], in order to control the congestion which may likely take place due to flooding of RREQ packets, the on-demand routing protocols—ad hoc on-demand distance vector routing (AODV) (in particular) employs a policy where a node does not forward a packet if it has seen the same packet in an earlier instance by comparing the sequence number field [4]. Nguyen and Nguyen [17] proposed the approach of duplicate suppression of RREQ packets allows rushing attackers to get into the route between source and destination and then possibly drop packets during data forwarding [14]. In rushing attack, the malicious upon receiving a RREQ packet will rush out the RREQ message along with its adjacent nodes without any delay (faster than all its neighbours) which ultimately ensures that the RREQ through the rushing attacker reaches the intended destination [10]. Accordingly, RREP is generated against this malicious RREQ packet [25]. Therefore, during route path establishment, the rushing attacker is placed in the data routing path which enables the attacker to drop the data packets thus disturbing the network operation [1]. The rushing attack can also be performed using other strategy such as flooding the network with bogus requests, forwarding the RREQ at higher transmission power and performing wormhole [5]. Figure 1 presents how rushing attack is carried out in a network that employs AODV as routing protocol [13]. In this paper, a simulation-based performance analysis on the AODV protocol has been performed using network simulator (NS-2.35) where the rushing attackers takes advantage of the duplicate removal mechanism by quickly forwarding the route discovery packet to the forwarding group [3]. Such attack can cause devastating effects on the network performance, and therefore, an approach to prevent rushing attack has also been presented in order to make the network robust and minimize the effect of rushing attack [2].

1.1 Effect of Rushing Attack

Rushing attack takes advantage of the duplicate removal mechanism by quickly forwarding the route discovery packet to the forwarding group [18]. This enables the malicious node to be in the route path between source and destination, thereby controlling the data forwarding [21]. This paper examines the impact of rushing attack in MANET routing protocol and analyses its performances during the presence of rushing attacker in the network. It also proposes a defence strategy by incorporating a mechanism to identify and isolate the rushing attacker from the network

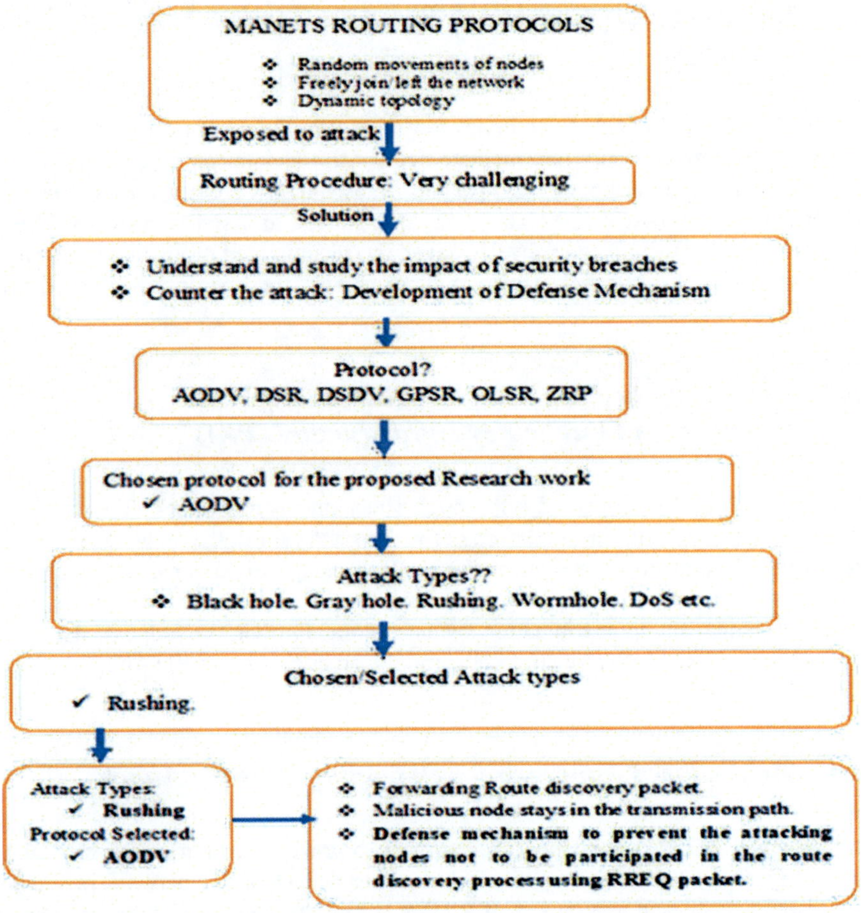

Fig. 1 Block diagram for the defence mechanism for rushing attack in MANETs

and, subsequently, measures the routing performance after incorporation of defence strategy [19]. The block diagram for the defence mechanism for rushing attack in MANETs is displayed in Fig. 1 [16].

2 Literature Survey

The following approaches given in Sects. 2.1, 2.2, and 2.3 have drawn interest among the research fraternity to counter the rushing attack.

2.1 Defending Rushing Attack Based on Random Selection of RREQs

As per Chinkit and Panchal [6], nodes receiving RREQs are not forwarded immediately upon arrival at any intermediate node. All nodes in the network are made to wait and receive multiple RREQ packets. Subsequently, a random RREQ packet is chosen and forwarded [20]. This technique is used to avoid forwarding RREQ packets from the rushing attackers. However, this technique of mitigation is defeated if the random packet selected in from an attacker. Also, this approach has high delay.

2.2 Defending Rushing Attack Based on Neighbour Detection and Public Key Infrastructure (PKI)

Ghoreishi et al. [9] performed a thorough study on rushing attack and presented the issues pertaining to the positioning of rushing attacker in the network. It further proposes a framework to defend rushing attack using neighbour detection, using high gain antennas and PKI along with border gateway protocol (BGP). However, no experimental set up or evaluation has been done. Further, the proposal demands high-computational infrastructure in the mobile ad hoc networks (MANETs) [22].

2.3 Timer-Based Approach for Defending Rushing Attack

Junaid and Iqbal [12] presented an approach to prevent rushing attack in AODV using timer-based approach. In this technique, each node in the network maintains a threshold timer, and any receipt of RREPs before the timer expires is considered to be malicious. Such nodes are tagged as rushing nodes and not considered for any future route establishments [24].

3 Delay-Based Approach for Preventing Rushing Attack

As it is well known that rushing attacker upon receipt of RREQ packet immediately forwards it without any processing or look up at the packet, thereby implying that it has minimal processing delay in forwarding the packet [8].

Therefore, in order to prevent rushing attack, the characteristic of processing delay is exploited, and any node receiving a RREQ in the network checks the processing delay of the sender, and if any RREQ has been received within a threshold time, that RREQ is discarded [23]. The threshold value is set based on experimental values obtained by calculating the average value of end-to-end delay of several simulations

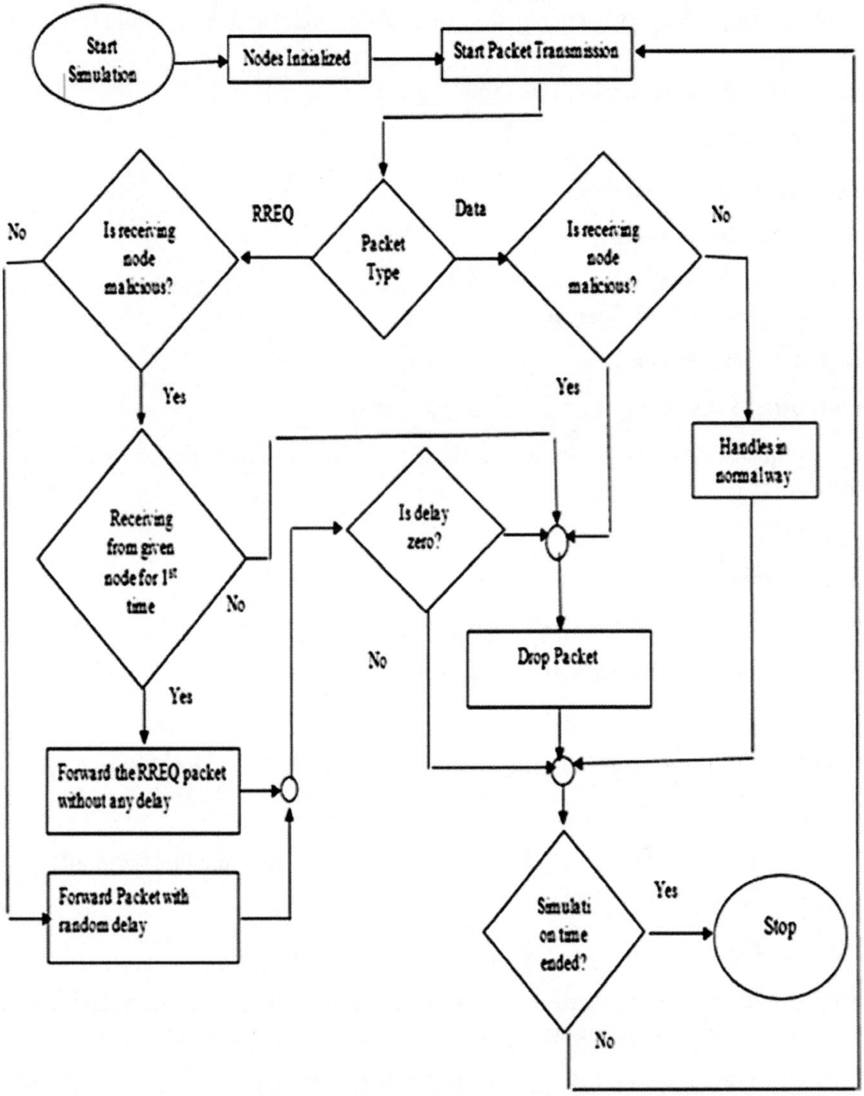

Fig. 2 Flowchart for prevention of rushing attack

of the network in its initial phase. This provides a measure of how the network performs normally in the absence of any malicious attacker. This average end-to-end delay thus obtained serves as a threshold for discovering anomalous in the network. Accordingly, any RREQ arriving before the threshold time is assumed to be from malicious nodes that, therefore, are not considered for further processing or forwarding. The corresponding flow diagram is depicted in Fig. 2 followed by the algorithm.

Algorithm for prevention of Rushing Attack

Algorithm Name:

Rushing Attack Prevention

Algorithm Description:

This algorithm is used to prevent the rushing attack by identifying the attacking nodes and dropping the RREQ packets.

Input(s): 1. RREQ packets.

Output(s): 1. Decision to drop the RREQ packets.

2. Obtain the attacking nodes which should not be participated in route discovery process.

Steps:

Step 0: Begin

Step 1: RREQ packets are sent by any nodes.

Step 2: Check the processing delay while packet transmission.

Step 2.1: Begin

Step 2.2: Calculate the processing delay of the packet forwarded by each node.

Step 2.3: End

Step 3: If RREQ packet delay obtained in step 2 is less than the threshold then

Step 3.1: Drop the packet

Step 3.2: Do not forward the packet.

Step 4: Check whether there is any node forwarding the packet much faster than other nodes, declare that particular node as the attacking node.

Step 5: If the result obtained in step 4 is positive then

Step 5.1: Disallow the attacking node in route discovery process.

Step 5.2: Drop the RREQ packets forwarded by attacking nodes.

Step 6: End

Table 1 Simulation parameters for rushing attack

Parameter	Value
Traffic type	Constant bit rate (CBR)
Number of nodes	25, 50
Transmission range (m)	250
Simulator	NS-2 (Version 2.35)
Simulation time	200 s
Area of network	1000 m × 1000 m
Pause time	0, 50 s, 100 s, 150 s
Maximum connection	1
Maximum speed of nodes (m/s)	2, 5, 10, 15, 20
Mobility model	Random waypoint

4 Simulation Environment

The approach to understand the effect of rushing attack in AODV has been simulated by using NS-2.35 simulator tool which is a discrete networking event simulator. It offers substantial support for TCP, routing and multi-cast protocol simulation across wired and wireless networks. In this simulation-based study, the performance of AODV is evaluated in terms of throughput when exposed to rushing attack in the network. The rushing attackers are varied from 0% (indicating absence of any malicious node) to 75% of the total nodes in the network. The node density considered in this study has considered two scenarios—25 nodes and 50 nodes. Thereafter, the rushing attack prevention mechanism described in Sects. 3 and 4 has been simulated to find out if this defence technique has any effect on the AODV protocol to make it more resistant against rushing attack. The simulation environment chosen is listed in Table 1.

5 Results and Discussions

Results have been obtained by using NS-2.35 simulator and analysing the trace files. Throughput has been calculated by using the following equation.

$$\text{Throughput} = \text{File Size}/\text{Transmission time(bps)} \tag{1}$$

5.1 Throughput Analysis for 25 Nodes Under Rushing Attack

Figure 3 depicts the AODV performance in terms of throughput during collaborative rushing attack for 25 nodes scenario where the rushing attackers vary from 5%, 15%, 25%, 50% and 75% of the total nodes.

The results show that the throughput performance is found to decline consistently with rise of rushing nodes in the network. A sharp decline in throughput is experienced when 25% nodes in the network are rushing attackers where upto 52% loss of throughput is incurred. It is also evident that the throughput reaches almost zero when 75% of the nodes in the network are rushing attackers. In addition, it can also be observed with the incorporation of processing delay-based prevention technique against rushing attack, and enhancement on throughput could be achieved with an improvement of upto 48% when 25% nodes in the network are malicious. The improvement on throughput performance in terms of percentage is presented in Fig. 4. The throughput improvement is achieved due to the prohibition of forwarding of rushed RREQ messages in the network, thereby eliminating rushing nodes to be in the routing path thus improving the throughput. The corresponding data pertaining to

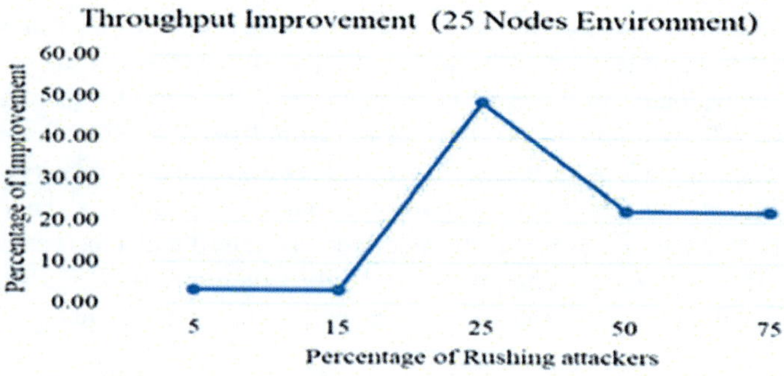

Fig. 3 Throughput under rushing attack for 25 nodes environment

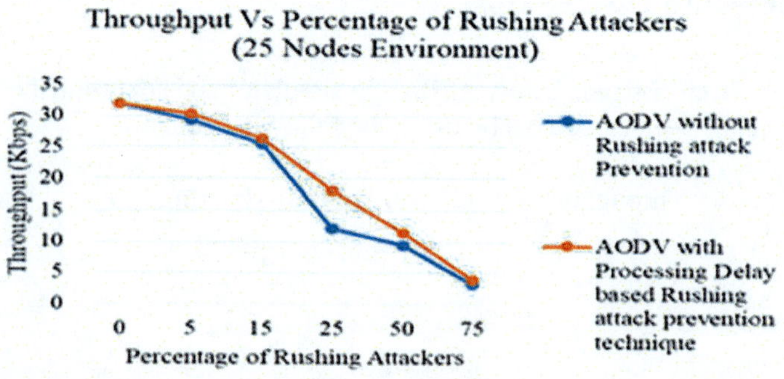

Fig. 4 Percentage of improvement on throughput (25 nodes)

throughput for 25 nodes is presented in Table 2, whereas the throughput improvement obtained after incorporation of the processing delay-based rushing attack prevention technique is presented in Table 3.

Table 2 Throughput comparison (rushing attack prevention)

No. of nodes	Rushing nodes (%)	Throughput without any prevention mechanism (kbps)	Throughput after incorporating prevention technique (kbps)
25	0	32	32
	5	29.34	30.31
	15	25.6	26.44
	25	12.2	18.14
	50	9.4	11.54
	75	3.2	3.92
50	0	48.96	48.96
	5	31.82	41.46
	15	28.51	37.53
	25	20.74	28.22
	50	17.57	26.60
	75	9.65	12.34

Table 3 Percentage of throughput improvement

No. of nodes	Percentage of rushing nodes	Percentage of improvement on throughput
25	5	3.31
	15	3.28
	25	48.69
	50	22.77
	75	22.50
50	5	30.27
	15	31.62
	25	36.11
	50	51.39
	75	27.91

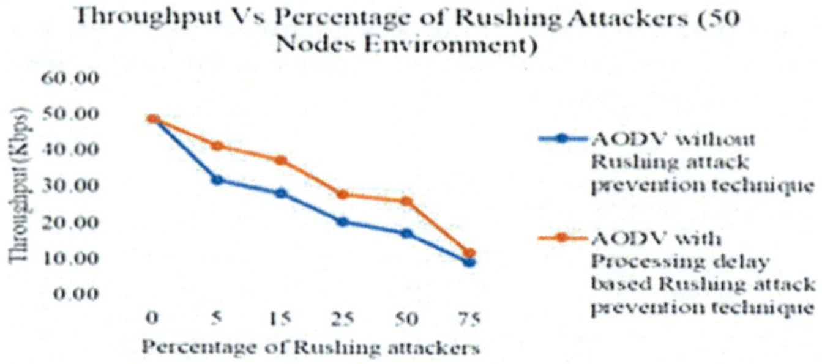

Fig. 5 Throughput under rushing attack for 50 nodes environment

5.2 Throughput Analysis for 50 Nodes Under Rushing Attack

Figure 5 depicts the AODV performance in terms of throughput during collaborative rushing attack for 50 nodes scenario where the rushing attackers vary from 5%, 15%, 25%, 50% and 75% of the total nodes. The results show that the throughput performance is found to decline consistently with rise of rushing nodes in the network. However, unlike 25 nodes environment, a sharp decline in throughput is not experienced, and the decline of throughput is found to be gradual due to the dense network environment where AODV is able to find alternative path during loss of data. It is also evident that with 75% nodes as rushing attacker, the throughput loss incurred is 80%. In addition, it can also be observed with the incorporation of processing delay-based prevention technique against rushing attack, and enhancement on throughput could be achieved with an improvement of upto 41% when 5% nodes in the network are malicious.

The improvement on throughput performance in terms of percentage for 50 nodes is presented in Fig. 6 where it can be observed that the processing delay-based rushing attack prevention technique is able to enhance throughput upto 52% even when the percentage of rushing attackers is 50%.

It can, therefore, be inferred that a highly dense network is more favourable for the prevention mechanism as explained in Sect. 3. The corresponding data pertaining to throughput for 50 nodes is presented in Table 2, whereas the throughput improvement

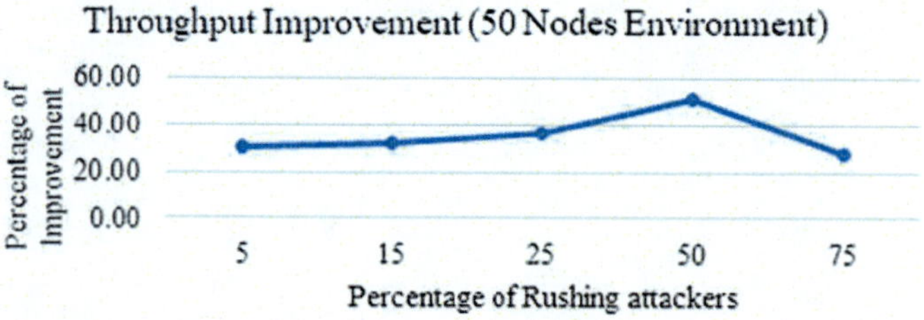

Fig. 6 Percentage of improvement on throughput (50 nodes)

obtained after incorporation of the processing delay-based rushing attack prevention technique is presented in Table 3.

6 Comparison with Other Model

The processing delay-based rushing attack prevention mechanism as discussed in Sect. 3 has been compared with an existing approach presented by Chinkit and Panchal [6]. While the proposal in Sect. 3 uses a strategy to prevent rushing attack by checking the delay in receipt of RREP messages, Chinkit and Panchal [6] adopt a different strategy where any node upon receiving RREQ messages will wait for more RREQ message to accumulate and then picking any random RREQ and forwarding for route discovery.

In order to compare the performance of both the rushing prevention strategy, Chinkit and Panchal [6] model has been simulated and compared with the prevention mechanism as proposed in Sect. 3 along with the simulation parameters discussed in Sect. 5 (as per Table 1) with percentage of rushing nodes being varied from 5%, 15%, 25%, 50% and 75%.

Figures 7 and 8 depict the average throughput under three circumstances—(a) When no prevention strategy has been incorporated, (b) Using processing delay-based rushing attack prevention technique and (c) using random selection of RREQs

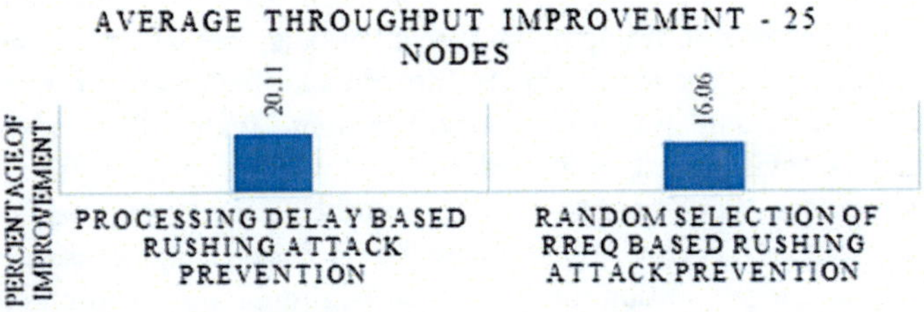

Fig. 7 Throughput comparison with 25 nodes

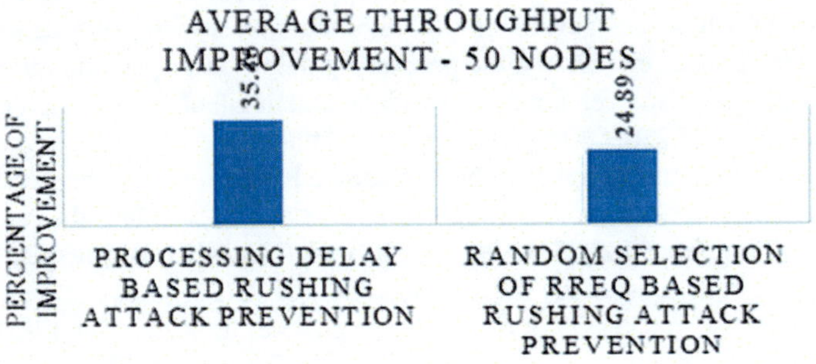

Fig. 8 Throughput comparison with 50 nodes

presented by Chinkit and Panchal [6], to prevent rushing attack. The results show that while both the prevention mechanisms improve the throughput performance, the prevention technique described in Sect. 3 performs better than the prevention technique presented by Chinkit and Panchal [6] for both 25 nodes as well as 50 nodes environment. It is found that in 25 nodes environment, the delay-based prevention technique is found to generate 4% more average throughput over the random selection of RREQ-based rushing attack prevention approach presented by Chinkit and Panchal [6]. Similarly, in denser environment of 50 nodes, delay-based prevention technique generates upto 11% more average throughput against the technique presented by Chinkit and Panchal. The reason for better performance in case of more dense network environment is due to availability of multiple routes which eventually helps in path finding process in case of frequent route changes. In addition, the approach adopted by Chinkit and Panchal [6] depends on accumulation of RREQs and then picking a random RREQ for path establishment. This random selection of RREQ may result in picking up of a RREQ which is from an attacker [7]. On the contrary, the proposed prevention discussed in Sect. 3 uses processing delays of neighbouring nodes as a key parameter to identify a malicious node and prohibit propagation of RREQs from those rushing nodes, thereby yielding better result.

7 Conclusions

In this paper, a study on the impact of rushing attack in the performance of AODV has been evaluated in terms of throughput where it has been found that rushing attackers affect the throughput. The simulation results obtained indicate that in low density environment, the throughput loss incurred was more than 90%, whereas in high density network environment, with 75% nodes as rushing attackers, the throughput delivery percentage was found to be 20%. Further, a novel approach to prevent rushing attack in AODV for MANETs, which is based on processing delay, has been presented. The prevention mechanism upon simulation was found to enhance throughput performance upto 20% and 35% for both low density as well as high density network environment viz. 25 nodes and 50 nodes, respectively. It can, therefore, be inferred that a highly dense network is more favourable for the prevention mechanism which is due to availability of alternate routes which may be exploited in the presence of attackers. Lastly, the processing delay-based rushing attack prevention technique has also been compared with an existing technique which is based on random selection of RREQs to prevent rushing attack. The performance comparison in terms of average throughput has been presented and found that processing delay-based prevention technique was found to be 4% and 11% more efficient in terms of average throughput for 25 nodes and 50 nodes environment, respectively. The reason for the same is that the random selection of RREQ does not guarantee that the selected RREQ packet is not from rushing attacker thus affecting the throughput performance.

References

1. Agrawal, S., Jain, S., Sharma, S.: A survey of routing attacks and security measures in mobile ad-hoc networks. arXiv preprint arXiv:1105.5623 (2011)
2. Aluvala, S., Sekhar, K.R., Vodnala, D.: An empirical study of routing attacks in mobile ad-hoc networks. Procedia Comput. Sci. **92**, 554–561 (2016)
3. Babakhouya, A., Challal, Y., Bouabdallah, A.: A simulation analysis of routing misbehaviour in mobile ad hoc networks. In: The Second International Conference on Next Generation Mobile Applications, Services, and Technologies pp. 592–597 (2008)
4. Bettstetter, C., Resta, G., Santi, P.: The node distribution of the random waypoint mobility model for wireless ad hoc networks. IEEE Trans. Mob. Comput. **2**(3), 257–269 (2003)
5. Bounpadith, K., Hidehisa, N., Yoshiaki, N., & Nei, K.: A survey of routing attacks in mobile ad hoc networks. Wireless Commun. **14**(5), 85–91 (2007). ISSN: 1536-1284
6. Chinkit, S., Panchal, B.: Rushing attack prevention with modified AODV in mobile ad-hoc network. IJEDR **2**(4), 3489–3493 (2014). ISSN: 2321-9939
7. Deshmukh, S.R., Chatur, P.N., Bhople, N.B.: AODV-based secure routing against black-hole attack in MANET. In: IEEE International Conference on Recent Trends in Electronics, Information & Communication Technology (RTEICT), pp. 1960–1964 (2016)
8. Geetha, J., Ganapathy, G.: Performance comparison of mobile ad-hoc network routing protocol. IJCSNS Int. J. Comput. Sci. Netw. Secur. **7**(11), 77–84 (2007)
9. Ghoreishi, S.M., Abd Razak, S., Isnin, I.F., Chizari, H.: Rushing attack against routing protocols in mobile ad-hoc networks. In: International Symposium on Biometrics and Security Technologies (ISBAST), pp. 220–224 (2014)
10. Hu, Y.C., Perrig, A., Johnson, D.B.: Rushing attacks and defense in wireless ad hoc network routing protocols. In: Proceedings of the 2nd ACM Workshop on Wireless Security, pp. 30–40 (2003)
11. Jamal, T., Butt, S.A.: Malicious node analysis in MANETS. Int. J. Inf. Technol. **11**(4), 859–867 (2019)
12. Junaid, W., Iqbal, A.: Prevention of multiple rushing attacks in mobile ad hoc network using AODV routing protocol. EVISA, Sci. Int. (Lahore) **30**(1), CODEN: SINTE 8, 173–177 (2018). ISSN 1013-5316
13. Kannhavong, B., Nakayama, H., Nemoto, Y., Kato, N., Jamalipour, A.: A survey of routing attacks in mobile ad hoc networks. IEEE Wirel. Commun. **14**(5), 85–91 (2007)
14. Mishra, M.K., Pattanayak, B.K., Jagadev, A.K., Nayak, M.: Measure of impact of node misbehavior in ad hoc routing: a comparative approach. IJCSI 10 (2010)
15. Mohapatra, P., Krishnamurthy, S. (eds.): Ad Hoc Networks: Technologies and Protocols. Springer Science & Business Media (2004)
16. Murthy, C.S.R., Manoj, B.S.: Ad Hoc Wireless Networks: Architectures and Protocols, Portable Documents. Pearson Education (2004)
17. Nguyen, H.L., Nguyen, U.T.: A study of different types of attacks on multicast in mobile ad hoc networks. Ad Hoc Netw. **6**(1), 32–46 (2008)
18. Panicker, A.V., Jisha, G.: Network layer attacks and protection in MANET—a survey. Int. J. Comput. Sci. Inf. Technol. **5**(3), 3437–3443 (2014)
19. Rampurkar, K., Lavande, N., Shilgire, S., Mane, S.N.: Study of routing overhead and its protocols (2017)
20. Rifquddin, M.R.: Performance of AOMDV routing protocol under rushing and flooding attacks in MANET. In: 2nd International Conference on Information Technology, Computer, and Electrical Engineering (ICITACEE), pp. 386–390 (2015)
21. Shrivastava, S.: Rushing attack and its prevention techniques. Int. J. Appl. Innov. Eng. Manage. **2**(4), 453–456 (2013)
22. Shrivastava, S., Mangal, D.: A new technique to prevent MANET against rushing attack. Int. J. Comput. Sci. Inf. Technol. **5**(3), 3460–3464 (2014)

23. Toubiana, V., Labiod, H., Reynaud, L., Gourhant, Y.: Performance comparison of multipath reactive ad hoc routing protocols. In: 2008 IEEE 19th International Symposium on Personal, Indoor and Mobile Radio Communications, pp. 1–6 (2008)
24. Verma, D.K., Jain, R., Kush, A.: Intrusion detection using RREP messages of AODV routing protocol. Int. J. Appl. Eng. Res. **12**(9), 1956–1961 (2017)
25. Wu, B., Chen, J., Wu, J., Cardei, M.: A survey of attacks and countermeasures in mobile ad hoc networks. In: Wireless network security, pp. 103–135. Springer, Boston (2007)

ASCTWNDN:A Simple Caching Tool for Wireless Named Data Networking

Dependra Dhakal, Mohit Rathor, Sudipta Dey, Prantik Dey, and Kalpana Sharma

Abstract Name Data Networking (NDN) is a new and proposed Future Internet Architecture inspired by unsolved problems in current Internet architectures like IP. There can be a multitude of applications realized through Information Centric Network (ICN) in various domains where data is the focal point of communication. One such domain is the Intelligent Transport System which can be realized through Vehicular Ad hoc Network (VANET) and Vehicular Sensor Network (VSN). The VSN has grown rapidly to exchange information, especially in the area of safe and efficient driving. Despite the broad interest, there is a shortage of publicly accessible resources to assess the efficiency of caching systems effectively. To address these issues, we have developed a python-based caching simulator for Wireless Named Data Networking considering vehicular network caching. This tool allows users to evaluate caching strategies and also provides modelling tools useful for caching research.

Keywords Named data networking · Vehicular sensor network · Caching · Simulation

1 Introduction

The Named Data Network (NDN) is dedicated to achieving new dimensions and improved Internet structures for knowledge distribution. This is the target of the NDN. The biggest distinction between the new Internet is the right to use names rather than the host data that addresses the standard Internet (IP) protocol. ICN key objective is to turn the current host-based connectivity model into a service-based

D. Dhakal (✉) · M. Rathor · S. Dey · P. Dey · K. Sharma
Sikkim Manipal Institute of Technology, Majitar, Rangpo, Sikkim, India
e-mail: dependra.d@smit.smu.edu.in

S. Dey
e-mail: sdptd20@gmail.com

K. Sharma
e-mail: kalpana.s@smit.smu.edu.in

H. K. D. Sarma et al. (eds.), *Contemporary Issues in Communication, Cloud and Big Data Analytics*, Lecture Notes in Networks and Systems 281,
https://doi.org/10.1007/978-981-16-4244-9_11

model to effectively deliver content [1] on a network-based basis. In recent years, this paradigm change has brought a tremendous amount of attention to the research community and launched several research projects worldwide to investigate and promote this thinking. For many of these modern network implementations, users need data objects, whether data is served by a server or network intermediate nodes [2] in particular. In reality, various network technologies (Content Delivery Networks, Load Balancers, etc.) that we rely upon have been specifically designed to fill the gap between the host centred IP and the data centre demands of the advanced web-based applications. This ground-breaking networking model, known as ICN, focuses content between any two users of the system, choosing to ignore place and host descriptions. The contents are identified specifically and exclusively in the NDN. Instead of identifying the node, contents are identified in the network. NDN caching is performed on routers where the previously used content is cached for future use. The caching strategy used by the vehicle network [3–5] is similarly aimed at a variety of study. There are very few tools to simulate caching strategy in named data network [6]. Our tools provide flexible and graphical analysis for any new or existing caching technique that can be implemented. ASCTWNDN supports scalability, extensibility, cache design modularity like cache partitioning to the user without any restriction to the structure of the caching algorithm. The tool can be used for generating NDN-based names or URL generators from the existing dataset. The tool can be basically used for testing existing or new caching techniques in NDN.

2 Related Work

The need for simulation tools for cache-based caching techniques in NDN has drawn several researchers. Different projects are already running in the field of information-centric networks and named data network. Many researchers are also working on a different type of caching approach. ndnSIM [7] is an ns3 based open source simulator for Named Data Networking that simulate all NDN protocol operations. It supports large scale simulation experiments and widely used by the researcher for simulating research based on Named Data Network. CCNx [8] is an emulation testbed, for emulation, covering many of the operations for content centric network. It proposes new architecture and focuses on a real simulation testbed. ccnSIM [9] is a Omnet++ based chunk level simulator with the content based architecture. It focuses on scalability and large scale especially in terms of important factors like cache sizes. ICARUS [10] is one of the first simulation tools focuses on caching in information centric network. It focuses on evaluating the caching technique effectively. It is a python based simulator and provides modelling tool for caching research.

3 Architecture and Design

ASCTWNDN has been designed to allow NDN based simulation of caching strategies with extensible and easy to use functionalities. It has been developed using a python-based programming language with a built-in graph analysis tool to analyse any caching strategies implemented using the tool. The tool has been developed considering the vehicular network caching but it can be used for other caching research related to NDN as well. This tool also allows the existing dataset to be converted into named based URLs and allows to create a random dataset that can be used to test the caching strategies. The modular approach is used to easily extend caching strategies of choice without changing the basic animator. The user can simply write the caching strategy and replace the existing caching strategy which includes LRU and FIFO. Unlike traditional network NDN uses named based interest which is missing in an existing dataset like accidental parameter [11] for the vehicular network which is in the form of columns can be converted into URL based names. The ASCTWNDN implements the workflow depicted in Fig. 1. The programme is executed in the following six steps:

- **NDN Based URL Generator**: One of the major problems in executing the NDN scenario is named data. Due to the lack of standardized data format till now in NDN, the data set has to be designed from the existing dataset. The data set which is column-based has to be converted into URL names. The generator generates URL names by selecting a desirable column to be used in the URL name as shown in Fig. 2. Columns from the raw data are listed based on which user can choose the key column and further the column which it wants to append to the named URL as shown in Fig. 2. Once the column is selected, it generates a CSV file with the named column as shown in Fig. 3.
- **Temporal and Space Validator**: Temporal and space validator can append time-based priorities and distance-based limits. These parameters can be used for the vehicular network for assigning priorities based on time and distance. These are optional and these steps can be skipped in keeping with the requirement of the user.

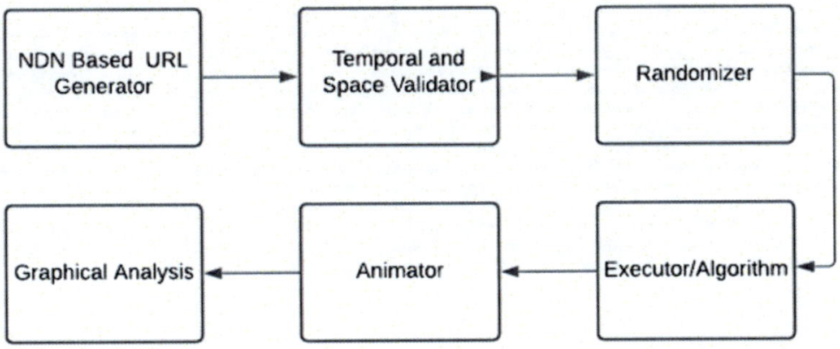

Fig. 1 Workflow of ASCTWNDN

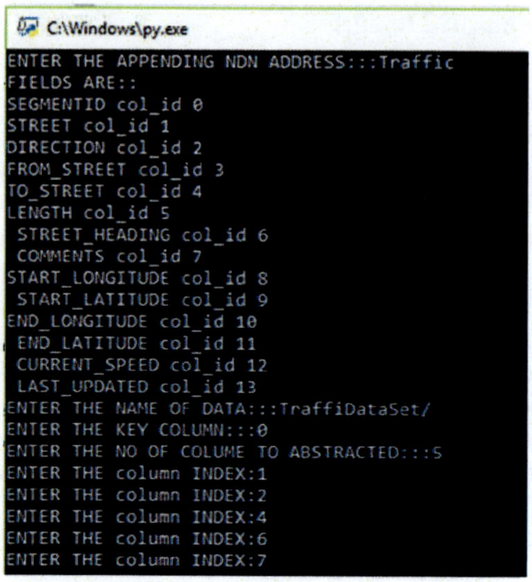

```
C:\Windows\py.exe
ENTER THE APPENDING NDN ADDRESS:::Traffic
FIELDS ARE::
SEGMENTID col_id 0
STREET col_id 1
DIRECTION col_id 2
FROM_STREET col_id 3
TO_STREET col_id 4
LENGTH col_id 5
 STREET_HEADING col_id 6
 COMMENTS col_id 7
START_LONGITUDE col_id 8
 START_LATITUDE col_id 9
END_LONGITUDE col_id 10
 END_LATITUDE col_id 11
 CURRENT_SPEED col_id 12
 LAST_UPDATED col_id 13
ENTER THE NAME OF DATA:::TraffiDataSet/
ENTER THE KEY COLUMN:::0
ENTER THE NO OF COLUME TO ABSTRACTED:::5
ENTER THE column INDEX:1
ENTER THE column INDEX:2
ENTER THE column INDEX:4
ENTER THE column INDEX:6
ENTER THE column INDEX:7
```

Fig. 2 Named URL generator from column dataset

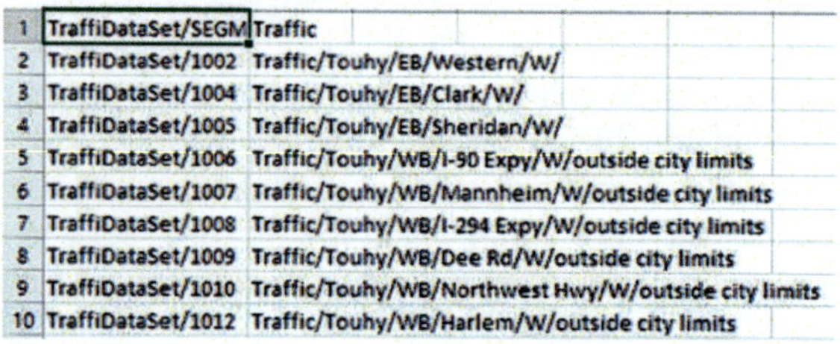

1	TraffiDataSet/SEGM	Traffic
2	TraffiDataSet/1002	Traffic/Touhy/EB/Western/W/
3	TraffiDataSet/1004	Traffic/Touhy/EB/Clark/W/
4	TraffiDataSet/1005	Traffic/Touhy/EB/Sheridan/W/
5	TraffiDataSet/1006	Traffic/Touhy/WB/I-90 Expy/W/outside city limits
6	TraffiDataSet/1007	Traffic/Touhy/WB/Mannheim/W/outside city limits
7	TraffiDataSet/1008	Traffic/Touhy/WB/I-294 Expy/W/outside city limits
8	TraffiDataSet/1009	Traffic/Touhy/WB/Dee Rd/W/outside city limits
9	TraffiDataSet/1010	Traffic/Touhy/WB/Northwest Hwy/W/outside city limits
10	TraffiDataSet/1012	Traffic/Touhy/WB/Harlem/W/outside city limits

Fig. 3 Named URL generated

	A	B
1	weather-data/2006-04-1(	infotainment/medium/data-world/weather-data/time/type/precip/temp/humidity/visibility/2006-04-10 03:00:00.000 +0200/Partly
2	facebook-data/1344	infotainment/low/kaggle/fk-data/id/userid/age/gender/1073384/41/female/1344/75818/1331/25
3	map-data/396	infotainment/medium/kaggle/map-data/id/location/lat/long/KAB. TORAJA UTARA/-2.8622/119.8352/396/1719/7014/13
4	facebook-data/1177	infotainment/low/kaggle/fk-data/id/userid/age/gender/1049406/35/female/1177/75186/1977/36
5	facebook-data/40	infotainment/low/kaggle/fk-data/id/userid/age/gender/1472999/14/female/40/81858/2020/43
6	twitter-data/3307078897:	infotainment/low/data-world/twitter-data/id/text/330707889729712131/b'@Vangsgutane Norway rocks!'/73064/1244/10
7	facebook-data/108	infotainment/low/kaggle/fk-data/id/userid/age/gender/1998492/15/female/108/73752/2628/10
8	facebook-data/273	infotainment/low/kaggle/fk-data/id/userid/age/gender/1632155/19/female/273/73951/4297/25
9	twitter-data/7319847390	infotainment/low/data-world/twitter-data/id/text/731984739012251648/b'Most recent rocket took max damage, due to v high ent
10	twitter-data/7546875250	infotainment/low/data-world/twitter-data/id/text/754687525034270720/b'@Mikey67Tang much appreciated'/77822/1634/8
11	twitter-data/7989846938	infotainment/low/data-world/twitter-data/id/text/798984693844541440/b'@ArminKeyani All P100D vehicles produced to date, as
12	news-data/	infotainment/medium/kaggle/news-data/id/category/headline/link/CRIME/Woman Alleges Rapper Young Lo Raped Her While She
13	twitter-data/8279256530	infotainment/low/data-world/twitter-data/id/text/827925653022453760/b'In addition, I again raised climate. I believe this is doing
14	accident-data/2139	sa/high/data-world/accident-data/id/crash-type/location/distance/injured/2139/PROPERTY DAMAGE ONLY/E WYOMING AVE/60/0,
15	traffic-data/117316	sa/high/data-world/traffic-data/id/location/117316/ALBANY AVENUE/1091/9833/10
16	twitter-data/6435387019	infotainment/low/data-world/twitter-data/id/text/643538701981716481/b"In expendable mode, Falcon Heavy can send a fully loa
17	twitter-data/1747512682	infotainment/low/data-world/twitter-data/id/text/174751268215066625/b"@TomMao1 Studied physics in college, read a lot of bo
18	weather-data/2006-04-0	infotainment/medium/data-world/weather-data/time/type/precip/temp/humidity/visibility/2006-04-04 16:00:00.000 +0200/Partly
19	weather-data/2006-08-1	infotainment/medium/data-world/weather-data/time/type/precip/temp/humidity/visibility/2006-08-13 21:00:00.000 +0200/Overc
20	twitter-data/3010818321	infotainment/low/data-world/twitter-data/id/text/301081832169013249/b'RT @A_C_R: @elonmusk My niece has one and has repo
21	facebook-data/1158	infotainment/low/kaggle/fk-data/id/userid/age/gender/1256256/34/male/1158/77935/3011/25

Fig. 4 Randomized data set

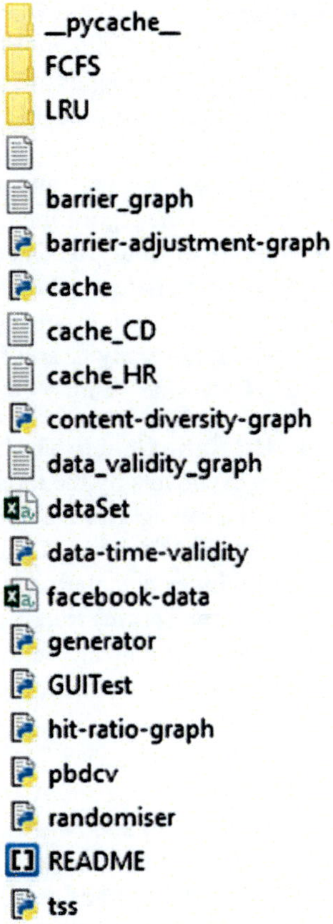

Fig. 5 Files in tool

- **Randomizer**: Once the number of datasets is prepared by the generator, a randomizer can be used to randomize this dataset. Each dataset can be combined manually into one CSV file and a randomizer can be used to randomize these data sets as shown in Fig. 4. The last four values in the URL are temporal and spatial values related to vehicular network space-time priorities.
- **Executor/Algorithm**: Executor is the code that calls the algorithm that is written as the caching strategies by the user. This algorithm is the independent module that is called by the executor. In Fig. 5 "pbdcv" is one of our algorithms which calls our cache algorithm "cache". Likewise, different algorithms like LRU and FCFS can be selected. The generator and tss are NDN URL generator and temporal and space validator, respectively.

- **Animator**: Animator has 3 parts which run after the executor, it contains an animation of how data are moving to cache and whether there are hit and miss. No forwarding strategies are used, and only the caching strategies are addressed. The other part contains the PIT table populated from the data set, the interest generated from the data set created by the user, and the command prompt showing hit ratio and content diversity and dynamic cache adjustment which can be configured with respect to the algorithm used as shown in Fig. 6.
- **Graphical Analysis**: Once the Executor executes, graphs are generated which are dynamic in nature. These graphs dynamically adjust with the values and for the time the algorithm is executed. Two graphs are generated, Hit ratio Graph and Content Diversity Graph for all the algorithms, whilst Barrier Adjustment Graph and Data time Validity Graph are related to vehicular network with priority.

1. **Hit Ratio Graph:** The hit ratio graph provides the hit ratio (which is the number of cache hits by the total number of cache hits and misses) with respect to time in a moving average dynamically as shown in Fig. 7.
2. **Content Diversity Graph:** To calculate the diversity in cache, Simpson diversity index calculation is used. Here, the population set is analogous to the data set. The species referred to as the priority of data, i.e. data belong to the same priority is considered to be belonging to the same species.

$$D = \sum n(n-1)/N(N-1) \tag{1}$$

   where, D is diversity of the cache and n refers to the no of data for a given priority (pi) and N is the total number of data present in cache.
3. **Barrier Adjustment Graph:** The barrier Adjustment graph is optional, but this graph can be used if you want a dynamic cache changing the size from one priority to another. Allocation of memory is adjusted into high, medium, and low priority tasks as per the requirement of the algorithm which is shown in Fig. 6 where we have a high barrier, medium barrier, and low barrier. Based on the size adjustment, graph is plotted in barrier adjustment graph as shown in Fig. 8. The size adjustment has been done in megabytes in all three barriers.
4. **Data Time Validity Graph:** Data Time Validity graph is the average time the data resides in different memory allocation where,

$$\text{AverageTime} = \frac{\text{TotalTimeDataWasInEachPriority}}{\text{NumberofDataforthatPriority}}$$

   From Fig. 9 it is clear the data that has been there in each cache partition based on its priority that is high, medium, and low. These graphs can be used for priority-based cache with partitioning.

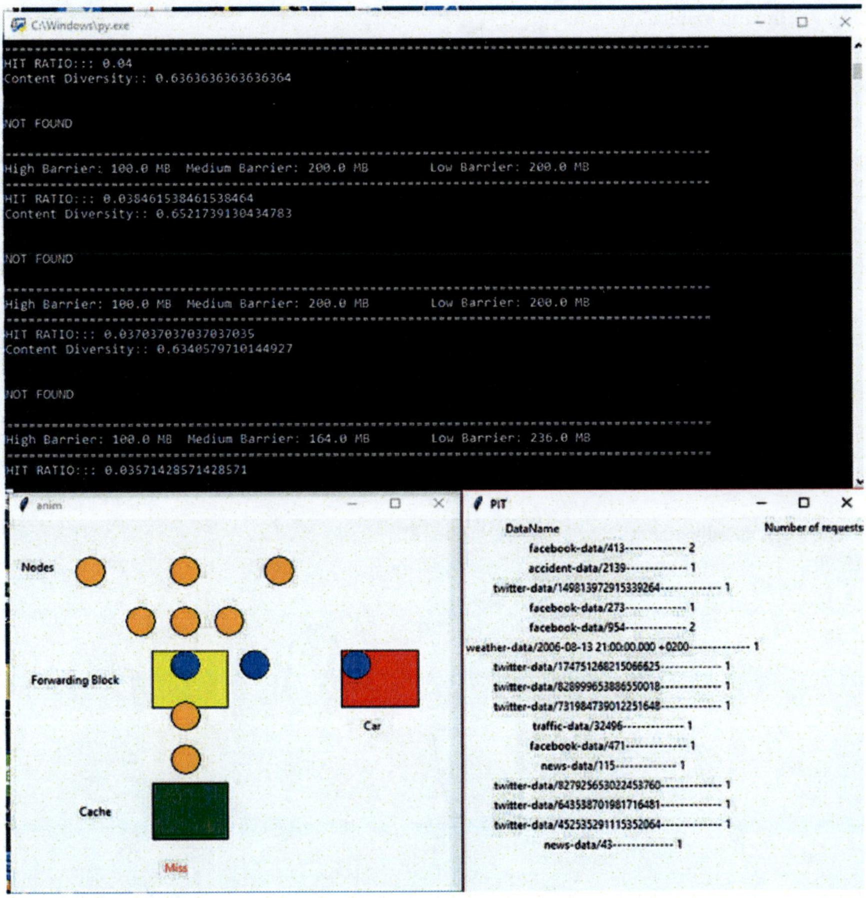

Fig. 6 Animator

4 Conclusion

In this paper, we presented ASCTWNDN, a simple caching tool specifically designed for caching in vehicular and other named data networks. It has been designed to have a flexible and dynamic in-network caching systems to test the existing and new replacement algorithms. The extensibility and open objective will give the user to innovate and create a new type of caching technique for named data networks. Our tool can generate ndn based names from the existing dataset. The tool is available in git hub and can be further improved as it is open for all. In the future, forwarding techniques can be added and a new caching algorithm using Quality of Service in Vehicular Network can be implemented using these tools.

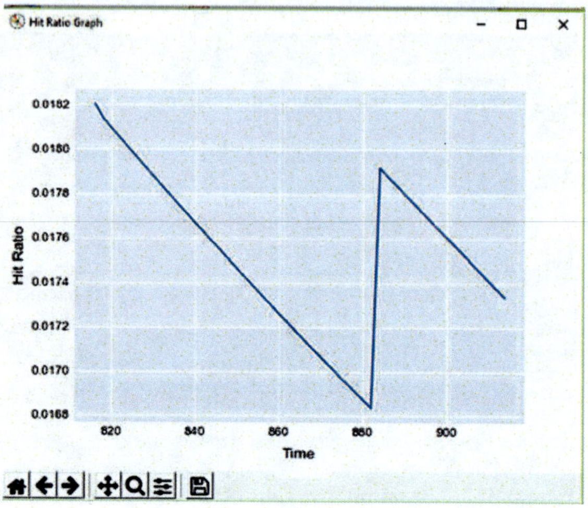

Fig. 7 Hit ratio graph

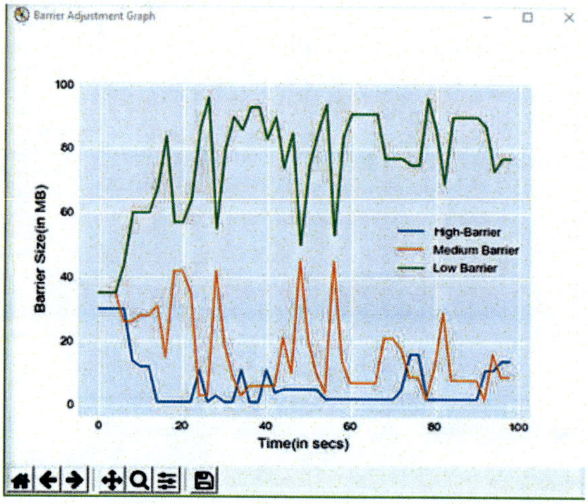

Fig. 8 Barrier adjustment graph

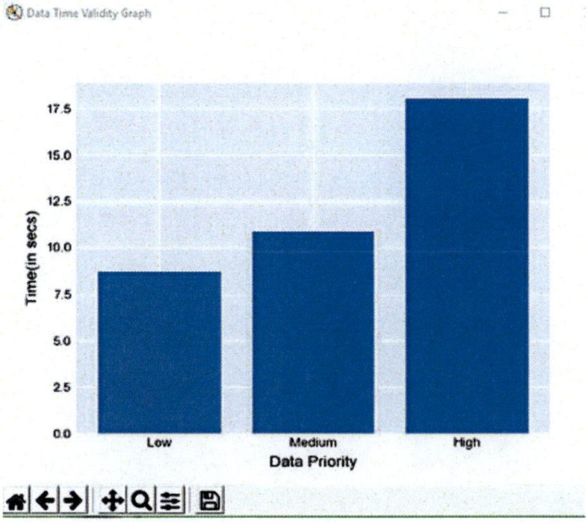

Fig. 9 Data time validity graph

References

1. Jacobson, V., Smetters, D.K., Thornton, J.D., Plass, M.F., Briggs, N.H., Braynard, R.L.: Networking named content. In: Proceedings of the 5th international conference on Emerging networking experiments and technologies, pp. 1–12, 2009
2. Naeem, M.A., Rehman, M.A.U., Ullah, R., Kim, B.-S.: A comparative performance analysis of popularity-based caching strategies in named data networking. IEEE Access **8**, 50057–50077 (2020)
3. Amadeo, M., Campolo, C., Ruggeri, G., Lia, G., Molinaro, A.: Caching transient contents in vehicular named data networking: a performance analysis. Sensors **20**(7), 1985 (2020)
4. Sampath, V., Karthik, S., Sabitha, R.: Position-Based adaptive clustering model (PACM) for efficient data caching in vehicular named data networks (VNDN). Wirel. Personal Commun. 1–17 (2020)
5. Meng, Y., Naeem, M.A., Ali, R., Kim, B.-S.: EHCP: an efficient hybrid content placement strategy in named data network caching. IEEE Access **7**, 155601–155611 (2019)
6. Zhang, M., Luo, H., Zhang, H.: A survey of caching mechanisms in information-centric networking. IEEE Commun. Surveys Tutor. **17**(3), 1473–1499 (2015)
7. Afanasyev, A., Moiseenko, I., Zhang, L.: ndnSIM:NDN simulator for NS-3. Technical Report NDN-0005,NDN, Oct 2012
8. Content-Centric Networking Packet Level Simulator. https://code.google.com/p/ccnpl-sim/
9. Chiocchetti, R., Rossi, D., Rossini, G.: ccnSim: An highly scalable CCN simulator. In: 2013 IEEE International Conference on Communications (ICC), Budapest, 2013, pp. 2309–2314. https://doi.org/10.1109/ICC.2013.6654874
10. Saino, L., Psaras, I., Pavlou, G.: Icarus: a caching simulator for information centric networking (ICN). In: SimuTools, vol. 7, pp. 66–75. ICST, 2014
11. United Kingdom Open Data Portal. Retrieved 27 Sept 2019. https://data.gov.uk/roadtraffic-accidents1

Design of MIMO Cylindrical DRA's Using Metalstrip for Enhanced Isolation with Improved Performance

A. Jayakumar, K. Suresh Kumar, T. Ananth Kumar, and S. Sundaresan

Abstract For evolving 5G mm-wave applications, an enhanced performance and amplification antenna for the MIMO cylindrical dielectric resonator (DRA) are designed. Here, the dielectric resonators of the cylindrical system (DRA) are centred on a substratum. Every DRA has a surface graved metal strip that unites the cylindrical DRA antenna elements to enhance isolation. A simulated impedance bandwidth of 26,928 GHz to 27,589 GHz (S11-10 dB) is obtained from the planned antenna for the 5G frequency applications of the Telecom Commission covering the 29 GHz array. Improved isolation is completed between 26,928 and 27,589 GHz. The mechanics for enhancing insulation and the process of construction can be contained in this article. The suggested detach approach is checked in the prototype to avoid the expulsion of crosstalks and interchannels.

Keywords 5Gmm-wave · Dielectric resonator antenna (DRA) · Decoupling · Antenna

1 Introduction

In recent year, communication technologies were increasing rapidly. As generation passes, we came across 2G, 3G, 4G generations of mobile communication technologies which improved voice quality. After 4G, the fifth generation of mobile communication as met an increased demand in mobile terminals for future benefits [1, 2]. The new generation network will have a greater bandwidth, high data speeds, up to 10 Gbps. Hence, it is a future of mobile technologies [3, 4]. Due to need of large sequential bandwidth a millimetre-wave and sub millimetre bands are keenly

A. Jayakumar (✉) · K. Suresh Kumar
Electronics and Communication Engineering, IFET College of Engineering, Villupuram, India

T. Ananth Kumar
Computer Science and Engineering, IFET College of Engineering, Villupuram, India

S. Sundaresan
Electronics and Communication Engineering, National Institute of Technology Puducherry, Puducherry, India

H. K. D. Sarma et al. (eds.), *Contemporary Issues in Communication, Cloud and Big Data Analytics*, Lecture Notes in Networks and Systems 281,
https://doi.org/10.1007/978-981-16-4244-9_12

noticed [3, 5]. It is a short range, high-frequency network technology for 5G, it is a step towards providing for 5G's potential with high speed and more capacity [3]. Different types of antenna for 5 g are microstrip antenna and their array in the millimetre wave [5, 6]. Likewise, due to metallic and surface losses in millimetres, the radiation performance of the microstrip antenna is limited. MIMO from a dielectric resonator antenna is also used for minimizing metal and ionic losses that are lacking in metal sections.

DRA is a radiation structure that can minimize dielectric loss of DRA to a lower level with a high-dielectric constant [7]. Although high radiation quality can be retained in the millimetre-wave band. MIMO technology will considerably boost the channel ability of a device and have specific advantages without increasing the propagation and spectrum of the antenna. And to reduce the reciprocal connections between close-packed antennas [8, 9]. The issues faced by MIMO 5G antenna's are antenna miniaturization, integration issues, decoupling and isolation between the antenna's [10]. For miniaturization and integration issues can be utilized cylindrical DRA's [11], which reduced its size 10 times by increasing the dielectric constant ε_r From (10–100 times). The decoupling [12] and isolation of the antenna number of techniques have been introduced to expand the isolation of MIMO DRA [13], such as preventing migration between antenna elements that develop orthogonal mechanisms such as hybrid feeding systems and parasitic structures [14]. For improving the isolation of MIMO DRA, a systematic approach based on the degeneration theory was introduced in the preparation of two orthogonal methods with the same resonant frequency [15].

In this paper, the isolation between an antenna element is improved by the metal strip which is installed on the top of DRA surface. The metal strip will lead a dominant part of the assembly area shifted away from the throttle slot next to the DRA [16], which is decoupling way [12, 17].

2 Antenna Geometry and Design

Figure 1 and Table 1, respectively, denote the MIMO DRA's geometry and dimensions proposed for enhanced isolation for use on 5G mm waves. Two cylindrical DRA's of dielectric material alumina (99.5%) lossy ceramic with a dielectric constant εr of 9.9, relative permittivity of 2. 2 s, tan δ 0.93 and a thickness t is 0.254 mm are attached to a Rogers 5880 substrate. For excitation purposes, there is a rectangular excitation space provided microstrips under each DR. A length lp and width Wp metal strip on top of each DR are printed to provide better isolation between the two antenna components. The MIMO DRA comprehensive design process is briefed as,

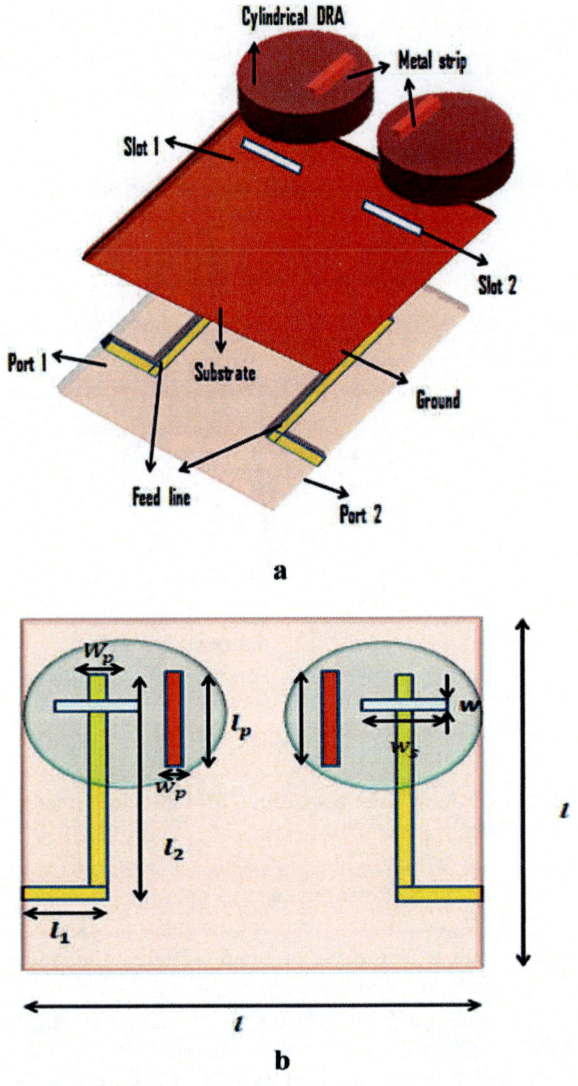

Fig. 1 Geometrical construction of the new MIMO DRA. **a** Exploded 3-D view. **b** Top view

Table 1 Parameters of the MIMO_DRA (units in mm)

Variables	Value	Variables	Value	Variable	Value
a	4.0	l_2	11.75	t	0.245
b	0	l_1	4.75	r	0.015
l	20	W_p	1.25	f	0.035
l_p	5.0	w_s	3.0	s	2.152

(continued)

Table 1 (continued)

Variables	Value	Variables	Value	Variable	Value
w_p	1.0	w	0.6	e	0.05

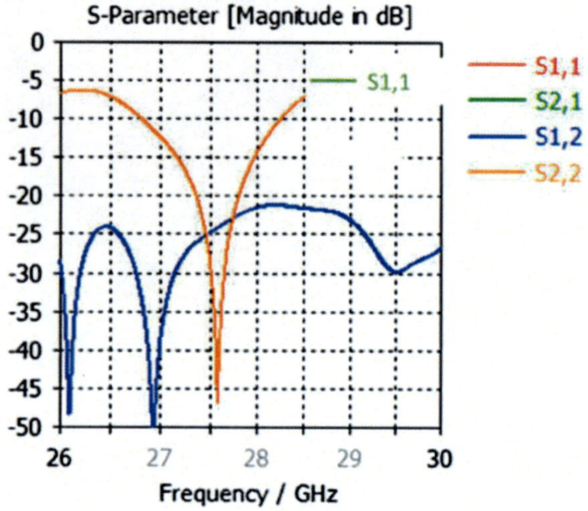

Fig. 2 Simulated output of *S* parameters in the DRA

2.1 *MIMO Cylindrical DRA*

Two ideal cylindrical DRAs with outer radius a of 4 mm and inner radius b of 0 mm are mounted on a metal base plane. The height *s* of cylindrical DRA is 2.152 mm. The antenna port is excited by a rectangular slot, which is powered by a microstrip. The Markatili approximation method can be used to measure wave networks. Figure 2 displays the simulated *S* parameter. We can see that at 27.00–28 GHz the S11 is stronger than −10 dB and at 28 GHz it's around −25 dB. The isolation needs to be further improved for meet the need for greater diversity.

The dielectric material used for structuring of dielectric resonator antenna is alumina (99.5%) lossy (ceramic) with dielectric constant of ε_r of 9.9 with the thermal conductivity of 30.3–35.0 W/mK has the extreme working temperature 1750 °C, and with the shock_resistance of 200 °C. It has the advantages of high-mechanical strength, excellent electrical insulation, good corrosion and dielectric properties.

2.2 *Substrate and Ground Plane*

The substrate used for the mechanical support and insulating properties of antenna is ROGER 5880 substrate which is placed below ground plane which is under the DRA's and above the rectangular microstrip line for feeding antenna with the dimension l of 20 × 20 × 20 mm each side. Likewise for the ground plane of dimension of 20 × 20 × 20 mm with the rectangular slot for the excitation of antenna.

The dominant part of the communication field lies on the rectangular slot with microstrip feed. Therefore, the strength of the contact field can be moved to port_a through the communication boosting slot. The antenna S parameters are shown in Fig. 2. The data reveal that the main portion of the communications area is not located immediately above the slot and must be put directly below the metal strip, improving separation from both components.

2.3 Microstrip Line

The rectangular microstrip line is used as a feed for cylindrical DRA for excitation of the antenna by means of slot present in the gound plane the microstrip feed line is made of pure copper which is port_a and port_b.

2.4 Metal Strip

The metal band is metal that is the same substance as the metal plane. The metal strip with 5.0 mm lp length and 1.0 mm wp width. The best thing about the pattern of radiation in the metal strip is for the metal level and the metal strip combination to be used as a plate condenser which absorbs the most power between the different substances in the coupling field (Fig. 3).

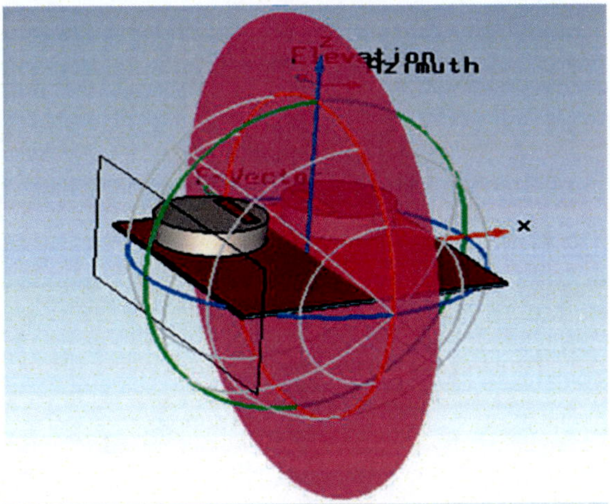

Fig. 3 Decoupling plane

3 Results and Discussions

3.1 Scattering Parameter

The *s* parameter for the cylindrical DRA is achieved in Fig. 4. The suggested MIMO DRA's reflections coefficient is stronger as −10 dB, whilst *s*1.1 is −46.67 dB at 26,928–27,589 GHz and covers the 28 GHz band for future 5G applications. The isolation measured in the 28 GHz band is good. It can be analysed that inserting metal strips improves isolation.

Parameter *S* of Fig. 4 of the DRA shows that if the mutual properties of parameter *S* are opposite, the voltage displayed on port_b is due to the current voltage on port_a. Similarly, when the same flow is applied. Port_b, which features a pair of two-way ports which is characterized by

$$S_{ij} = S_{ji} \tag{1}$$

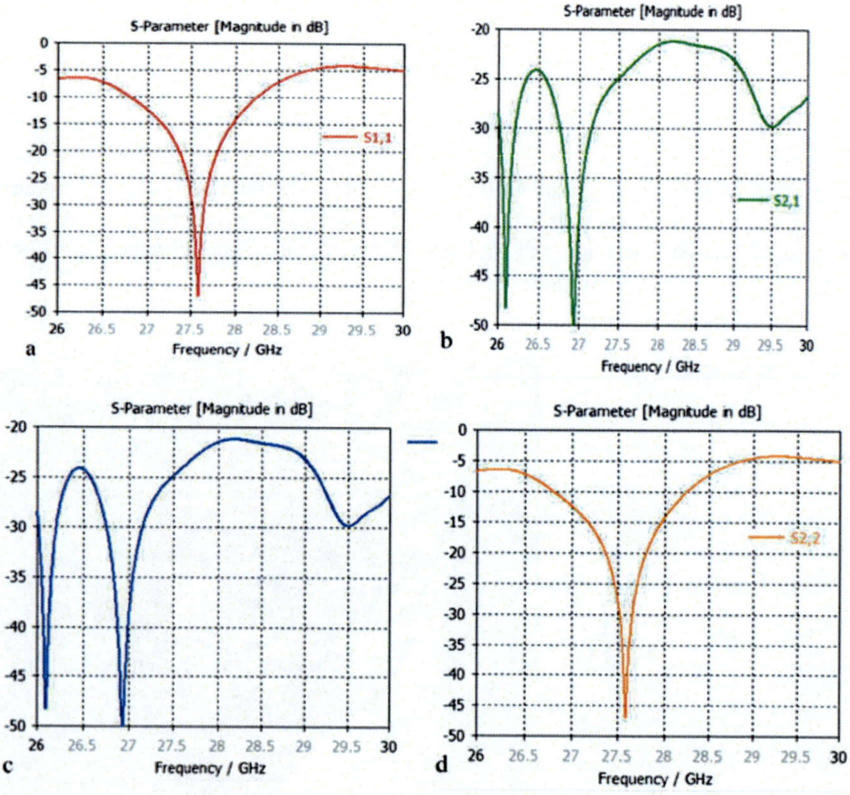

Fig. 4 *s*-parameter of DRA. **a** *S*1.1, **b** *S*2.1, **c** S1.2, **d** S2.2

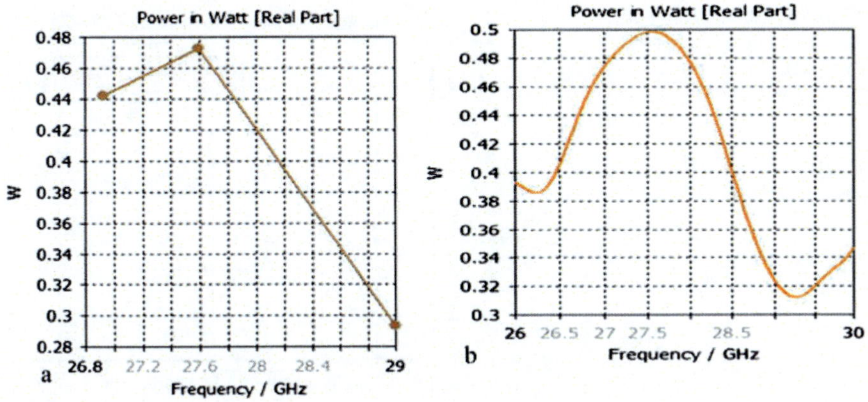

Fig. 5 **a** Power radiated, **b** power accepted

$$|S_{11}| = |S_{22}| \tag{2}$$

$$S_{21} = S_{12} \tag{3}$$

3.2 Power and Loss

It supplies a power point to the antenna terminal. Antenna power loss does not include energy lost when the joule is heated in the power line and reflections behind the power line because antennas/lines with impedance are not compatible. The power radiated and the power accepted by the antenna are shown in Fig. 5 is.

The power radiated from DRA is 0.44–0.48 W for 26.928–27.589 GHz. The power accepted by an antenna for 5G frequency is 0.5 W. The power loss in metals is 0.0186–s0.0189 W, and power loss in dielectrics is 1.02e−007 W.

3.3 VSWR

The smaller the VSWR, the transmission line antenna will get better bandwidth and the better than the other antenna. In the real system, it is very difficult to get a perfect match, so VSWR < 2 is defined as a good match system. The correct combination of VSWR = 1.009 is obtained and revealed in Fig. 6.

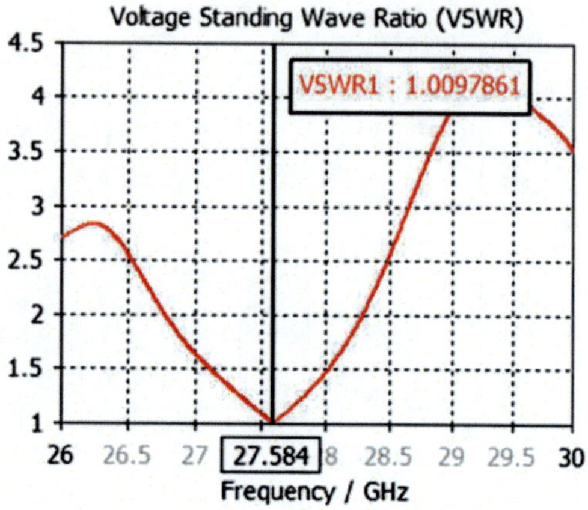

Fig. 6 Simulated graph of voltage standing wave ratio (VSWR)

3.4 Radiation Pattern

The radiated fields of antenna are used to analyse the radiation efficiency of antenna by means of farfield measurement. The farfield pattern with angle phi(Φ) and azimuth angle of DRA is shown in Figs. 7 and 8.

1. *When $\Phi = 90°$ at 27.589 GHz*

2. *When (azimuth = 90°) angle at 27.589 GHz*

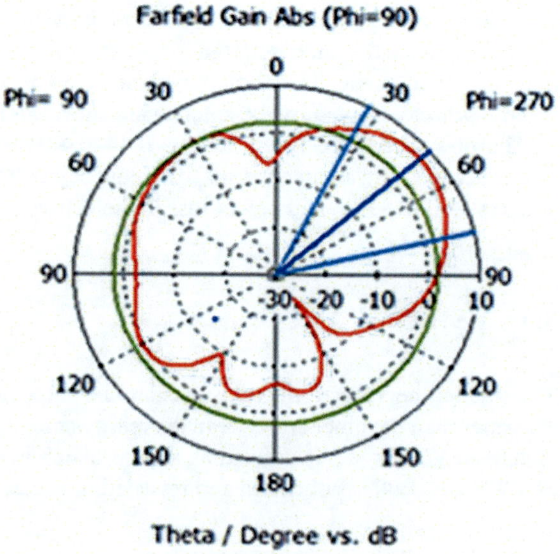

Fig. 7 Simulated graph of farfield pattern for 27.589 GHz, $\Phi = 90°$ at 27.589 GHz

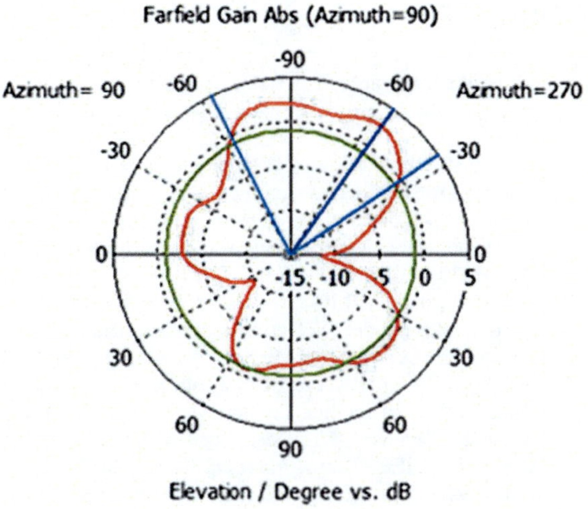

Fig. 8 Simulated graph of farfield pattern for 27.589 GHz (1) at azimuth = 90°

The farfield pattern of the angle ($\Phi = 90$) at 27.589 GHz frequency has the main lobe magnitude of 7.38 dB and main lobe direction 49.0°. It has the angular width of 49.5° and side lobe level of −2.1 dB. The analysis of azimuth = 90° angle at 27.589 GHz frequency achieved the main lobe magnitude of 1.82 dB and main lobe direction of −8.5°. It has the angular width (3 dB) of 87.2° and side lobe level −2.3 dB. The MIMO DRA achieves a radiation efficiency of −0.2280 dB and directivity of 7.853 dBi. The gain and diversity gain obtained are 7.796 and 9.99997 dB. Envelope coefficient (ECC) is an important statistic in MIMO communication systems. For a 2-element MIMO antenna, the antenna's radiation pattern will determine the ECC. As shown in Fig. 8, the proposed MIMO DRA in ECC 28 GHz is less than 0.00362, contributing the MIMO's broad channel potential and diversity advantage to the communication system. The ECC obtained is −6.6417308e−008.

4 Conclusion

This paper designs, analyzes, and measures advanced isolation MIMO DRA for future 5G mm-wave applications. By positioning one metal strip over each dielectric resonator, the proposed isolated antenna is reinforced to hold the best part of the interconnecting field away from the excitation window. The greatest improvement in isolation compared to 26.928–27.589 GHz. With an improved VSWR of 1.009 and a directivity of 7.853 dB, a gain of 7.796 dB is achieved. With this antenna, the reflection coefficient (parameter *s*) −6.641737308e−008 is obtained. The DRA of MIMO is simulated, measured and tested. The simulation and measurement results show that the proposed decoupling approach is efficient.

References

1. Adimoolam, M., John, A., Balamurugan, N.M., Ananth Kumar, T.: Green ICT communication, networking and data processing. In: Green Computing in Smart Cities: Simulation and Techniques, pp. 95–124. Springer, Cham (2020)
2. Chin, W.H., Fan, Z., Haines, R.: Emerging technologies and research challenges for 5G wireless networks. IEEE Wireless Commun. **21**(2), 106–112 (2014)
3. Rappaport, T.S., et al.: Overview of millimeter wave communications for fifth-generation (5G) wireless networks—with a focus on propagation models. IEEE Trans. Antennas Propag. **65**(12), 6213–6230 (2017)
4. Shoaib, N., et al.: MIMO antennas for smart 5G devices. IEEE Access 6 (2018): 77014–77021.
5. Sharawi, M.S., et al.: Dielectric resonator based MIMO antenna system enabling millimetre-wave mobile devices. IET Microw., Antennas Propag. **11**(2), 287–293 (2017)
6. Revathy, P., Ananth Kumar, T., Rajesh, R.S.: Design of highly efficient dipole antenna using HFSS. Asian J. Appl. Sci. Technol. (AJAST) **3**(1), 01–09 (2019)
7. Yan, J.-B., Bernhard, J.T.: Design of a MIMO dielectric resonator antenna for LTE femtocell base stations. IEEE Trans. Antennas Propag. **60**(2), 438–444 (2011)
8. Dadgarpour, A., et al.: Mutual coupling reduction in dielectric resonator antennas using metasurface shield for 60-GHz MIMO systems. IEEE Antennas Wireless Propag. Lett. **16**, 477–480 (2016)
9. Chiu, C.-Y., et al.: Reduction of mutual coupling between closely-packed antenna elements. IEEE Trans. Antennas Propag. **55**(6), 1732–1738 (2007)
10. Jamil, A., Yusoff, M.Z., Yahya, N.: Current issues and challenges of MIMO antenna designs. In: 2010 International Conference on Intelligent and Advanced Systems. IEEE (2010)
11. Huynh, A.P.M.: Investigations of cylindrical dielectric resonator antennas with improved impedance and radiation performance. University of Houston (2011)
12. Chen, W.-J., Lin, H.-H.: LTE700/WWAN MIMO antenna system integrated with decoupling structure for isolation improvement. In: 2014 IEEE Antennas and Propagation Society International Symposium (APSURSI). IEEE (2014)
13. Lee, Y., Ga, D., Choi, J.: Design of a MIMO antenna with improved isolation using MNG metamaterial. Int. J. Antennas Propag. **2012** (2012)
14. Payandehjoo, K., Abhari, R.: Investigation of parasitic elements for coupling reduction in multiantenna hand-set devices. Int. J. RF Microw. Comput.-Aided Eng. **24**(1), 1–10 (2014)
15. Wang, F., et al.: High isolation millimeter-wave wideband MIMO antenna for 5G communication. Int. J. Antennas Propag. **2019** (2019)
16. Zhang, Y., et al.: A MIMO dielectric resonator antenna with improved isolation for 5G mm-wave applications. IEEE Antennas Wireless Propag. Lett. **18**(4), 747–751 (2019)
17. Abdalrazik, A., Abd El-Hameed, A.S., Abdel-Rahman, A.B.: A three-port MIMO dielectric resonator antenna using decoupled modes. IEEE Antennas Wireless Propag. Lett. **16**, 3104–3107 (2017)

Cloud

A Robust BSP Scheduler for Bioinformatics Application on Public Cloud

Leena I. Sakri and K. S. Jagadeeshgowda

Abstract Aligning genomic sequences is a simple method for handling and analyzing the information in Bioinformatics. Here, the author expresses the problems related to BLAST algorithm which is used for the alignment of the sequences of biological data as the data grows huge. The current BLAST algorithm provided by National Centre of Biotechnology Information (stand-alone) cannot address biological data dynamically that are in terabytes. Many schedulers have been proposed to address this issue. The existing sequencing methodology is the one which is based on Hadoop MapReduce framework that executes in a sequential manner by consuming more time and is cost expensive. Hence, the author in this paper has achieved the BLAST-BSPMAPREDUCE (Bulk Synchronization Parallel MapReduce) algorithm by further improving the BLAST (Basic Local Alignment Search Tool) algorithm that is on the basis of Azure Cloud principles to tackle the case of serial execution in Hadoop MapReduce framework. The exploratory study which has been presented in this work demonstrates that the proposed BSPMAPREDUCE has a better coordinating speed of bioinformatics genomic sequences and achieves much higher speed over existing Hadoop-BLAST algorithm and the experiment result shows that the proposed customized scheduler is highly robust and scalable.

Keywords Bioinformatics · Genomic sequence · Scheduler · Parallelization · Hadoop · Microsoft azure

1 Introduction

Cloud computing has arisen as the famous foundation of information escalated processing standard [1, 2]. Cloud computing stage give on-request admittance to

L. I. Sakri (✉)
Department of Information Science and Engineering, SDM College of Engineering and Technology, Dharwad, India

K. S. Jagadeeshgowda
Department of Computer Science, Sri Krishna Institute of Technology, Bengaluru, India

H. K. D. Sarma et al. (eds.), *Contemporary Issues in Communication, Cloud and Big Data Analytics*, Lecture Notes in Networks and Systems 281,
https://doi.org/10.1007/978-981-16-4244-9_13

shared, versatile, flaw lenient, and configurable processing resources with ostensible administration endeavors and sensibly estimated [3, 4]. The big data application or high-performance computation (HPC) is a generally selected and worthy arrangement whenever contrasted with free private processing groups [5, 6]. The Cloud offers highlights like asset control, virtual processing stages and flexibility which empower simple relocation of data escalated applications in the Cloud environment. The systems needed to empower effective utilization of Cloud assets at insignificant expenses to run data intensive application needs to be addressed [7, 8]. The MapReduce structure created by Google [9] is generally famous created for circulated calculation implementations on the Cloud. Hadoop [10, 11] embraces the MapReduce system, and it fulfills the necessities in supporting big data applications. In this work, a double-stage implementation technique is embraced. In the underlying stage, the input file to be handled is divided in the form of pieces of data. Each piece is related with a Mapper or a Map worker. This worker gives the ⟨Key, Value⟩ set as the output that are arranged based on the key and values. The arranged key values are given to the Reduce worker for instance ⟨Key and Sorted List (Value)⟩. The outcomes within the Hadoop distributed file system (HDFS) are stored in the reduce worker. The virtual machines, the Map and Reduce are within the Cloud environments which is public.

Research analysts have perceived the drawbacks of the Hadoop MapReduce and have thought about different enhancement strategies. In [12], Hadoop with CUDA (Compute Unified Device Architecture) is talked about improving the calculation limit and upgrade execution time. The exploration framework in [12] empowers viable use of graphical processing units (GPU). A Cloud-based MapReduce model is introduced in [13]. To upgrade the execution and to accomplish the help for flexible valuing, the model uses pipelining. Kondikoppa et al. [14] have proposed a Replica Exchange Statistical Temperature Molecular Dynamics dependent on equal treating structure for logical analysis. To address the non-advanced asset arrangement issue in Hadoop, a novel dynamic Hadoop slot assignment conspire is introduced in [15]. The outcome introduced in [15] gives a methodology that proficiently upgrades the speed of Hadoop in taking care of different MapReduce occupations. Improving execution adequacy utilizing booking rehearses is the most generally utilized technique and is introduced in [16–19].

In [20], the authors have presented the computation of 'Imprecise Applications' on MapReduce frameworks. In predictable MapReduce applications, the Reduce stage is implemented after the Map phase. In such case, the Reduce stage can be started depending on the partial results retrieved from the Map stage. Applications like frequency count of words, detection of hot word, etc., are taken as imprecise applications. Computation of imprecise applications on conventional MapReduce frameworks such as Hadoop tends to present execution latency. In public Cloud, the user is charged for all the computations and the storage services used, and execution of imprecise applications can lead to additional costs. To overcome this drawback in [20], the authors have introduced the Check phase in the MapReduce framework to reduce the costs and reduce execution latencies. In MapReduce frameworks, the performance of applications that were iterative in nature and certain graph-based application provided unsatisfactory results. In [21], Google proposed the Pregel framework Facebook or Cloud computations for such applications. The Pregel framework is on the basis of the Bulk Synchronous Parallel (BSP) computation model presented in [22]. In [23], the execution proficiency of the Pregel framework in computing graph-based applications when compared to the MapReduce is proved. Apart from the serial execution strategy adopted in MapReduce, virtualized computing environment node failure handling capabilities, scheduling strategies, and multiway join techniques are the problems that still exist which are need to be addressed. In this paper, the authors introduce a Parallel Computing MapReduce framework (MapReduce) for public Cloud environments. **The BLAST-BSP MapReduce is based on the MapReduce architecture and incorporates the parallel computation of the BSP model to reduce the execution time. The BLAST-BSPMapReduce considers the virtualized computing environment available in the public Clouds to realize the Map and Reduce workers by using Microsoft Azure VM computing environments which consists of multi-core systems that enable parallel computation. The BLAST-BSPMapReduce proposed exploits this feature of parallel execution to reduce calculations times of the Map and Reduce worker nodes. In the BLAST-BSP MapReduce framework, the Reduce phase is initialized when two or more worker nodes of Map have completed their tasks in order to deduct the calculation time at the Reduce phase unlike the serial approach adopted in the existing MapReduce frameworks such as Hadoop. The computation of bioinformatics BLASTx applications is considered in the BLAST-BSP MapReduce presented here.**

The arrangement of the paper is as follows. In section two, the BLAST-BSP MapReduce framework with the parallel execution strategy is presented. The last section discusses the experimental study conducted to compare the performance of the BLAST-BSPMapReduce with Hadoop-BLAST MapReduce for bioinformatics applications. The conclusions drawn and the work anticipated in the future are considered in the last section.

2 Proposed System

BLAST-BSPMap Reduce proposes the calculation that executes order in alignment coordinating and considering genome information saved through NCBI. BLAST-BSP Map Reduce accepts a Cloud computing framework for MapReduce processing. To help adaptability in BLAST-BSPMapReduce, the Map and Reduce azure worker nodes are directed on a Microsoft azure Cloud cluster of VM's. The BLASTx calculation is used for genomic sequence alignment matching. The utilization of blastx calculation for alignment matching and its points of interest over the current aligners with its profits are found in [24]. The BLAST-BSPMapReduce comprises of two stages, Map and Reduce stage for sequence alignment. The existing setup for alignment on the Cloud platform consumes Hadoop structure. In the frame work of Hadoop-based arrangement, the Reduce will work only when the Map phase is finalized. To defeat the downsides or burden of Reduce stage to stand by until Map stage finishes, an arrangement to implement in parallel of the both stages has been introduced. Optimization of the BLASTx is an additional part thought to be in BLAST-BSPMapReduce. Execution of both stages' limits are planned to run parallel and feasibly make use of the cores in the virtual machine.

BLAST Algorithm: The Seed-and-Extend Approach

BLAST comprises of various searching sequence through different programs (counting BLASTn, tBLASTn, BLASTx, BLASTp, tBLASTp, and tBLASTx). Every application alteration is now of its performance; however, the center of calculations for every one of them is practically same. The BLAST algorithmic program parts a query sequence into expressions of affixed size, discovers matches for the words (which square measure alluded to as hits) during an information genomic sequencing, so stretches out the hits to get longer comparable area sets. Here, the creator considers blastx to clarify the algorithmic program. Blastx straightforwardly analyzes the genomic query sequencing of nucleotides with data super molecule arrangements. The blastx proceeds an active interesting method known as seed-and-extend which examines to get the gene sequence alignment with the highest score between the query sequence and the database. It is made up of 3 stages: In the first stage, it constructs the list of words that holds all the nearby w-mers in the query sequence, most of the time with w being 11. A list of $n - \mathrm{w} + 1$ word is constructed where n is the length of the query sequence. Figure 1 gives the outline of the system discussed above.

In the second stage, BLASTx analyzes the given data as examining for the hits (precise matching pair of words). To trigger an increase, the BLAST grasps a 'two-hit' system which needs the presence of two non-overlapping pair of words with a distance D from one another [25]. In the last stage, to find out the calibration called a high-scoring segment pair (HSP), a gap free extension is done in two ways. This is done when the condition of two-hit is satisfied. Figure 2 gives an outline of gap-free extension. If score of HSP is over the limit T, by then a gap extension is generated and the alignment is represented. In order to construct this gap local alignment, a dynamic programming calculation is used [25, 26].

Fig. 1 Build word List

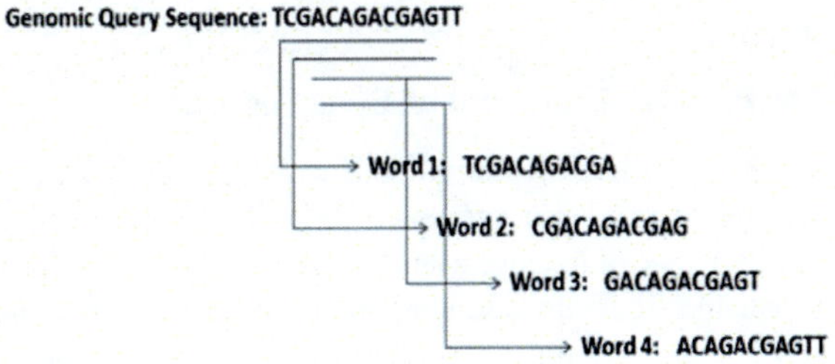

Fig. 2 Gap-free extension

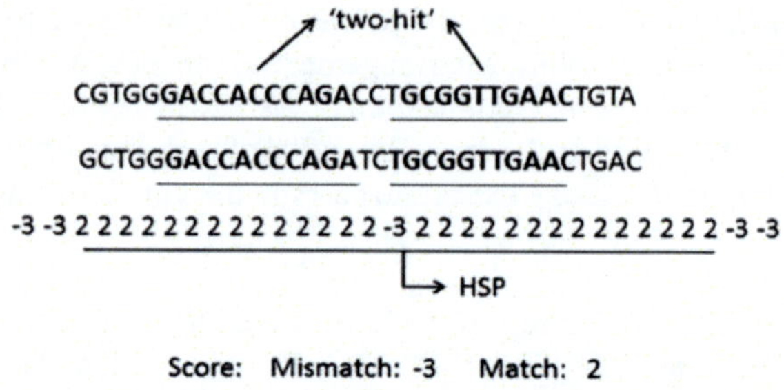

Proposed BLAST on the Parallelized MapReduce Model BLAST-BSPMapReduce

In our work, at first, the author fragments the genomic grouping database into minor pieces of static size and schedule them among the virtual azure processing hubs. The pieces can be parallelly handled by various nodes of computing. At that point, we redesign the BLAST three principle steps (assemble the list of the words, examine for the strike, and augmentation to handle the most computationally concentrated parts in parallel.)

BLAST (Basic Local Alignment Search Tool) [27] is the genome sequencing technique which compares biological DNA sequence of different proteins. It allows us to identify database sequences that look like the query sequence above a certain stipulated threshold and compare a query sequence with a database of sequences. The most intensive computing task includes two stages according to this algorithm. The first stage is searching and matching seeds in the database and second stage is extending the seeds. Since it faces significant challenges in scalability due its sequential process to match the sequence alignment, the proposed BLAST-BSPMapReduce framework can be used to achieve parallelization of BLAST algorithm to improve the overall efficiency of computation. The BLAST-BSPMapReduce algorithm is divided into three stages. At first, the gene sequence data is stored on azure blob storage. In the second stage, genomic sequence of each computing node is pre-processed. In the third stage, further accurate matches, extension and statistics are made for sequence that are pre-processed from these on stage. The data obtained is given as input to the proposed algorithm. The stages of the BLAST-BSPMapReduce algorithm are described below.

The Parallel Alignment of Gene Sequence

The Scanner is built using list of words which is built in accordance with the preprocessing of genomic sequence data. The parallel alignment of genomic sequence through a word list and a scanner mainly involves three steps, exact match of words, expansion of seeds, and statistical during the expansion process. The built word list and scanner should be sent to the multiple virtual computing nodes to speed up parallelization of the alignment to genomic sequence. The flow chart of proposed algorithm is shown in Fig. 3.

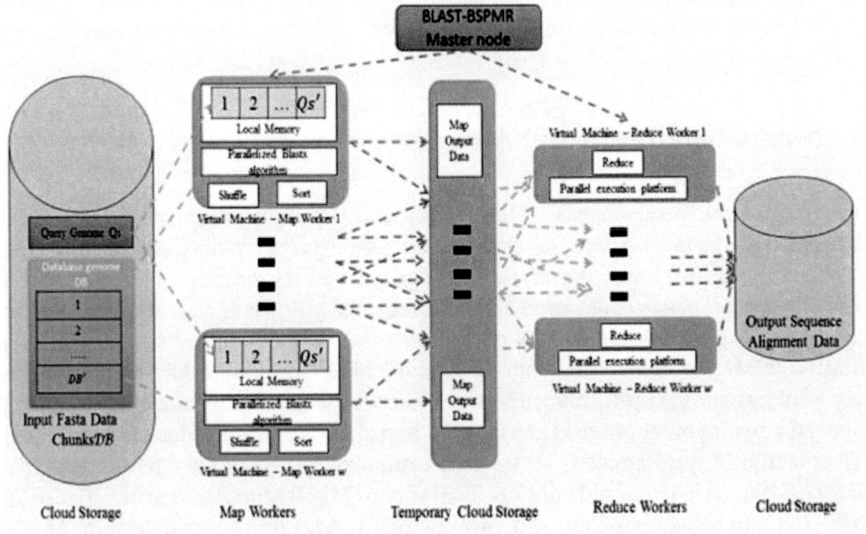

Fig. 3 The parallelized BLAST-BSPMapReduce model

Steps involved for Parallel Alignment of BLAST-BSPMapReduce Algorithm is Shown as Follows

To begin with, the genomic sequence database is divided into little lumps of data whose size is fixed. Then this data is distributed among the different worker nodes. The lumps have to be handled in parallelized design in various computing nodes. At this point, the fundamental three phases of BLAST algorithm are reorganized, i.e., build the list of words, scan, and the extension. Later, run or cycle most computationally concentrated data in the parallelized manner. The general progression of the BLAST-BSPMapReduce is given in Fig. 3. Once an input query is given, the BLAST-BSPMapReduce technique will work as follows: (1) Word list is formed using input query sequence by the BLAST program, (2) and then from this word list, a scanner is constructed. (3) Scanner and query sequence are transferred to processing worker nodes and thereby initiating the MapReduce job. (4) In the Map stage, the scanner and the query are stacked by each of the Map worker from cached records and genomic database is scanned on local memory for word hits. The 'two-hit' list is put aside for the future cycle. (5) Expansion starts after the sweep cycle. The hole-free augmentation is first finished for every two-hit. In an event that the accomplished score is over an edge T, a gapped DP (dynamic programming) expansion is set off. It is considered to put both sweep and expansion steps in the Map stage because of the following two reasons: Initially, it stays away from the huge data of irrelevant information over the setup; also, the output and augmentation assignments on various genomic data set pieces can completely run-in parallel design. (6) In the shuffle stage, results obtained for alignment are naturally arranged depending on the obtained scores. (7) The cycle in the Reduce stage collects this result and stores in the azure blob container storage.

3 Simulation Result and Analysis

The framework makes use of Home Windows 10 Enterprises 64-bit Operating System, 8 GB RAM, i-5 Quad Core processor. C# 6.0 programming language and .net framework, and 4.0 java programing language are used for the current Hadoop framework. A research was considered at the variables like linear acceleration and execution time for series alignment. A comparative study was finished on the proposed blast–BSPMapReduce model with the existing Hadoop–blast model with the intention to determine the performance between the two fashions. With one VM processing node, the deployment of the Hadoop–blast and blast-BSPMapReduce is considered. The blast–BSPMapReduce is set up on the microsoft azure cloud environment. The Hadoop-blast is constructed utilizing the Hadoop–MapReduce framework. To deploy SW-Hadoop, apache Hadoop and yarn 2.4.0 is used. Similar configuration of VM processing nodes is considered in the arrangements. The nr (non-redundant) protein genomic database (nr.01) is considered for assessment [28]. Trials are led utilizing a steady database genomic arrangement and four query sequence of different sizes are thought of. The nr.01is considered for the experiment purpose (this is the bigger non-redundant protein database that is accessible openly). Table 1 contains the summary of the database and query genomic sequences. The execution time is observed by the computations on the BLAST-BSPMapReduce and Hadoop-BLAST considering the BLASTx sequence alignment. The complete time taken for the sequence alignment is observed as shown in Fig. 4. The proposed BLAST-BSPMapReduce aligner deployed on Azure surpasses the Hadoop-BLAST. As the query document length expands, the execution time likewise increments for both BLAST-BSPMapReduce and Hadoop-BLAST. In trial 1 (sequence file size of 16 kb), the accelerate accomplished for BLAST-BSPMapReduce is about 4.1. For sequence alignments in trial 2 (sequence file size of length 32 kb which is shown in Fig. 5), the speedup was seen to be 2.6. For sequence alignments trial 3 (sequence file size of length 64 kb which is shown in Fig. 5), the speedup was seen to be 2.14. For example, in trial 4 (sequence file size of length 128 kb which is shown in Fig. 5), the speedup was seen

Table 1 Information taken to compare the performance of BLAST-BSPMR with the BLAST-Hadoop

No.	Database genome	Size (GB)	Query genome	Size (kb)
1	nr.01 (non-redundant protein)	2.99	Nucleotide sequence	16
2	nr.01 (non-redundant protein)	2.99	Nucleotide sequence	32
3	nr.01 (non-redundant protein)	2.99	Nucleotide sequence	64
4	nr.01 (non-redundant protein)	2.99	Nucleotide sequence	128

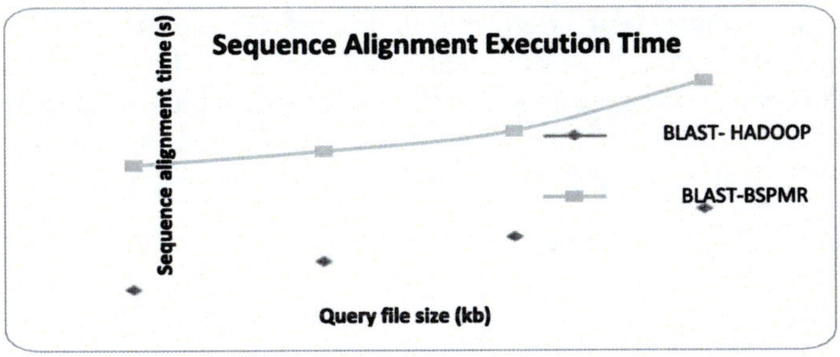

Fig. 4 Sequence alignment execution time

Fig. 5 Linear speedup and throughput

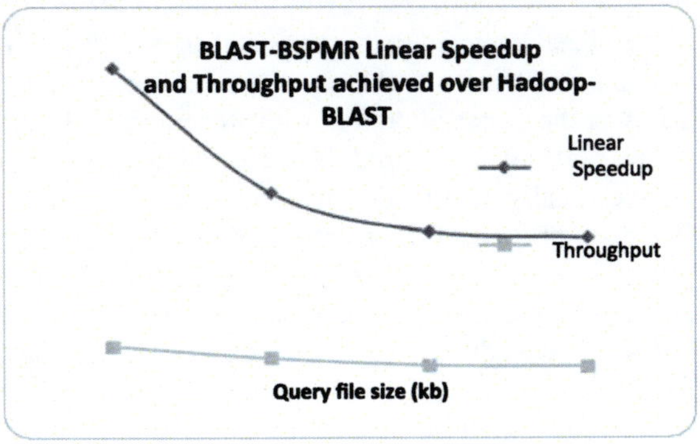

to be 2.07. A normal accelerate of 2.72 is accomplished which is depicted in Fig. 6 considering BLAST–BSPMapReduce when contrasted with the Hadoop–BLAST. In trial 1 (sequence file size of 16 kb), the throughput accomplished for BLAST–BSPMapReduce is about 76%. For grouping arrangements in trial 2 (sequence file size of length 32 kb which is shown in Fig. 5), the throughput was seen to be 62%.

Fig. 6 Average speedup and throughput

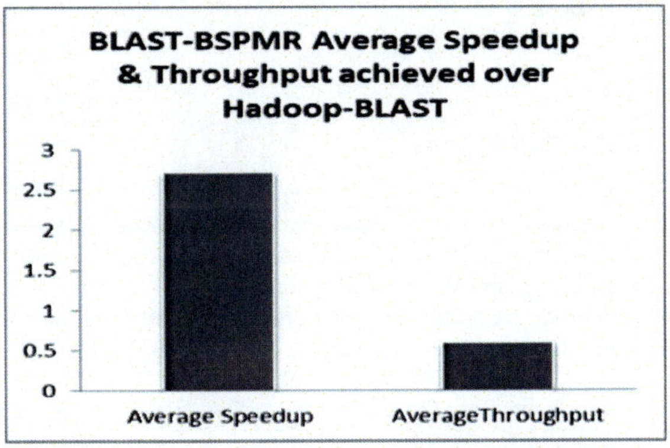

For sequence alignment trial 3 (sequence file size of length 64 kb which is appeared in Fig. 5), the throughput was seen to be 53%, and in trial 4 (sequence file size of length 128 kb which is appeared in Fig. 5), the throughput was seen to be 51%. An average throughput of 60.8% is accomplished which is appeared in Fig. 6 considering BLAST–BSPMapReduce when contrasted with the Hadoop–BLAST.

4 Conclusion

Genomic sequence alignment is a straightforward technique which handles and examines the data in bioinformatics. The suitable issues of sequence alignment BLAST calculation which is the most well-known local sequence alignment calculations is been depicted in this paper. As of now, the BLAST calculation gave by NCBI (independent) cannot address organic information dynamic which are in terabytes. To settle the issues identified with storing the data and parallel computation of data, Cloud platform is utilized. The current sequence aligners scheduler shows inadequacies in alignment sequencing in cost-effective manner. And also experience the ill effects of issues that are examined in this work. The BLAST-BSPMapReduce to align sequences is proposed on this paper. The BLASTx calculation is been utilized for bio-sequence alignment inside the BLAST-BSPMapReduce cloud degree and a parallel MapReduce execution method depending on azure cloud is utilized. In order to raise using the digital machine-based totally cloud computing platform, a parallel implementation of the map and reduce shape is executed. The paper likewise features the correlation of the proposed BLAST-BSPMapReduce with the current frameworks sequence arrangement. Trials to illustrate the effectiveness of the BLASTx calculation are introduced. Correlation with the Hadoop-BLAST for collection alignment is brought through trial study. The consequence received suggests massive improvement thinking about the BLAST-BSPMapReduce when contrasted with the Hadoop-BLAST. Later on, the direct analysis on differed database is proposed. Furthermore, run fluctuated application that are accessible in NCBI, for example, blastn, blastp etc.…, is proposed to run on BSPMapReduce Scheduler to further analyze the robustness.

References

1. Marinescu, D.C.: Parallel and distributed computing: memories of time past and a glimpse at the future. In: 2014 IEEE 13th International Symposium on Parallel and Distributed Computing (ISPDC), pp. 14, 15, 24–27 June 2014
2. Gartner, Inc.: Gartner says worldwide Cloud services market to surpass $68 billion in 2010. http://www.gartner.com/it/page.jsp?id=1389313
3. Mell, P., Grance, T.: The NIST Definition of Cloud Computing, US National Institute of Science and Technology Std., 2011 [Online]. Available http://csrc.nist.gov/publications/nistpubs/800-145/SP800-145.pdf

4. Bera, S., Misra, S., Rodrigues, J.J.P.C.: Cloud computing applications for smart grid: a survey. IEEE Trans. Parall. Distrib. Syst. **99**, 1. https://doi.org/10.1109/TPDS.2014.2321378
5. Mehrotra, P., Djomehri, J., Heistand, S., Hood, R., Jin, H., Lazanoff, A., Saini, S., Biswas, R.: Performance evaluation of Amazon EC2 for NASA HPC applications. In: Proceedings of the 3rd Workshop on Scientific Cloud Computing. ACM, New York, NY, USA, 2012
6. Suen, C.H.: Evaluating and improving the performance and scheduling of HPC applications in Cloud. IEEE Trans. Cloud Comput. **1**, 1. https://doi.org/10.1109/TCC.2014.2339858
7. Cai, Z., Li, X., Gupta, J.N.D.: Heuristics for provisioning services to workflows in XaaS Clouds. IEEE Trans Serv Comput **99**, 1. https://doi.org/10.1109/TSC.2014.2361320
8. Okur, M.C., Buyukkececi, M.: Big data challenges in information engineering curriculum. In: EAEEIE (EAEEIE), 2014 25th Annual Conference, pp. 1, 4, May 30 2014–June 1 2014
9. Dean, J., Ghemawat, S.: MapReduce: simplified data processing on large clusters. In: OSDI, 2004, pp. 137–150
10. White, T.: Hadoop: The Definitive Guide. M. Loukides, Ed. O'Reilly, 2010
11. Hadoop, 2010. http://Hadoop.apache.org
12. Zhu, J., Li, J., Hardesty, E., Jiang, H., Li, K.: GPU-in-Hadoop: enabling MapReduce across distributed heterogeneous platforms. In: 2014 IEEE/ACIS 13th International Conference on Computer and Information Science (ICIS), pp. 321, 326, 4–6 June 2014
13. Dahiphale, D., Karve, R., Vasilakos, A.V., Liu, H., Yu, Z., Chhajer, A., Wang, J., Wang, C.: An advanced MapReduce: Cloud MapReduce, enhancements and applications. IEEE Trans. Netw. Serv. Manage. **11**(1), 101, 115 (2014)
14. Kondikoppa, P., Platania, R., Park, S.-J., Keyes, T., Kim, J., Kim, N., Kim, J.H., Bai, S.: MapReduce-based RESTMD: enabling large-scale sampling tasks with distributed HPC systems. 2014 6th International Workshop on Science Gateways (IWSG), pp. 30, 35, 3–5 June 2014
15. Tang, S., Lee, B., He, B.: Dynamic MAPREDUCE: a dynamic slot allocation optimization framework for MapReduce clusters. IEEE Trans. Cloud Comput. **2**(3), 333, 347, 1 July–Sept 2014
16. Zaharia, M., Konwinski, A., Joseph, A.D., Katz, R.H., Stoica, I.: Improving MapReduce performance in heterogeneous environments. In: OSDI. USENIX, pp. 29–42, 2008
17. Tao, Y., Zhang, Q., Shi, L., Chen, P.: Job scheduling optimization for multi-user MapReduce clusters. In: 2011 Fourth International Symposium onParallel Architectures, Algorithms and Programming (PAAP), pp. 213, 217, 9–11 Dec 2011
18. Rasooli, A., Down, D.G.: An adaptive scheduling algorithm for dynamic heterogeneous Hadoop systems. In: Proceedings of the 2011 Conference of the Center for Advanced Studies on Collaborative Research, ser. CASCON '11. IBM Corp., Toronto, Ontario, Canada, pp. 30–44, 2011
19. Zaharia, M., Borthakur, D., Sen Sarma, J., Elmeleegy, K., Shenker, S., Stoica, I.: Delay scheduling: a simple technique for achieving locality and fairness in cluster scheduling. In: EuroSys 2010, pp. 265–278, Apr 2010
20. Wang, C., Peng, Y., Tang, M., Li, D., Li, S., You, P.: Map check reduce: an improved MapReduce computing model for imprecise applications. In: 2014 IEEE International Congress on Big data (BigData Congress), pp. 366, 373, June 27–July 2 2014
21. Malewicz, G., Austern, M., Bik, A., Dehnert, J., Horn, I., Leiser, N., Czajkowski, G.: Pregel: a system for Large scale graph processing. In: Proceedings of the 2010 International Conference on Management of Data, SIGMOD'10. ACM, New York, NY, USA, pp. 135–146, 2010
22. Valiant, L.G.: A bridging model for parallel computation. Commun. ACM **33**(8), 103–111 (1990)
23. Kajdanowicz, T., Indyk, W., Kazienko, P., Kukul, J.: Comparison of the efficiency of MapReduce and bulk synchronous parallel approaches to large network processing. In: 2012 IEEE 12thInternational Conference on Data Mining Workshops (ICDMW), pp. 218, 225, 10–10 Dec 2012
24. Job, T.N., Park, J.H.: Exploiting High Performance on Bioinformatics Applications in a Cloud System, vol. 22, no. 2, pp. 22–24, 2014

25. Altschul, S.F., Madden, T.L., Schäffer, A.A., Zhang, J., Zhang, Z., Miller, W., Lipman, D.J.: Gapped BLAST and PSI-BLAST: a new generation of protein database search programs. Nucl. Acids Res. **25**(17), 3389–3402 (1997)
26. Zhang, Z., Schwartx, S., Wagner, L., Miller, W.: A greedy algorithm for aligning DNA sequences. J. Comput. Biol. **7**(1/2), 203–214 (2000)
27. Altschul, S., Gish, W., Miller, W., Myers, E., Lipman, D.: Basic local alignment search tool. J. MolBiol. 215(3), 403–410. PMID 2231712. https://doi.org/10.1016/S0022-2836(05)80360-2
28. ftp://ftp.ncbi.nlm.nih.gov/blast/db/

Mobile Cloud-Based Framework for Health Monitoring with Real-Time Analysis Using Machine Learning Algorithms

Suman Mohanty, Ravi Anand, Ambarish Dutta, Venktesh Kumar, Utsav Kumar, and Md. Ruhul Islam

Abstract Cloud computing in the field medical sciences has made remarkable progress and has been a boon for medical firms. The motive being to provide health consultancy remotely as well as quickly. It mostly emphasizes on proper diagnosis of the patient as and when required. In contrast to the existing system which is at times prone to errors leading to many deaths due to faulty diagnosis and monitoring. Cloud-based system provides much fluidity by providing quick assistance for patient irrespective of their location. Cloud infrastructure has greater computational power and can analyze patient's data remotely helping the medical practitioner to provide diagnosis rapidly, and greater precision is achieved by deploying machine learning algorithms. To make this system accessible from anywhere, in our paper, we propose the use of mobile cloud computing-based architecture for health monitoring. Mobile cloud computing relies on cloud computing to deliver applications to monitoring devices. Real-time monitoring is possible where data can be fetched with the use of mobile cloud applications. Mobile computing provides a platform for making the use of high-end cloud infrastructure to use powerful computation ability to deploy forecast models. Mobile cloud computing plays a vital role since it extracts the advantages of integrating both cloud and mobile computing to provide healthcare assistance. The proposed architecture is scalable as data storage can be increased/decreased by health institutions, reliable as it implements MCC and affordable as it works as a subscription model.

Keywords Mobile cloud · Gene · Machine learning techniques · Time-series analysis · Artificial neural networks · Support vector (SVM)

1 Introduction

The widespread of medical devices along with the Internet has revolutionized patient monitoring and medical assistance systems. Different acquired information such as

S. Mohanty (✉) · R. Anand · A. Dutta · V. Kumar · U. Kumar · Md. R. Islam
Department of Computer Science and Engineering, Sikkim Manipal Institute of Technology, Sikkim Manipal University, Majitar, India

H. K. D. Sarma et al. (eds.), *Contemporary Issues in Communication, Cloud and Big Data Analytics*, Lecture Notes in Networks and Systems 281,
https://doi.org/10.1007/978-981-16-4244-9_14

blood sugar level, heart rate, and other signals which are transmitted to the doctor's end through cellular networks and wireless media automated medical forecasting and analytical tools enable remote diagnosis [1]. Cloud computing is an efficient platform for monitoring and managing services. It provides secure transmission, and retrieving medical information using private cloud, secure storage, sharing, and medical information exchanges is achieved through private cloud. The use of machine learning algorithms and deep learning models for clinical applications is capable of transforming the delivery traditional health monitoring services.

2 Literature Review and Existing System

Extensive work on health monitoring, diagnosis, and detection is being carried by many researchers. Computer scientists and researchers tried to blend health monitoring along with remote diagnosis. A work by Bhosale et al. [2], Patient Management System For Doctors Using Cloud Computing, helped us to understand how management systems can be deployed in the cloud and can be recursively used for future reference of patients health.

Another good literature was by Mamun [3], and his Cloud-Based Framework for and Monitoring System for Remote Healthcare Applications idea is on how the cloud technology can be used in a similar viewpoint and how it can improve the detection and monitoring method. This is a literature that helped us to build a solid foundation.

This literature research on [4] Mobile Cloud Computing: Review, Trend and Perspectives by Mr Han Qi and his team of Malaysia University helped us to understand the importance of Mobile Cloud Computing.

One of the other notable work we used for our reference was by Qayyum et al. [5] Secure and Robust Machine Learning for Healthcare: A Survey. It helped us to how data processing is important as a strategic resource and how mobile cloud is an extension of cloud computing with high mobility and scalability

3 Cloud Computing Services

Cloud computing is a hardware–software package, which is the storage and computing power and also the software required to utilize the hardware very efficiently. Cloud computing looks after the weighted tasks involved in computation and processing data away from devices we use. It migrates all computation to greater block of computer in the Web. The Internet tends to be the cloud, the information, applications, and work which is accessed through any device irrespective of geographical location. [6]. Cloud computing can both be private and public. Public cloud devices charge some fee for rendering service. Private cloud services are confined to a group of people and provide services to a specific organization which is confined to the people associated. These are framed networks which supply hosted

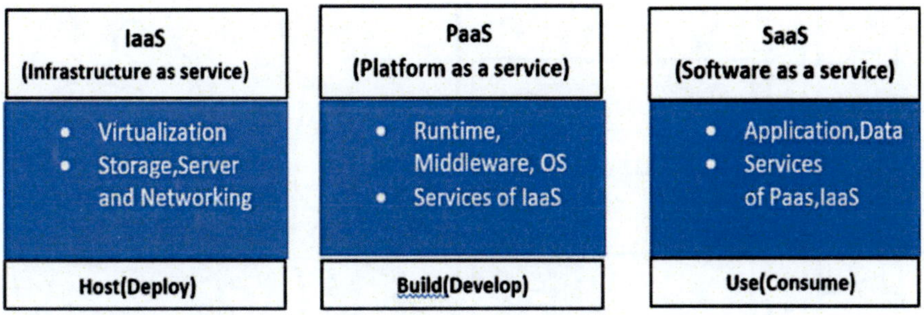

Fig. 1 Cloud services

services. Hybrid cloud is a technology that incorporates one or more public cloud services with a private cloud, with customized software that allows for communication within each separate service. When demands and costs vary wildly, a hybrid cloud approach provides organizations with greater versatility by shifting workloads across cloud solutions (Fig. 1).

4 Mobile Cloud Computing

We can say that mobile cloud computing is a mixture of mobile computing and the expansion of the cloud computing. It is somewhat similar to cloud computing. With the help of mobile cloud computation, users of the mobile can run applications on their mobile without completely relying on the mobile operating system and the computing or the memory capacity of the smartphone [7]. Mobile cloud computing is complete mixture of mobile development and cloud computing. With the help of mobile cloud computing, the mobile user can use applications with many options and functional capabilities delivered over the Internet and powered by infrastructure of cloud backend.

Architecture

In mobile cloud computing, the mobile devices and the mobile networks are connected with the help of satellite and access point. The main part of this satellite or access point is to establish the connections and control functional interfaces between the networks and mobile devices. The requests of the mobile users and their information such as their ID and location are transmitted to the central processor. These central processors are connected to the servers; the server provides the mobile network services. The mobile device users who are connected in the mobile cloud network can access services such as authorization and authentication based on the user or subscriber's data stored in the databases. Then, the subscriber's data and the subscriber's request are transferred to cloud through the Internet. In the cloud, there is cloud controllers, and this cloud controller is responsible to process the subscriber's

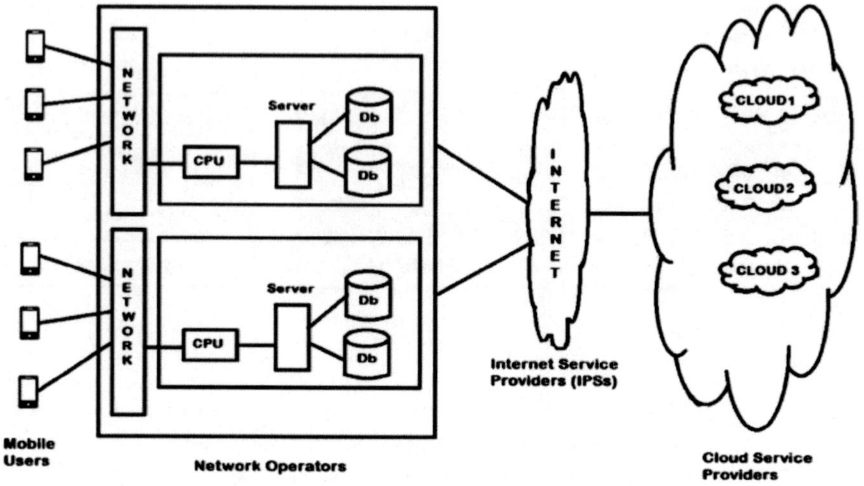

Fig. 2 Mobile cloud architecture health monitoring systems

request and provide the subscriber the corresponding cloud services. Figure 2 shows the architecture of the mobile cloud computing.

5 Health Monitoring System and Sensors

The patients end comprises of IoT module. The module consists of multiple biomedical sensors which enable the measurement of vital health data such as heart rate, saturation of oxygen in blood, and body temperature. The considered sensors include electromyography sensors which measure response of muscles and reads electrical activity due to simulation of muscle nerves. It helps to detect muscle and nerve problems. It also includes electrooculography which is a method for measuring the potential difference between the retina and cornea which is the front and back of human eye. The signal generated due to this is termed as electrooculogram. It mainly focuses on diagnosis of eye and other ophthalmological treatment. It does not take in account the response to individual visual stimulus.

An embedded blood sugar is to measure the glucose concentrated in the blood. It also exists in form of strips of glucose paper which is dipped to some substance and measured with respect to standard glucose chart for monitoring patients with hypoglycaemia/diabetes mellitus. We have taken in account electroencephalography (EEG), which measures electrical activity in the brain, as cells in brain communicate through electrical impulses. It helps to detect brain disorders such as seizures, stroke, dementia, and disorder in sleep. One of the most vital sensors used is the heart rate monitor which is a common sensor module used to monitor heart rate during various physical activities.

- This enables close monitoring of patient's health at real time using mobile cloud architecture.
- The data triggered after being read through sensors helps the doctor to remotely diagnose the patient as well as maintaining a health record in the proposed system

6 Machine Learning Algorithms for Deployment in Cloud

6.1 *Support Vector Machine*

SVM is one of the most used algorithms that are being deployed for classification and regression problems. Not only that, its extensive use is also found in the problems, specially the classifications were dataset that needs to be divided into two different classes using a hyperplane. For the new data to be classified correctly, the selection of hyperplane should be in such a way that the distance between any point within the training set and the hyperplane should be maximum [8]. Support vectors are the most essential and most difficult data points that need to be classified. These are the one that is nearest to the decision surface or the hyperplane and hence should be attended carefully (Fig. 3).

$$\text{the function of line, } y = mx + c$$
$$mx + c - y = 0, \text{Let vector } X = (x, y) \text{ where } x, y \tag{1}$$

$$w = (m, -1), \text{Now hyperplane is } w.X + b = 0$$
$$\text{the hypotheshis function } h = \begin{cases} +1, & \text{if } w.X + b \geq 0 \\ -1, & \text{if } w.X + b < 0 \end{cases} \tag{2}$$

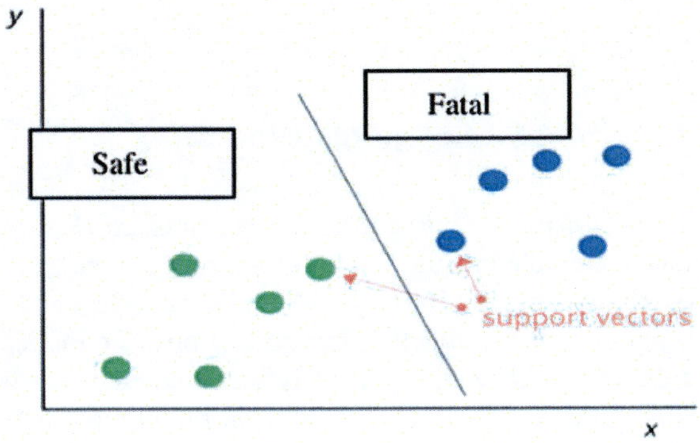

Fig. 3 2D support vector graphs

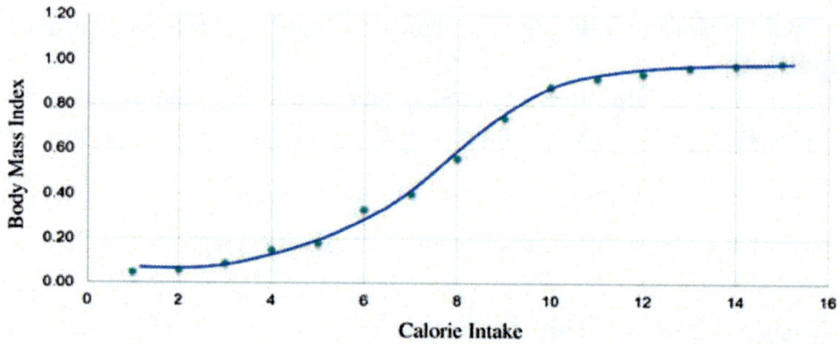

Fig. 4 Logistic function

6.2 *Logistic Regression*

Another supervised learning algorithm is logistic regression which is grouped under classification algorithms and are mainly deployed for predicting the probability of an output variable. Multiple linear regression is also akin to this, having an exception that response variable is binomial. In case of healthcare, where logistic regression is broadly deployed for predicting the risk of a disease and to further ameliorate in making decisions, predicting the probability of output variable and classification problems is important [8]. It forecasts the likelihood of a given case, making it a valuable method for determining disease risk and enhance clinical choices (Fig. 4).

$$\sigma(x) = \frac{1}{1 + e^{-x}}$$

the weighted sum of inputs, $x = \Theta.\text{weight} + b$

Logistic equation i.e., Obese Probablity

$$= \frac{1}{1 + e^{-x}} = \frac{1}{1 + e^{-\Theta.\text{weight}+b}} \tag{3}$$

6.3 *Natural Language Processing*

Taking healthcare into consideration, substantial fraction of data is unstructured and indecipherable like discharge summaries, patient's health reports, operational procedure, etc., thereby making it difficult to understand by computers without special programs or methods [9]. The collections of enriched and well-structured data that are available from the previously encountered disease-related keywords help in identifying different unseen words from the clinical notes and hence overcoming this problem.

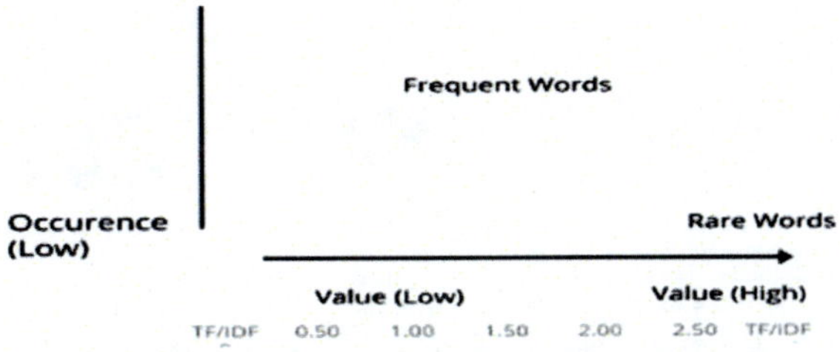

Fig. 5 Automated health report data mining

TF-IDF Term frequency-inverse document frequency measures the importance of a word under a document in the collection or corpus. Term frequency is the frequency measure of words in the document, and inverse document frequency is the measure of commonness in the words among the different text corpus. The frequency count of the words helps in understanding the most significant words and the least significant words as well [9]. TF-IDF is also important in situations where disease correlation between patients are high as well as in searching of databases for sequential patterns (Fig. 5).

7 Architecture and Working

7.1 Cloud Services

Why Azure over alternative Cloud services?

Microsoft has produced a number of the industry's leading advanced security technologies, so you can be guaranteed that your information is securely protected [10]. Microsoft has taken significant measures to ensure high standards of protection around the cloud space. With tools consisting of threat intelligence, threat analytics, Azure Info Security, and two-factor authentication, your Azure environment is secure. These methods help you assess, create, and re-establish threats in real time and respond to suspicious activity of the server and the user on your network and have the best protection levels. With Azure, the collection, use, and distribution of your client data are owned and managed by yourself. Microsoft has provided detailed information about their expertise to handle your knowledge

Azure Virtual Machine

Virtual machine allows you to produce and manage a gaggle of similar, scalable, and load-balanced VMs. It is a computer file usually referred to as an image that functions like an actual computer. It is one of the files holding it all. It operates on Windows, Linux, and so on. This gives you the ability to run multiple computers on a physical machine. You can have a different operating system on each system. It solves the following purposes

Collect	Process	Store	Analyze
Collect Device Data in variety of formats and frequencies	Transform and enrich messages with external sources	Data is stored in time-series data store for analysis	Run SQL queries, use pre-built models, and your custom analysis to perform machine learning and make predictions

Fig. 6 Cloud services working diagram

- Development and testing
- Applications within the cloud
- Augmented data/knowledge center.

Azure Cosmos DB

It is a completely NoSQL information management service for parallel app development Microsoft's internationally distributed, multi-structured database service which is Azure Cosmos DB. Cosmos DB helps you to scale throughput and storage elastically. Using your chosen API, including: SQL, MongoDB, you can elastically scale and take advantage of quick, millisecond data access.

Azure Time Series Insights

End-to-end IoT analytics platform observes, analyzes, and visualizes your industrial IoT knowledge at scale. Turn IoT knowledge into unjust insights. Improve operations and decision making with decades of IoT knowledge delivered with made image and a jailer expertise. Use period of time information insights and interactive analytics accelerates IoT information (Fig. 6).

7.2 *Working of Time-Series Forecasting Using Trained Model*

Time-series analysis is a mode which helps in recognizing patterns in historical data, to detect occurrence of sequences and forecasting or predicting future values on the basis of past trends [11]. In this paper, our motive was to visualize assumed patient's data who is a prolonged hypertension patient using various machine learning algorithms such as support vector machines (SVMs) and artificial neural networks (ANNs). We were interested to explore the established time-series forecasting methodologies to have insight on health data trends and discover early signs and symptoms of deteriorating health conditions on the basis stored sets of health

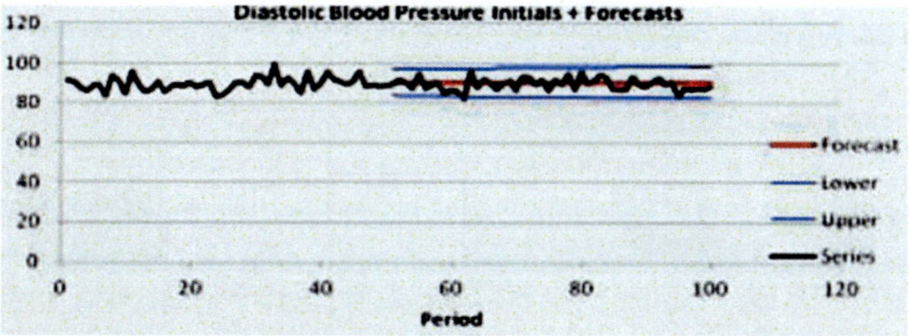

Fig. 7 Forecast range versus actual values [11]

data and events. The autoregressive integrated moving average (ARIMA) model is established for its detailed statistical features and as an effective process for linear modeling [11]. In case of nonlinear modeling, SVM and ANN models can help in capturing historical info by nonlinear function and hence be proved to be robust time-series forecasting methods due to their flexibility in nonlinear mapping ability and are tolerant to twisted forecasting data (Fig. 7).

8 Proposed Architecture of Cloud-Based Framework for Health Monitoring System

The proposed system works in the following manner. Sensor at the patient's end fetches necessary data through the services available via mobile network to the cloud which acts as a federal point of entry enabling full access to capabilities inherent in the mobile network platform. APIs are used to transfer data TO Microsoft VM and then to the Cosmo DB. After storing the data, it is sent to the monitoring system for detecting any fatalities or alarming condition. Real-time data is used for time-series forecasting and is stored as historic data. In case of any abnormalities based on the monitoring and time-series forecasting, doctor is notified in order to provide assistance and diagnose. Doctors can view patient records using API.

A model is built using ML approaches mentioned above; once the model is trained, real-time data along with recorded data is transferred to Azure Time Series insights for prediction. It can be visualized as an infrastructure, providing centralized and powerful platforms in the cloud via base stations of mobile network (Fig. 8).

The proposed architecture in the given figure above can be deployed, and it is an add-on to the existing system as stated in Table 1 for comparison.

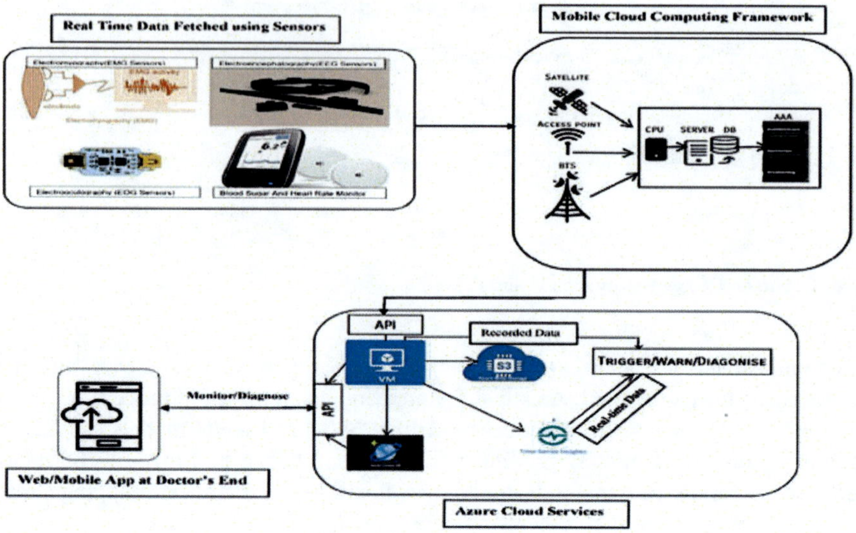

Fig. 8 Proposed architecture with MCC framework and cloud services

Table 1 Proposed architecture with MCC framework and cloud services

	Existing system	Proposed system
Architecture	Cloud based with Internet connectivity	Mobile cloud-based framework using base stations
Algorithms	Support vector machines, neural networks	SVM, logistic regression, natural language processing
Reliability and accuracy	Lower reliability, average accuracy	Enhanced reliability with greater accuracy
Monitoring	Monitoring using forecasting models	Real-time monitoring using time-series insights

9 Conclusion

The above-stated work is a theoretical approach, the architectural framework is a robust model for practical application in health monitoring systems, and it aims to efficiently use mobile cloud and machine learning approach for remote monitoring and real-time analysis. It will be very useful in areas with less resources and connectivity issues, when immediate assistance is not available or with patients who need constant monitoring. For future scope, this architectural framework can be developed further to make software which can integrate with the framework to facilitate real-time application

References

1. Shahzad, A., Lee, Y.S., Xiong, N.: Real-Time cloud-based health tracking and monitoring system in designed boundary for cardiology patients. J. Sens. **2018** (2018)
2. Bhosale, C., Narkar, N., Bhat, V., Maste, D.: Patient management system for doctors using cloud. Comput. IOSR J. Eng. (IOSRJEN)
3. Malhotra, L., Agarwal, D., Jaiswal, A.: Virtualization in cloud computing. J. Inf. Technol. Softw. Eng. https://doi.org/10.4172/2165-7866.1000136
4. Qi, H., Gani, A.: Research on mobile cloud computing: review, trend and perspectives. In: 2012 Second International Conference on Digital Information and Communication Technology and it's Applications (DICTAP), Bangkok, 2012, pp. 195–202. https://doi.org/10.1109/DICTAP.2012.6215350
5. Qayyum, A., Qadir, J., Bilal, M., Al-Fuqaha, A.: Secure and Robust Machine Learning for Healthcare: A Survey (2020)
6. Gill, S.S., Arya, R.C., Wander, G.S., Buyya, R.: Based smart healthcare as a big data and cloud service nature Switzerland AG 2019. In: Hemanth, J. et al. (eds.) ICICI 2018, LNDECT 26, pp. 1376–1381, 2019. https://doi.org/10.1007/978-3-03
7. Dinh, H.T., Lee, C., Niyato, D., Wang, P.: A survey of mobile cloud computing: architecture, applications, and approaches. Wirel. Commun. Mob. Comput. **13**, 1587–1611 (2013). https://doi.org/10.1002/wcm.1203
8. Salazar, D., Velez, J., Salazar Uribe, J.: Comparison between SVM and logistic regression **35**, 223–237 (2012)
9. Yu, Z., Wallace, B.C., Johnson, T.R.: Healthcare data analytics challenge. In: 2015 International Conference on Healthcare Informatics, Dallas, TX
10. Tatum, W.O.: Handbook of EEG interpretation. Demos Medical Publishing, pp. 155–190. ISBN 9781617051807. OCLC 874563370 (2014)
11. Billis, A., Bamidis, P.D.: Employing time-series forecasting to historical medical data: an application towards early prognosis within elderly health monitoring environments. In: AI-AM/NetMed@ ECAI, pp. 31–35, 2014
12. Microsoft azure v/s Amazon AWS cloud services: a comparative study. Int. J. Innov. Res. Sci. Eng. Technol. (An ISO 3297: 2007 Certified Organization) **5**(3) (2016)
13. Ren, H., Xu, B., Wang, Y., Yi, C., Huang, C., Kou, X., Xing, T., Yang, M., Tong, J., Zhang, Q.: Time-Series Anomaly Detection Service at Microsoft. Microsoft China

Big Data Analytics

Genomic Data and Big Data Analytics

Hiren Kumar Deva Sarma

Abstract Genomic research has been highly prominent in recent times. Society has witnessed huge progress in genomic research in the last decade. The amount of data generated due activities like genome sequencing is huge. It is important to analyse such huge amount of data for acquiring meaningful insight so that such knowledge finds application in real-life scenarios. However, analysing such huge volume of data is extremely difficult because of the unique characteristics and complexities of these data. Big data analytic approaches are possible to explore for analytic purpose, and there have been researching efforts in that direction. In this paper, the relationship between genomic data and big data analytics has been explored. Challenges in processing of genomic data are analysed. The issue like how big data analytics concepts can be applied in genomic data processing is addressed. Future trends in combined research direction in the area of genomics and big data analytics are outlined.

Keywords Genomic data · Big data analytics · Gene editing · Next generation sequencing

1 Introduction

Gene editing mechanisms allow to bring changes in an organism's DNA by adding, deleting, or altering genetic information or genetic material at certain specific locations within the genome. The benefits which are possible to draw through gene editing are as mentioned below [1].

Gene editing finds applications in different areas such as disease management including research in bio-medical science, and also in agriculture and environment sciences. Gene and cellular therapies used for treatment are going to be enhanced. Various areas going to be benefited out of gene editing are infectiology, oncology,

H. K. D. Sarma (✉)
Department of Information Technology, Sikkim Manipal Institute of Technology, Majitar, Sikkim 737136, India

H. K. D. Sarma et al. (eds.), *Contemporary Issues in Communication, Cloud and Big Data Analytics*, Lecture Notes in Networks and Systems 281,
https://doi.org/10.1007/978-981-16-4244-9_15

hepatology, neurology, haematology, dermatology, ophthalmology, pneumology, and organ transplantation [1].

Whole genome sequencing, whole-exome sequencing, targeted sequencing are the next generation sequencing (NGS) technologies. Such technologies are applied to medical practices which can identify diseases or/and may be applied to advance precision medicine [2]. Precision medicine allows prediction of more accurate therapeutic and preventive measures to certain specific illness. Precision medicine approaches consider genetic make-up of patients, their lifestyles, and also environmental factors [2].

As per [2], more than 6000 Mendelian disorders have already been studied considering genetic level. Even more than 1500 clinically relevant complex traits were studied with various approaches based on Genome-Wide Association Study (GWAS) [2]. Electronic Health Records (EHR) are also available now a days [2]. Modern studies combine genomic and EHR data for better results of clinical and healthcare research.

Genomic data that are available in large scale with clinical data derived from EHR can facilitate individual diagnosis and therapy. Thus, next generation sequencing (NGS) combining with EHR will certainly lead towards patient-centred precision medicine and it is expected that in the days to come it will be a common clinical practice.

However, there are challenges in handling genomic data generated by NGS technologies because of its volume and associated complexities [3]. Again such data are very important for supporting genomic medicine. As per [4], a single whole genome sequencing generates more than 100 gigabytes of data. Thus, associating NGS with clinical practice in future will demand novel technologies in place to handle such big volume of data and associated complexities. There is need of powerful and more advanced analysis techniques in order to exploit the utility of such genomic data [5].

The success of genomic industry is in providing huge amount of sequence data. These sequence data if utilized properly can bring revolution to the medicine industry with smarter approaches for drug discovery. In fact, in the last decade, genomics has played a vital role in drug discovery. Genomics have paved the way to develop targeted therapies. Genomics data have capacity to show the path towards personalized drug discovery and recommendations. Personalized treatments against various diseases may be possible based on genomics data.

Researchers have argued and even concluded that different types of cancers have link with the disease of the genome. Genes are being extensively studied that cause cancer. Better understanding of the genes is highly important in order to address the issues with cancer. Now basic cancer research demands better understanding of the genes. Genome sequence data are available in huge volume due to extensive study in the field in last two decades and shall be useful for cancer research as well.

Big data analytics essentially involve storage and processing of huge volume of data and discovering of knowledge otherwise hidden inside the huge volume of data. Big data processing demands novel techniques and algorithms in order to discover meaningful information. Data generated out of human genome sequencing (i.e. genomic data) resembles the attributes of big data [6] and therefore, processing techniques and algorithms developed for big data are also applicable for processing of human genome sequence data. There are several instances where principles of big data analytics and associated techniques have been exploited for processing of sequence data [7, 8]. Thus, this is an open area of research where novel tools and techniques may be developed for processing of human genome sequence data by using approaches developed or applicable for big data analytics.

In this article, basics of genomics and big data analytics are reviewed in Sect. 3. Challenges and scope with the processing of genomic data are highlighted in Sect. 4. Various big data analytics task in the context of genomic data processing are mentioned in Sect. 5. Applications of the principles of big data analytics in genomic data processing are explored and reviewed in Sect. 6. Future trends in using big data analytics in genomics are summarized in Sect. 7. Finally, the paper is concluded in Sect. 8.

2 Related Work

There are significant research efforts in the domain of genomics across the world. In this section, few relevant literature with respect to this article is mentioned. Stefano Ceri and Pietor Pinoli share their experiences while using data science for managing demonic data [9]. The authors highlight various challenges to be faced by data scientists while using public datasets for solving biological and clinical problems. Hirak Kashyap et al. describe various aspects of big data analytics in bioinformatics from machine learning perspective as a whole [10]. Use of various machine learning techniques in big data analytics has been elaborated in this work. Various tools suitable for bioinformatics applications that have been developed using big data framework are also highlighted. Karen Y. He et al. elaborated applications of big data analytics in genomic medicine [2]. Applications of big data analytics in health research are also highlighted in this work. Tim Hulsen et al. describe how big data is associated with precision medicine [11]. Association of big data with bio-medical research has been elaborated in this work from different perspectives. Fabio C. P. Navarro et al elaborates the relationship between genomics and data science [6]. How genomics fits into the modern data science framework has been detailed in this work.

3 Background

3.1 Genomic Data

There are more than 6000 Mendelian disorders already studied at the genetic level, however, there is lack of clear understanding about the roles of the majority of such disorders in human health and diseases [2]. Due to the development of next generation sequencing (NSG) technologies, it has become easier to sequence a whole genome or exome, but the tasks like handling, analysing, and interpreting the genomic information that is generated by NSG are considerably challenging. A human genome has more than three billion base pairs (sites). Thus, sequencing a whole genome shall generate more than 100 gigabytes of data. For example, the file formats used for storing such data are BAM and VCF. BAM stands for the binary version of sequence alignment or map. On the other hand, VCF stands for Variant Call Format.

What is Mendelian Disorder?
By analysing chromosomes or through bio-chemical studies, in earlier days, it was not possible to have precise diagnosis in the field of genetics. But due to advancements in the field of genetics, direct analysis of the defects in gene level has become possible. This also gives hope to correct such defects. The reason behind the Mendelian or monogenic diseases is the process of mutation in one gene. Mendelian disorder is resulted by mutation at a single genetic locus. This is a kind of genome abnormality. Such disorders are visible in child since his or her birth and can also be predicted considering the family history, i.e. through pedigree analysis. Good news is that genetic disorders are highly rare and can affect one out of million individuals. These disorders may be heritable or non-heritable.

What is Gene Mutation?
Gene is made up of DNA sequences. A gene mutation indicates permanent alteration in DNA sequence. As a result of such alteration, the DNA sequence differs from that pattern what is normally found in most people. Mutation can affect anywhere in the DNA sequence; thus, it may be in a single DNA block or even to a large extent in the chromosome including multiple genes.

Mutations are of two major types. Hereditary mutations are a result of inheritance from a parent. Such mutations are present in virtually every cell of the body and persistent throughout the life of the person. This type of mutation is also known as germline mutation. Such mutations are present in the parents' egg or sperm cells. At the time of union of an egg and a sperm cell, DNA from both the parents are received by the resultant fertilized egg cell. If this DNA suffers from mutation, then the child grown out of this fertilized egg cell carries this mutation in each of the cells of his or her body. On the other hand, there is another type of mutation known as somatic mutations. Somatic mutations happen in a single cell. It happens in the early stage of embryonic development. Finally, this leads to a situation known as mosaicism. This type of genetic changes does not occur at parents' egg or sperm cells. It is not present even in the fertilized egg. It occurs a bit later when embryo contains several cells. During growth and the development, all the cells are divided. The cells that are created from the cells with the altered gene are going to have the mutation. And other cells will not have that mutation. Now depending on the extent of mutation and the number of cells affected, the situation called mosaicism will cause or not cause health problems.

Gene mutations that cause diseases are not common in general population. But there are genetic alterations which occur in more than 1% of the population. Such alterations are known as polymorphisms. Differences between people in terms of eye colour, hair colour, blood type, etc. occur due to such polymorphisms. It is important to know that many polymorphisms do not have negative effects on human health, however, some of such variations in the DNA may escalate the risk of developing disorders of certain types.

3.2 Big Data Analytics

Big data is characterized by its volume, velocity, variety, veracity, and value [12]. These are five Vs that exhibit essential characteristics of big data. In present scenario, there are various sources of big data that include social networks, various mobile devices, Internet of Things (IoT), healthcare networks, human genome project, Internet as a whole, and many more. If such huge volume of data can be analysed properly, it is possible to have significant insight in a particular subject matter, which in turn may be helpful for organizations, business houses, and the society as a whole, significantly.

There are various architectures proposed for having application in big data analytic systems. For example, MapReduce architecture, fault tolerant graph architecture, streaming graph architecture, etc. MapReduce is a parallel processing architecture and was originally developed by Google [13]. Parallel processing is achieved through execution of tasks by multiple machines or nodes. Apache Hadoop is an implementation of MapReduce framework through open-source platform.

Motivated by the necessity of fault tolerance, Low et al. [14] proposed GraphLab. It is a fault tolerant graph-based architecture in which computation is divided among various machines or nodes. Each of the nodes perform assigned particular tasks. The data model has two parts, first a graph having computing nodes and second, a globally shared memory. Pregel [7] and Giraph [8] are two other big data solutions based on graph architecture.

Mainly due to high overhead for disc read/write operations, the graph architecture as mentioned above is not efficient for stream data. Therefore, graph-based architecture for distributed processing of large scale data is designed. MPI (Message Passing Interface) [15] is a suitable solution to this problem.

4 Challenges and Scope with Genomic Data Processing

Genomic data processing suffers from several problems. Genomic data are dispersed over many data repositories therefore, it is challenging to process the desired data spread across in order to achieve meaningful insight. There is lack of documentation against available data, as a result of which interpretation of the data becomes challenging. Moreover, data formats used to store such data are heterogeneous and not comparable. In summary, there is lack of standardization in genomic data storage.

Again general challenges faced by most data science scenarios such as necessity of data mapping, normalization, data cleaning are also present in genomic data scenarios. In [9], it is showed that in spite of next generation sequencing which is considered to be a technological revolution in DNA processing, there is not adequate genomic data available for particular applications such as precision medicine, just as an example. It is elaborated in the paper with an example of cancer research from genomic data perspective. Good volume of data is available for cancer research in Genomic Data Commons (GDC) [16] or in the International Cancer Genome Consortium (ICGC) [17]. Genomic data sets available in The Cancer Genome Atlas (TCGA) [18] is of high volume and it provides more than 10,000 better quality samples. But when the matter of specific cancer type and that too within a targeted patient population comes into the picture, the data volume that is available becomes small. This puts challenges in front of data scientists in addressing crucial problems. It is already mentioned that it is quite challenging in integrating data that come from different collections because the data collections use incompatible types of measures [9]. It is again challenging to compare gene expressions produced by different technologies. For example, technologies like microarray and next generation sequencing use different principles while producing genomic data. It is difficult to compare gene expressions processed by different laboratories as the laboratories may adopt different protocols for analysis of the data. From computational aspect, lack of adequate data (i.e. data scarcity) offers strong limitation in adopting machine learning (ML)-based approaches (which is very common in classical data science scenario) in addressing specific research problems or research questions. This is so because there is mismatch between the huge number of features and a very limited number of samples. This fact also discourages use of deep learning (DL)-based approaches for the same purpose. Finally, computational experts having desire to support biologists and clinicians end up using statistical tests instead of using computationally efficient ML or DL-based approaches.

There are challenging problems, for example, developing complete model in order to describe the details regarding how transcription factor (TF) contributes in gene regulation. TFs are proteins responsible for contributing to many different molecular functions. TFs enter the nucleus of the cell and bind to DNA. Molecular functions such as chromosome organization, regulation of gene expression, etc. are influenced by TFs.

Scope: Genomic data processing has tremendous scope in the line of precision medicine, and personalized treatment.

It is expected that adequate volume of relevant genomic data will be available in the near future. NGS and other associated technologies may be the causes behind it.

5 Big Data Analytics Tasks in the Context of Genomic Data Processing

The volume of data related to genomic studies is increasing rapidly. Genomic data are being made available by different research groups across the globe continuously. Next generation sequencing (NGS) technologies are facilitating in the dramatic growth of genomic data volume. Best part about it is that high volume of data facilitates efficient analytics in terms of accuracy, especially in the filed like genomics as well as precision medicine which is highly sensitive. In [10], authors elaborate how bioinformatics data are different in the context of big data analytics from other big data such as particle physics data captured at CERN, or data (high resolution) received from satellites. As per the authors, bioinformatics data are different in two major aspects: first, bioinformatics data are **highly heterogeneous** in the sense that such data are generated by different organizations in different formats (i.e. there is lack of standardization in data formats) although the nature of the data is same. At the same time, multiple heterogeneous and independent databases are necessary for drawing inference and also for validation. Second, bioinformatics data are **geographically distributed** across the world. Transfer of these data in totality may not be possible due to their volume, privacy and other ethical issues [19]. Thus necessary analysis of these data is to be done remotely and results may be shared. Considering the advancements in cloud computing technologies, analytics of such data nowadays may be planned to be carried over cloud platforms, although various other issues shall still persist.

While the entire bioinformatics arena is considered, there are five major types of data used intensively in bioinformatics research [10]. These are, gene expression data; protein-protein interaction data (PPI); DNA, RNA, and protein sequence data; pathway data; and finally, gene ontology (GO).

Using gene expression data the expression levels of large numbers of genes are analysed. Generally, microarray-based gene expression profiling is adopted in order to record the various expression levels. Gene-sample, gene-time, and gene-sample-time are the three different types of microarray data [10]. Analysis of such data may be helpful in disease diagnosis, disease prevention, and also in precision medicine.

Protein-protein interaction data (PPI) analysis may be helpful in understanding protein functions. It is important to understand that defective PPI is fundamental causes behind various diseases such as cancer, Alzheimer, etc.

Analysis of DNA, RNA, and protein sequences are helpful in understanding their associations with various diseases also in identification of potential drugs for those diseases. Such type of sequence data is massive in volume and therefore, it needs sophisticated tools for analysis and also high-end computing infrastructure.

Pathway data analysis can reveal molecular basis of a disease. Such analysis can also be used to discover the genes and proteins associated with a particular type of disease, and therefore, may also be useful for drug discovery.

Table 1 Example databases available for various types of data used in bioinformatics

Type of data	Example databases
Gene expression data	Array Express (www.ebi.ac.uk/arrayexpress) Gene Expression Omnibus (www.ncbi.nlm.nih.gov/geo)
Protein–protein interaction data (PPI)	DIP (dip.doe-mbi.ucla.edu) STRING (string.embl.de) BioGRID (thebiogrid.org)
DNA, RNA, and protein sequence data	DNA Data Bank of Japan (www.ddbj.nig.ac.jp) RDP (rdp.cme.msu.edu) miRBase (www.mirbase.org)
Pathway data	KEGG [20], Reactome [21], Pathway Commons [22]
Gene ontology	www.geneontology.org

GO database maintains gene ontologies for different biological processes, molecular functions and also cellular components. It is species-independent. Analysis of GO databases may foster new biological discoveries (Table 1).

With respect to various types of data associated with genomic studies, and as a whole in the area of bioinformatics, there are several big data analytics-based tasks which are to be accomplished in order to derive meaningful inferences and for the discovery of new knowledge. In [10], Kashyap et.al. categorizes seven such analytics problems relevant to bioinformatics. Those are microarray data analysis, gene-gene network analysis, PPI data analysis, sequence analysis, evolutionary research, pathway analysis, and disease network analysis.

6 Genomic Data and Big Data Analytics

In this introductory part of this section, we see how the 5 V framework of big data infrastructure is relevant to the genomic data processing.

As per [6] genomics emerged during 1980s and it was the result of convergence of genetics, statistics, and large scale datasets. There has been significant advancement in nucleic acid sequencing and as a result of that huge volume of data is getting generated. Such a big scale of generated raw data having promises of huge knowledge discovery that is going to be tremendously beneficial for the mankind made this discipline a prominent one. Similarly, there are several other data rich areas associated with biological sciences such as medical image processing, neuroimaging, health record maintenance and analysis, proteomics, macromolecular structure, etc. and these areas are also rapidly increasing. As already mentioned, big data are characterized by 5 Vs, namely volume, velocity, variety, veracity, and value, there is a natural mapping of these 5 Vs that exists with the genomic data. Thus, genomics probably shall become a sub-discipline of data science.

Volume: As a result of the availability of next generation sequencing technologies, there has been tremendous growth in the genomic data volume in recent times. Data generated in other disciplines such as earth science [23], astronomy, or social networks is voluminous and the same in the field of genomics is also becoming at par. As per [24], the data generation due to genomic studies shall surpass the other applications such as social media, earth sciences, and astronomy in the near future. Moreover, within biological sciences also genomic data volume surpasses other data intensive and data driven sub-disciplines.

Velocity: The speed at which data is getting generated is an important characteristics of big data. The human genome sequencing techniques are more advanced nowadays. Human Genome Project (HGP) [25] has shown the pathways for such studies. Thus, sequencing of human genome may take even less than 24 hours due to advanced technologies. Other associated technologies such as diagnostic imaging, microarrays, etc. are also advanced nowadays and as a result data are possible to generate faster.

Variety: There are two aspects involved in genomic data. In one side, genomic data is the monolithic sequencing data which is an ordered list of nucleotides. These are mapped to human genome. Different sequencing methods or protocols essentially produce varieties. On the other side, genomic data is the complex phenotypic data. Phenotypic data are composed of diverse entities such as electronic health records which are simple and unstructured texts, then images, and then measurements from various bio-medical instruments, sensors, etc. This nature of phenotypic data in terms of variety is a complicated matter. Standardization of such data in handling data of large scale is an important issue.

Veracity: Accuracy or truthfulness of the data set under consideration is indicated by the term veracity. Veracity in the context of big data essentially indicates the factors like trustworthiness of the data source, type of it, and also, the processing algorithms involved in it. Data must be free from inconsistencies, duplication, noise, etc. In the context of genomic data processing, the accuracy aspect plays a big role. The conclusions drawn out of the analysis of the data shall be correct only when the input data is accurate.

Value: In the context of big data analytics, value indicates the worth that can be generated for the organization by the data under consideration. In healthcare sector, bio-medical data as well as genomic data are having great value for the organizations, companies, and individuals too. In the context of genomic data processing, this *V*, (i.e., value) of 5 V framework of big data infrastructure is undoubtedly extremely relevant.

6.1 Issues with Genomic Data Science

Privacy: Privacy of data has always been a prime concern. In current scenario, privacy in e-mail, bank transactions, credit card information, users' data necessary for online transactions, even surveillance cameras deployed for public place monitoring or private property monitoring, etc. are major concerns. Genomics-related privacy concerns are also similar to general data science-related privacy concerns; however, genomic-related privacy has some unique aspects considering the fact that genome passes down through the generations. Genomic data is fundamentally important for the public from all perspectives. Leaking of genomic data may be very dangerous and the possible damage that can be caused by such data leakage may be more significant than other types of data. Considering the current state of the art in genomic studies, it is true that the amount of available knowledge with respect to genomics has not yet reached the peak, but in another 50 years from now this volume is certainly going to increase a lot. Thus, any genomic data leaked today will have severe privacy and security concerns in future. This is so because any genomic data that flows across the generations, leaked today will remain relevant even after 50 years when it is expected to have more domain knowledge in the field of genomics. Moreover, genomic data is always voluminous. It carries lot of individual information and it is much more than a credit card number or bank account number. These aspects of genomic data make its privacy a problematic issue.

At the same time, in order to get more insights in the field of genomics, it is necessary to share and aggregate more genomic data. Such activities shall bring more social benefits although individual privacy may be compromised to some extent. The Global Alliance for Genomics and Health (GA4GH) (https://www.ga4gh.org) works for developing mechanisms in order to balance the individual privacy and social benefits considering genomic studies.

Ownership of Data: Data ownership and control over genomic data is a tricky issue. Question is, "who owns the genomic data?". In general, the individual or the patient is thought to be the owner of their personal data. However, in bio-medical research, there is an idea that the researcher who generates the dataset also owns it. Thus, data generated out of patients or individuals are used by researchers for analysis purpose in search of discovery and new knowledge. Again health-related data has medical as well as commercial value inside it. Thus, ownership and control of these data are another tricky issue.

There is a need to decide on the openness of such data along with its ownership and control aspects. Worldwide applicable regulations are necessary for recognizing and rewarding the efforts behind generation and analysis of such data for discovery of new knowledge. Deciding on ownership, control, and access of genomic data remains an issue that needs attention.

6.2 *Genomic Data Processing Using Big Data Analytics*

In this sub-section, utilization of big data frameworks for various activities involved in genomic data processing is reviewed. Few practical systems in place, which exploit big data infrastructure for analysis of genomic data have been referred.

Next Generation Sequencing Read Alignment: DNA is broken into large number of segments which is essential part of next generation sequencing (NGS). Each segment is known as a "*read*". The length of the reads across the genome may be uneven [26, 27]. Therefore, some genomic regions may be covered with more reads, whereas some are covered with less number of reads. *Read depth* is a measure to indicate the average number of times each base has been read [2]. Similarly, for RNAseq, read depth is in terms of number of millions of reads. *Read alignment* is a parameter that indicates the lining up of the sequence reads to a reference or benchmark sequence. Read alignment permits comparison of a sample sequence data with some reference genome. Thus, read alignment is a computing task with high computational complexity. There are few tools available for reading alignment, for example, CloudBurst [28], Crossbow [29], and SEAL [30]. These tools are developed using big data infrastructures. Quality control in this computational task is a concern.

Variant Calling: Variant calling is the process of identifying the variants from sequence data. For example, in the first step, whole genome sequence is stored in appropriate file format. Then with respect to a reference genome, the sequence is aligned. Finally, the locations where the aligned reads are different from the reference genome are identified (i.e., variants are located). Higher read depth ensures more reliable variant calling, as a result of which rare genetic variants can be detected with higher confidence. There are few programmes available for germline variant calling, for example, SAMtools [31], GATK [32], FreeBayes [33], and Atlas2 [34].

Variant Annotation: In analysis of genome sequencing data, variant annotation is considered to be a crucial step. In drawing the ultimate conclusions in disease studies, annotation results always have significant influence. Inappropriate annotations may lead to false positives. During the process of variant annotation, functional information is assigned to DNA variants [35]. Variant annotation relies on the biological knowledge with which it is possible to provide information related to the impact of variants on protein function or gene regulation [2]. Next generation sequencing (NGS) generates large amounts of sequence data. There are several annotation software widely used such as ANNOVAR [36], snpEff [37], Ensembl Variant Effect Predictor (VEP) [38]. As reported in [2], authors have developed a cloud-based version of ANNOVAR using Hadoop framework.

Genomic Data and Statistical Analysis: Genomic data may be analysed in order to discover various mutations in genetic level that may contribute towards various diseases. The analysis may be carried out using following approaches.

Family-based analysis: Family-based data analysis and research may discover many mutations which may be the cause behind recessive, and inherited diseases. SeqHBase [39] is a software developed using family-based analysis approach that can identify genes causing diseases. This tool is based on big data framework developed for analysing large scale family-based gene sequence data.

Population-based analysis: Population-based sequencing in large scale is an undergoing effort [2]. The volume and other characteristics of gene sequence data over a particular population, e.g. population from a country like USA or India, essentially promote the necessity of developing big data infrastructure-based tools for analysing genomic data of huge population in terms of millions.

6.3 Genomic Data Security

Security of genomic data is of high priority. Privacy and confidentiality of genomic data are important as mentioned in the previous section. Considering the nature and volume of genomic data, cloud platforms are utilized for its storage and also for distribution. Moreover, big data frameworks are exploited for necessary processing and analysis purpose. Therefore, all security concerns of cloud platforms and big data frameworks are applicable to such genomic data [19]. Access control and security measures for such genomic data is an important aspect that needs urgent attention from researchers as well as various regulatory bodies and authorities.

7 Future Trends

Genomic research shall facilitate in meeting many challenges in the days to come. Genomic research produces huge data. However, in order to exploit the power of genomic research, analysis of this huge data is important. If we want to infuse genomic information into day-to-day medical practice, there are many issues to be faced. For example, confidentiality of genomic data, electronic medical records, education for masses in this direction, also, making genomic information available to the physicians are some of the issues.

Data analysis is a big challenge in the field of genomics, and it is envisioned that there will be research effort in developing novel algorithms and practices for analysing huge genomic data. Big data approaches combined with machine learning techniques shall bring out novel techniques. Machine learning techniques may find extensive applications in genomic data analysis [40].

Security of genomic data is a big concern. There is research effort already seen in the direction of developing novel techniques for ensuring security to the genomic data considering its unique characteristics. In future, more research effort shall be visible in this area.

8 Conclusion

Genomic research generates huge volume of data. Next generation sequencing gives rise to huge volume of data. Unless these data are analysed properly, the power of such data remains untapped. While processing these huge volume of data, the task becomes highly challenging due to the unique nature and complexities of these data. Big data analytics-based approaches can be explored for the analysis of such data. Basics of genomic data and big data analytics are reviewed. Challenges to be faced in genomic data processing are highlighted. Interrelations between genomic data and big data are explored in detail. Finally, the future research directions in the combined area of big data analytics and genomic data processing are outlined. Privacy and security issues of genomic data are a major concern and they need novel techniques in order to make secure. Proper policies are needed in place to have a balance between privacy matter concerned with genomic data and applicability of the genomic data for betterment of the human society.

References

1. Furtado, R.N.: Gene editing: the risks and benefits of modifying human DNA. Rev. Bioét. **27**(2) (2019). https://doi.org/10.1590/1983-80422019272304; On-line version ISSN 1983–8034
2. He, K.Y., Ge, D., He, M.M.: Big data analytics for genomic medicine. Int. J. Mol. Sci. **18**(2), 412 (2017). https://doi.org/10.3390/ijms18020412
3. Gullapalli, R.R., Lyons-Weiler, M., Petrosko, P., Dhir, R., Becich, M.J., LaFramboise, W.A.: Clinical integration of next-generation sequencing technology. Clinics Laborat. Med. **32**(4), 585–599 (2012)
4. Robison, R.J.: How big is the human genome? Precision Med (2014)
5. Ritchie, M.D., Holzinger, E.R., Li, R., Pendergrass, S.A., Kim, D.: Methods of integrating data to uncover genotype-phenotype interactions. Nat Rev Genet. **16**, 85–97 (2015). https://doi.org/10.1038/nrg3868
6. Navarro, F.C.P., Mohsen, H., Yan, C., et al.: Genomics and data science: an application within an umbrella. Genome Biol **20**, 109 (2019). https://doi.org/10.1186/s13059-019-1724-1
7. Malewicz, G., Austern, M. H., Bik, A. J., Dehnert, J. C., Horn, I., Leiser, N., Czajkowski, G.: Pregel: A System for Large-Scale Graph Processing, SIGMOD'10, June 6–11, 2010,, pp. 135–145. Indianapolis, Indiana, USA (2010)
8. Sakr, S., Orakzai, F. M., Abdelaziz, I., Khayyat, Z.: Large-Scale Graph Processing Using Apache Giraph. Springer (2016). ISBN 978-3-319-47430-4
9. Ceri, S., Pinoli, P.: Data science for genomic data management: challenges, resources experiences. SN Comput. Sci. **1**, 5 (2020). https://doi.org/10.1007/s42979-019-0005-0
10. Kashyap, H., Ahmed, H.A., Hoque, N., Roy, S., Bhattacharyya, D.K.: Big Data Analytics in Bioinformatics: A Machine Learning Perspective. (2015) arXiv preprint arXiv:1506.05101
11. Hulsen, T., Jamuar, S.S., Moody, A.R., Karnes, J.H., Varga, O., Hedensted, S., Spreafico, R., Hafler, D.A., McKinney, E.F.: From big data to precision medicine. Front. Med. **6**, 34 (2019). https://doi.org/10.3389/fmed.2019.00034
12. Sarma, H.K.D., Dwivedi Y.K., Rana N.P., Slade E.L.: A MapReduce based distributed framework for similarity search in healthcare big data environment. In: Janssen, M., et al. (eds.) Open and Big Data Management and Innovation. I3E 2015. Lecture Notes in Computer Science, vol. 9373. Springer, Cham (2015). https://doi.org/10.1007/978-3-319-25013-7_14

13. Dean, J., Ghemawat, S.: Mapreduce: simplified data processing on large clusters. Commun. ACM **51**(1), 107–113 (2008)
14. Low, Y., Gonzalez, J.E., Kyrola, A., Bickson, D., Guestrin, C.E., Hellerstein, J.: Graphlab: a New Framework for Parallel Machine Learning (2014). arXiv preprint arXiv:1408.2041, 2014.
15. Gropp, W., Lusk, E., Doss, N., Skjellum, A.: A high performance, portable implementation of the mpi message passing interface standard. Parallel Comput. **22**(6), 789–828 (1996)
16. Grossman, R.L., Heath, A.P., Ferretti, V., Varmus, H.E., Lowy, D.R., Kibbe, W.A., Staudt, L.M.: Toward a shared vision for cancer genomic data. N. Engl. J. Med. **375**(12), 1109–1112 (2016)
17. Zhang, J., Baran, J., Cros, A., Guberman, J.M., Haider, S., Hsu, J., Liang, Y., Rivkin, E., Wang, J., Whitty, B., Wong-Erasmus, M., Yao, L., Kasprzyk, A.: International Cancer Genome Consortium Data Portal—A One-Stop Shop for Cancer Genomics Data. Database (2011); 2011:bar026.
18. Weinstein, J.N., Collisson, E.A., Mills, G.B., Shaw, K.R., Ozenberger, B.A., Ellrott, K., Shmulevich, I., Sander, C., Stuart, J.M.: Cancer Genome Atlas Research Network. The cancer genome atlas pan-cancer analysis project. Nat. Genet. **45**(10), 1113 (2013)
19. Sarma H.K.D.: Security issues in big data. In: Sarma H.K.D., Bhuyan B., Borah S., Dutta N. (eds.) Trends in Communication, Cloud, and Big Data. Lecture Notes in Networks and Systems, vol. 99. Springer, Singapore (2020). https://doi.org/10.1007/978-981-15-1624-5_7
20. Kanehisa, M., Goto, S.: KEGG: kyoto encyclopedia of genes and genomes. Nucleic Acids Res. **28**(1), 27–30 (2000)
21. Croft, D., OKelly, G., Wu, G., Haw, R., Gillespie, M., Matthews, L., Caudy, M., Garapati, P., Gopinath, G., Jassal, B., et al.: Reactome: a database of reactions, pathways and biological processes. Nucleic Acids Res. gkq1018 (2010)
22. Cerami, E.C., Gross, B.E., Demir, E., Rodchenkov, I., Babur, O., Anwar, N., Schultz, N., Bader, G.D., Sander, C.: Pathway commons, a web resource for biological pathway data. Nucleic Acids Res. **39**(1). D685–D690 (2011)
23. NASA. https://earthdata.nasa.gov
24. Stephens, Z.D., Lee, S.Y., Faghri, F., Campbell, R.H., Zhai, C., Efron, M.J., et al. (2015) Big data: astronomical or genomical? PLoS Biol. **13**(7), e1002195. https://doi.org/10.1371/journal.pbio.1002195
25. Lander, E., et al.: Initial sequencing and analysis of the human genome". Nature **409**, 860–921 (2001). https://doi.org/10.1038/35057062. International Human Genome Sequencing Consortium, Whitehead Institute for Biomedical Research, Center for Genome Research
26. Lander, E.S., Waterman, M.S.: Genomic mapping by fingerprinting random clones: a mathematical analysis. Genomics **2**, 231–239 (1988). https://doi.org/10.1016/0888-7543(88)90007-9
27. Sims, D., Sudbery, I., Ilott, N.E., Heger, A., Ponting, C.P.: Sequencing depth and coverage: Key considerations in genomic analyses. Nat. Rev. Genet. **15**, 121–132 (2014). https://doi.org/10.1038/nrg3642
28. Schatz, M.C.: Cloudburst: Highly sensitive read mapping with mapreduce. Bioinformatics **25**, 1363–1369 (2009). https://doi.org/10.1093/bioinformatics/btp236
29. Langmead, B., Schatz, M.C., Lin, J., Pop, M., Salzberg, S.L.: Searching for SNPS with cloud computing. Genome Biol. **10**, R134 (2009). https://doi.org/10.1186/gb-2009-10-11-r134
30. Pireddu, L., Leo, S., Zanetti, G.: Seal: A distributed short read mapping and duplicate removal tool. Bioinformatics **27**, 2159–2160 (2011). https://doi.org/10.1093/bioinformatics/btr325
31. Li, H., Handsaker, B., Wysoker, A., Fennell, T., Ruan, J., Homer, N., Marth, G., Abecasis, G., Durbin, R.: The sequence alignment/map format and samtools. Bioinformatics **25**, 2078–2079 (2009). https://doi.org/10.1093/bioinformatics/btp352
32. De Pristo, M.A., Banks, E., Poplin, R., Garimella, K.V., Maguire, J.R., Hartl, C., Philippakis, A.A., del Angel, G., Rivas, M.A., Hanna, M., et al.: A framework for variation discovery and genotyping using next-generation DNA sequencing data. Nat. Genet. **43**, 491–498 (2011). https://doi.org/10.1038/ng.806

33. Garrison, E., Marth, G.: Haplotype-based variant detection from short-read sequencing. Available online: http://arxiv.org/abs/1207.3907
34. Evani, U.S., Challis, D., Yu, J., Jackson, A.R., Paithankar, S., Bainbridge, M.N., Jakkamsetti, A., Pham, P., Coarfa, C., Milosavljevic, A., et al.: Atlas2 Cloud: a framework for personal genome analysis in the cloud. BMC Genom. **13**(Suppl. 6), S19 (2012). https://doi.org/10.1186/1471-2164-13-S6-S19
35. McCarthy, D.J., Humburg, P., Kanapin, A., et al.: Choice of transcripts and software has a large effect on variant annotation. Genome Med. **6**, 26 (2014). https://doi.org/10.1186/gm543
36. Wang, K., Li, M., Hakonarson, H.: Annovar: functional annotation of genetic variants from high-throughput sequencing data. Nucleic Acids Res. **38**, 164 (2010). https://doi.org/10.1093/nar/gkq603
37. Cingolani, P., Platts, A., le Wang, L., Coon, M., Nguyen, T., Wang, L., Land, S.J., Lu, X., Ruden, D.M.: A program for annotating and predicting the effects of single nucleotide polymorphisms, SnpEff: SNPs in the genome of *Drosophila melanogaster* strain w1118; iso-2; iso-3. Fly **6**, 80–92 (2012). https://doi.org/10.4161/fly.19695
38. McLaren, W., Gil, L., Hunt, S.E., Riat, H.S., Ritchie, G.R., Thormann, A., Flicek, P., Cunningham, F.: The ensemble variant effect predictor. Genome Biol. **17**, 122 (2016). https://doi.org/10.1186/s13059-016-0974-4
39. He, M., Person, T.N., Hebbring, S.J., Heinzen, E., Ye, Z., Schrodi, S.J., McPherson, E.W., Lin, S.M., Peissig, P.L., Brilliant, M.H., et al.: Seqhbase: A big data toolset for family based sequencing data analysis. J. Med. Genet. **52**, 282–288 (2015). https://doi.org/10.1136/jmedgenet-2014-102907
40. Libbrecht, M.W., Noble, W.S.: Machine learning applications in genetics and genomics. Nat. Rev. Genet. **16**, 321–332 (2015). https://doi.org/10.1038/nrg3920

Image Processing

Feature Extraction Techniques for Shape-Based CBIR—A Survey

Naiwrita Borah and Udayan Baruah

Abstract Content-based image retrieval or CBIR is a method used to retrieve images from large databases with the help of local features. These local features are mainly colour, shape and texture. These features can either be used individually or in conjunction with one another to perform CBIR. Here, several shape descriptors, benchmark as well as author developed techniques, used for shape-based CBIR have been discussed. The shape data is primarily obtained from the edge detected from images. The experimentations performed were mostly on benchmark data like COREL-1000, OLIVIA, COIL-100, PRODUCE-1400, etc. It has been observed that the methods developed by authors have outperformed the benchmark methods in terms of precision and recall.

Keywords Content-based image retrieval · Shape features · Edge histogram

1 Introduction

It is believed that "A picture is worth a thousand words". This is so because a picture can represent complex and multiple ideas and meanings at a given moment. Representing multiple complex ideas can be difficult with the help of mere words or sentences. With the ever-growing digital world, it has become imperative to represent this information in some digital format for the purposes of storing, sharing, etc. This stored information finds its use in application areas like architecture, engineering, fashion industry entertainment and so on. As World Wide Web found its niche, it was observed that a large number of images well uploaded, processed and shared across platforms be it in social networking sites or other aforementioned industries. It is found that not all images are indexed properly to be retrieved as needed for regular text-based searching [1]. Thus, content-based image retrieval (CBIR) was a solution for this. In CBIR, search is performed using the visual content or information of the

N. Borah (✉) · U. Baruah
Department of Information Technology, Sikkim Manipal Institute of Technology, Sikkim Manipal University, Majitar, Sikkim, India

H. K. D. Sarma et al. (eds.), *Contemporary Issues in Communication, Cloud and Big Data Analytics*, Lecture Notes in Networks and Systems 281,
https://doi.org/10.1007/978-981-16-4244-9_16

images [2]. The earliest CBIR system that was developed and used on a commercial scale was Query-Based Image Retrieval (QBIC) by IBM [3].

In CBIR, the objective is to retrieve, from the image database (DB), all the images that are similar to the Query Image (QI) based on the visual information. Now, we need to know the actual manifestation of the word "visual information". By visual information, we represent the visual features of the image like colour, texture, shape, etc. Together, these visual features are able to represent the information rendered by the image. CBIR has been an emerging means for image recognition and retrieval. This paper primarily focuses on the techniques of extraction of the shape feature for CBIR. Good shape features have certain properties associated with them. Some of them are mentioned below [2]:

a. They should be identifiable: shapes which are considered as same by the human eye should also be considered as same by the shape features.
b. Shape data should be translation, rotation and scaling (TRS) invariant, i.e. shape features should remain the same event if the images have undergone any translation rotation or scaling.
c. Noise resistant: features must not be affected by the noise present in the images.
d. Occlusion resistant: at times, features may have to be obtained from images having parts that are hidden, so the features obtained from the visible part should not be affected in any way by the occlusion.

2 Shape-Based CBIR

Every object in the real world can be represented with the help of shape. The human cognitive ability is such that the moment any object is mentioned to us, a picture of that object comes to our mind having a particular shape and a possible size and colour or even texture. But most commonly the shape and colour are the most immediate things that come to one's mind. Similarly, in CBIR also, shape, colour and texture play an integral role in image recognition and retrieval. In shape-based CBIR, the shape features are extracted from the QI to form the feature vector, and then these feature vectors are compared with the feature set that is extracted from the database of images. From this comparison, several similarity matching techniques are employed and those images that are highly similar are to be displayed by the system as the retrieved images. Figure 1 gives the general schematic of a CBIR framework. Mentioned below are some of the recent related works to shape-based CBIR.

Kumar et al. [4] in their CBIR model used canny edge detection to identify the key points from the edges in the objects of images of the Wang dataset. The features extracted were then classified using Support Vector Machine (SVM), and Euclidian distance was used for similarity matching. The proposed methods outperformed existing methods of Multilayer Perceptron (MLP), Hierarchical Neural Network (HNN), Fuzzy and Neural Network and k-Nearest Neighbour (K-NN) in terms of specificity, sensitivity, precision and recall.

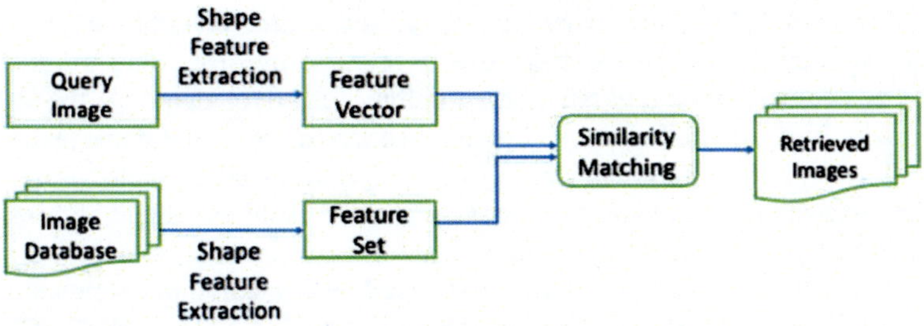

Fig. 1 CBIR framework with reference to shape feature

Alsmadi, 2020 [5] in his work has used all three features of colour, shape and texture in his CBIR system. The dataset used was the Corel dataset which contained 1000 different image of size 384 × 256 or 256 × 384. The shape descriptors were mostly based on moment on region as well as boundary. To achieve the pre-processing step, he made use of neutrosophic clustering in order to separate the pixels having close values from the insignificant pixels in grey level images. Once the pre-processing was performed, he used canny edge detection method in order to retrieve the edge data from the surrounding pixels. After the boundary details were obtained from the QI, the shape content indices were extracted to form the feature vector. The author then used a metaheuristic algorithm in order to retrieve the precision values. It was seen that the author's method outperformed other existing techniques in terms of precision and recall.

Ali and Sadoon [6] in their proposed method used a combination of three processes, namely Single Value Decomposition (SVD), Edge Histogram Detection (EHD) and colour auto correlogram for feature extraction. They performed their experimentation on a dataset containing four types of image classes with 100 images of each type. EHD was used for shape feature extraction with which the authors described the relative frequency five types of edges in the image. They preferred EHD as it is resistant to TRS. A total of 150 features were extracted using EHD. The similarity matching was performed using Euclidian distance. A comparative analysis was performed between the classification results obtained with the different features used individually, and all features used together. It was observed that the classification results obtained with all features used together outperformed all other individual feature classification.

Surendrenadh and Rao in their method [7] used three moments-based CBIR system on two benchmark image databases, namely Georgia Tech Face and COIL—100. Their shape-based feature extraction was performed by using exponential Fourier moments [8] improved exponential Fourier moments [9] and a fast exponential Fourier moment [10]. After the feature extraction, a comparative analysis of the similarity matching was done using different distance measures like Euclidian distance, Manhattan distance, Chi-square distance, etc. After the experimentation, the authors found that average and fast exponential Fourier moments results outperformed the other moment techniques.

Pradhan et al. [11] in their work have suggested a three-layer hierarchical CBIR method. At each layer, the database size is reduced by discarding the irrelevant images by comparing them based on their colour, feature or shape. At the last level of hierarchy, only the similar images remain. In order to extract the shape features, the authors have proposed a system where the position of the edge pixels as well as the distance between them is taken into consideration to build an edge joint histogram. This is done by using the canny edge map and fuzzy edge map to obtain the edge-related data. The experimentation was performed on five commonly known dataset, namely COREL-1000 [12], 32 GHIM-10 K [13], COIL-100[14], PRODUCE-1400 [15], OLIVIA [16] and OUTEX [17].

3 Shape Description Techniques

The previous section mentions some of the most recent feature extraction techniques that have been proposed by authors which are based on some form of enhancement to existing or traditional methods of shape feature extraction. This section shall contain details of some of the classical shape extraction techniques developed over the years.

3.1 Shape Descriptors

In a computer system, shape data is represented in the form of a set of numbers which represent shape. These numbers try to quantify shape data the way humans interpret shape data. The matching of feature vectors with feature sets on the basis of shape can be done in the following ways:

a. Similarity matching is usually done on the basis of some distance measure. If the concerned DB image has a distance within a specific range, it can be considered as similar to the QI.
b. Shapes obtained from the DB images can even at times be affine transformed to get shape data similar to that of the QI.
c. Generating shape data from lesser elements can also be done in order to match the shape data of the QI.

Aktas in his work has categorised the several shape descriptors into broad categories [33]. Some of these categories have been noted in Fig. 2. A brief mention of the same can be found in the following section.

a. Area-Based Shape Descriptors

One of the most common area-based shape descriptors is the region-based moment [31]. The moment invariant method was first introduced by Hu in 1962 [18, 32], wherein the author was able to extract the shape data which was robust to transformations due to translation, scaling, rotation and reflection. It was observed that

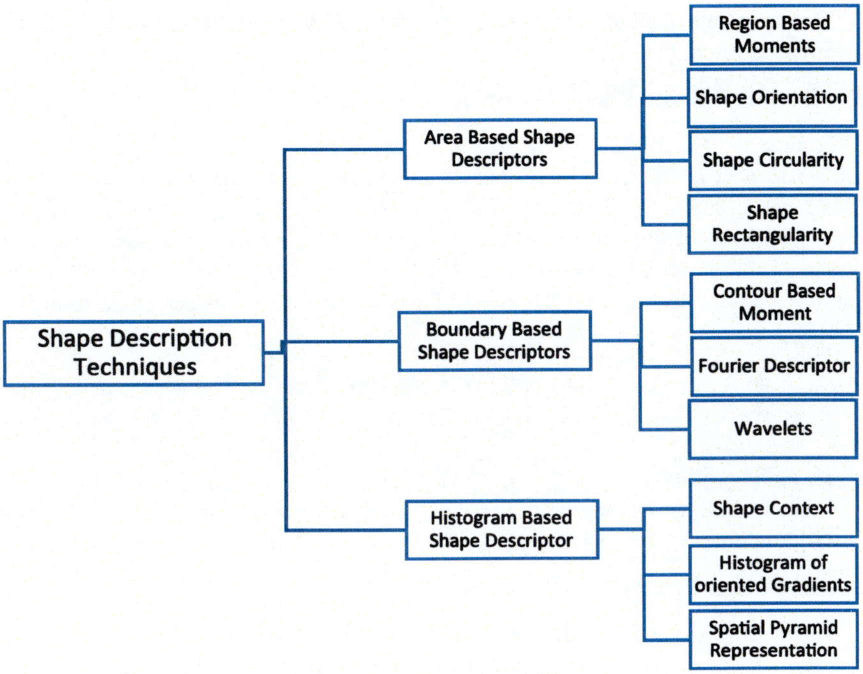

Fig. 2 Overview of some of these shape feature descriptor techniques

when shape based matching was to be performed, there were certain discrepancies when translation, scaling, or even rotation occurred. Thus moment based techniques were preferred. However an issue was observed. Images are of discrete nature but shape based matching based on moments do not perform well if the original images undergo some geometric transformation. Huang and Leng [19] in their research have performed some experimentation on how moment invariants perform under geometric transformations. The authors have also provided some solutions to deal with the fluctuating moment invariants. Moments can be described by the Eq. (1):

$$M_{pq} = \int_{-\infty}^{\infty}\int_{-\infty}^{\infty} x^p y^q f(x, y)\mathrm{d}x\,\mathrm{d}y \tag{1}$$

where $p, q = 1, 2, 3, 4 \ldots \infty$ and $f(x, y)$ is the intensity function of the pixels in image I (say). Thus, M_{pq} represents the shape S of the object in the image based on moment value. Thus, Eq. (1) can also be re-written as Eq. (2)

$$M_{pq}(S) = \int_{-\infty}^{\infty}\int_{-\infty}^{\infty} x^p y^q f(x, y)dxdy \tag{2}$$

There are several other types of moments like scale invariant moment [20], rotation invariant moment [21], etc.

b. Boundary-Based Shape Descriptors

Equations (1) and (2) are based on moments that are extracted from the area information. It is also possible to extract this information from the boundary data. This is called as line moments [22]. Let B is the boundary of any given shape S, with P as the length of the boundary. Let the boundary be a curve or an arc. Then the points on $B(S)$ can be represented as $B(S) = (x(s), y(s))$ for $0 \leq s \leq P$. If the curve is a complete one, then $B(0) = B(P)$. Then line moments can be described as

$$M_{pq}^{l}(B) = \int_{B} x(s)^{p} y(s)^{q} \mathrm{d}s \tag{3}$$

where $s\varepsilon 0$, length(B),

Other techniques belonging to this category are convexity measure [23], Fourier descriptors [24], Wavelets [25], etc.

c. Histogram-Based Shape descriptor.

Shape context is one of the methods used for describing shape features that can be used for similarity matching eventually. The technique works in the following way let P be a set of n points from an edge detected image such that $P = \{p_1, p_2, p_3, p_4, \ldots p_n\}$. Thus, at least $(n - 1)$ vectors are present which connect the n points to one another. These collections of vectors give ample information regarding the shape description. However, it is a very detailed set, and not all the shape description is needed for shape matching. Thus, a subset of the vector collection would be enough. The authors off the paper [26] have identified a process using only the relative positions of points in P. Using only the selected points, which the relative positions of $(n - 1)$ vectors a course histogram $h(i)$ is made with the uniform number of bins. This histogram is the shape context of P. When comparing the shape context of an image having say points Q with points of P, the authors have mentioned the cost incurred in matching p_i with q_i as

$$C_{ij} \equiv C(p_i, q_i) = \frac{1}{2}\sum_{k}^{K} \frac{\left[h_i(k) - h_j(k)\right]^2}{\left[h_i(k) + h_j(k)\right]} \tag{4}$$

Other techniques belonging to this category include Histogram of Orientation Gradients [27, 28], Spatial Pyramid Representation [29, 30] to name a few.

4 Summary of Findings

Section 2 mentions some recent works associated with feature extraction based on benchmarked techniques. The intent of this section is to provide a summary of their findings in terms of accuracy scores and metrics. Table 1 gives a summary of the findings.

5 Conclusion

This paper has been an attempt to identify some of the recent techniques that have been prevalent in shape-based content-based image retrieval. Apart from contemporary techniques, mentions of some of the benchmark techniques have also been made. It has been observed that the contemporary works have achieved very high accuracy rates as compared to the benchmark methods. Some authors have combined features of colour and texture along with shape to have an optimally inclusive feature set, while others have performed comparative analysis between features of colour, texture and shape to test their proposed methods. The works have mostly been done on benchmark data like Corel 1000 produce, 1400 Olivia, etc. Attempts have been made to perform shape feature extraction by replicating the human cognitive ability, which has proved to be a very difficult as the human brain is a complex system. Furthermore, it has been observed that different shape descriptors work differently depending on the type of dataset. The same shape descriptor may work brilliantly on one type of data but fail on some other. There is no fixed process as to which shape descriptor is the best in performance. Even so, much has been achieved in recent years through proper experimentation and examination.

Table 1 Summary of results achieved by recent authors on shape-based CBIR

Authors	Method	Dataset used	Benchmarked method compared with:	Metrics used			
				Specificity (%)	Sensitivity (%)	Precision (%)	Recall (%)
Kumar et al	Proposed [SF + SVM]	Wang dataset	MLP, HNN, Fuzzy and Neural N/W, K-NN, rough set	99.2	96.48	98.04	96.62
Alsmadi	Proposed CBIR method	Corel dataset	GCCL, WATH, spatial level-2, spatial LGH, GMM spatiogram, colour fusion	NA[a]	NA	90.15	18.03
Surendrenadh and Rao	A&F EFM-based CBIR	COIL-100 face database	EFM-based CBIR, IEFM-based CBIR	NA	NA	64.2	NA
		GT face database		NA	NA	65.35	NA
Pradhan et. al	Proposed hierarchical system using probability edge joint histogram (Existing benchmark technique for shape feature extraction)	Corel 1000 image dataset	Not compared with any benchmarked process. Comparison performed using the proposed CBIR method with colour vs texture vs shape features	NA	NA	64.5	12.9
		GHIM-10 K image dataset		NA	NA	61	2.44
		PRODUCE-1400		NA	NA	64.64	12.93
		OLIVIA image dataset		NA	NA	72.5	4.41

[a]NA refers to not applicable as not all works have used certain metrics for evaluation

References

1. Beall, J.: The weaknesses of full-text searching. J. Acad. Librarianship **34**(5), 438–444 (2008)
2. Mingqiang, Y., Kidiyo, K., Joseph, R.: A survey of shape feature extraction techniques. Pattern Recogn. **15**(7), 43–90 (2008)
3. Flickner, M., Sawhney, H., Niblack, W., Ashley, J., Huang, Q., Dom, B., Gorkani, M., Hafner, J., Lee, D., Petkovic, D., Steele, D., Yanker, P.: Query by image and video content: The QBIC system. Computer, **28**(9), 23–32 (1995).
4. Kumar, P.S., Kumar, N.U., Ushasree, A., Sumalata, G.L.: Key point oriented shape features and SVM classifier for content based image retrieval. Mater. Today Proc. (2020).
5. Alsmadi, M.K.: Content-based image retrieval using color, shape and texture descriptors and features. Arabian J. Sci. Eng. 1–14 (2020)
6. Ali, S.K., Sadoon, S.A.: Image classification based on CBIR. J. Phys. Conf. Series **1591**(1), 012022 (2020). IOP Publishing.
7. Surendranadh, J., Rao, C.S.: Exponential fourier moment-based CBIR system: a comparative study. In: Microelectronics, Electromagnetics and Telecommunications, pp. 757–767. Springer, Singapore (2020)
8. Xiao, B., Li, W.S., Wang, G.Y.: Errata and comments on Orthogonal moments based on exponent functions: Exponent-fourier moments. Pattern Recogn. **48**(4), 1571–1573 (2015)
9. Hu, H.T., Ju, Q., Shao, C.: Errata and comments on Errata and comments on Orthogonal moments based on exponent functions: Exponent-fourier moments. Pattern Recogn. **52**, 471–476 (2016)
10. Singh, S.P., Urooj, S.: Accurate and fast computation of exponent Fourier moment. Arabian J. Sci. Eng. **42**(8), 3299–3306 (2017)
11. Pradhan, J., Kumar, S., Pal, A.K., Banka, H.: A hierarchical CBIR framework using adaptive tetrolet transform and novel histograms from color and shape features. Digital Signal Process. **82**, 258–281 (2018)
12. Wang, J.Z., Li, J., Wiederhold, G.: SIMPLIcity: semantics-sensitive integrated matching for picture libraries. IEEE Trans. Pattern Anal. Mach. Intell. **23**(9), 947–963 (2001)
13. Liu, G.H., Yang, J.Y., Li, Z.: Content-based image retrieval using computational visual attention model. Pattern Recognition, **48**(8), 2554–2566 (2015)
14. Nene, S.A., Nayar, S.K., Murase, H.: Columbia object image library (coil-100) (1996)
15. tropical-fruits-db-1024x768.tar.gz, http://www.ic.unicamp.br/~rocha/pub/downloads/tropical-fruits-DB-1024x768.tar.gz/. Accessed 18 Aug 2017.
16. Oliva, A., Torralba, A.: Modeling the shape of the scene: a holistic representation 83 of the spatial envelope. Int. J. Comput. Vis. **42**(3),145–175 (2001). https://doi.org/10.1023/A:1011139631724
17. "site www, vision & image", lagis-vi.univ-lille1.fr, http://lagis-vi.univlille1.fr/85datasets/outex.html. Accessed 18 Aug 2017.
18. Ming-Kuei, Hu.: Visual pattern recognition by moment invariants. IRE Trans. Inf. Theory **8**(2), 179–187 (1962). https://doi.org/10.1109/TIT.1962.1057692
19. Huang, Z., Leng, J.: Analysis of Hu's moment invariants on image scaling and rotation. In 2010 2nd International Conference on Computer Engineering and Technology (Vol. 7, pp. V7–476). IEEE.
20. Srivastava, P., & Khare, A. (2016, December). Content-based image retrieval using scale invariant feature transform and moments. In: 2016 IEEE Uttar Pradesh Section International Conference on Electrical, Computer and Electronics Engineering (UPCON), pp. 162–166. IEEE (2016).
21. Mehtre, B.M., Kankanhalli, M.S., Lee, W.F.: Shape measures for content based image retrieval: a comparison. Inf. Process. Manage. **33**(3), 319–337 (1997)
22. Lambert, G., Gao, H.: Line moments and invariants for real time processing of vectorized contour data. In: International Conference on Image Analysis and Processing, pp. 347–352. Springer, Berlin, Heidelberg (1995)

23. Padraig Corcoran, P.M., Winstanley, A.: A convexity measure for open and closed contours. In: Proceedings of the British Machine Vision Conference, pp. 81.1–81.11. BMVA Press (2011). http://dx.doi.org/10.5244/C.25.81
24. Folkers, A., Samet, H.: Content-based image retrieval using Fourier descriptors on a logo database. In: Object Recognition Supported by User Interaction for Service Robots, vol. 3, pp. 521–524. IEEE (2002)
25. Ashraf, R., Ahmed, M., Jabbar, S., Khalid, S., Ahmad, A., Din, S., Jeon, G.: Content based image retrieval by using colour descriptor and discrete wavelet transform. J. Med. Syst. **42**(3), 44 (2018)
26. Belongie, S., Mori, G., Malik, J.: Matching with shape contexts. In: Statistics and Analysis of Shapes, pp. 81–105. Birkhäuser Boston (2006)
27. Dalal, N., Triggs, B.: Histograms of oriented gradients for human detection. In: 2005 IEEE Computer Society Conference on Computer Vision and Pattern Recognition (CVPR'05), vol. 1, pp. 886–893. IEEE (2005)
28. Tian, S., Bhattacharya, U., Lu, S., Su, B., Wang, Q., Wei, X., et al.: Multilingual scene character recognition with co-occurrence of histogram of oriented gradients. Pattern Recognition, **51**, 125–134 (2016).
29. Lazebnik, S., Schmid, C., Ponce, J.: Beyond bags of features: Spatial pyramid matching for recognizing natural scene categories. In: 2006 IEEE Computer Society Conference on Computer Vision and Pattern Recognition (CVPR'06), vol. 2, pp. 2169–2178. IEEE (2006)
30. Elfiky, N.: A novel spatial layout representation for object recognition. In: Joint European-US Workshop on Applications of Invariance in Computer Vision, pp. 553–563. Springer, Cham (2020)
31. Kim, W.Y., Kim, Y.S.: A region-based shape descriptor using Zernike moments. Signal Process Image Commun. **16**(1–2), 95–102 (2000)
32. Leu, J.G.: Computing a shape's moments from its boundary. Pattern Recogn. **24**(10), 949–957 (1991)
33. Aktas, M.A.: Shape Descriptors. PhD Thesis, University of Exeter, UK (2012)

Object Detection Under Occlusion in Aerial Images: A Review

Praveen Kumar Pradhan and Udayan Baruah

Abstract For processing of an image for the applications of object detection used in various fields such as vehicle detection, traffic monitoring, human detection, surveillance systems, object mapping, etc., the occurrence of occlusion is predominant especially in an aerial image having wide angle coverage and variety of objects under the screenshot other than object of interest. Various works have been done for addressing the occlusion issues with a good degree of success and these variety of approaches have yield promising results. From optimal hardware utilization, segmentation, and mathematical modeling that have been explored in this paper, the current relevance and development of Artificial Neural Networks of type convolutional neural networks (CNNs) have caught predominance. It is, however, found that a viable Self-Organizing Maps (SOMs) model development has not taken place but has a great potential. This paper reviews various studies, experiments, and approaches on occlusion detection and handling for providing a comprehensive scenario for applications in aerial images.

Keywords Object tracking · Occlusion handling · Aerial images

1 Introduction

A typical field of research in the area of computer vision has been object tracking and detection. A lot of applications depend in the area of object detection such as vehicle tracking, person tracking, driving assistance system for vehicles, self-directed navigation, traffic monitoring, etc. Various challenges, however, still exist for spotting an object, such as illusion, low visibility, shadows, and, importantly, occlusions of object [1]. Object tracking involves the process of finding and tracking objects which may be one or more in a setting (Fig. 1). This helps in detecting the activities of a

P. K. Pradhan (✉)
Centre for Computers and Communication Technology, Chisopani, South Sikkim, India

U. Baruah
Department of Information Technology, Sikkim Manipal Institute of Technology, Sikkim Manipal University, Majitar, East Sikkim, India

H. K. D. Sarma et al. (eds.), *Contemporary Issues in Communication, Cloud and Big Data Analytics*, Lecture Notes in Networks and Systems 281,
https://doi.org/10.1007/978-981-16-4244-9_17

Fig. 1 Object and person detection. (*Source* hackerearth.com)

target object such as people on a street or vehicles on the road and many of its likes (Fig. 2).

However, occlusion is a pertinent problem during object tracking in a setting with multiple objects. Occlusion happens when an object is overlaid or is blocking the image information of an object by another object placed in the same projection of the camera (Figs. 3 and 4).

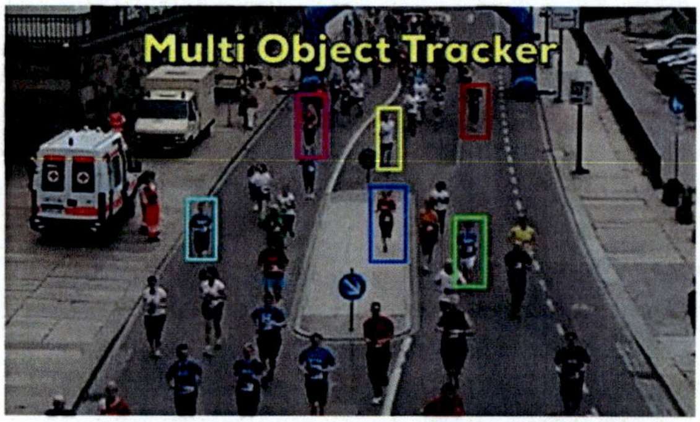

Fig. 2 Multiple object tracking. (*Source* medium.com)

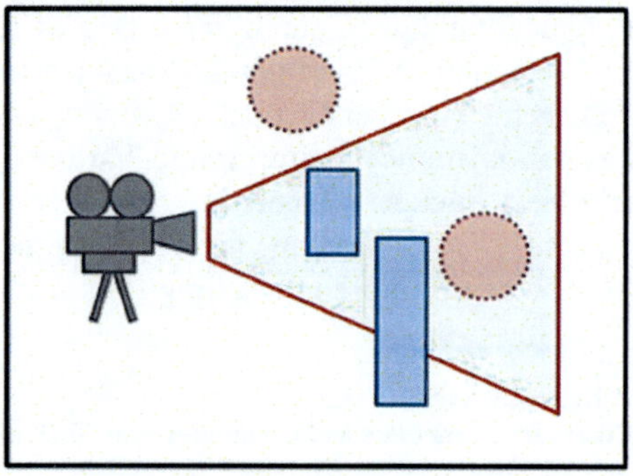

Fig. 3 Occlusion. (*Source* medium.com)

Fig. 4 Occlusion detection. (*Source* stackoverflow.com)

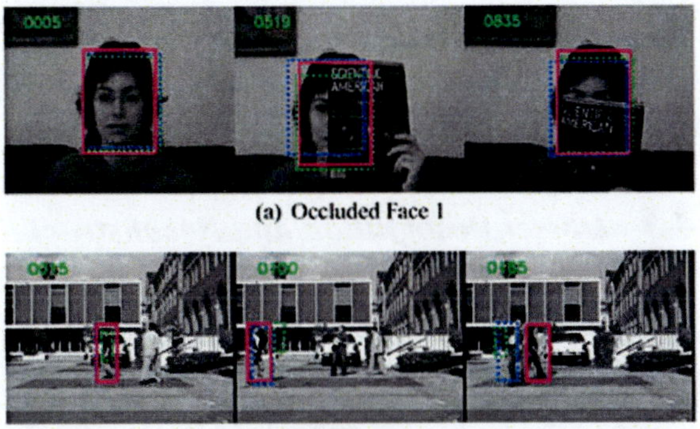

(a) Occluded Face I

(c) Person

1.1 Occlusion

Occlusion happens when multiple objects come very near to each other and superficially merge in an image (Fig. 4). A system can sometimes lose trail of the object or the incorrect object may be trailed after being overlaid due to occlusion [2].

There are two types of occlusion scenarios:

1. Self-occlusion: A part of an object get occluded by part of the same object from a perspective.
2. Inter-object occlusion: More than one objects being tracked overlay each other occluding one another.

Occlusions is one of the major hurdles for automatically deriving conjugate point pairs in overlapping images. They lead to discrepancies or uncorrelated areas of features in an image. This is largely grave in urban areas, where many inconsistencies often occur, produced by relief shift of the buildings [1–3].

1.2 Occlusion Handling

Occlusion issues occur when a tracked object gets placed behind real objects in an image. Usually, the area taken up by the real objects being tracked gets masked such that no computer objects is able to be drawn inside the occluded area. This mask basically captures the outlines of the obstructing real objects, and the mask edges dictate how unified the merging turns out to be. The effect of the mask can be attained in 2D by labeling a 2D pixel area manually in the real image [4]. Occlusion boundaries contain rich intuitive information about the original scene. They also provide critical information in many visual perceptions works such as scene understanding, object recognition, and segmentation [5]. System detecting occlusion with boundary patterns from a single image or multiple images of a video has been the way forward.

Various mathematical models, neural networks training, and algorithms have been worked upon for not only accurately and automatically detecting occlusion boundary, but also for replacement of objects, virtual, or real.

1.3 Object Recognition and Reconstruction

There exist challenges in identifying objects with incomplete presence in images, which can have appearance variance, orientation or pose disparity, and occlusion. Self-Organizing Maps (SOMs) using Artificial Neural Networks are known to be very effective in image segmentation and classification [6]. Methods based on region convolutional neural network (R-CNN) are also widely applied for object detection as well as for image classification. However, they mostly either use the full objects for training, even take occluded objects as disparities. They do not recuperate the object pose info or the entire occluded object outline. Variations in the model-based approaches have proved to mitigate some extent of occlusion. However, they have hitches in handling object appearance differences. Lately, a merger of deformable part model (DPM) alongside CNN has been anticipated for improved object detection. However, they have to use pyramid of images as input. Further, constituents may not look like in the object hierarchy structure or calculate the object pose info. R-CNN can also be used for object detection and pose estimation in one network. However, it has not been proved to handle occluded objects reliably [7] (Figs. 5 and 6).

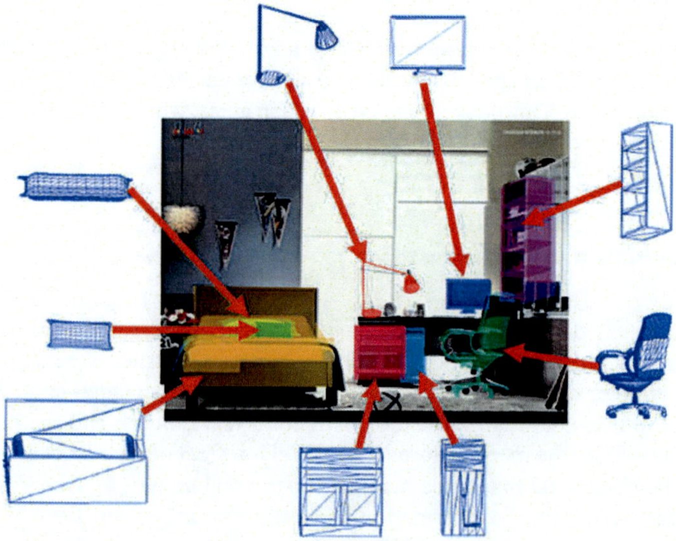

Fig. 5 Recognition and reconstruction. (*Source* http://cvgl)

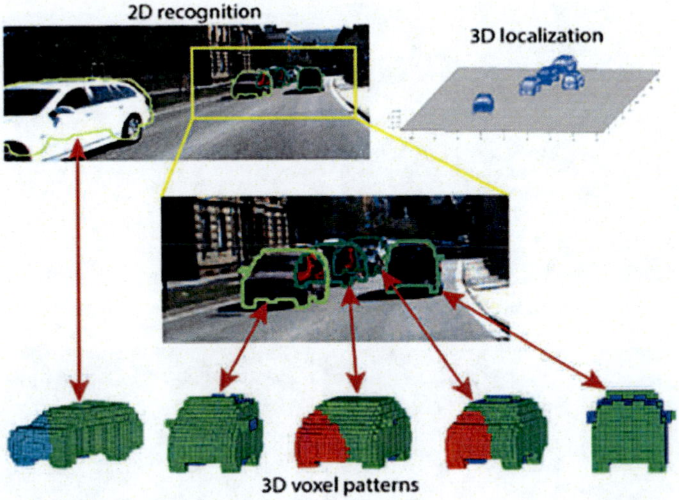

Fig. 6 Recognition and reconstruction. (*Source* http://cvgl)

1.4 Aerial Images

Aerial images are similar to normal images taken by cameras, however, not from ground but at an elevation such as satellites, aircrafts, drones, or UAV. Their typical application is of remote sensing, surveillance, and traffic monitoring and management. Articulation of geography, topography, and maps can also be improvised with accuracy and timeline study. The advantage of aerial images is that it provides a bird's eye view of the landscape with multiple real-time objects included in the same frame including the ground surface. Depending if the camera angle is vertical, oblique or 3D, augments application of 3D transformation can have a better understanding of the topology of the surface, object identification as well as other applications (Fig. 7).

2 Approaches in Occlusion Handling

Image processing and handling of occlusion have been the key area of study. Occlusion detection, boundary detection, and reconstruction have been the key area of development in object detection as well as reconstruction. Various scientific methodologies and algorithms have been developed for the same which has come to a decent level for application of object tracking. However, one system fitting all scenario is not available. Occlusion detection involves the field of pattern recognition and utilizing computation for accurate detection. Vehicle and human tracking, face recognition, UAV navigation, and other applications of the likes have faced challenges posed by

(a) Google Earth

(b) UAV *(source: www.gettyimages.in)*

(c) Hilly Terrain *(source: indiatoday.com)*

(d) Urban City *(Source: https://unsplash.com)*

Fig. 7 Samples of aerial images

occlusion and also its solutions. Aerial or satellite imagery for remote sensing applications, which are even used for GIS applications, poses a good field for exploration for object detection under occlusion imposed by terrain, shadows, natural objects like trees, forest, manmade structures, etc.

2.1 Comparative of Approaches for Occlusion Detection

Shravya et al. [8] explored different methods doing object tracking under occlusion. It assimilates different performance methods available to assess the tracking results under occlusion, widely available datasets list which are used to test algorithm for object tracking. For the various techniques/approaches mentioned in the paper, it lists particle filter approach whose results are probabilistic, depth-based approaches using pseudo depth information, and deep learning-based approaches which track and also update the appearance together. Re3, long short-term memory (LSTM), YOLO deep convolutional neural networks, ROLO (recurrent YOLO), and recurrent autoregressive network (RAN) are compared. It mentions that learning-based methods are more suitable for tracking objects in aerial images. For metrics used for evaluating the trackers, it mentions Frame-based metrics: True Negative or TN, True Positive or TP, False Negative or FN, and False Positive or FP are used. Tracker Detection

Rate or TRDR which is True Positive by TG, False Alarm Rate or FAR is equal to False Positive by sum of True Positive and False Positive. For object-based metrics, it categorizes in detection centered metrics, tracking centered, and perimeter-imposed detection metrics. It, however, mentions specific performance metric for occlusion detection is not yet available. It suggests that a viable metric could be the percentage of object occluded. For datasets availability, the paper mentions ALOV300++ which is the Amsterdam Library of Ordinary videos, OTB which classifies the structures by marking them in eleven attributes, VOT2014 which is a fully marked set with visual properties, Need For Speed (NFS) having high frame rate videos, AMP having fifteen omnidirectional sequences with known twelve motion types, and UAV123 which is taken from low altitude UAV's having 123 video classifications. The TLP dataset consists of fifty videos of which half are indoor type and other half outdoor sequences. Each dataset has their own challenges. However, there are no aerial dataset for occlusion detection purposes.

Chandel et al. [1] provided summarizing occlusion handling study of various algorithms and systems under different scenarios. Car occlusion in street scenes for vehicle detection, pedestrian occlusion in diverse scenes for human detection, and diverse real-world occluded scene were compared. It exhibits to show several image processing, machine learning, and computer vision techniques for the same occlusion problem. Shape has been identified as the best feature to match occluded portion of an object for the same in the subsequent frames. Various methods under stereo vision exhibited good development. 100% accuracy in handling occlusions under various scenarios was not available. It concludes that there exists a large gap on occlusion reasoning and human perception. If features for detection improves, the occlusion estimation could be better.

2.2 *Occlusion Detection in Video Sequences*

Cheong et al. [2] used algorithms to handle multiple object tracking (MOT) using segmentation of foreground object, tracking of color, specification of object, and handling occlusion. They used mean color values and segmentation of differentiate objects. A short video was the system's input and every subsequent frame analyzed. Results showed that it was able to uninterruptedly track the object selected and preserve the identity despite occlusion by another object in the same frame.

2.3 *Occlusion Handling Using Artificial Neural Networks*

Li et al. [9] used a convolution neural network (CNN). The Net was left–right symmetric which together learn binocular occlusion by cooperatively mining the binocular information(s). The experiments showed that their model achieved better results on detecting the stereo and motion occlusion as compared to KZ and MC-CNN + LCR for various parameters of precision, recall, and F-score. The learning and Depth order of the NN was reported to have been improved.

Nedyalko et al. [6] used Self-Organizing Maps (SOMs) a type of Artificial Neural Network (ANN) for image classification to several classes based on their features and texture. Three main experiments were conducted: The classifiers were trained without any statistical pre-processing of dataset; the classifiers were trained after normalization of the available data; the SOMs after training use linear transformations of the earlier features acquired after pre-processing with Principal Component Analysis (PCA). Tests were conducted 50 times, and classification results evaluated for accuracy rate, sensitivity, and specificity metrics. Best results were obtained when statistical pre-processing (normalization) and PCA were implemented.

2.4 Accelerated Computation for Occlusion Handling

Hu et al. [10] used orthophoto and LiDAR for automatic image occlusion detection with shadow detection. Z-Buffer or depth buffer technique was used to detect occlusion. The algorithm for Z-Buffer computation was expensive especially when handled by the CPU. Using the common GPU in modern graphic hardware, they use techniques to optimize speed for the Z-Buffer computation by the GPU. Two ways explored were:

1. OpenGL (a cross language, cross platform API for rendering 2D and 3D vector graphics)
2. CUDA which is a parallel computing platform API from NVIDIA.

A 37.8 Megapixels aerial image with 0.13 m/pixel resolution was used. Experiments were executed on a personal computer with Intel Core i7 CPU- 2.67 Hz, 4 GB RAM, NVIDIA GeForce GTX285 GPU with memory of 1 GB, Windows-7 64 bit. Detection of occlusion took for CPU- 24.2 s, OpenGL as 4.3 s and using CUDA in GPU as less than 3.4 s. Seven fold speedup time was attained.

3 Collation of Study

Of the papers taken up for the review, a collation is provided in the table below for providing a compilation of factors that were used. The category of image used for the compilations has been listed. As some paper provide a comparative study and others utilized specific structure or architecture for conducting the experiment and findings, it has also been listed. The performance evaluation and deductions provided by the authors have also been comprehended in the table to provide the diversity of approaches and findings of the various studies conducted in the field of occlusion handling erstwhile object detection for images providing focus on aerial image type (Table 1).

4 Challenges in Occlusion Detection and Handling in Aerial Images

Occlusion as an imperative composition of an aerial image is faced with a challenge of being detected and handled effectively. From still imagery, sequence of frames in videos and online tracking and monitoring applications the method(s) used for handling occlusion needs high accuracy for application to be effective. Though CNN in recent times has shown great promises, however, a reliable system is still not developed. Augmented that in an aerial image, the objects can be of varying sizes within the frame, and the small dimensions can lead to false detection or even omission. Efficient feature selection for object is still a wild guess, and it varies on application. The dynamic nature of aerial images also possesses a challenge as in some a method may prove to be effective but not in all. Standardization of parameters is also such challenge apart from approach and datasets.

5 Conclusion

From the study undertaken, it is clear that occlusion detection challenges have been addressed in vision system for quite some time. Occlusion scenarios pose hindrance in many image processing applications by posing a false information than in reality. A sufficiently large approaches and sophisticated systems line CNN, SOM have been experimented for handling the occlusion scenario effectively to a certain degree. Few of the approaches use mathematical models providing improvements in existing system, segmentation approaches have shown promising results and the utility if ANN of specific type CNN and SOM have shown significant alignment to handle the issue. Dataset to execute experiments is limited or has been developed for some other applications. The need for such dataset is also exhibited which can then lead for experimenting various approaches to yield 100% accuracy which no system has been able to achieve. The most promising approaches have been CNN. However, newer facility of cloud services can greatly reduce these providing greater reach for newer approaches and experiments.

Table 1 Collation

Category of images	Structure/architecture	Performance evaluation	Deductions
Aerial images of various dataset	Comparative study	For aerial object tracking the CNN learning method is more viable	A viable metric could be the percentage of object being occluded. There is no aerial dataset for occlusion detection purposes specifically [8]
1. Car occlusion in street scenes 2. Pedestrian occlusion in diverse scenes 3. Human detection and diverse real-world occluded scene	Comparative study	Performance when same tracking algorithm used for both occluded and occlude: Poor Best feature selection: Shape 100% accuracy: Not got gap between human perception and occlusion reasoning: Large requirement for better estimation of occlusion: Feature	Conducted study for handling occlusions under different scenarios for the same problem of occlusion, exhibited multiple image processing, computer vision and machine learning techniques [1]
3. Different videos apiece with different occlusion problems	Segmentation of object and color tracking algorithms	2 objects overlapping each other: handled occlude identified as 1 object despite it being obstructed from the mid	Future improvements could be done on real-life scenarios with compound colored objects like pedestrian, cars, etc. [2]

(continued)

Table 1 (continued)

Category of images	Structure/architecture	Performance evaluation	Deductions
SceneFlow and Middlebury dataset	Different model of SymmNet was used keeping same parameter numbers other than for the input and output layers	SceneFlow/middlebury Precision: 0.799/0.810 Recall: 0.919/0.849 F-score: 0.873/0.849	Demonstrated good ability for stereo and motion occlusion detection that can be extended to stereo and optical flow applications [9]
335 greyscale images of size 230 × 340 pixels of cork tile samples	Self-Organizing Maps (SOM) using 60 and 120 neurons		SOM	N-SOM	PCA Accuracy	45.4–100	63.6–100	63.3–100 Sensitivity	66.7–90.9	76.9–100	84.6–100 Specificity %	92.3–100	94.7–100	94.8–100	Statistical pre-processing and PCA gave best results. Results were competitive and in few cases more for mean accuracy, sensitivity, and specificity metrics [6]
1.2 million LIDAR points having density of 1.4 points/m^2, 7228 *5228, 0.13 m/pixel resolution aerial image	Z-Buffer Algo on a PC: Intel(R) Core(TM) i7-920 @ 2.67 Hz CPU, 4 GB RAM, NVIDIA GeForce GTX285 GPU (1 GB) on Windows-7—64 bit OS. OpenGL and CUDA applications		seconds CPU	24.2 OpenGL	4.3 CUDA in GPU	< 3.4 TIN creation CUDA in GPU	2.1	Using OpenGL accelerated by CUDA in GPU gave about 7 times faster execution for occlusion detection. Using GPU, the execution time can be greatly enhanced [10]

References

1. Himanshu, C., Vatta, S.: Occlusion detection and handling: a review. Int. J. Comput. Appl. (0975–8887) **120**(10) (2015)
2. Cheong, Y.Z., Chew, W.J.: The application of image processing to solve occlusion issue in object tracking. MATEC Web Conf. **152**, 03001 (2018)
3. Brito, J.: Occlusion detection in digital images through bayesian networks. In: International Archives of Photogrammetry and Remote Sensing, vol. XXXIII, Part B3. Amsterdam (2000)
4. Ong, K.C., Teh, H.C., Tan, T.S.: Resolving occlusion in image sequence made easy. Visual Comput. **14**(4), 153–165 (1998)
5. Huan Fu, Chaohui Wang, Dacheng Tao, Michael J. Black, "Occlusion Boundary Detection via Deep Exploration of Context", IEEE Conference on Computer Vision and Pattern Recognition (CVPR), 2016.
6. Nedyalko, P., Jordanov, I.: Unsupervised texture image classification using self-organizing maps. In: Proceedings on the International Conference on Artificial Intelligence (ICAI). The Steering Committee of the World Congress in Computer Science, Computer Engineering and Applied Computing (WorldComp) (2011)
7. Liu, L., Yin, X., Zhu, Y., Zhang, J., Li, J.: Object recognition and reconstruction with partial appearance. In: Scene Understanding Workshop Hawaii 2017, (CVPR) (2017)
8. Shravya, A.R., Monika, K.M., Malagi, V., Krishnan, R.: A comprehensive survey on multi object tracking under occlusion in aerial image sequences. In: 2019 1st International Conference on Advanced Technologies in Intelligent Control, Environment, Computing & Communication Engineering (ICATIECE), pp. 225–230. Bangalore, India (2019). https://doi.org/10.1109/ICATIECE45860.2019.9063778
9. Li, A., Yuan, Z.: A symmetric convolutional neural network for occlusion detection. British Mach. Vision Conf. (2018)
10. Hu, X., Li, X.: Fast occlusion and shadow detection for high resolution remote sensing image combined with LIDAR point cloud. In: ISPRS—International Archives of the Photogrammetry, Remote Sensing and Spatial Information Sciences, vol. XXXIX-B7 (2012)
11. Hoiem, D., Stein, A.N., Efros, A.A., Hebert, M.: Recovering occlusion boundaries from a single image. In: IEEE 11th International Conference on Computer Vision (2007). https://doi.org/10.1109/ICCV.2007.4408985
12. Huang, Y., Essa, I.: Tracking multiple objects through occlusions. In: IEEE Computer Society Conference on Computer Vision and Pattern Recognition (CVPR'05) (2005). https://doi.org 10.1109/CVPR.2005.351
13. Wang, J., Xie, C., Zhang, Z., Zhu, J., Xie, L., Yuille, A.: Detecting semantic parts on partially occluded objects. British Mach. Vision Conf. (BMVC) (2017)
14. Humayun, A., Mac Aodha, O., Brostow, G.J.: Learning to find occlusion regions. In: IEEE Computer Vision and Pattern Recognition (CVPR) (2011)
15. Pepik, B., Stark, M., Gehler, P., Schiele, B.: Occlusion patterns for object class detection. In: IEEE Computer Society Conference on Computer Vision and Pattern Recognition (CVPR) (2013)
16. Sundberg, P., Brox, T., Maire, M., Arbeláez, P., Malik, J.: Occlusion boundary detection and figure/ground assignment from optical flow. In: IEEE Conference on Computer Vision and Pattern Recognition (CVPR) (2011)
17. Fatema, T.Z., Gavrilova, M.: Face recognition using occluded area localization method. In: Gavrilova, M.L. et al. (eds.) Transactions on Computational Science XXX, LNCS 10560, pp. 12–28. Springer-Verlag GmbH Germany (2017)
18. Wang, T.-C., Alexei, A.E., Ravi, R.: Occlusion-aware depth estimation using light-field cameras. In: IEEE International Conference on Computer Vision (ICCV) (2015)
19. Ozgur, Y.: Classification of occluded objects using fast recurrent processing. In: IEEE 14th International Conference on Machine Learning and Applications (ICMLA) (2015)

20. Op het Veld, R.M., Wijnhoven, R.G.J., Bondarev, Y.:. Detection and handling of occlusion in an object detection system. In: Video Surveillance and Transportation Imaging Applications 2015, vol. 9407, p. 94070N. International Society for Optics and Photonics (2015)
21. Lin, Y.-Y., Liu, T.-L., Fuh, C.-S.: Fast object detection with occlusions. In: The 8th European Conference on Computer Vision (ECCV) (2004)
22. Hsiao, E., Hebert, M.: Occlusion reasoning for object detectionunder arbitrary viewpoint. IEEE Trans. Pattern Anal. Mach. Intell. **36**(9), 1803–1815.
23. Bailloeul, T., Duan, J., Prinet, V., Serra, B.: Urban digital map updating from satellite high resolution images using GIS data as a priori knowledge. InL 2003 2nd GRSS/ISPRS Joint Workshop on Remote Sensing and Data Fusion over Urban Areas, pp. 283–287. IEEE (2003)
24. Ardeshir, S., Zamir, A. R., Torroella, A., Shah, M.: GIS-assisted object detection and geospatial localization. In: European Conference on Computer Vision, pp. 602–617. Springer, Cham (2014)
25. Oliveira, H.C., Habib, A.F., Dal Poz, A.P., Galo, M.: Height gradient approach for occlusion detection In: UAV Imagery International Conference on Unmanned Aerial Vehicles in Geomatics (2015)
26. Miljković, D.: Brief review of self-organizing maps. In: International Convention on Information and Communication Technology, Electronics and Microelectronics (MIPRO) (2017)
27. Reyes-Aldasoro, C.C.: Image segmentation with Kohonen neural network self-organising maps. In: International Conference on Telecommunications ICT (2000)

Comprehensive Comparative Study on Several Image Captioning Techniques Based on Deep Learning Algorithm

Chitrapriya Ningthoujam and Tejbanta S. Chingtham

Abstract Image captioning is evolving as an interesting area of research that involves generating a caption or describing the content in the image automatically. The idea behind image captioning is to make the computer perceive a given image like a human mind leading to automatic description. Image captioning is a challenging task that involves capturing semantically correct information and expressing in a simple sentence. A large number of methods have been proposed in the recent past, and we aim to do a comprehensive survey in the different deep learning algorithms used in image captioning based on the method framework.

Keywords Captioning · Annotation · Image · Deep learning · Encoder · Decoder · Neural image generator · LSTM · Language method

1 Introduction

In recent years, image captioning has received huge attention. It involves observing the contents in an image and then describing it. It has a broad application area with a wide range of scenarios. Areas of research in Natural Language Processing (NLP) and also in Computer Vision (CV) fields are achieving immense advancements; larger datasets have been made available while generating text of images and videos leading to implementation of deep neural network-based methods acquiring more and more accurate results on image captioning. It involves the task of capturing an image, analyzing the video contents, recognizing the most important features of the image, and then generating the textual description based on it. Deep learning algorithms have shown better results in handling many complex and challenges of an image captioning task [1]. The image processing can be categories into three different approaches based on: retrieval, text, and novel. Retrieval-based approach caption an image from a collection of already existing captions [2]. In template based, captions are generated based on the templates which identify a set of visual notions

C. Ningthoujam (✉) · T. S. Chingtham
Department of Computer Science and Engineering, Sikkim Manipal Institute of Technology, Sikkim Manipal University, Majitar, Sikkim, India

H. K. D. Sarma et al. (eds.), *Contemporary Issues in Communication, Cloud and Big Data Analytics*, Lecture Notes in Networks and Systems 281,
https://doi.org/10.1007/978-981-16-4244-9_18

first then connected through the sentence template to compose a sentence used by [3]. Novel based on the other hand, generates captions of an image from both visual spaces as well as from multimodal space.

This paper starts with the discussion of different image captioning methods categories into two different frameworks in Sect. 1.1. Section 1.1.1 discusses the encoder-decoder framework along with five different methods under it. References [4–8] methods are based on encoder-decoder architecture to generate a caption. Similarly, Sect. 1.1.2 discusses the compositional architecture-based image captioning and five other different methods another same. References [9–13] methods are based on the second type of framework where captions are generated by extracting components from relevant captions and later combined for describing the image. In Sect. 1.2 summarizes the various image captioning methods based on deep learning method on two different frameworks.

1.1 Image Captioning Methods

Among the various methods based on deep learning model, this paper has considered the framework used to build a model that can generate a caption or describe a given image trained and tested on some of the benchmark datasets. The architecture considered are: encoder-decoder-based framework and compositional-based framework.

1.1.1 Encoder-Decoder Framework

(a) ***Encoder-Decoder pipeline***: The main idea of this method is to translate a sentence from one language into another language by supplying an input as an image and the output as a sentence illustrated in Fig. 1 [4]. This method has been adopted from the neural translation concept as given by [14].

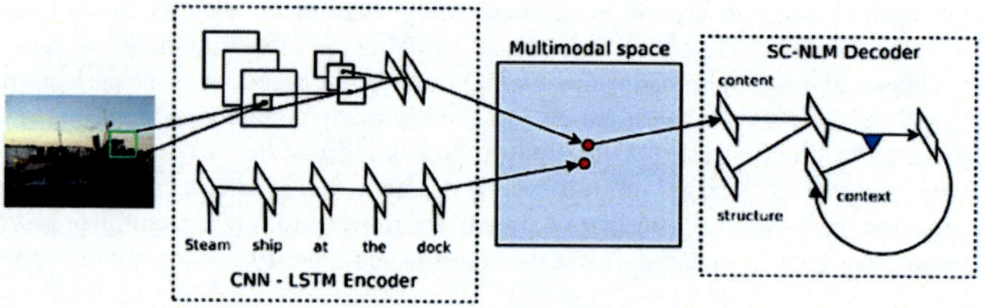

Fig. 1 The encoder-decoder method proposed by [4]

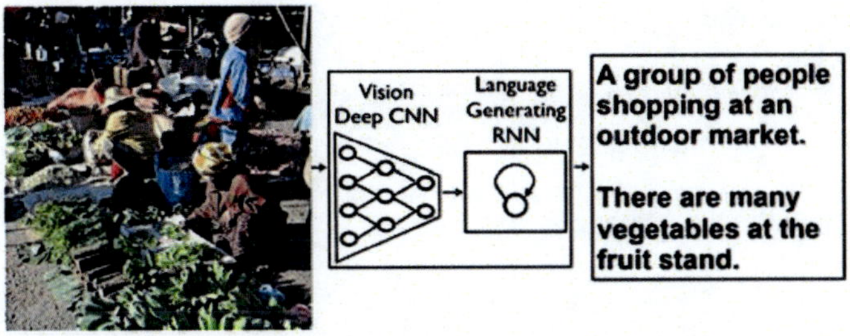

Fig. 2 Neural image caption generator [5]

Working

It contains two stages: ***encoder*** and ***decoder.*** Firstly, the encoder phase makes a combined multimodal space which is used to order the images along with its descriptions. This encoder encodes the sentences by using the idea of machine translation using LSTM model [15]. Features of an image are embedded using a CNN. The encoder tries to minimize the pairwise ranking loss that will help to learn the ranking of images and along with its descriptions. In the second stage, the method uses the multimodal representation so that it can generate novel descriptions. The decoder part uses a new type of a neural network-based language method named as Structure-Content Neural Language Method by [4] and can generate novel descriptions.

(b) ***Neural Image Caption (NIC) Generator***

This method is proposed by [5] which uses a CNN as an encoder for image representations and RNN as a decoder for generating captions of an image shown in Fig. 2. The encoder in this method follows a novel approach where the last hidden layer in the model is fed as an input to the decoder [16].

Working

The encoder (RNN) translates the input of variable length into a fixed dimensional vector [5] and decodes this representation into required output which is the description. The probability of the right caption is calculated using Eq. 1 [17], where I is an image, and S sentence of its length is unbounded.

$$\theta^* = \underset{\theta}{\arg\max} \sum_{(I,S)} \log p(S|I;\theta) \tag{1}$$

Sampling is one of the approaches used in [17] where the first word was sampled according to p_1, equivalent embed was supplied as an input for sample p_2, continuing in the same order like this until all the samples reach a special end-of-sentence token or it has reached with some maximum length. Second approach uses a search

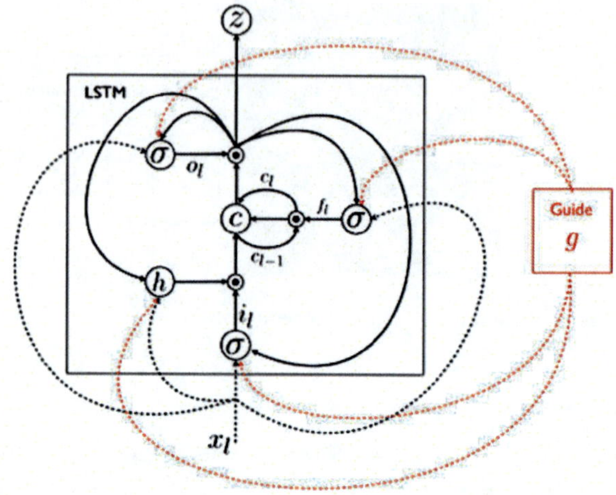

Fig. 3 gLSTM network proposed by [6]

technique called a Beam Search, by iteratively taking all the *k*-best description till some time *t*.

(c) *gLSTM*

This method is an extension of LSTM proposed by Jia et al. [6]. The author [6] used a concept known as guided LSTM [6] which could create long description in form of sentence by adding global semantic info. This information was then added to each LSTM's gates and cells shown in Fig. 3. It also takes into consideration various normalization strategies to manage the caption length.

Working

Firstly, in this method, for describing an image and extracting the semantic information from an image a Cross-Modal Retrieval (CRM) is used. Multimodal embedding space can also be used to extract the information of the image. Secondly, semantic information is added to the computation of each gates and cell state. Thus, the information is obtained together from the images and its descriptions, aiding as a guide in the procedure of generating a word sequence. In the LSTM method, the generation of a word generally depends on the embedding word at the present time step and the previous hidden state.

$$i_l' = \sigma\left(W_{ix}x_l + W_{im}m_{l-1} + W_{iq}g\right) \quad (2)$$

$$f_l' = \sigma\left(W_{fx}x_l + W_{fm}m_{l-1} + W_{fq}g\right) \quad (3)$$

$$o_l' = \sigma\left(W_{ox}x_l + W_{om}m_{l-1} + W_{oq}g\right) \quad (4)$$

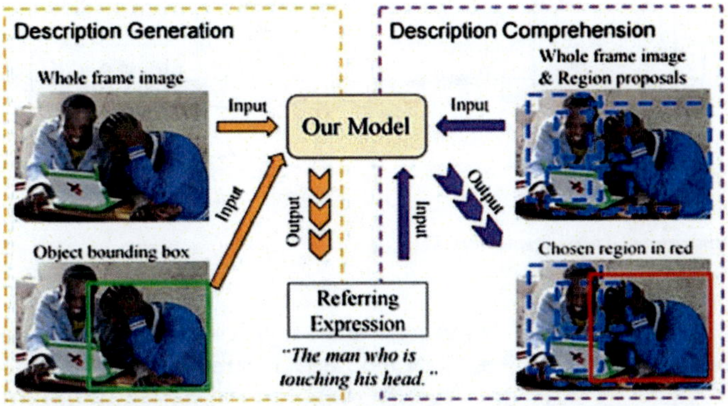

Fig. 4 Illustration of referring expression proposed by [7]

$$c_l' = f_l' \odot c_l' + i_l' \odot h(W_{cx}x_l + W_{cm}m_{l-1} + W_{cq}g) \quad (5)$$

$$m_l = o_l' \odot c_l' \quad (6)$$

In the above equations by [6], vector representation of semantic information is denoted by g. While $\odot$ denotes the element-wise multiplication, σ represents the sigmoid function, and h is the hyperbolic tangent function. The variable i_l represents the input gate, f_l, is forget gate o_l is output gate, c_l, and m_l memory cell unit and hidden layer.

(d) ***Referring expression***

Mao et al. [7] proposed a new method called as a referring expression. The expression determines a unique description for a particular object or may be an area specified in a given image shown in Fig. 4. Further, the expression can be interpreted to infer which object is being described [7]. This method used well-defined performance metric which gives more detailed than just image captions and as a result provides more helpful.

Working

This method generally considers two characteristics: description generation and description comprehension. In first characteristics, a text expression is generated that exclusively identifies object which is highlighted or some specific region emphasized in the image. In the second characteristics, the method inevitably chooses an object from a given expression that refers to this chosen object. Alike other image captioning methods this method uses a CNN model to represent the image latter followed by an LSTM. The method computes feature for the whole image, to serve as context [7]. It is considered a novel dataset which they termed as ReferIt dataset [18].

(e) ***Variational Auto Encoder (VAE)***

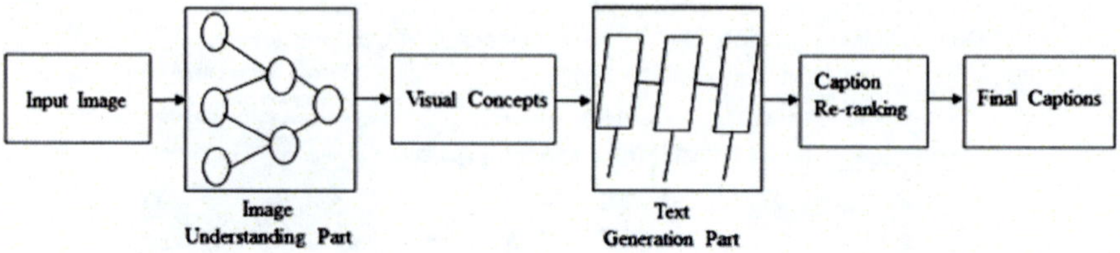

Fig. 5 Compositional-based framework [1]

This method is proposed by [8] using a semi-supervised learning technique. The encoder considered here is a deep CNN and Deep Generative Deconvolutional Neural Network (DGDN) as a decoder. The framework may also even allow unsupervised CNN learning, based on an image [8].

Working

CNN is used as an image encoder for captioning, whereas a recognition method was established for the DGDN as a decoder which decodes the latent image features [8]. The encoder supplies an approximation of distribution for the latent DGDN features which is then related to generative methods for labels or captions. They used Bayesian Support Vector Machine for generating the labels of an image and RNN for giving captions. In the process of generating a label or a caption for any new images, the task of calculating the average across the distribution of latent codes is performed.

1.1.2 Compositional-Based Framework

The second type of architecture is mainly composed of several individual functional components [1]. This approach used CNN that extracts the meaning from an image using a language method illustrated in Fig. 5.

This framework performs the following steps:

(i) Extraction of unique visual features from the image.
(ii) Derived visual attributes from the extracted features.
(iii) Use the visual features and the visual attributes in a language method to generate probable captions.
(iv) Provides ranking for the probable caption using a deep multimodal similarity method, to determine the best suitable captions.

(a) ***Generation-based image captioning from sample***

This method is proposed by [15], which is composed of several components: (i) visual detectors, (ii) a language method, and (iii) multimodal similarity method so as to train the method on dataset of an image captioning.

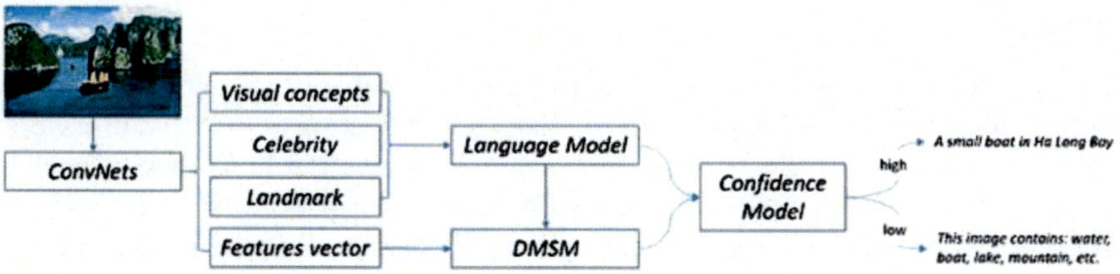

Fig. 6 Illustration of image caption pipeline [10]

Working

A Multiple Instance Learning (MIL) [19] for training the visual detectors for word that are commonly occurs in a caption. This includes several parts of speech like a noun, verb, and adjectives. An image sub-region was considered by this method rather than the complete image. Outputs of a word detector act as conditional inputs to a maximum-entropy language method. The features extracted from the sub-regions are represented with the words probably present in the image captions. Maron and Lozano-Pérez [19] performed re-ranking of the captions using sentence-level features and a deep multimodal similarity method to acquire the semantic information of an image.

(b) ***Generation of description from wild***

This method is introduced by [10] for different image captioning that automatically describes images in the wild. It used the compositional framework like [9], where in [10] the image caption systems are established using different components which are trained independently and latter combined in the main structure shown in Fig 6.

Working

In this method, for identifying a comprehensive visual concept the authors have considered a deep residual network-based vision method. On the other hand, to identify images of celebrities and landmarks; an entity recognition method is used. A classifier for estimating the confidence score for each output caption [10] for generating candidates a language method and for ranking the caption deep multimodal semantic method are considered.

(c) ***Generation of descriptions with structural words***

A compositional network-based image captioning method is proposed by [11]. This method follows some structural words format as: *<object, attribute, activity, scene>* [11]. These structural words are used to generate semantically meaningful descriptions using multi-task method which is comparable to MIL [19] method. Then LSTM [20] machine translation method is used to translate the structural words into image captions.

Working

Figure 7 describes the framework with two stages capable of identifying structural words and generating descriptions from image. Identification of the structural words sequence <objects, attributes, activities, scene> is carried out in the first stage. The image captions that contain comparatively larger information by deep RNN are translated from the word sequence (recognized in first stage) in the second stage.

(d) ***Parallel-fusion RNN-LSTM architecture***

Wang et al. [12] proposed a method based on deep convolutional networks and recurrent neural networks shown in Fig. 8. The main idea is to combine the benefits of RNN and LSTM which leads to decrease in complexity and increase in performance. RNN hidden units are composed of several equal dimension components that work parallel. The outputs are then merged with corresponding ratios to generate final output.

Working

This method follows the following strategies:

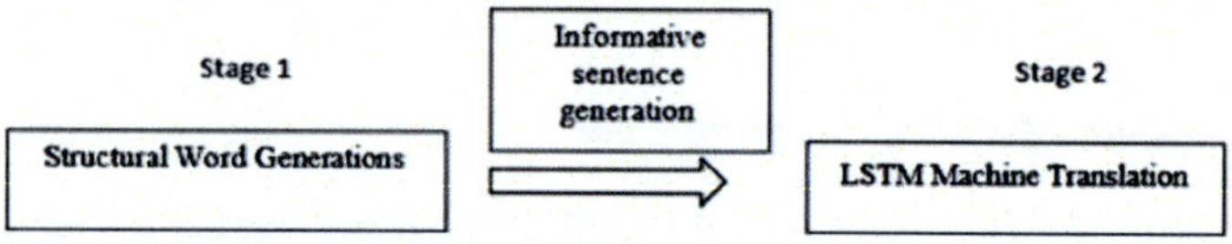

Fig. 7 Stages of generating sentence [11]

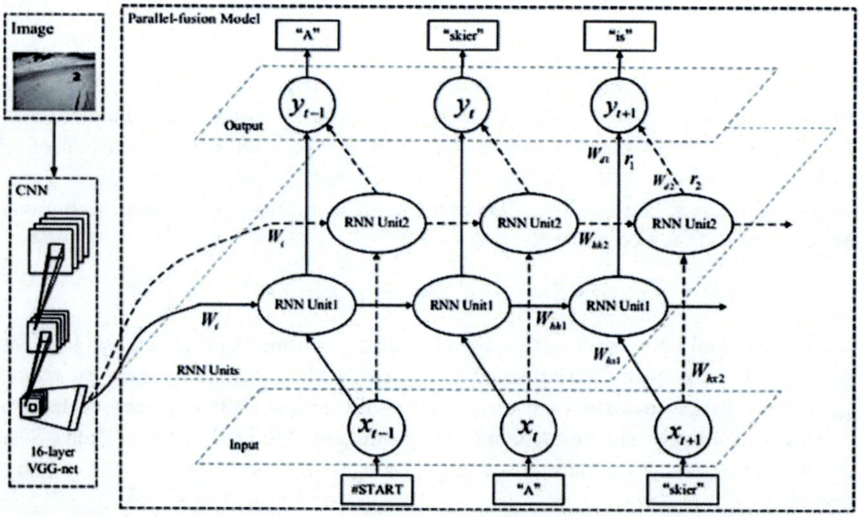

Fig. 8 Method proposed by [12]

1. Splits the hidden layer into 2 parts with both the parts remaining uncorrelated until the output unit.
2. From source data, identical features vectors are fed to the hidden layers along with feedback outputs of the respective hidden layers from the past.
3. Send the generated output of the RNN unit to y_t component of the overall output module.

$$h_{1_t} = \max\left(W_{hx1}x_t + W_{hh1}h_{1_{t-1}} + b_{h1}, 0\right) \tag{7}$$

$$h_{2_t} = \max\left(W_{hx2}x_t + W_{hh2}h_{2_{t-1}} + b_{h2}, 0\right) \tag{8}$$

$$y_t = \text{softmax}\left(r_1 W_{d1}h_{1_t} + r_2 W_{d2}h_{2_t} + b_d\right) \tag{9}$$

$$dy_1 = r_1 \times dy \tag{10}$$

$$dy_2 = r_2 \times dy \tag{11}$$

As per [12], the parameters considered are $\{h1, h2\}$ as the hidden units, $\{W_{hh1}, W_{hh2}, b_{h1}, b_{h2}, W_{d1}, W_{d2}\}$ denotes the weighted parameters while dy is the Matrix for a softmax derivative. The ratios considered are r_1, r_2: ratios.

(e) ***Fusion-based Recurrent Multimodal (FRMM) method***

This method proposed by [13] introduced an end-to-end trainable Fusion-based Recurrent MultiModal (FRMM) method that can address multimodal applications which allow each input modality to be independent w.r.t architecture, parameters and length of input sequences shown in Fig. 9.

Working

The method has separate stages whose outputs are mapped to a common description so that it can be associated with one another during the fusion stage. The outputs are predicted by the fusion stage based on the association. Supervised learning occurs in each all stage. Figure 9 describes how the FRMM method works by taking a video description method as an example. The FRMM method learns the behavior in separate stages.

1.2 Summary

Tables 1 and 2 show the different image captioning methods using deep neural algorithms. The table is categorized into two parts based on the framework used to generate a caption for an image experimented on datasets such as *Flickr8K*,

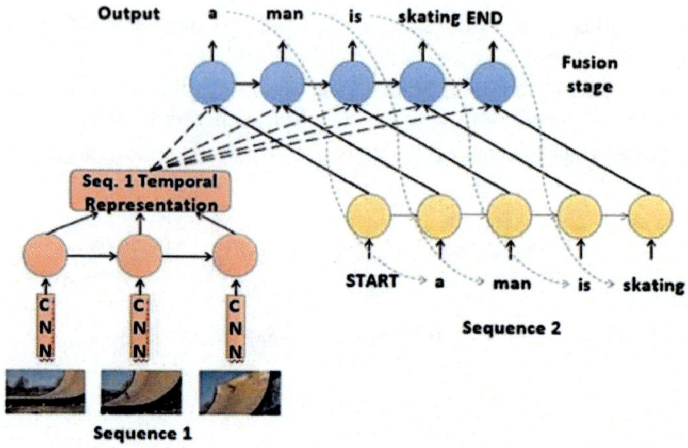

Fig. 9 Method proposed by [13]

Table 1 Summary of generating image caption based on encoder-decoder framework on the dataset using evaluation metric

Method	Dataset	BLEU-1	R@10
Encoder-decoder pipeline [4]	Flickr8K Flickr30K	–	0.55 0.629
NIC [5]	PASCAL Flickr8K Flickr30K	0.59 0.63 0.66	0.61 0.56
gLSTM [6]	Flickr8K Flickr30K MS COCO	0.647 0.646 0.67	–
Referring expression [7]	MSCOCO	–	–
Variational autoencoder [8]	Flickr8K Flickr30K MSCOCO	0.70 0.69 0.71	–

Table 2 Summary of generating image caption based on compositional-based framework on the dataset

Method	Dataset	BLEU-1	BLEU-4
Generation-based image captioning from sample [9]	MSCOCO	–	0.291
Generation of description from wild [10]	MSCOCO Adobe-MIT FiveK Instagram Image	–	–
Generation of descriptions with structural words [11]	UIUC Pascal Flickr8k	0.40 0.621	–
Parallel-fusion RNN-LSTM [12] architecture	Flickr8k	0.647	–
FRMM [13]	Flickr30k MSCOCO	0.589 0.702	0.177 0.276

Flickr30K, PASCAL, UIUC PASCAL, and *MSCOCO* considering only **BLEU** and **Recall@k** evaluation metrics. The first category, encoder-decoder framework uses to generate a caption which got inspired from the concepts of translating sentences from one language into another language. Under this framework, various authors [4–8] have been successful in generating captions of images. Contrary to the encoder-decoder framework, the second category that has been considered is a Compositional Architecture proposed by [9–13] image captioning.

In Table 1, among the method that follows encoder-decoder framework, variational autoencoder [8] has shown higher BLEU-1 value compared with the other methods experimented in MS COCO datasets. In Table 2, among the compositional-based architecture: FRMM [13] has higher BLEU-1 evaluation value experimented on MSCOCO dataset.

1.3 Conclusion

On carrying out a comprehensive survey on image captioning methods based on some deep learning methods. The following ideas were derived from: encoder-decoder framework and compositional architecture. The encoder-decoder framework is used to generate various captions from images. It first encodes an image to an intermediate representation and then generates a sentence word by word from the representation using the decoder. The compositional image captioning uses a method to detect concepts that visually appear in the input image. The detected concepts are then forwarded to the language method to generate various candidate captions from where one probable caption is chosen as the final caption or description for a given input image.

References

1. Zakir Hossain, M., Sohel, F., Shiratuddin, M.F., Laga, H.: A comprehensive survey of deep learning for image captioning. ACM Comput. Survey **51**(6), 1–36 (2019)
2. Bal, S., An, S.: A survey on automatic image caption generation. Neurocomputing **311**, 291–304 (2018)
3. Kulkarni, G., Premraj, V., Dhar, S., Li, S., Choi, Y., Berg, A.C., Berg, T.L.: Baby talk: understanding and generating simple image descriptions. CVPR (2011)
4. Kiros, R., Salakhutdinov, R., Zemel, R.: Unifying visual-semantic embeddings with multimodal neural language models. arXiv:1411.2539 (2018)
5. Vinyals, O., Toshev, A., Bengio, S., Erhan, D.: Show and tell: a neural image caption generator. In: Proceedings of the IEEE Conference on Computer Vision and Pattern Recognition, pp. 3156—3164 (2015)
6. Jia, X., Gavves, E., Fernando, B., Tuytelaars, T.: Guiding long-short term memory for image caption generation. In: Computer Vision (ICCV), IEEE International Conference on Computer Vision (ICCV) (2015)

7. Mao, J., Huang, J., Toshev, A., Camburu, O., Yuille, A.L., Murphy, K.: Generation and comprehension of unambiguous object descriptions. In: Proceedings of the IEEE Conference on Computer Vision and Pattern Recognition, pp. 11–20 (2016)
8. Pu, Y., Gan, Z., Henao, R., Yuan, X., Li, C., Stevens, A., Carin, R.: Variational autoencoder for deep learning of images, labels and captions. In: Proceedings of the Advances in Neural Information Processing Systems, pp. 2352–2360 (2016)
9. Fang, H., Gupta, S., Iandola, F., Srivastava, R.K., Deng, L., Dollãr, P., Gao, J., He, X., Mitchell, M., Platt, J.C.: From captions to visual concepts and back. In: Proceedings of the IEEE Conference on Computer Vision and Pattern Recognition, pp. 1473–1482 (2015)
10. Tran, K., He, X., Zhang, L., Sun, J., Carapcea, C., Thrasher, C., Buehler, C., Sienkiewicz, C.: Rich image captioning in the wild. In: Proceedings of the IEEE Conference on Computer Vision and Pattern Recognition Workshops, pp. 49–56 (2016)
11. Ma, S., Han, Y.: Describing images by feeding LSTM with structural words. In: 2016 IEEE International Conference on Multimedia and Expo (ICME), pp. 1–6. IEEE (2016)
12. Wang, M., Song, L., Yang, X., Luo, C.: A parallel-fusion RNN-LSTM architecture for image caption generation. In: IEEE International Conference on Image Processing (ICIP), pp. 4448–4452. IEEE (2016)
13. Oruganti, R.M., Sah, S., Pillai, S., Ptucha, R.: Image description through fusion based recurrent multi-modal learning. In: IEEE International Conference on Image Processing (ICIP), pp. 3613–3617 (2016)
14. Xu, K., Ba, J., Kiros, R., Cho, K., Courville, A., Salakhudinov, R., Zemel, R., Bengio, Y.: Show, attend and tell: Neural image caption generation with visual attention. In: International Conference on Machine Learning. PMLR, pp. 2048–2057 (2015)
15. Hochreiter, S., Schmidhuber, J.: Long short-term memory. Neural Comput. **9**, 1735–1780 (1997)
16. Donahue, J., Hendricks, L., Guadarrama, S., Rohrbach, M., Venugopalan, S.: Long-term recurrent convolutional networks for visual recognition and description. In: Proceedings of the IEEE Conference on Computer Vision and Pattern Recognition, pp. 2625–2634 (2015)
17. Cho, K., Merrinboer, B.V., Gulcehre, C.: Learning phrase representations using RNN encoder—decoder for statistical machine translation. arXiv:1406.1078v3 (2014)
18. Kazemzadeh, S., Ordonez, V., Matten, M., Berg, T.L.: ReferItGame: referring to objects in photographs of natural scenes. EMNLP, pp. 787–798 (2014)
19. Maron, O., Lozano-Pérez, T.: A framework for multiple-instance learning. In: Advances in Neural Information Processing Systems, pp. 570–576 (1998)
20. Wang, C., Yang, H., Bartz, C., Meinel, C.: Image captioning with deep bidirectional LSTMs. In: Proceedings of the 2016 ACM on Multimedia Conference, pp. 988–997. ACM (2016)

A Comparison of Image Compression Techniques Over Wireless Fading Channel

Priyanka Hazowary, Sushanta Kabir Dutta, and Rupaban Subadar

Abstract In today's life, the multimedia transmission is becoming an integral part of wireless communication. However, the transmission of a multimedia file requires sending a large amount of data which consume a huge amount of bandwidth. Due to the limitations in bandwidth utilization in wireless mobile communication, different compression techniques are used, and they reduce the amount of information to be transmitted. However, the compression techniques very often reduce the quality of a multimedia file because of their inherent nature. In addition, the quality may further be deteriorated after wireless transmission due to the presence of fading. In this work, we compared two compression techniques JPEG and JPEG2000 that are widely used in still image compression. From the study, it has been observed that the JPEG2000 performs better in low SNR environment, while JPEG performs better in high SNR environment. In addition, the performance of both the techniques improves, while using maximal ratio diversity combining technique as compared to selection combining technique.

Keywords Image compression · Fading channels · Diversity technique · PSNR

1 Introduction

Wireless communication is the fastest growing and the most vibrant technological areas in the communication area. However, due to unavoidable fading effects, the performance of the wireless communication system degrades to a large extent. The fading is a phenomenon that depends on the communication scenario to a large extent. It is modeled by the different statistical distributions in different scenarios. Nakagami-*m* [1] models provide a very good matching with the practically obtained data in urban scenarios. Therefore, it is regarded as a very popular model.

Reasonable amount of literature is available describing the works that have been carried out in this model to analyze the performance of the system with and without

P. Hazowary (✉) · S. K. Dutta · R. Subadar
Department of Electronics and Communication Engineering, North Eastern Hill University, Shillong, India

H. K. D. Sarma et al. (eds.), *Contemporary Issues in Communication, Cloud and Big Data Analytics*, Lecture Notes in Networks and Systems 281,
https://doi.org/10.1007/978-981-16-4244-9_19

a diversity system [2–5]. However, a very limited works are available on the image transmission over fading channels. We have discussed here a few of them.

In [6], secure image source coding was formed by utilizing compression feature of JPEG2000 called chaotic JPEG2000. It was robustified to image communication by turbo channel coding. And later with Gaussian impulsive noisy channels, the efficacy of the proposed scheme was experienced through the implementation of the result for an image communication. ZainEldin et al. [7] studied and evaluated pertinent research route as well as went through the latest image coding algorithms over WMSN. This survey outlines the pros and cons of the latest efforts of these algorithms. In [8], a joint compression-encryption and symbol scrambling techniques were proposed, and the performance of the system was analyzed based on the peak signal-to-noise ratio (PSNR) and bit error rate (BER) values. A joint source and channel coding techniques have been proposed in [9] for wireless image transmission.

From the literature available, it is observed that although there are some works carried out on the transmission of images over fading channels, and a performance comparison of the standard compression techniques has been missing. Moreover, no work is found on the use of diversity techniques in image transmission. In this work, we have considered selection combining (SC) and maximum ratio combining (MRC) receivers over Nakagami-*m* fading channels and compared the received images that were compressed using JPEG and JPEG2000 compression standards.

2 Communication Model

A simple block diagram of the system considered has been provided in Fig. 1. In this study, we have considered a SIMO system with SC and MRC diversity receiver. The wireless channels have been considered as the Nakagami-*m*. This model is generally observed in the urban scenario where the numbers of obstacles present are comparatively high. The obtained data from this model are the best fit with the practically obtained data.

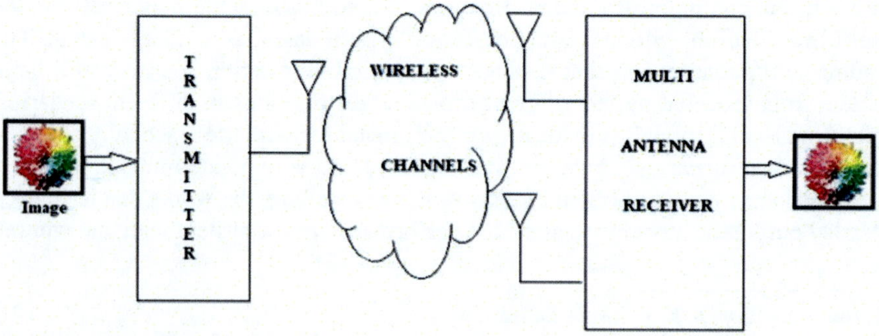

Fig. 1 Block diagram of the communication system

The received signal through wireless channel at the lth received antenna can be given as

$$r_l(t) = \alpha_l e^{j\phi l} s(t) + n_l(t), \quad l = 1, 2, \ldots, L \tag{1}$$

where, $s(t)$ is the transmitted information, and $n_l(t)$ is the additive Gaussian noise. $\alpha_l e^{j\phi l}$ is the complex wireless channel with α_l is the Nakagami-m distributed [10].

$$P_\alpha(\alpha) = \frac{2m^{m\alpha^{2m-1}}}{\Omega^m \Gamma(m)} \exp\left(\frac{-m\alpha^2}{\Omega}\right), \quad \alpha \geq 0 \tag{2}$$

where $m \geq 0$ is the parameter which determines the severity of fading.

Diversity is a widely used concept to get rid of the effect of fading. In this technique, the receiver obtained the transmitted signal through the different fading paths to avoid the occurrence of deep fading. The selection diversity which selects the best-received signal based on the SNR is the simplest one and easy to implement. The maximal ratio combining is the optimal one, where the output SNR equals to the sum of all the input SNR.

3 Image Compression Techniques Used

Here, we have exclusively used two image compression algorithms JPEG and JPEG2000 for our study. The basic principles of both are available in various standard textbooks. However, an optimal description is append here for clarity of understanding without referring to any standard literature.

3.1 JPEG

It is a commonly used method of lossy compression [11] for digital images, specially for the images resulting from digital photography. The degree of compression can be adjusted, which usually allows a selectable tradeoff between storage size and image quality. JPEG typically achieves 10:1 compression with negligible perception loss in the image quality. Since introduction in 1992, JPEG has been the most widely used image compression standard in the world and also the most widely used digital image format. The term 'JPEG' is an acronym for the Joint Photographic Experts Group, which created the standard in 1992. The basis for JPEG is the discrete cosine transform (DCT) which is a lossy image compression technique.

3.2 JPEG2000

It is an improved version [12] of JPEG image compression standard and coding system. It was developed from 1997 to 2000 by a Joint Photographic Experts Group committee. The improvement was in superseding original DCT based JPEG standard with a newly designed, wavelet-based method. In JPEG2000, it is possible to store different parts of the same picture using different quality.

JPEG2000 is thus a discrete wavelet transform (DWT) based compression standard that could be adapted for motion imaging video compression with the Motion JPEG2000 extension. JPEG2000 technology was selected as the video coding standard for digital cinema in 2004.

4 Experimental Results and Discussion

4.1 SC Receiver

In SC, the receiver uses multiple antennas to capture the transmitted signal. We have considered that the antennas are sufficiently apart from one another and the signals received by them are independently faded. The SC diversity combiner selects the signal from the receiving antennas with the highest SNR for decoding. It is the simplest among all the diversity techniques available and hence very much important from the practical point of view. In this study, we have transmitted both JPEG and JPEG2000 compressed images and received with the help of SC diversity receiver and evaluated their performances.

(1) *JPEG*: In Figs. 2 and 3, the compressed JPEG transmitted and received images are shown for SC receiver systems for different power levels for the channel parameter $m = 1$. It can be easily observed that as the power and diversity

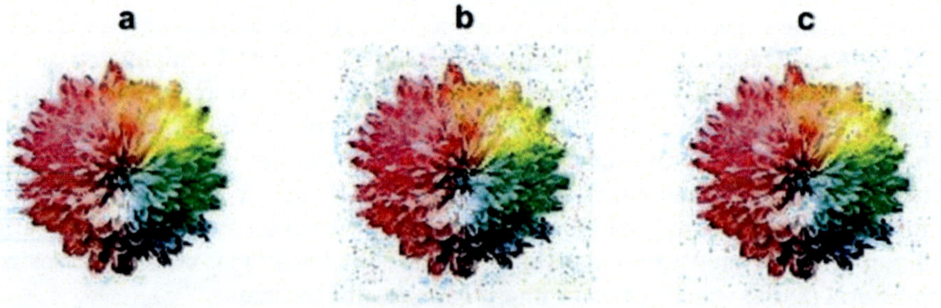

Fig. 2 **a** Original image (JPEG). **b** Received image through 2-branch SC with 5 dB transmitted power. **c** Received image through 3-branch SC with 5 dB transmitted power

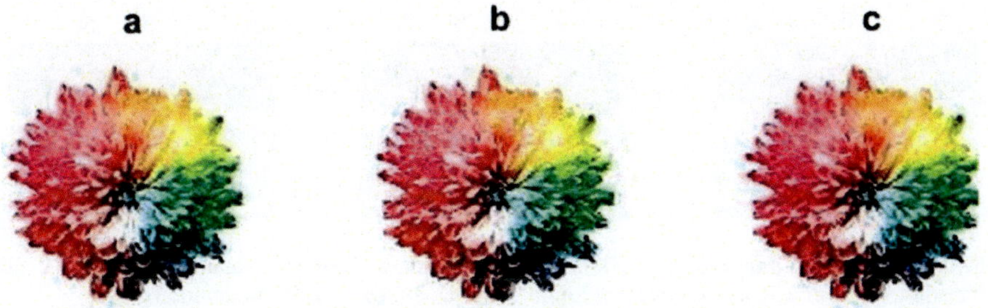

Fig. 3 **a** Original image (JPEG). **b** Received image through 2-branch SC with 12 dB transmitted power. **c** Received image through 3-branch SC with 12 dB transmitted power

branches are increased the quality of the received image gets improved. This is observed from the PSNR and BER parameters mentioned in Table 1.

(2) *JPEG2000*: In Figs. 4 and 5, the compressed JPEG2000 transmitted and received images are shown for SC receiver systems for different power level for the channel parameter $m = 1$. It can be easily observed that as the power and diversity branches are increased the quality of the received image is also improved. This is also seen from the PSNR and BER parameters mentioned in Table 2.

Now, while comparing the two compression standards as above, it is observed from Tables 1 and 2 that JPEG2000 performs better in a noisy the channel. That is when the SNR is low, it provides a low BER at the output. However, JPEG outperforms JPEG2000, while the channel is less noisy, that is, the SNR is high. This is evident from the BERs at various noise levels as in Tables 1 and 2. Again, as we

Table 1 Comparison of image parameters in 2 branch and 3 branch SC

Power	2 dB		5 dB		12 dB	
Parameter	PSNR	BER	PSNR	BER	PSNR	BER
2 branch SC	17.55	0.3964	21.67	0.1578	33.9414	0.0101
3 branch SC	20.03	0.2315	25.32	0.067	43.3484	0.0011

Fig. 4 **a** Original image (JPEG2000). **b** Received image through 2-branch SC with 5 dB transmitted power. **c** Received image through 3-branch SC with 5 dB transmitted power

Fig. 5 **a** Original image (JPEG2000). **b** Received image through 2-branch SC with 12 dB transmitted power. **c** Received image through 3-branch SC with 12 dB transmitted power

Table 2 Comparison of image parameters in 2 branch and 3 branch SC

Power	2 dB		5 dB		12 dB	
Parameter	PSNR	BER	PSNR	BER	PSNR	BER
2 branch SC	17.82	0.3961	21.79	0.157	33.66	0.0102
3 branch SC	20.13	0.231	25.48	0.0671	42.49	0.0014

increase the number of diversity branches, the performance increases in both the cases. Consequently, the BER decreases.

4.2 *MRC Receiver*

Similar to SC receiver MRC receiver also uses multiple antennas to capture the transmitted signal. In this case also, we have considered an independent fading environment. MRC receiver combines all the received signal after weighting them according to SNR and phase cancelation. It performs optimally among all the diversity techniques available. In this study, we have transmitted both JPEG and JPEG2000 images and received with the help of MRC diversity receiver and evaluated the performances.

(1) *JPEG*: In Figs. 6 and 7, the compressed JPEG transmitted and received images are shown for MRC receiver systems for different power level for the channel parameter $m = 1$. It can be easily observed that as the power and diversity branches are increased, and the quality of the received image is also enhanced. This is also seen from the PSNR and BER parameters tabulated in Table 3.

(2) *JPEG2000*: In Figs. 8 and 9, the compressed JPEG2000 transmitted and received images are shown for MRC receiver systems for different power level for the channel parameter $m = 1$. It can be easily observed that as the power and diversity branches are increased the quality of the received image is also enhanced. This is also seen from the PSNR and BER parameters tabulated in Table 4.

It is observed from Tables 3 and 4 that JPEG2000 compression scheme performs better when the channel is noisy, although JPEG compression technique supersedes

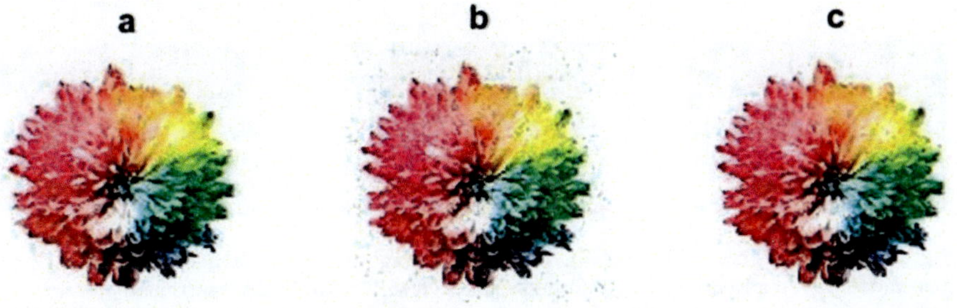

Fig. 6 **a** Original image (JPEG). **b** Received image through 2-branch MRC with 5 dB transmitted power. **c** Received image through 3-branch MRC with 5 dB transmitted power

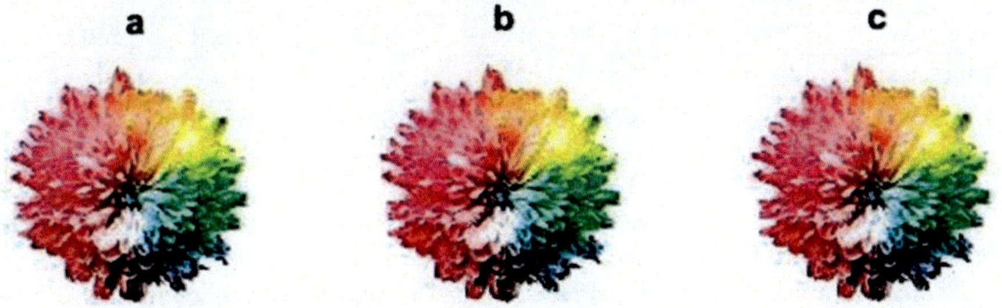

Fig. 7 **a** Original image (JPEG). **b** Received image through 2-branch MRC with 12 dB transmitted power. **c** Received image through 3-branch MRC with 12 dB transmitted power

Table 3 Comparison of image parameters in 2 branch and 3 branch MRC

Power	2 dB		5 dB		12 dB	
Parameter	PSNR	BER	PSNR	BER	PSNR	BER
2 branch SC	19.35	0.263	24.11	0.095	36.78	0.0054
3 branch SC	24.19	0.086	30.89	0.019	50.12	0.00025

Fig. 8 **a** Original image (JPEG2000). **b** Received image through 2-branch MRC with 5 dB transmitted power. **c** Received image through 3-branch MRC with 5 dB transmitted power

Fig. 9 **a** Original image (JPEG2000). **b** Received image through 2-branch MRC with 12 dB transmitted power. **c** Received image through 3-branch MRC with 12 dB transmitted power

Table 4 Comparison of image parameters in 2 branch and 3 branch MRC

Power	2 dB		5 dB		12 dB	
Parameter	PSNR	BER	PSNR	BER	PSNR	BER
2 branch SC	19.57	0.2621	24.00	0.0947	36.36	0.0054
3 branch SC	24.44	0.0858	30.91	0.0191	50.55	0.0002

JPEG2000 when the channel noises are sufficiently low. This is similar to what has been observed with SC receiver.

Therefore, it is observed that JPEG2000 has a better performance in noisy channels, while JPEG shows more efficiency in low SNR environments. This may be attributed to the observations in Tables 3 and 4. However, it is also interesting to note that with the increase in number of diversity branches here the BERs falls in both cases.

Upon comparing the two diversity schemes, the MRC technique provides better results with both the compression standards than SC, which is quite evident.

5 Conclusions

While using both the compression standards in transmitting the images in fading channels, we found that the JPEG standard has a superior performance in non-noisy channels and JPEG2000 performs better in noisy channels. This may be attributed to the signal processing effects in them. In JPEG2000 DWT is used unlike DCT in JPEG. Use of DWT provides better compression ratio but introduces Spackle noise. Therefore, performance is just below JPEG in noise free channels. However, JPEG does not have noise resiliency which is available JPEG2000 and therefore, in noisy channel the later has a superior performance to the former. This observations may be very informative in certain applications like medical image transmission over noisy and noise free channels. Here, the proper selection of image compression standard would be of a great help to get a high-quality image in the receiver side. As a part of future work, a few more hybrid compression techniques may be explored.

References

1. Nakagami, M.: The m-distribution—a general formula of intensity distribution of rapid fading. In: Hoffman, W.C. (ed.) Statistical Methods of Radio Wave Propagation. Pergamon, Oxford (1960)
2. Zhang, Q.T.: A decomposition technique for efficient generation of correlated Nakagami fading channels. IEEE J. Sele. Areas Commun. **18**(11), 2385–2392 (2000)
3. Karagiannidis, G.K., Zogas, D.A., Kotsopoulos, S.A.: On the multivariate Nakagami-m distribution with exponential correlation. IEEE Trans. Commun. **51**, 1240–1244 (2003)
4. Peppas, K., Sagias, N.C.: A trivariate Nakagami-m distribution with arbitrary covariance matrix and applications to generalized-selection diversity receivers. IEEE Trans. Commun. **57**, 1896–1902 (2009)
5. Ryu, H.-S., Lee, J.-S., Kang, C.-G.: BER analysis of dual-carrier modulation (DCM) over Nakagami-m fading channel. IEICE Trans. Commun. **E94-B**(7), 2123–2126 (2011)
6. Deergha Rao, K.: A robust and secure scheme for image communication over wireless channels. In: IEEE 7th CAS Symposium on Emerging Technologies (2005)
7. ZainEldin, H., Elhosseini, M.A., Ali, H.A.: Image compression algorithms in wireless multimidia sensor networks: a survey. Ain Shams Eng. J. **6**(2) (2015)
8. Wang, Z.: Secure image transmission in wireless OFDM systems using secure block compression-encryption and symbol scrambling. IEEE Access **7**, 126985–126997 (2019)
9. Bourtsoulatze, E., Burth Kurka, D., Gündüz, D.: Deep joint source-channel coding for wireless image transmission. IEEE Trans. Cogn. Commun. Netw. **5**(3), 567–579 (2019)
10. Simon, M.K., Alouini, M.S.: Digital Communication Over Fading Channels, 2nd edn. Wiley (2005)
11. Pennebaker, W.B., Mitchell, J.L.: JPEG Still Image Data Compression Standard. Springer (1993)
12. Taubman, D., Marcellin, M.: JPEG2000 Image Compression Fundamentals, Standards and Practice. Springer Science and Business Media (2012)

Image Transformation into Cartoon Using OpenCV

Aaditaa Soni and Anand Sharma

Abstract The objective of this paper is to give a solution regarding transformation of images into cartoon images. The previous methods of image transformation require convoluted computer graphics and programming skills. An image transformation system that can produce a adapted cartoon face from the input image is presented in this paper. This proposed system is easy to use and requires less programming skills with little user interaction. The basic concept of this paper is particularly on nominated images that are changed to cartoons. This focuses on using OpenCV to transform input image to cartoon image, which will be used for more image processing systems for various applications.

Keywords Image processing · Cartoon · Cartoonify · Generative adversarial network (GAN) · OpenCV

1 Introduction

Image cartoons are generally worn in number of implementations. The cartoons are creatively formed, and it is necessary graceful and plus man creative abilities. The depictions of cartoons in large number for any creative films it could be tedious for craftsman as all have to characterize the drawing of cartoon appropriately to bring a decent outcome. We as a whole realize that animation assumes a significant function in the realm of film, so to defeat the issue looked by the craftsman a program with the assistance of generative adversarial network (GAN) was created, which converts images as well as change video into animation.

In past times, the designing of photos comprises in specific space called "non-photorealistic rendering." A conventional method has been created particularly for space for designing of images, and they are fruitful in designing any image by attaching texture, styles, impacts, and so on. With the assistance of the this method,

A. Soni (✉)
MUST, Lakshmangarh, Sikar 332311, India

A. Sharma
Computer Science and Engineering Department, MUST, Lakshmangarh, Sikar 332311, India

H. K. D. Sarma et al. (eds.), *Contemporary Issues in Communication, Cloud and Big Data Analytics*, Lecture Notes in Networks and Systems 281,
https://doi.org/10.1007/978-981-16-4244-9_20

numerous product was created to change over genuine images into cartoon a portion of the techniques fizzled, while a portion of the strategies gave outcome, however, did not fulfill all necessities.

Additionally, animated pictures are byzantine contrasted with genuine original pictures. To fulfill all the necessities of changing over genuine original pictures for example preview into cartoon, the cartoon GAN [1] is taken. This will make the work easy for experts.

In stated above, a genuine original picture is changed over into an animated picture with the assistance of the implementations created known as "Cartoon GAN" is smaller in length of time. Thus, time is spared, fine work is produced inside an abundant measure of time, which will give an extraordinary open door for activity businesses to make as, and for example, film or movement cuts. This technique gives a more precise after effect of changing over pictures in addition to it changes over video into an activity cut contrasted with the past strategies.

The coming section will give an overview of previous work. Proposed method is presented in next section. After that the implementation and experimental result are discussed. Finally, the conclusion part is there at the end.

2 Related Work

In recent years, there had been colossal development in the exploration of GAN [2]. Figure 1 shows the GAN model architecture. GAN was advanced in the year

Fig. 1 GAN model architecture

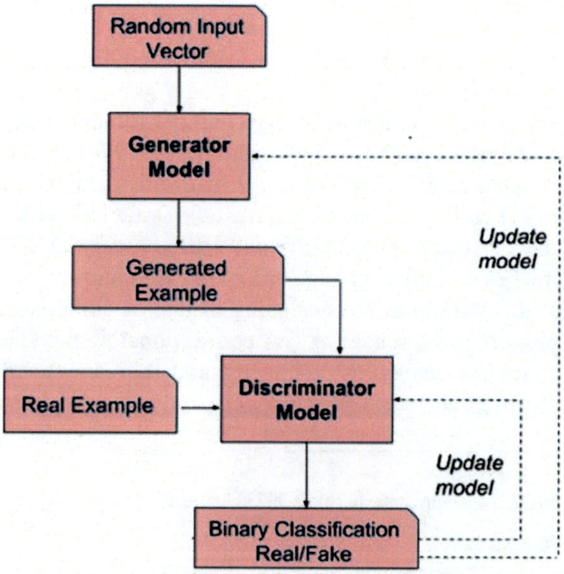

2014 where it was presented in different applications, for example, natural language processing (NLP) and deep learning. In research article [3], authors investigated the various strategies for picture synthesis, for example, direct strategy, iterative method and hierarchal technique. They talked around two strategies for picture amalgamation those are "image-to-image translation" and "text-to-image change".

In text-to-image change, the latest strategies functioned admirably on a pre-characterized information where every picture having one article, for example, Caltech-UCSD Birds [4] and Oxford102 [5], yet the presentation on complicated dataset, for example, MSCOCO [6] is a lot of poor performer. That was the only restriction this model had, and it was important to learn various ideas of the article. To get better, the exhibition of GAN and upgrade yield in assignment they prepared various models that would create a solitary article and train different model which would figure out how to join different items as per text depictions, and that CapsNet [7].

Presently coming to picture interpretation, they examined some broad models from supervised mechanism to unsupervised mechanism that are pixel loss [8], and self-distance loss [9]. Beside that they additionally developed some image–image interpretation model for the face altering and video expectation. It has been observed that picture interpretation was a snappy use of GAN which had an incredible breadth for versatile application. Despite the fact that during the examination unsupervised technique apparently was more main stream contrasted with the supervised strategy.

Many carton making frameworks [10] center on low-level editing and manage devices for simple and adaptable cartoon activity and drawing. As of late, motivated by the intensity of CNN, the spearheading work of author Gatys et al. [11] present an overall answer for moving the style of an offered fine art to any picture naturally. Inkwell [12], for instance, built up some compelling methods on abusing layers, gathering/progression of components and permitting the control of movement capacities. Additionally, the Cartoon framework [13] gave extraordinary skeleton-driven components, a broad arrangement of building-blocks to configuration faces by re-using segments and bits of activities. Interestingly, our framework permits clients to make a customized cartoon and movement from an input face picture. There is a progression of tasks dependent on the GAN projected for an overall image-to-image interpretation. Isola et al. [14] build up the pix2pix network prepared with the oversight of pictures matches and accomplish sensible outcomes on numerous interpretation errands, for example, photograph-to-label, photograph-to-sketch, and photograph-to-map.

3 Proposed Method

To make a customized cartoon, we receive a picture based methodology. In particular, we developed an example-based learning way to deal with produce a cartoonified image drawing from a giving face picture, by seeing how a specific craftsman would sketches the cartoons shows from preparing face pictures. Since this is hard to portray

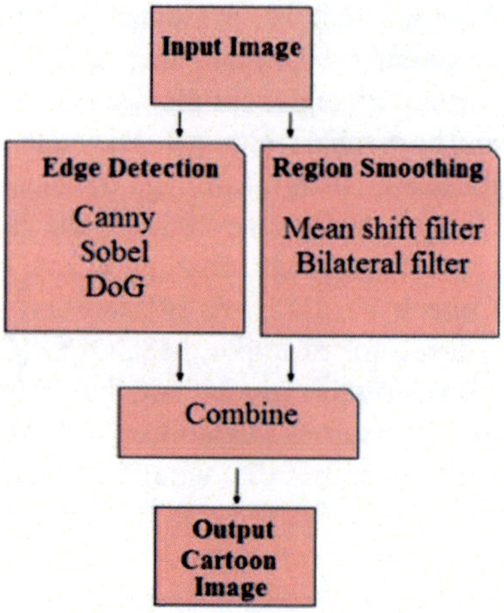

Fig. 2 Proposed model for image transformation into cartoon

the standards for how a craftsman sketches, we utilize a sampling approach which is non-parametric that catches the connection between the training pictures and their relating outlines drawn by a similar craftsman. Figure 2 shows the proposed model for image transformation into cartoon.

In this proposed model, we are using edge detection techniques and region smoothing filtering for the cartoon creation.

i. Edge Detection: There are a number of edge detection implementations in the literature but for the image transformation we can use Canny or Sobel or DoG (Difference of Gaussian). This edge detection technique (any one) is applied on the input image for enhancing the edges. As per the results obtained, we can say that the performance of DoG is much better than the Canny and Sobel.
ii. Region Smoothing: By using region smoothing, we are removing all the shades and color boundaries (like shades due to variation of light) which can be measured as inconsequential for the cartoon making process. We are having mean shift filter and bilateral filter for the region smoothing purpose. In case of mean shift filter, the color regions are significantly smoothened but it still contains shades and features which can give negative impact for cartoon making. So we can prefer bilateral over the mean shift filter.
iii. Combine: After doing region smoothing and edge detection we will combine the images and finally get the cartoon image as output.

4 Implementation and Experimental Result

Computer vision is perhaps the sultriest field in Artificial Intelligence with a wide assortment of utilizations that pre-owned Python. Python is the pool of libraries. It has various libraries for genuine applications. One such library is OpenCV [15]. OpenCV is the most well-known library utilized in computer vision with a ton of intriguing stuff. OpenCV is a cross-platform library that incorporates applications like video and picture capturing and preparing. It is straightforward and actualizes by everybody. It is significantly utilized in image transformation, face recognition, object detection, and numerous other shocking applications. We will follow the accompanying steps in this article to convert the image to cartoon:

i. Importing libraries
ii. Reading input image

It should be obvious that the image read by OpenCV is being appeared as Blue-Green-Red (BGR) picture so we have to change it over to the Red-Green-Blue (RGB).

iii. Detecting edges in the image

Here, we will detect the edges in the image utilizing versatile thresholding strategies.

iv. Cartoonifying the image

In this progression, we will be cartoonifying the image by utilizing bilateral filter method.

v. Final output

At last, we will visualize the final output as cartoon image (Fig. 3).

5 Conclusion

A systematic approach for image transformation into cartoon is proposed in this paper. This system will design a personalized animated photo from the input image. With the help of edge detection and region smoothing we have developed the system for image transformation into cartoon. The system is implemented in OpenCV. The proposed system is anything but difficult to use for common individuals. When an exact expressive cartoon image is produced, it very well may be misrepresented with the animation supervisor and enlivened by the speech-driven animator. This framework further be utilized in an assortment of utilizations, like online visiting and customized e-cards.

The selection of best edge detection techniques and region smoothing filtering may be improved in coming couple of years. Color distance and spatial radius might be influencing factor for the image transformation into cartoon.

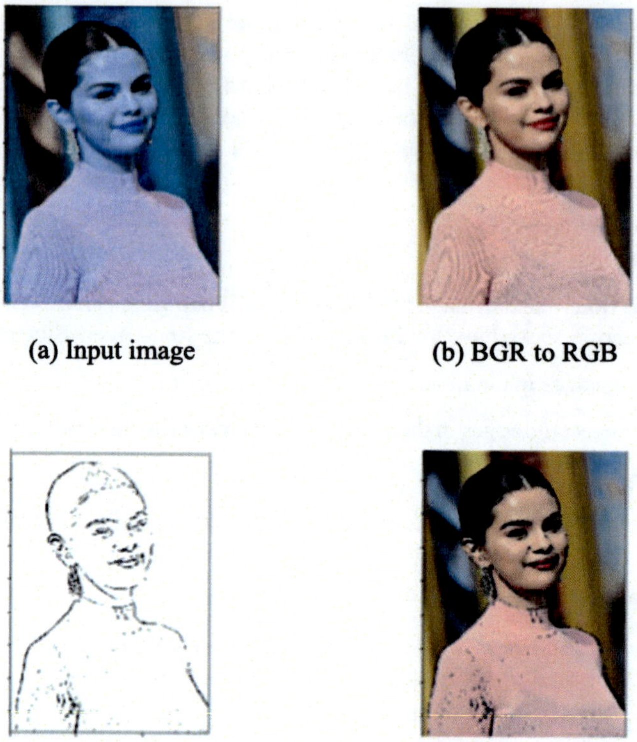

(a) Input image (b) BGR to RGB

(c) Edge detection (d) Output cartoon image

Fig. 3 Experimental results

References

1. Chen, Y., Lai, Y.-K., Liu, Y.-J.: CartoonGAN: generative adversarial network for photo cartoonization. In: International Conference on Image Processing (2018)
2. Goodfellow, I., Pouget-Abadie, J., Mirza, M., Xu, B., WardeFarley, D., Ozair, S., Courville, A., Bengio, Y.: Generative adversarial nets. Adv. Neural Inf. Process. Syst. 2672–2680 (2014)
3. Qin, Z., Luo, Z., Wang, H.: Autopainter: cartoon image generation from sketch by using conditional generative adversarial networks. In: International Conference on Image Processing (2017)
4. Wah, C., Branson, S., Welinder, P., Perona, P., Belongie, S.: The Caltech-UCSD birds-200-2011 dataset (2011)
5. Nilsback, M.-E., Zisserman, A.: Automated flower classification over a large number of classes. In: Proceedings of the Indian Conference on Computer Vision, Graphics and Image Processing, Dec 2008
6. Zhu, J.-Y., Park, T., Isola, P., Efros, A.A.: Unpaired image-to-image translation using cycle-consistent adversarial networks. arXiv preprint arXiv:1703.10593 (2017)
7. Sabour, S., Frosst, N., Hinton, G.E.: Dynamic routing between capsules. arXiv preprint arXiv: 1710.09829v2 (2017)

8. Isola, P., Zhu, J.-Y., Zhou, T., Efros, A.A.: Image-to-image translation with conditional adversarial networks. arXiv preprint arXiv:1611.07004 (2016)
9. Benaim, S., Wolf, L.: One-sided unsupervised domain mapping. arXiv preprint arXiv:1706.00826 (2017)
10. http://www.macromedia.com/software/
11. Gatys, L.A., Ecker, A.S., Bethge, M.: A neural algorithm of artistic style. arXiv preprint arXiv:1508.06576 (2015)
12. Litwinowicz, P.: Inkwell: a 21/2 D animation system. Comput. Graph. **25**(4), 113–122 (1991)
13. Ruttkay, Z., Noot, H.: Animated CharToon faces. In: The First International Symposium on Non-Photorealistic Animation and Rendering, pp. 91–100 (2000)
14. Isola, P., Zhu, J.-Y., Zhou, T., Efros, A.A.: Image-to-image translation with conditional adversarial networks. In: Proceedings of CVPR (2017)
15. Baggio, D.L.: Mastering OpenCV with Practical Computer Vision Projects. Packt Publishing Ltd (2012)

An Advanced Revealing and Classification System for Plant Illnesses Using Unsupervised Bayesian-based SVM Classifier and Modified HOG-ROI Algorithm

P. Pugazhendiran, K. Suresh Kumar, T. Ananth Kumar, and S. Sundaresan

Abstract Identification of crop diseases is a challenging task in the agricultural field. Various methods are proposed for classification and disease identification. Considering all the researches done so far, accuracy is the major problem identified. It is due to the lack of clarity in obtaining the image of the crop from different viewpoints. A preprocessing method is implemented for the captured image and is imperiled to an improved clustering algorithm, specifically k-means algorithm, to acquire the disease-ridden part of the leaf. The classification of the diseases in the leaf is done by using Bayesian-based SVM classifier. The identified diseased portion might be given on the way to a processing method in order to expand the most affected area. Some part like the plague-ridden part of the leaf might get exposed toward a combined histogram algorithm called histogram of oriented gradient (HOG) algorithm and region of interest (ROI) algorithm to segregate the features. For obtaining the extracted features and to classify and identify the plant diseases, the proposed unsupervised machine learning method (SVM) is better than previous techniques.

Keywords HOG · ROI · K-means · Clustering algorithm · Unsupervised SVM classifier

1 Introduction

India grades second position in agriculture in the form of output. Still, agriculture is considered to be the most significant financial division, which assumes a noteworthy job being developed. It is considered to be the life booming opportunities for 66% of

P. Pugazhendiran (✉)
Electrical and Electronics Engineering, IFET College of Engineering, Villupuram, India

K. Suresh Kumar
Electronics and Communication Engineering, IFET College of Engineering, Villupuram, India

T. Ananth Kumar
Computer Science and Engineering, IFET College of Engineering, Villupuram, India

S. Sundaresan
Electronics and Communication Engineering, National Institute of Technology, Puducherry, India

H. K. D. Sarma et al. (eds.), *Contemporary Issues in Communication, Cloud and Big Data Analytics*, Lecture Notes in Networks and Systems 281,
https://doi.org/10.1007/978-981-16-4244-9_21

the people groups in India. India consumes more than 84 million hectares of agrarian lands. Rice and wheat are the principal cereals. Fruits like banana and sapota are the furthermost extensively recognized organic products. Sugarcane, cotton, bean stew, and groundnuts are the real business crops. Harvest development relies upon regular precipitation, soil quality, soil supplement, and climate conditions. In South India, rubber cultivation is the most common business involved. Any deviation in these prompts' extreme misfortune. Ailments are significant reasons for the loss of harvests, and the controlling of diseases is again a difficult task. The techniques used for the classification are based on random transform and other machine learning methods [1]. The economic development of the country is based on agriculture development. A momentous involvement in the commercial affluence of the forward-looking countries is based on agriculture. It is multidirectional, having run rapidly and fast reach out regarding reality. Green unrest made the ranchers expand their social practices and agrarian contributions to serious trimming frameworks with escalated projects to improve the assembling prospective per unit land, time, and information.

The process of agriculture pushes India to seventh in the world. Industries need raw materials such as cotton, sugar, and tobacco, which is driven by agriculture. Exporting of such products is commonly occurring in India because of its soil nature and fertile land. Crop diseases are the foremost problems facing by the farmers now a day. Crop diseases are of various types, and the identification and classification of such conditions are considered to be a difficult task; no such methods are there, which provide better accuracy. The portion of the recognized plant ailments that influence the various pieces of the plant parts is due to fungal infection, viruses, bacteria, nematodes, and plant pests [2]. The use of agriculture in the field is cast off for cultivating the soil's erection, and it diminishes the contamination. The crop sicknesses are related to an increase in agriculture loss. Compared to earlier detection of diseases already available, some of the new techniques are implemented in image processing, which will be helpful for the farmers. Thus, a method for the classification and revealing of the illness of the plant using various algorithms is discussed in this paper.

2 Related Works

A lot many algorithms and modern techniques are proposed for the determination of the leaf diseases and its classification. According to Sushil R. Kamlapurkar, crop disease identification is done by feature extraction algorithm after extracting features, and the crop leaf infections are classified using an artificial neural network (ANN). Computational complexity is considered to be the most important disadvantages mentioned in this article. They have applied SVM course of action calculation, calculation using decision tree like Chaid algorithm [3], and calculation using logistic regression might have discovered that the grouping of the SVM model provides improved exactness when contrasted with different predictions [4].

Dr. N. Sasirekha et al. identified a new approach utilizing information mining methods. Information mining system has different inconveniences. There are protections that give that happen in information mining procedures. The information provided by this is not much accurate as it utilizes a few calculations in the field of picture preparing and the effective neural system to distinguish crop illnesses and creepy crawly bugs. So as to more readily distinguish sicknesses and creepy crawly bugs, this paper utilizes an assortment of plant ailments and bug bugs for examination and research, at that point grows an assortment of yield illnesses and bug bugs on the first premise, extends the common sense field by altering model parameters and different techniques, which spares remaining task at hand, yet in addition, spares a great deal of time, and can be founded on the first [5]. The infected part is identified and is subjected to segmentation based on the threshold value, and the feature is removed utilizing the shading event strategy and dim level co-event grids, and the classification is done by using the linear and nonlinear filters as proposed. Malvika ranjan et al. proposed a thought for recognizing and characterizing the leaf infections by extracting the element by changing over RGB design in tone immersion. Here also, ANN is used for the classification [6]. Another method to separate the infected some portion of the leaf is distinguished utilizing k-implies bunching calculation, and the grouping of the malady is again done by utilizing a feature extraction algorithm along with SVM classifier as proposed by Pinto et al. [7]. They proposed a methodology where the element is removed utilizing the shading co-event technique, and assessment of diseases is done by using surface figures. Hence, it is utilized to adjust RGB picture into HSI picture and, in the past, fragmented the sick part using morphological activity. The examination of the preparation set is used for arranging the harvest illnesses. Al Bashish et al. anticipated a technique to provide fast and accurate in segmenting and classifying the plant diseases. Here, ANN is used along with the k-means algorithm which is pushed off to isolate the tainted piece of the leaf utilizing edge-based division strategy, and order of the illnesses is finished by utilizing self-organizing map (SOM) and neural system classifier [8]. Jaisakthi et al. initiated with an advanced approach that consists of a programmed framework in order to recognize the maladies in the grapevines utilizing picture preparing and AI system and acquired an exactness of 93%. This approach is used to extract the feature using a dim dimension-event grid, and classification of the crop diseases is done by using the nearest neighborhood classifier in classification algorithm. The segmentation of the diseased part of the image is done using the detection of the image, and the classification of the infections is done by utilizing the homogeneous pixel checking method for cotton infection recognition [9].

The examination of different AI procedures for distinguishing proof and grouping of the plant malady designs from the leaf pictures is finished by Akhtar et al. [10]. Here, a three-stage system is actualized for identifying the diseased regions. Another method is used for identifying the diseases which affect the pomegranate plant. Here, some acute conditions are taken into consideration like bacterial blight, fruit spot, fruit spoil, and leaf spot. Here, clustering algorithm and ANN are used along with the GLCM method. Considering the accuracy, the various methods for classification and identification of crop diseases are not sufficient. So, in this proposed method, the PSNR estimation of the picture is expanded to provide better quality to find the diseases.

O. Kulkarni proposed an artificial intelligence (AI)-based model, which is set up for using an open dataset containing pictures of sound and tainted gather leaves. This prototypical serves its objective by masterminding pictures of leaves into a debilitated class subject to the case of distortion [11]. The various applications that are coated in the detection and tracking of the disease identified regions using an artificial intelligence method is discussed in [12]. The combination of HOG and the SVM in addition with the convolutional neural networks is discussed in this study.

The changing of the section that leads to the deep learning approach that includes the convolutional neural networks is discussed in [13] for the effective identification of plant diseases. The improved module for inception is then subjected for generating the corresponding network. The classification and extraction of the feature for the high-dimensional images are used for the extension of the layers. The remaining portion of this paper describes the proposed system with its methodologies and the ways for discriminating the plant diseases identification process.

3 Proposed Work

Preprocessing the image is a significant task in the proposed system. The picture is taken and is subjected to the histogram equalization method. The diagram given below describes the various steps tangled in the classification of the leaf syndromes. Initially, the picture was caught by the computerized camera, and the obtained image is subjected to the preprocessor. The preprocessor uses two steps; using image enhancement method, certain features of the image is modified and is converted from the patterns of RGB to some other, and this will uniformly distribute the intensity or color and afterward segment the infested fragment by means of k-means bunching procedure. The morphological processing method is used to enhance the diseased part. The diagram shown below gives a clear sketch on the processing methods used in this new approach (Fig. 1).

After performing the morphological processing, the size of the infected part is increased and is extracted by using the modified HOG algorithm [14]. This will be used for extracting the feature from the expanded image. Subsequently, the crop disease is classified, and it is made for contrasting the preparing database into the test picture by utilizing the unsupervised support vector machine classifier. The identified diseased part is given to a processing method to expand the disease-ridden area. The tainted piece of leaf is oppressed to a combined histogram of the oriented gradient (HOG) algorithm and region of interest (ROI) algorithm to extract the affected leaf features. The unsupervised classifier using support vector machine is utilized to categorize the sicknesses and to identify the disease-affected area founded on the mined feature.

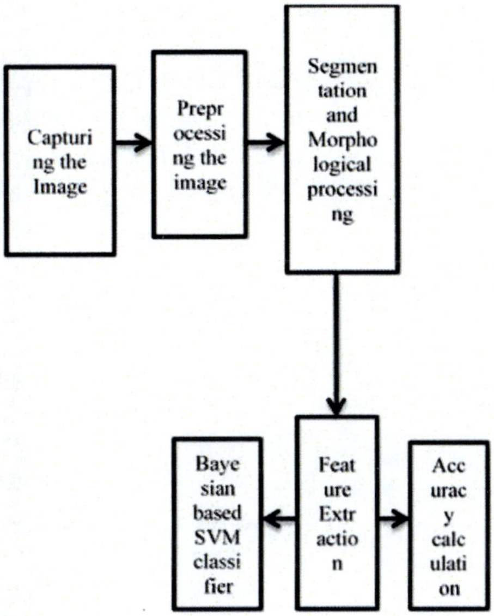

Fig. 1 Proposed system implementation

4 Methodologies Involved in Proposed System

It includes two strategies for the procedures of change of the image. The test is taken and is incorporated into two procedures: image capturing and image preprocessing. In the image capturing, the picture of the infected leaves was caught utilizing digital camera. The captured picture is in RGB layout. For reference, the image is taken for rubber leaf, and it is then preprocessed. The preprocessing of the image is done through image enhancement and image transformation. The captured image is enhanced using an enhancement method that uses contract stretching. It is utilized to feature the specific element present in that picture. Propelled picture upgrade method is utilized to improve the precision of activity by disseminating the power esteems. The obtained image is then converted to the CIELAB image (Figs. 2 and 3).

The converted CIELAB image is transformed from the RGB image by using an advanced method of image transformation. The glowing part of the image is obtained from the picture taken. A piece of the picture will offer data about the shading range of the image from green to red; other part gives data of the shading range of the image from blue on the way to yellow.

$$\text{Luminosity} = 1.2426 \times R + 1.8152 \times G + 1.1722 \times B \tag{1}$$

$$X = 1.5689 \times (0.4213 \times R - 0.5290 \times G + 0.2177 \times B) + (1 \times 128) \tag{2}$$

$$Y = 0.4245 \times (0.3949 \times R + 1.8131 \times G - 1.8106 \times B) + (1 \times 128) \tag{3}$$

Fig. 2 Input image

Fig. 3 Preprocessing image

The luminescence portion will not provide flawless statistics. So, the luminescence part is removed, and the content related to the *X* and *Y* parts is taken into a justification for additional processing. *X* and *Y* elements are reserved into interpretation for processing further. After transforming the image, it is then subjected to segmentation and morphological processing. The framework sections the leaf (region of interest) from the foundation picture utilizing get cut division technique. From the fragmented leaf part, the sick district is further sectioned dependent on two distinct strategies, for example worldwide thresholding and utilizing a semi-regulated system. The highlights are separated from the sectioned ailing part, and it has been delegated solid, spoil, esca, and leaf curse utilizing distinctive artificial intelligence strategies like modified unsupervised support vector machine (SVM) algorithm.

The infected part is then separated from the leaves and segmented using k-means clustering algorithm. Two points are fixed for clustering the bunching point and are gathered independently from the contaminated part. The steps tangled in the implementation of the k-means clustering algorithm are:

(i) To establish the value of the cluster according to the information taken.
(ii) Each pixel of the picture is allotted with the least separation between the clusters.
(iii) The recomputed cluster value is subjected to the device by taking a pixel value average.
(iv) Keep repeating the previous steps and check for the change in the value of the cluster, and if it is found to be idle without changing, then stop it.

The above steps are used to segment only the infected parts from the result obtained by k-means clustering, as shown in Figs. 4 and 5.

The tainted piece of the leaf picture is then extended by utilizing a widening strategy in the morphological handling technique. The enlargement process is utilized

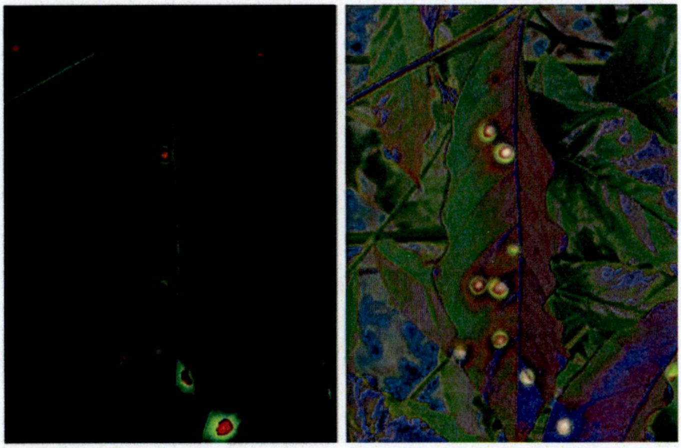

Fig. 4 Segmented image

Fig. 5 Separate the contaminated part utilizing k-means clustering

to analyze the closest pixel with the original pixel. The concentration value of the original pixel is replaced with the nearest pixel.

4.1 Feature Extraction and Final Processing

After the segmentation process, it is necessary for the feature which is extracted using histogram of the oriented gradient (HOG) algorithm and region of interest algorithm (ROI). The steps intricate in finding HOG and ROI are shown below.

(i) Divide the contaminated picture into discrete cells, and portion the equivalent into the picture.
(ii) Calculate the inclination level and extent of every cell.
(iii) Map the binning-based estimation of inclination worth and extent esteem.
(iv) Normalize the outcome acquired in Eq. 3 utilizing block standardization.
(v) Calculate the element like mean, kurtosis, differential entropy, distortion, and homogeneity (Figs. 6 and 7).

4.2 Classification of the Diseases

Support vector machine and region of interest are used for classifying and identifying the diseases of the rubber leaves and sunflower leaves, as both the leaves have the same infections as noted. The hyperplane technique is cast off in the SVM classifier to divide the clusters into two classes based on the obtained dataset. This technique

Fig. 6 Contrast enhanced

Fig. 7 Using ROI and HOG algorithms

is better in the classification and identification of leaf diseases in an accurate manner. The precision is distinguished to be more when contrasted with the past works.

4.3 Accuracy Calculation

Accuracy calculation is one of the most critical needs in disease identification as it involves a lot of methods that bring the threshold value to reach the limit. The prediction of the disease is made with maximum accuracy by the benchmarking process. For calculating the accuracy, the following formula can be used.

$$F(x) = \frac{B(x)}{T(x)} \times 100 \tag{4}$$

The prediction of each disease is given as the input, and while comparing with the existing system, the accuracy is improved here. The accuracy is spotted to be 99.037%, as it is quite more when compared to the previous methods.

5 Results and Discussion

The parameters like distortion, differentiation, entropy, consistency, and kurtosis are estimated in addition with the mean value by arranging the ailments. This is implemented by using MATLAB 2012a. For analysis purpose, rubber tree leaf and sunflower leaf are taken. The obtained result shows the accuracy as 99.037%, which is more than the previous one. The skewness value is extracted as 1.602; the homogeneity value is 1.222; the contrast is 0.6712; mean is 42.119; entropy is 3.432; and the Kurtosis value is 4.1112.

6 Conclusion

The utilization of picture handling in recognition and grouping of plant illnesses is acquired in an extremely precise way by the use of the HOG and ROI algorithms in combination with the SVM classifier. The proposed system can provide a better result in the form of accuracy when compared with existing ones. The identified accuracy is 98.34% in the previous works, and now, it is obtained as 99.03%. It can be enhanced by taking the video proofs from a surveillance camera.

References

1. Pujari, J.D., Yakkundimath, R., Byadgi, A.S.: Detection and classification of fungal disease with Radon transform and support vector machine affected on cereals. Int. J. Comput. Vis. Robot. **4**(4), 261–280 (2014)

2. Adimoolam, M., John, A., Balamurugan, N.M., Ananth Kumar, T.: Green ICT communication, networking and data processing. In: Green Computing in Smart Cities: Simulation and Techniques, pp. 95–124. Springer, Cham, 2020
3. Song, T.-M., Song, J.: Prediction of risk factors of cyberbullying-related words in Korea: application of data mining using social big data. Telematics Inform. 101524 (2020)
4. Kumar, A., Sarkar, S., Pradhan, C.: Recommendation system for crop identification and pest control technique in agriculture. In: 2019 International Conference on Communication and Signal Processing (ICCSP), pp. 0185–0189. IEEE, 2019
5. Hu, H., Su, C., Yu, P.: Research on pest and disease recognition algorithms based on convolutional neural network. In: 2019 International Conference on Virtual Reality and Intelligent Systems (ICVRIS), pp. 166–168. IEEE, 2019
6. Ranjan, M., Weginwar, M.R., Joshi, N., Ingole, A.B.: Detection and classification of leaf disease using artificial neural network. Int. J. Techn. Res. Appl. **3**(3), 331–333 (2015)
7. Pinto, L.S., Ray, A., Udhayeswar Reddy, M., Perumal, P., Aishwarya, P.: Crop disease classification using texture analysis. In: 2016 IEEE International Conference on Recent Trends in Electronics, Information & Communication Technology (RTEICT), pp. 825–828. IEEE, 2016
8. Naikwadi, S., Amoda, N.: Advances in image processing for detection of plant diseases. Int. J. Appl. Innov. Eng. Manage. (IJAIEM) **2**(11) (2013)
9. Revathi, P., Hemalatha, M.: Classification of cotton leaf spot diseases using image processing edge detection techniques. In: 2012 International Conference on Emerging Trends in Science, Engineering and Technology (INCOSET), pp. 169–173. IEEE, 2012
10. Akhtar, A., Khanum, A., Khan, S.A., Shaukat, A.: Automated plant disease analysis (APDA): performance comparison of machine learning techniques. In: 2013 11th International Conference on Frontiers of Information Technology, pp. 60–65. IEEE, 2013
11. Kulkarni, O.: Crop disease detection using deep learning. In: 2018 Fourth International Conference on Computing Communication Control and Automation (ICCUBEA), pp. 1–4. IEEE, 2018
12. Aslan, M.F., Durdu, A., Sabanci, K., Mutluer, M.A.: CNN and HOG based comparison study for complete occlusion handling in human tracking. Measurement, 107704 (2020)
13. Chen, J., Chen, J., Zhang, D., Sun, Y., Nanehkaran, Y.A.: Using deep transfer learning for image-based plant disease identification. Comput. Electron. Agriculture **173**, 105393 (2020)
14. Priyankha, J.J., Suresh Kumar, K.: Crop disease identification using a feature extraction HOG algorithm. Asian J. Appl. Sci. Technol. **1**, 35–39

Computation

Advanced FET-Based Biosensors—A Detailed Review

M. Suryaganesh, T. S. Arun Samuel, T. Ananth Kumar, and M. Navaneetha Velammal

Abstract This review paper has done a detailed performance analysis for advanced field-effect transistor (FET) based biosensors which shall be distinguished by their outstanding features, such as mass-production capability, sensitivity, ultra-sensitivity detection, and low-cost manufacturing, within a range of advanced biosensing technologies. In order to encourage the understanding of FET-based biosensing technology and its sensing analyte, major FET-based biosensing devices are presented in this review: Dielectric modulated FET, impact ionization MOSFET, double gate -dielectric modulated tunnel FET, junction less electrolyte insulator-semiconductor FET and nanowire FET, etc. This work is also designed to provide a state-of-the-art analysis of biosensors, based on an advanced field-gate field-effect transistor in the area of bioanalytical applications. Besides, a connection will be made between the various FET structures, with particular attention paid to materials and technologies.

Keywords Field-effect transistor · MOSFET · In-vitro diagnosis (IVD) · Biosensors

1 Introduction

Micro/Nanotechnologies are emerged with ultra-sensitive biosensors. Due to early detection potential, the advent of specificity medicine, genetic testing, and gene sequencing attention is drawn. Academic biosensing explores practical tools for in-vitro diagnosis (IVD) as distinct as you are, sensing methods are developed, and these advances translate. Simultaneously, analysis strips and instruments are being developed in a ground-breaking transition from electrochemical / optical to nano-electronic technologies. It is also worth looking at electronic-based biosensing instruments, one

M. Suryaganesh (✉) · T. S. Arun Samuel
Department of ECE, National Engineering College, Kovilpatti, Tamil Nadu, India

T. Ananth Kumar
Department of CSE, IFET College of Engineering, Villupuram, Tamil Nadu, India

M. Navaneetha Velammal
Department of ECE, Francis Xavier Engineering College, Tirunelveli, Tamil Nadu, India

H. K. D. Sarma et al. (eds.), *Contemporary Issues in Communication, Cloud and Big Data Analytics*, Lecture Notes in Networks and Systems 281,
https://doi.org/10.1007/978-981-16-4244-9_22

of the leading medical diagnostic sensing technologies. FET-based biosensors have been proposed and have become an emerging field due to solid-state technologies' rapid growth. Biomolecules are used to carry electrostatic charges where the bioactivities are needed for electrostatic potential adjustments. A FET-based biosensor would be a healthy choice for ultra-sensitivity and rapid reaction conditions.

2 Biosensors Using FETs-Related Works

The study of embedded field-effect transistor underlap channel submitted by Jee-Yeon Kim for biosensor application [1]. For both aqueous and dry electrical label-free biosensors, a low-powered FET channel is suggested, and the voltage/current properties are compared and evaluated in each collection. Avian influenza (AI) antigen–antibody binding is used to research the underlap framework's efficacy as a biosensor for both settings. The underlap-FET proved to be label-free; biomolecules were electrically defined, and the two conditions contrasted in both aqueous and dry conditions were quantitatively compared. When the anti-AI is spring to the underlap channel region enclosed by SBP-AI, the decrease in drain current is calculated using the negative charges of anti-AI molecules. This effect has been seen for both conditions, and the drain current change was greater in dry conditions than in aquatic environments.

"Study the performance of Dielectric modulated FET" was presented by Kannan, member IEEE [2]. In this paper, using computer-aided simulation technology as a label-free biosensor, analyses the outcome of a dielectric modulated I-MOS transistor (DIMOS). By inserting a nanogap into the I-MOS device by mixing chromium and gold gate structure, DIMOS provides the extremely sensitive with potential biomedical and biomolecule sensing applications. Impact ionization dielectric modulated based biosensor FET has been used to study the nanogap immobilization biomolecular detection and instrument's sensitivity. In the simulation, the nanogap's dielectric constant is modeled for immobilization, and even the dialectical constant of biomolecules would be separate from the air. A value of $K = 2$ and 12 is used to define a category of biomolecules with low- and high-dielectric constants. The notion of a DIMOSS has been suggested and studied. With a biosensor. The recommended structure shows high sensitivity. Dielectric modulation is the dominant effect of biomolecule detection. It retains dielectric modulation compared to conventional dielectric modulated FETs, where the effect on interface sensitivity is essential for biomolecule charges.

Ultra-sensitive sensing performs by using vertical strained impact ionization MOSFET presented by Ismael Saad [3]. Interestingly, the vertical strained impact ionization MOS suffers from an impressive high-supply voltage (VDS) and hysteresis. Hence, the idea of strained SiGe vertical I-MOS is implemented to reduce supply voltage. As the strained SiGe layer is inserted into the framework, both the supply and threshold voltage decreased significantly. After all, the system continues to experience low-breakdown voltages caused by parasitic bipolar transistors (PBT) that affect the device's reliability. Developing novel system designs to replace traditional I-MOS is a potential solution to address these limits. For potential applications biosensing, the performance of three attractive candidates was investigated and defined: vertical strained impact ionisation MOS dual channel (DC-VESIMOS), vertical straight-in ionisation MOS single-channel (SC-VESIMOS), and vertical stress ionisation MOS Dielectric Pocket (VESIMOS-DP) integration. Comparison of transfer characteristics (IDS-VGS) with SCVESIMOS for a candidate with the best possible biosensor for characterizing the respective unit dual channel vertical stressed ionisation MOS's performance and vertical stressed impact ionisation MOS-Channel length dual pocket devices Lg = 50 nm. The voltage sub-risk of VESIMOSDPP increases Ioff = 10^{-16} A/μm leaking currents and ON current of Ion = 10^{-4}A/μm single-channel-VESIMOS. Therefore, a low-leakage current gives low subthreshold advantages. The system offers high stability and low-energy consumption. The DC-VESIM/OS method is the best candidate for the potential lowest biosensor sensitivity software.

Saiyan Kanungo [4] presents the efficiency comparison of the full and short gate TFET with different dielectric values. In this paper, they studied the efficiency and fundamental operational physics of both FG-DMTFET and SG-DMTFET dependent biosensors. The schemes of DG Dielectric Modulated Tunnel FET biosensors are shown in Fig. 1a, b.

Silicon–Germanium acts as the basic material for the higher tunneling current value in both device structures with a germanium composition of 0.5. The gate length of the full gate-dielectric modulated tunnel FETs is known to be 42 nm, while the design of the short gate-dielectric modulated tunnel FET is 20 nm. Each of the drain and source regions is 20 nm long, and the channel thickness is considered to be 10 nm. The source (*p*+), channel (*P*), and drain (*N*+) regions have uniform doping concentrations are range between 10^{19} cm^{-3} and 10^{12} cm^{-3}. The bias dependency of tunneling junction electrostatics from such biosensors should be undertaken with a comparative analysis better to understand dielectric modulated tunnel FET biosensor's relative performance. The implementation of the shorter gate tunnel FET with dielectric modulated and full gate-dielectric modulated tunnel FET biosensing components has been extensively studied. Compared to the traditional FG architecture, the integration of the short gate architecture into the dielectric modulated tunnel FET framework provides a substantial increase in sensitivity without substantial manufacturing. The short gate biosensor can work within a precise range of biases. The short gate-dielectric modulated tunnel FET architecture will generate drain current in the simulation context similar to the streptavidin–biotin binding system (i.e. $k = 2.1$ and son = 0) and its found that approximately seven times improved sensitivity of short gate TEFT compared to Full gate DM Tunnel FET.

"Study of SiGe and pocket-doped sensing activity effects" was presented by Partha Sarathi Guptha [5]. Compared to its dielectrically modulated FET counterpart, biosensors based on dielectrically modulated tunnel FET (DMTFET) exhibit greater sensitivity, but it has low subthreshold characteristics. The impact of the use of the source silicon–germanium (SiGe) and the channel *n*+-pocket-doped is

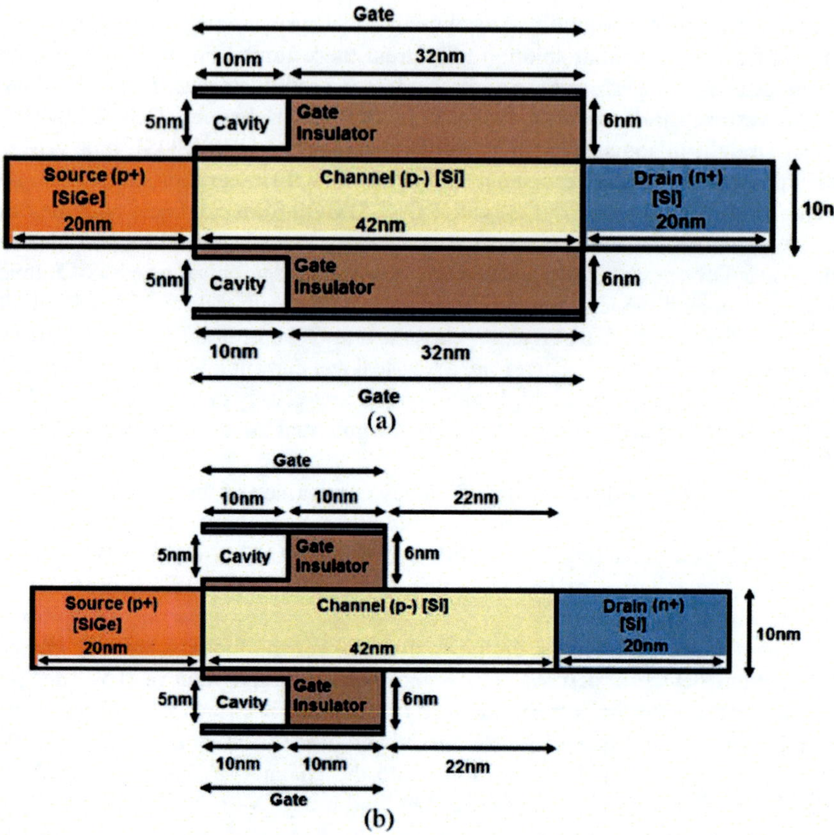

Fig. 1 **a** DGDM TFET biosensors with full Gate and **b** short gate DGDM tunnel FET biosensors

discussed in this context. For biomolecule conjugation, variation in doping concentration in the *n*+-pocket region, the source region, and germanium composition's fundamental mechanics are discussed. The decrease in source side bandgap and its effect on band-to-band tunneling has been studied in a reduced tunneling distance. The sensing efficiency of dielectric modulated tunnel FETs was measured subsequently. Research shows that SiGe source DMTFET is much higher than *n*+ pockets to achieve higher current subthreshold levels, while retaining the necessary sensitivity. This sensitivity-current optimization was investigated, and the required bias function was indicated for different gate voltages and drain variations. Each source and drain area should be 50 nm. The insulator channel and gate thicknesses are 20 and 10 nm, respectively. Suppose to optimize the electrical reaction of the conjugation, SiO_2 and aluminum (work function = 4.1 eV) were gated. Doping concentrations are consistent, respectively, of $1 - 1020$, $1 - 1016$, and $5 - 1018\ \text{cm} - 3$ in the $p+$ source, p-channel and *n*+-drain regions. The incorporation of SiGe source into the dielectric modulated FET device provides a major advantage over the pocket-doped

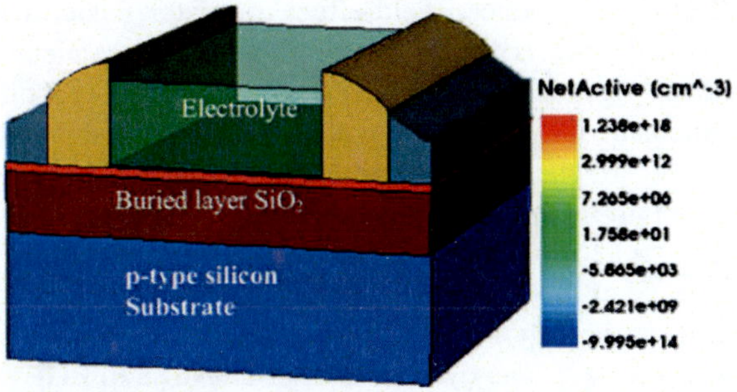

Fig. 2 Simulated *n*-type junction less electrolyte insulator-semiconductor FET

architecture for sensitivity optimization. The use of SiGe source offers more than one order of improvement for a wide variety of biomolecule samples without substantial trade-off sensitivity, while the $n+$ pocket dielectric module tunnel FET has seen a consequence of more than 5% sensitivity degradation to achieve a proportional current improvement.

"Junction less electrolyte insulator-semiconductor FET details" was presented by Ajay [6]. The TCAD simulation of n-type junction less electrolyte insulator-semiconductor FET is shown in Fig. 2. In typical MOSFET, the gate material is directly connected to the gate oxide (SiO_2 insulator). In ion-sensitive FET, the measuring electrode, also called the reference electrode (Vref), is placed around electrolytes into the dielectric layer (oxide layer). The ions present in the electrolyte and its charged molecules influence the gate terminal's electrostatic potential, so they can adjust the threshold voltage and drain current. An essential element of the ion-sensitive FET is the electrode. It provides the test electrolyte with a stable electrical contact and allocates the sensing liquid's electrical potential or the sensor's operating point depending on the FET. They may also exchange information with electrodes and electrolytes, and the drainage current of the system depends on the electrolyte arrangement. Thus, by changing the drain current is a change in the gate's surface potentials at the gate isolator and electrolyte interface.

The conductance of the junction less silicon-on-insulator electrolyte semiconductor FET channel depends not only on the device operating, but also , on electrolyte status and solution's ion concentration. Different pH values have been demonstrated to affect the conductance of the junction less FET. As the electrolyte's hydrogen ion concentration increases, the pH value of the electrolyte is also modified. This also affects the junction transfers less silicone electrolyte semiconductor FET. The Silicon-on—insulator junction less (JL) electrolyte semiconductor field effect transistor (EISFET), a well-known 3-D commercial TCAD semiconductor, was investigated as a simulation method. The area of the electrolyte consists of altering the properties of the pure semiconductor material as the electrolyte is the same as the pure semi conductive material. A solution determined by the necessary carrier load-densities of the intrinsic semiconductor material with efficient phosphate-buffered Salin (PBS) and ionic concentrations was implemented. Michael Kroon [7] presents fast detections of the Ebola antigen with FET. For real-time Ebola virus antigen detection, a reduced field-effect transistor device based on graphene oxide. This method utilizes the graphene-based smart semiconductor features and instantaneously produces highly sensitive and reliable Ebola glycoprotein detection. The schematic diagram of the graphene oxide-based FET biosensor is shown in Fig. 3.

The graphene oxide layer was dumped on the instrument to link the source and the drain regions electrode. A thin layer Al_2O_3 is coated for surface passivation on the graphene oxide layer. The transistor measurement was performed to describe the essence of the semiconductor and the start-up current ratio. The electrodes of drain and source region were given a constant 0.01 V of Vds with Vgs between −40.0 and +40.0 V. The dynamic sensor response to EGP was investigated with constant Vds of 0.01 V across drainage and supply electrodes in order to examine the sensor output and Vg set at zero. Higher Vds leads to a louder reaction and even damage to the rGO sheet. EGP was taken in the form of an industrial supplier's purified protein (IBT Bio Services) and interrupted into 0.01PBS /human/human plasma. As PBS/serum/plasma dilution ensures that the surface load transmitted by EGP is not screened the Debye duration increases. The respective length of Debye is around 7.4 nm 35 for PBS, while the gold NPs are less than 5 nm and the provider reports that Ebola and EGP antibodies are about 3 and 10 nm.

Yu-cheng-syu [8] is addressed with a field-effect transistor analysis with a clinical use. A biosensor is a device used by the International Union of Pure and Applied Chemistry (IUPAC) to detect chemical compounds by electrical, optical signals or thermal using biochemical reactions intercede by immune systems, isolated enzymes, tissues, and entire cells. In other words, to track the dynamics and interactions of the immune system, isolated enzymes, organelles, tissues, and whole cells, a biosensor is an analytical tool. DNA hybridization transfers the results of tracking into electrical signals. The FET-based biosensors were targeted to be the right candidate for POCT testing of the next generation. FET-based biosensors are planned to be an acceptable candidate for POCT (Point-of-Care) research of future generations and an IVD subdivision. It is characterized as a medical diagnosis at the point where patient care is required. For example, a glucometer is used as a central laboratory in the home of the patient instead of in health centers. It should be simple enough for less qualified people to work for POCT or not even trained. In addition, it greatly reduces the time taken for test results, providing doctors with immediate diagnostic information. Diamond field-effect transistor (FET) a solid surface sensing feature

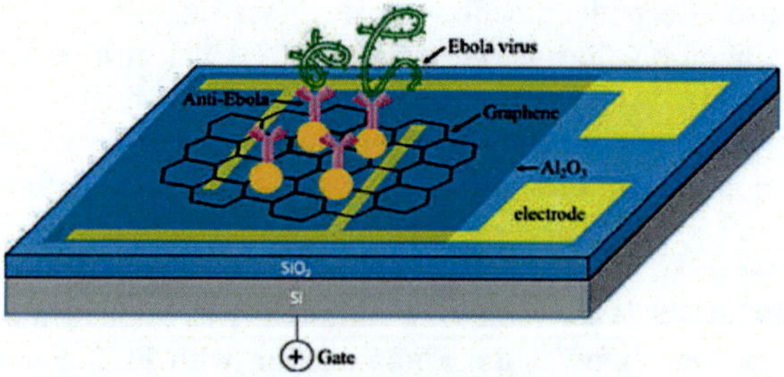

Fig. 3 The reduced graphene oxide-based FET biosensor schematic diagram

using ribonucleic acid aptamers (RNA). The results show a possible change at the 91.6 mV gate, whereby 8 μA shifts occur in the negative path when HIV-1 Tat proteins are present in the source-drain connected with RNAA protein. In addition, it was first demonstrated that aptamer-FET has been able to rely on the actual HIV-1 sample of Tat proteins, which demonstrated a good approach to the development, through diamond bio-interfaces, of relevant clinical biosensor applications.

"New dual pocket vertical heterostructure TFET performance" was presented by Amit Bhattacharyya [9]. The evolution of FET-based biosensors could be successfully realized by high-throughput biomolecule recognition technology. Medical diagnosis is helpful, but implementations often have certain drawbacks. Due to the effective detection, area saturation, size, and molecular concentration issues would restrict the response limit among various types of biomarkers. The cavity with a low-binding probability, surface functionality was included in the proposed architecture. So on, the collection of bio analytes inside the cavity is not smooth and complex. A more practical strategy would be this one. Also, the filled cavity is not allocated to the steric hindrance problem. Thus, different non-uniform phase patterns are considered inside the cavity.

The following International Technologies, the incorporation of regular/irregular biomolecules arrangements was proposed as a semiconductor roadmap. The proposed dual-gate TFET structure would be used to analyze the level of the well-set-up unit easily. Channel length (L_{ch}) = 42 nm the source and drain regions are fixed as 20 nm, and the cavity length is fixed as 15 nm are the simulation parameters. This paper shows a new FET-based biosensor with a double-pocket-dielectric modulated heterostructure tunnel involving lateral and vertical tunneling events. In comparison to other recognized TFET structures, the efficiency of double pocket-hetero-TFET (HTFET) was assessed.

Sandeep Kumar [10] is provided for the dual-gate operations conducted using an extended source double gate (ESDG)-FET, and the schematics are shown in Fig. 4. Also, the authors analyses the comparative studies for single and dual-metal gate DMTFET. Besides, a double gate-dielectric modulated tunnel FET-based biosensor

Fig. 4 Schematic view of ESDG-DMTFET

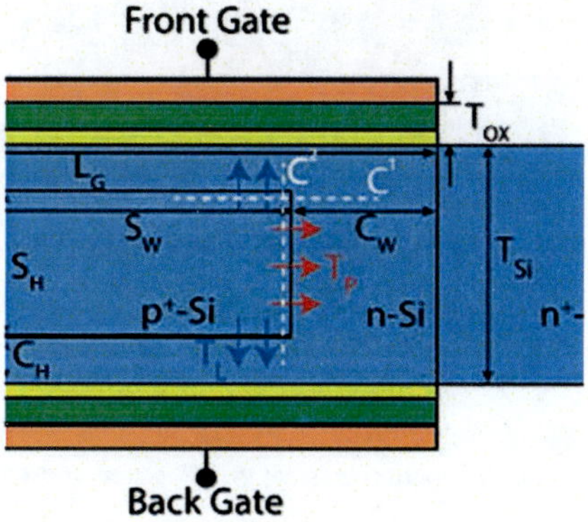

was recorded in the source region over which a single-sided *n*+-pocket DG-DMTFET-based biosensor was applied. In order to achieve improved sensitivity, the cavity was extended. A staggered DMTFET heterojunctions biosensor was suggested to give the threshold voltage sensitivity (Vth) of 450 mV. The source is enlarged further into n-channel so that both sides overlap the source. The role of gate metal work (φm) and length of the gate (L_G) 3.8 eV and 90 nm, respectively, are recognized. Field plate is used to increase tunneling and reduce channel resistance at source-channel junction. The (L_{ch}) value is chosen so that the extended source double gate -DMTFET configuration enables line and point tunneling at the source-channel junction. It boosts current sensitivity, and biosensor threshold voltage, drain current.

This presents the ESDG-DMTFET biosensor in the channel region with an extended source. Biomolecule immobilization cavities are formed in drain side gate-oxide. Changes in *k* of biomolecules allow for the full change of the Ion and Vth system. From the source-channel interface with an Increase in *k*, the electrical field increases, this reduces the tunneling width and increases the Ion and I_{ON}/I_{OFF} sensitivity of the above device structure. However, from the other performance, Vth reduce and I_{ON} rises due to the increase in *k*.

Chattopadhyay [11] presents a wide range of oxide stack detection for junction less FET (Fig. 5). The performance of a dielectrically modulated low and high-k oxide stack junction dual-metal double-gate was investigated with low HK-S JL-MOSFET and DM-DG-LK biosensor system for the effective recognition of various protein molecules in a dry condition, in terms of absolute and relative change, called Vth-responsiveness and threshold voltage (Vth) Vth-sensitivity.

For JL-MOSFET-based biosensing applications, the sensor with a long channel length is desired in the present situation, as it is preferable to connect sufficient numbers of biomolecules to the sensing device. The nanoscale regimen's surface is relatively thin. Unless otherwise reported, all our research considered a JL-MOSFET 1 μm channel length. However, shorter channel lengths are necessary for better system efficiency, especially for digital applications. Two different channel length

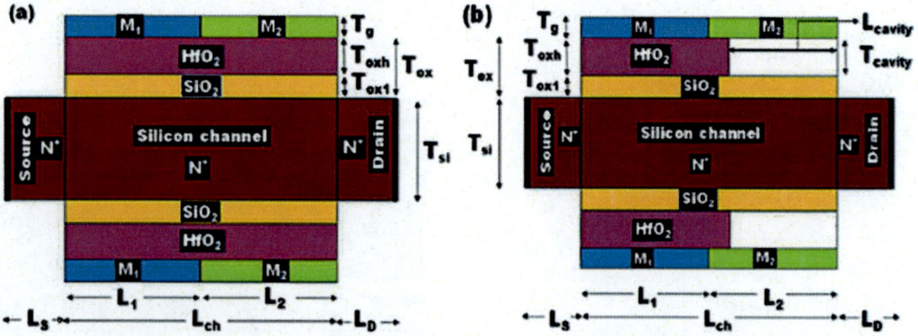

Fig. 5 **a** 2-D DM-DG-LK/HK-Stack JL-MOSFET cross-sectional view, **b** DMDG-Low-K/High-K-Stack Junction less-MOSFET schematic with Nanogap cavity

values, i.e., 1 μm and 50 nm, were performed in the dielectric modulated-double gate-low-K/high-K-stack junction less-MOSFET as a sensor system for the detection of protein molecules.

Louisa sellami introduces fluid-biosensor efficiency [12]. This paper consists of a gap-sensor with a bridge contains three regular PMOS devices attached as resistors and condensers, three bridge brackets. The fourth arm is a same PMOS transistor, and it distinguishes between a channel and a high polycap where a fluid like creatinine or gas is located. The circuit method uses spice simulations and spice extractions. In this, a sensor-sensitive is given to a dielectric fluid constant over a constant dielectric range of 11–1. This is done with a normal VLSI analog processing, which partially removes the thin dielectric gate of one of the four bridge bin transistors for which additional oxide grafting is applied. The Vset value of the bridge, representing the relatively dielectric constant of the fluid in silicon dioxide, can be compared by adjusting the Vset diagonal transistor gate.

Pranav Amborkar [13] describes the growth of the nano wire technology and in Fig. 6 the scheme is shown. High-sensitivity biomedical sensors may make it possible to identify diseases in their early stage, dramatically raising the likelihood of diagnosis and action that could potentially save lives. For example, five years' survival rate is higher than 90% when breast cancer is diagnosed and treated with current therapies at an early stage (local condition), but drops to about 20% when the late stage (distant disease) happens. Early detection of cancer also requires highly sensitive biosensors. Although much research has been carried out to boost the sensitivity of biomedical sensors, recent development in the field of nanotechnology will deliver the most promising biomedical sensor solutions. Nanotechnology encompasses large fields

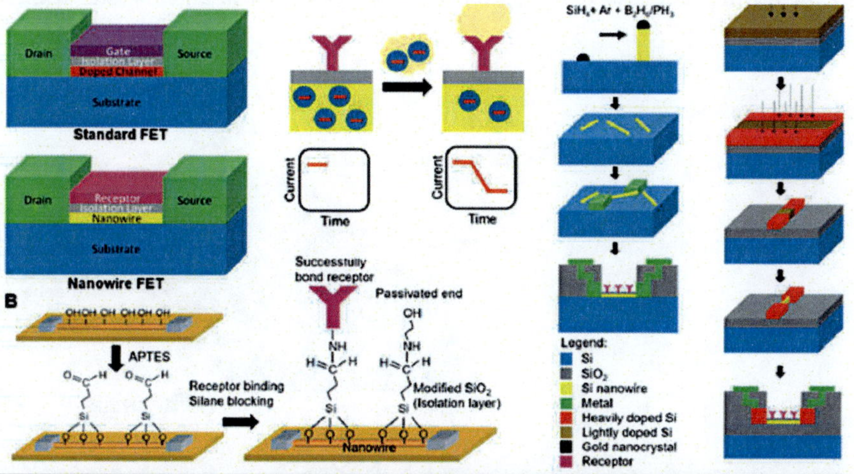

[redacted]

Fig. 6 Nanowire technology for bio medical sensors

of study in science and engineering in order to investigate materials and structures below about 100 nm.

Nanowire sensors have shown great dedication to being available sensors—the biological and medical recognition network. The devices offered to offer many advantages, such as high-sensitivity electrical signal transduction in real-time and label-free detection feasibility. While it is remarkably sensitive to its analytical signal capabilities, it is still too low for other methods, often used in vivo settings, to be polluted by high-background noise. This greater sensitivity and more fundamental problem of development can be solved by improved receptor binding methods. Moreover, the higher yield of recent up-to-date processing techniques allows lower prices of marketed products. However, nanowire sensor efficiency is based on their advances in simplicity, sensitivity, precision, and durability compared to existing standards in the goldfield, such as enzyme-linked immunoassay (ELISA) and polymerase chain reaction (PCR).

"The Chemically functionalized graphene FET" was presented by Clare watts [14]. Graphene FET (GFET) was covalently functionalized with 1-pyrene butyric acid and paired with anti-CD63 antibodies for N-hydroxy succinimide ester exosome-free detection. The microfluid tube solution revealed a portion of the graphene film. In addition to the original Dirac point in the background voltage (Vg) curve, the electrical properties of the diagram were further reduced. In the pipe, a minimum of one Vg was less than the initial Dirac stage when the phosphate saline (PBS) was present and, over time, changed by injecting exosomes into the tube. This minimum change from PBS was saturated after 30 min, and many exosome levels were observed. The highest concentration of exosome in combination with the anti-CD63 antibody means that functionally functioned GFET can directly detect exosomes low to a minimum of 0.1 μg/mL and susceptible to concentration, was low in combination with the isotype regulation (Fig. 7).

Fig. 7 Microfluidic integration biosensor

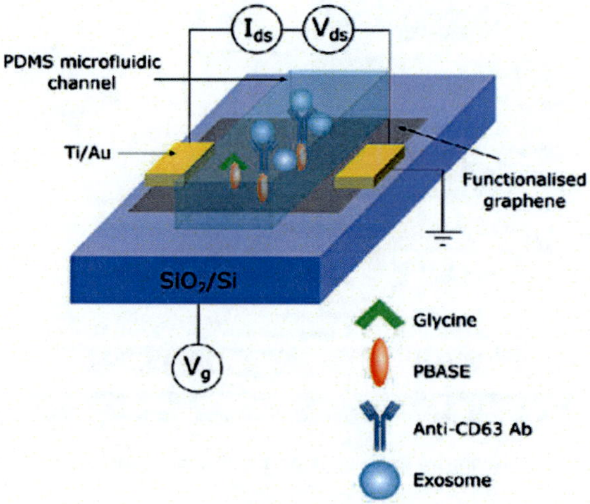

Table 1 Performance analysis of different FET based

Reference number	Different types of FET	Channel length (L_{ch}) (nm)	source -Drain voltage (V_{ds}) (V)	Source-gate voltage (V_{gs}) (V)
[2]	DIMOS	200	–	–
[3]	VESIMOS	50	1.75	1.75
[4]	DM-TFET(FG&SG)	–	0.4	1.2
[5]	DMTFET	100	0.5	6
[6]	JL-EISFET	5	0.5	1.3
[9]	HTFET	42	0.8	1.3
[10]	ES-DG-TFET	15–40	1	1.5

3 Major Findings

It has been inferred from the presented literature survey that, HTFET, ES-DG-FET privilege optimum performance for nano scale biosensor. The relationship between VDS, VGS has been plotted in Table 1.

4 Conclusion and Future Scope

The broad advancement of nanotechnology has rendered many possibilities feasible in the area of nano biosensors. In this paper, several FET structures with respective design specifications and performance metrics have been discussed. The biosensor should discuss settling time, sensitivity, and selectivity to ensure optimum operation. However, in terms of sensors' development, many aspects should be carefully examined, including biomolecular binding energy or the screening issue. There should be a compromise between the sensitivity of the sensor and stability of receptor-target molecules. In future, nano wire, extended gate FET, nanodot sensor, flexure FET sensor are may be used to design the biosensor for reduce the channel length and low power application.

References

1. Kim, J.-Y., Ahn, J.-H., Choi, S.-J., Im, M., Kim, S., Duarte, J.P., Kim, C.-H., Park, T.J., Lee, S.Y., Choi, Y.-K.: An underlap channel-embedded field-effect transistor for biosensor application in watery and dry environment. IEEE Trans. Nanotechnol. **11**(2), 390–394 (2012). https://doi.org/10.1109/TNANO.2011.2175006
2. Kannan, N., Kumar, M.J.: Dielectric-modulated impact-ionization MOS transistor as a label-free biosensor. IEEE Electron Device Lett. **34**(12), 1575–1577 (2013). https://doi.org/10.1109/LED.2013.2283858

3. Saad, I., Syazana, A.H.B., Zuhir, M.H., Seng, B.C., Bolong, N.: Equivalent circuit model analysis of vertical impact ionization MOSFET (IMOS). In: 2015 3rd International Conference on Artificial Intelligence, Modelling and Simulation (AIMS), Kota Kinabalu, pp. 451–456 (2015). https://doi.org/10.1109/AIMS.2015.77
4. Kanungo, S., Chattopadhyay, S., Gupta, P.S., Rahaman, H.: Comparative performance analysis of the dielectrically modulated full-gate and short-gate tunnel FET-based biosensors. IEEE Trans. Electron Devices **62**(3), 994–1001 (2015). https://doi.org/10.1109/TED.2015.2390774
5. Kanungo, S., Chattopadhyay, S., Gupta, P.S., Sinha, K., Rahaman, H.: Study and analysis of the effects of SiGe source and pocket-doped channel on sensing performance of dielectrically modulated tunnel FET-based biosensors. IEEE Trans. Electron Devices **63**(6), 2589–2596 (2016). https://doi.org/10.1109/TED.2016.2556081
6. Ajay, Narang, R., Saxena, M., et al.: Novel junctionless electrolyte-insulator-semiconductor field-effect transistor (JL EISFET) and its application as pH/biosensor. Microsyst. Technol. **23**, 3149–3159 (2017). https://doi.org/10.1007/s00542-016-3013-1
7. Chen, Y., Ren, R., Pu, H., et al.: Field-effect transistor biosensor for rapid detection of Ebola Antigen. Sci Rep **7**, 10974 (2017). https://doi.org/10.1038/s41598-017-11387-7
8. Syu, Y.-C., Hsu, W.-E., Lin, C.-T.: Review—field-effect transistor biosensing: devices and clinical applications. ECS J Solid State Sci Technol **7**(7), Q3196–Q3207 (2018)
9. Bhattacharyya, A., Chanda, M., De, D.: Performance assessment of new dual-pocket vertical heterostructure tunnel FET-based biosensor considering steric hindrance issue. IEEE Trans. Electron Devices **66**(9), 3988–3993 (2019). https://doi.org/10.1109/TED.2019.2928850
10. Kumar, S., Singh, Y., Singh, B.: Extended source double-gate tunnel FET based biosensor with dual sensing capabilities. SILICON (2020). https://doi.org/10.1007/s12633-020-00565-4
11. Chattopadhyay, A., Tewari, S., Gupta, P.S.: Dual-metal double-gate with low-k/high-k oxide stack junctionless MOSFET for a wide range of protein detection: a fully electrostatic based numerical approach. SILICON (2020). https://doi.org/10.1007/s12633-020-00430-4
12. Sellami'r, L., Newcomb, R.W.: Electrical Engineering Department, U.S. Naval Academy 105 Maryland Ave, Annapolis, MD 21402, USAE-mail: sellami@eng.umd.edu, http://web.usna.navy.mil/'sellami
13. Ambhorkar, P., Wang, Z., Ko, H., Lee, S., Koo, K.-I., Kim, K., Cho, D.-I.D.: Nanowire-based biosensors: from growth to applications. Micromachines **9**, 679 (2018)
14. Kwong Hong Tsang, D., Lieberthal, T.J., Watts, C., et al.: Chemically functionalised graphene FET biosensor for the label-free sensing of exosomes. Sci. Rep. **9**, 13946 (2019). https://doi.org/10.1038/s41598-019-50412-9

A Survey on Trends of Two-Factor Authentication

Dereje Tirfe and Vivek Kumar Anand

Abstract The rapid increase of cybercriminals on identity theft and data breaches due to weak authentication schemes, user's poor password management experience, and attacks such as phishing and man-in-the-middle attacks firmly shifts the traditional way of authentication method (i.e., user name/passwords) to multifactor authentication security control. However, different organization starts implementing different ways of two-factor authentication (2FA) prevention mechanism which is the simplest form of multifactor authentication (MFA). This 2FA is mostly achieved by combining different factors such as what you know (i.e., PIN, password, etc.), what you have (i.e., one-time password (OTP), token, digital certificate, etc.) and what you are (i.e., fingerprint, iris, etc.) with traditional username and password. Even if an organization such as banking and e-commerce sectors started using this solution still multifactor authentication, not in impenetrable stage and multiple cases, has come out that highlights some of the weaknesses of these security measures. Bad guys will use a variety of attack methods such as social engineering (phishing), man-in-the-middle attacks (MITA), and breaching of weak credential to gain authentication access of a secure network infrastructure. The most effective methods for mitigating the risk of authentication process from being compromised by bad guys are to find out a solution that reduces the process of the user and builds the authentication process with strong cryptography which follows standards mainly by using digital certificates. In this survey study, we will briefly analyze different types of two-factor authentication based on their performance and we will suggest the best strong methods of multifactor authentication accordingly.

Keywords 2FA · MFA · Digital certificate · Authentication · OTP · MITM · Mutual authentication

D. Tirfe (✉) · V. K. Anand
Marwadi University, Rajkot, India
e-mail: derejetirfe.astatike108745@marwadiuniversity.ac.in

V. K. Anand
e-mail: vivekkumar.anand@marwadieducation.edu.in

H. K. D. Sarma et al. (eds.), *Contemporary Issues in Communication, Cloud and Big Data Analytics*, Lecture Notes in Networks and Systems 281,
https://doi.org/10.1007/978-981-16-4244-9_23

1 Introduction

Networked information dissemination is one of the extensively progressing area in the current era. The concept of networking is exploring in many directions including Internet of Things [1], cognitive radio networks [2, 3], dynamic networking [4], and information centric networks [5]. Irrespective of the networking strategies adopted, the authenticity of the sender and receiver in the process of communication is one of the vital concerns. To get started with authentication, it is one of the security control methods for identifying user or to ensure clients are who they state they are. The most popular, oldest, widely recognized, and easiest forms of authentication are based on username and password phrase. User name and password have been used for long time. Due to the rise of variety of cyber-crimes, just using only username and password were not enough to secure sensitive information. It has proved that hackers are able to break username and password authentication controls. Authentication schemes are one of the core target areas for cybercriminals. As the result of dramatic increase of daily data breaches, new customer trust toward online organization became eroded. Such companies are taking measures to restore their name to the market. They should focus on strengthening their security controls for the sake of online infrastructures by exploring secured, cost-effective, easy deployable Web application security, and other ways of security authentication controls. Based on Verizon Data Breach Investigations Report (VDBIR) [6], there were around 157,525 security incidents and 3950 confirmed data breaches in 2020, with 80% of those incidents are in connection with weak credential. Nowadays, implementing two-factor authentication (2FA) which is the simplest form of multifactor authentication (MFA) has become the best solution to secure authentication.

Currently, most companies and organizations have shifted their attention on securing their authentication wall for the sake of customer trust because two-factor authentication gives a solution for data breaches. We can say that it is a solution for worst scenario because even if username and password have been hacked or obtained by different attacks techniques, 2FA defends and stops them from accessing the system. Two-factor authentication has now became a standard framework which companies make interest of adopting. In general, we can have an advantage of two-factor authentication because it increases customer trust toward an organization or business company, it also reduces the risks of data and information leakage, and finally, it minimizes operational and security cost because two-factor authentication is less expensive method of security control compared to other types.

As a whole, two-factor authentication is achieved by combining known authentication factors; the first is known as knowledge factor; this includes password, PIN, and code of words. Anything that you can remember and then type, say, do, perform, or otherwise recall when needed falls into this category. The second one is known as possession factor; this includes physical and software items that belong to mobile phones, SIM card, smart cards, USB drives, and token devices. The third and the final one is known as inheritance factor which is given to someone by nature or is

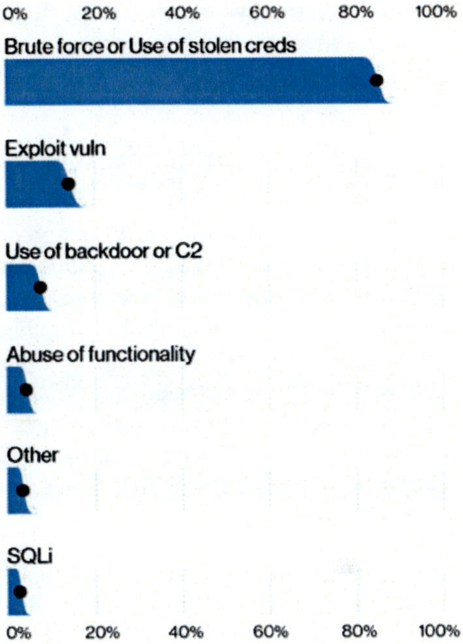

Fig. 1 Top hacking varieties in breaches

born with it. This factor includes biometric data of human such as fingerprint, iris retina voice, and so on (Figs. 1 and 2).

In addition to those three universally accepted factors, there are also some researchers which show different factors such as somebody you know (such as human

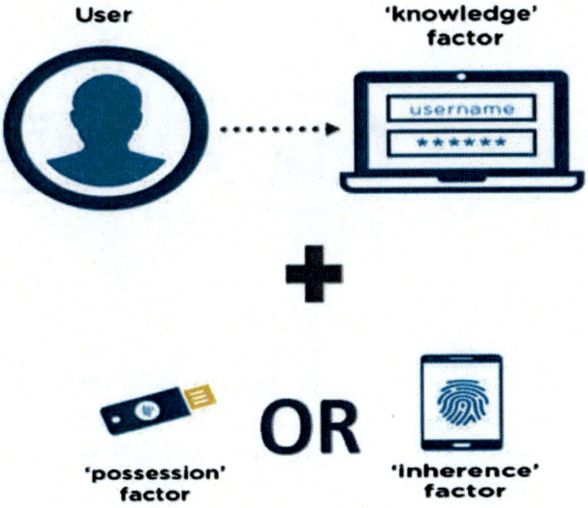

Fig. 2 2FA methods

relations) [7], somewhere you are (such as location) [8], something you do (such as gesture), and something you process (such computing mathematical operation) [9].

2 Two-Factor Authentication Methods

Two-factor authentication (2FA) is a guarantee against various cyber-attacks that target mainly to breach authentication walls for further usage of passwords and username. Two-factor authentication has the ability to prevent infrastructure from phishing, brute force attacks, key logging, and credential stolen attacks. It is the simplest, less step and most effective types of multifactor authentication (MFA) method of verifying that users are who he/she say he/she is by providing additional security evidence. As we seen in the figure [10] there are various forms of 2FA, but here we divided 2FA into three main categories, such as OTP-based, biometric-based and digital certificate-based 2FA. These categories are further divided in to sub-types which are discussed below.

2.1 One-Time Password (OTP)

A one-time password refers to one-time PIN or dynamic password which is a password that is only valid for a certain defined time by the algorithm and one login term on a computer application or other digital physical device. It falls under the second commonly used factor called something you have. It is often used in combination with a regular password as a layer of authentication mechanism to provide extra security (Fig. 3).

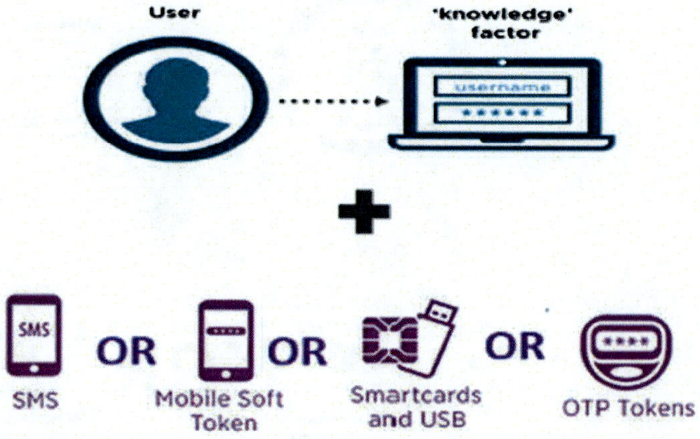

Fig. 3 OTP methods

OTP can be divided in two types; the first is event-based (**HOTP**) OTP tokens; this generates unique code when the user presses the pin of the hardware. The generated code becomes valid until used by the client and server. The second category is time-based OTP (**TOTP**) tokens; this generates a unique code which is only valid for specific period of time usually 30 min. Therefore, **TOTP** can be generally considered as more secure one-time password solution than **HOTP**. OTP authentication tokens are also further classified based on two delivery mechanisms: hardware based also known as 'hard tokens' and software based also known 'soft tokens.' A hardware based is a physical device that is attached to computer to generate an OTP such as USB device token and smart card token. And, they are considered as secured compared to software-based token. But this type of medium client should carry the device as they move. Software token, on the other hand, enables users to access OTP tokens through different means such as SMS, voice calling, push app, app based, e-mail delivery usually through user's personal device such as mobile phone and computer. OTP based two-factor authentication has become a choice for small and medium startup companies as it reduces the cost of building security control method [11].

2.2 *Biometric*

Biometric authentication can be categorized into the third category of commonly used factor which is something you are through inheritance. Biometrics are a category of authentication methods that use a unique biological trait of individual either physical pattern also known as static or behavioral pattern which is dynamic to verify the identity of user. Example of physical pattern are fingerprint, facial, iris, and retina scans, including hand geometry of individual; on the other hand, voice and speech patterns, typing rhythm, and body resonance fall under behavioral pattern which is dynamic behavior. The most acknowledged and commonly practiced biometric method is fingerprint authentication. Figure 4 shows how biometric finger print is combined with user name and password. In biometric-based two-factor authentication (2FA), just like other 2FA a client or a user is asked to prompt first factor evidence such as password or QR smart card and then the system scans input of biometric data to authenticate their unique which is taken as a second factor. Doing this in school or organization will allow them to enjoy the benefits of layered security. During the process of authentication, high technology medium is used such as computer, light, sound and biological sensors.

2.3 *Digital Certificate*

Digital certificate is equivalent to electronic password that enables a person, device, and organization to authenticate before granting a resource. To enable two-factor

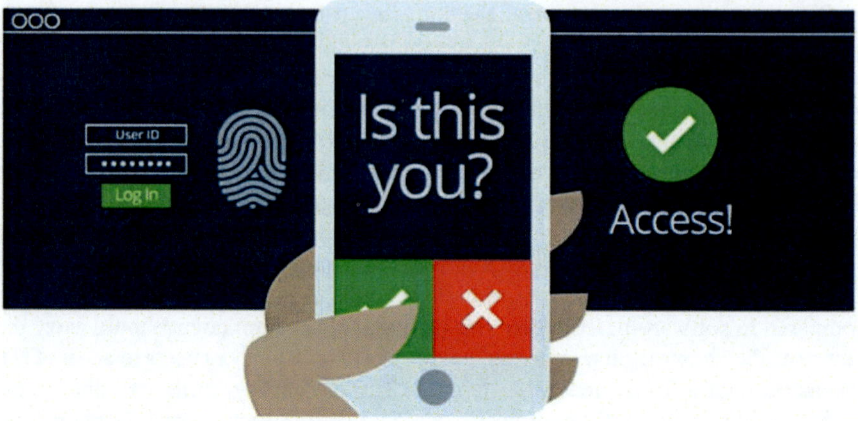

Fig. 4 2FA with biometric fingerprint

authentication with digital certificate, one must have an identity-based digital certificate issued by certificate authority (CA). Identity based means that the identity of the individual must be verified by an accredited Certification Authority (CA) before the certificate has been given to the user client. The goal of the digital certificate is to confirm that the public key attached in the certificate linked to the person to which the certificate was provided, digital certificate-based two-factor authentication what makes unique from other types of 2FA methods is that it is not limited only to authenticate a person but also used for machines, device and things on Internet of Things (IoT). So, it gives dual benefit at a time which is authenticating the person along with its device. This will be good at adding a value on the layer of security. It is based on asymmetric cryptography technique which uses public and private keys supported by a public key infrastructure (PKI) for the distribution and identification of public keys. **Public key infrastructure (PKI)** plays a significant role in facilitating the way of authentication process by using digital certificate with the aid of protocols, standards, and services. **X.509 format is** international standard which is very important for formatting and structuring the information within the certificate in a standard way so that it can send, retrieve, and interpret without any concern of who issued the certificate. Client certificate authentication mainly deals with process of identifying a user used as a proof before granting an access to network infrastructure. In paper [1–3], different schemes have been discussed regarding network protocols.

We can use digital certificates for different purpose; the first is for secure communication and authentication which are mainly achieved by using transport layer security or secured socket layer in short TLS/SSL certificate, the second is code signing certificate authentication used by vendors to authenticate software embedded in hardware or in the Internet, and the third one is client certificate. From these types of digital certificate, client certificate is used for two-factor authentication purpose. There are two ways of digital certificate authentication; the first is one-way SSL authentication in which only client authenticates the server to allow permission to retrieve data

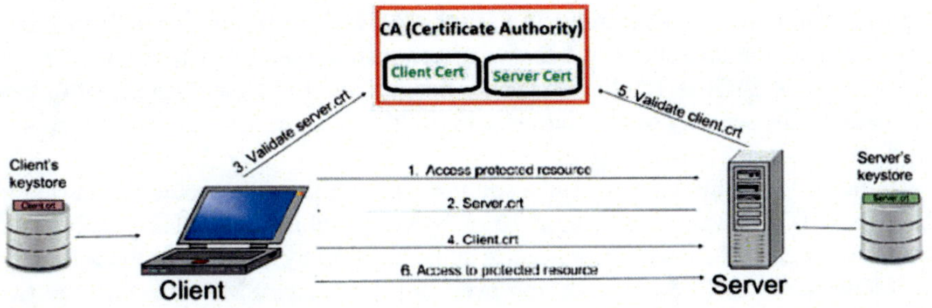

Fig. 5 Mutual SSL authentication process

from the given server. Therefore, in this case, only the server presents its public certificate to the clients for authentication. There is a trust store on the client side usually browsers and a KeyStore on the server side. The second one is two-way SSL authentication in which both client and server verify their identity before granting a resource. Two-way SSL authentication is also considered as mutual authentication which provides more level of trust between parties (Fig. 5).

This two-way authentication enables digital certificate a unique and a more secured ways of multifactor authentication, helps to mitigate different attacks, such as man-in-the-middle (MITM), phishing credential stealing, unauthorized access, and also prevents breaches of sensitive data transmission.

3 Literature Review

Most of the authentication scheme today depends on *static passwords* to authenticate and identify user. But, using password became traditional as it raised major security concerns [10]. Due to weak password management experience, most users prefer to use weak password, use same password for multiple accounts, put passwords on paper notes and on their personal devices [10]. Moreover, bad guys have the variety of techniques to steal credential such as shoulder surfing, snooping, sniffing, guessing, man-in-the-middle attack, phishing. The most common and widely used two-factor authentications is short message service (SMS)-based one-time password (OTP) which needs user involvement to enter an OTP after logging in with a user name and password, or the OTP is required to authorize a transaction [12]. Another paper in [11] outlines an architecture to build OTP-based two-factor authentication which acts as OTP as a service like other service provided in the cloud such as software as a service, infrastructure as a service, hardware as a service, and so on. The paper states that providing OTP as a service which is secured and confidential in the cloud to enterprise will help them to mitigate attacks by using the second factor of authentication. However, using SMS OTP sent from the cloud has many security problems such as SIM swap (impersonating victims), malware (for interception of forwarding text message), and social engineering attacks [6]. Moreover, nowadays SMS-based OTP has become unsecured two-factor authentication method [12] because of that the security of SMS OTP relies on the confidentiality of SMS messages that further depends on the security of cellular networks and for the reason that the wide spread of different attacks cybercriminal.

In paper [13], authors proposed a solution to solve possible problem occurred in SMS-based OTP two-factor authentication by using mobile phone. The paper tries to improve the SMS-based OTP two-factor authentication by eliminating the following problems such as SMS cost—during every login request, SMS lateness (due two network issue), international roaming (traveling overseas), and SMS security. In addition to the above paper [13], another author proposes a solution of mobile phone as software token for one-time password generation [10]. The proposed solution [10] tries to secure authentication gate for service such as online banking and ATM machines. The system works with the help of mobile phone as a software token in order to generate one-time password in the user side. The generated one-time password expires after a short period of time usually 30 min and is generated by factors that are unique to both the user and the mobile device itself. In addition to the above-proposed mechanism, the authors also gave an alternative option of SMS-based authentication for the purpose of a backup and synchronization feature. Time-based OTP is also proposed in paper [14] to prevent authentication vulnerability of Claim Token method of Membership Service provide in Private Blockchain which makes secured by enforcing two-factor authentication. OTP-based authentication is also applicable in transportation area to facilitate ticketing system [15].

Another paper related to two-factor authentication is two-factor verification which uses QR-code. This paper [16] states that OTP-based 2FA is vulnerable since they are compromised by man-in-the-middle attack, phishing, and spoofing attack and proposes a solution by displaying an encrypted IMEI number of a user in the form of QR codes for authenticating purpose. The first factor is QR code, and second factor is mobile phones to form two-factor authentication; in addition to this, paper in [8] proposed a new way of two-factor authentication by using the current GPS location of the users. Paper [17] scheme named S-Mbank also proposes other ways of mechanism by replacing the authentication using one-time password with the contactless smart card to prevent attackers to use the unencrypted message which is sent to the user's mobile phone and also removes the vulnerability of unauthorized users to act as a legitimate user to exploit the mobile banking user's account. Besides that, the author uses public–private key pair and PIN to provide two-factor authentication and mutual authentication. Furthermore, to paper [17], a paper in [18] called 2FMA-NetBank also uses a public–private key pair and smartphone IMEI number as a second factor for mutual Internet banking authentication and uses QR code application in smartphone for efficient implementation of the scheme. Each of the proposed solution for safe and secure multifactor authentication has their own advantage and disadvantage; Table 1 summarizes them accordingly.

Table 1 Observation of different 2FA proposed solutions

Paper and method	Advantages	Disadvantages
Short message service (SMS)-based OTP [12]	Better security than traditional username and password Simple to use Every user can avail Easy to implement	SIM swap, SMS interception, malware, social engineering attacks, SMS cost, SMS lateness
Mobile phone-based OTP [10, 13]	User friendly Easy to implement and cross-platform	Phone stealing Not available to every user Could account compromise Can be intercepted
QR code 2FA [16]	Support user mobility and convenience Affordable technology and easy to use	Unencrypted data Vulnerable to phishing and pharming attack Computing power
Smart card based [17]	Flexibility to use	Can physically stolen Need extra resource
Asymmetric-based-2FMA NetBank [18]	Mutual authentication Less cost	Have extra step Need more computing Not work for all phone types

4 Security Analysis

Due to the rise of business to reach their customer through digital channel, enterprises are more digital-oriented than ever. Weak authentication and traditional authentication (username and password) mechanism followed by business sector let the customer to compromise their account and lose trust among them. That is why enterprises deploying strong authentication methods are crucial. Mobile phone malicious software, such as Trojans malware, that are look legitimate but designed to forward SMS messages from user phone that containing OTPs, are one of the big threats. This kind of malware is developed intentionally by Internet criminals. The other security threats of SMS-based OTP is, since this kind of method relies on cellular networks service providers by default, they are vulnerable to different attacks due to lack of strong security control. Moreover, SIM swap and cloning activities give a good opportunity for hackers to steal the phone number. Despite this flaws SMS OTP still became one of the most common second factor authentication mechanisms in financial and e-commerce industries due to its easiness to use and platform independent, but experts suggest that SMS OTP 2FA is not any more secure and strongly recommend to observe options of available high-secured multifactor authentication solution such as hardware, software, and biometric authentication. One of the alternatives which can replace SMS OTP is application-based generated codes. These methods solve the issue of network requirement to deliver the OTP because it can work offline without cellular network and Internet. However, since the method works on smart phone, the application or user smart phone can be hacked. On the other hand, hardware-based tokens which are considered as one of the secured 2FAs also has a

Table 2 Ability to prevent the following attacks

Paper	Phishing attacks	Device theft	MiTA (man-in-the-middle attack)	Brute force
SMS-based OTP [12]	No	No	No	Yes
Mobile phone-based OTP [10, 13]	No	No	No	Yes
QR code 2FA [16]	No	No	No	Yes
Smart card based [17]	No	No	No	Yes
Asymmetric-based 2FMA NetBank [18]	Yes	Yes	Yes	Yes

chance of infection by virus and different attacks. The user may also lose the device. Furthermore, due to its uniqueness at individual level, biometric has become one of these easy and more reliable two-factor authentication mechanisms. Nevertheless, if a password is intercepted, we can recover it within couple of minutes; if we get hacked of our hardware 2FA, it takes a bit more time, but still it is possible to recover, but in biometric 2FA, once the image is compromised by attackers, it is impossible to recover back. So, biometric authentication is not recommended for the purpose of sensitive area such financial sector. Another method of authentication which is not widely used but considered as strong authentication is asymmetric based which uses private and public keys under public key infrastructure (PKI). Since the mechanism works by giving digital certificate which is stored on user's computer, it can be accidental deleted and can be copied by criminals if they got access to corporate network. User should take care of managing their private key (Table 2).

5 Conclusion and Recommendation

Two-factor authentication has become a mandatory solution by different organizations to reduce credential stealing by cybercriminals. It is playing a significant role in controlling attacker who seeks to get into personal login account and access personal data by having user name and password of users by any means. However, this technology is imperfect. It has its own advantages and disadvantages which vary on a particular type. Though, if a company provides one of the available methods of two-factor authentication, it is a good move for their client to create a trust. Even if organization begins to implement different ways of two-factor authentication mechanism, still different types of attacks came up by changing their ways of attacking so in response to that security measures should be much stronger than before. One of the most effective methods I suggest for minimizing the risk of authentication process from being compromised is to find out a solution that reduces the process of the user and builds the authentication process with strong cryptography which is based on international standards and framework mainly by using digital certificates.

Apart from security concern, this will help organization to implement trusted, easier, manageable, and less cheap methods authentication method.

References

1. Sathwara, S., Dutta, N., Pricop, E.: IoT forensic a digital investigation framework for IoT systems. In: 10th IEEE International Conference on Electronics, Computers and Artificial Intelligence (ECAI), Romania, pp. 1–5 (2018)
2. Dutta, N., Sarma, H.K.D., Polkowski, Z.: Cluster based routing in cognitive radio adhoc networks: reconnoitering SINR and ETT impact on clustering. Comput. Commun. **115**, 10–20 (2018)
3. Dutta, N., Sarma, H.K.D.: A probability based stable routing for cognitive radio adhoc networks. Wireless Netw. **23**(1), 65–78 (2017)
4. Dutta, N., Sarma, H.K.D.: A scheme for dynamic MAP selection in HMIPv6. Proc. Natl. Acad. Sci. (J.) India Sect. A Phys. Sci. (NASA) **90**, 371–382 (2020)
5. Delvadia, K., Dutta, N., Ghinea, G.: An efficient routing strategy for information centric networks. In: IEEE ANTS, Goa, India, pp. 1–6 (2019)
6. Available accessed 10/2/2020. https://enterprise.verizon.com/resources/reports/2020-data-breach-investigations-report.pdf
7. Brainard, J.G., et al.: Fourth-factor authentication: somebody you know. In: CCS '06 (2006)
8. Kumar, D., Agrawal, A., Goyal, P.: Efficiently improving the security of OTP. In: 2015 International Conference on Advances in Computer Engineering and Applications, Ghaziabad, pp. 912–915 (2015)
9. Shah, S.U., Fazl-e-Hadi, Minhas, A.A.: New factor of authentication: something you process. In: 2009 International Conference on Future Computer and Communication, Kuala Lumpur, pp. 102–106 (2009). https://doi.org/10.1109/ICFCC.2009.79
10. Aloul, F., Zahidi, S., El-Hajj, W.: Two factor authentication using mobile phones. In: International Conference on Computer Systems and Applications IEEE/ACS, pp. 641–644 (2009)
11. Erdem, E., Sandıkkaya, M.T.: OTPaaS—one time password as a service. IEEE Trans. Inf. Forensics Secur. **14**(3), 743–756 (2019)
12. Mulliner, C., Borgaonkar, R., Stewin, P., Seifert, J.P.: SMS-based one-time passwords: attacks and defense. In: Rieck, K., Stewin, P., Seifert, J,P. (eds.) Detection of Intrusions and Malware, and Vulnerability Assessment. DIMVA 2013. Lecture Notes in Computer Science, vol. 7967. Springer, Berlin, Heidelberg (2013)
13. Eldefrawy, M.H., Alghathbar, K., Khan, M.K.: OTP-based two-factor authentication using mobile phones. In: 2011 Eighth International Conference on Information Technology: New Generations, Las Vegas, NV, pp. 327–331 (2011)
14. Park, W., Hwang, D., Kim, K.: A TOTP-based two factor authentication scheme for hyperledger fabric blockchain. In: 2018 Tenth International Conference on Ubiquitous and Future Networks (ICUFN), Prague, pp. 817–819 (2018)
15. Edna Elizabeth, N., Nivetha, S.: Design of a two-factor authentication ticketing system for transit applications. In: 2016 IEEE Region 10 Conference (TENCON), Singapore, pp. 2496–2502 (2016)
16. Rodrigues, B., Chaudhari, A., More, S.: Two factor verification using QR-code: a unique authentication system for Android smartphone users. In: 2016 2nd International Conference
17. Putra, D.S.K., Sadikin, M.A., Windarta, S.: S-Mbank: secure mobile banking authentication scheme using signcryption, pair based text authentication, and contactless smart card. In: 2017 15th International Conference on Quality in Research (QiR): International Symposium on Electrical and Computer Engineering, Nusa Dua, pp. 230–234 (2017)

18. Pratama, A., Prima, E.: 2FMA-NetBank: a proposed two factor and mutual authentication scheme for efficient and secure internet banking. In: 2016 8th International Conference on Information Technology and Electrical Engineering (ICITEE), Yogyakarta, pp. 1–4 (2016). https://doi.org/10.1109/ICITEED.2016.7863247
19. Chavan, J.: Internet banking-benefits and challenges in an emerging economy. Int. J. Res. Bus. Manag. **1**(1), 19–26 (2013)
20. Hiltgen, A., Kramp, T., Wigold, T.: Secure internet banking authentication. IEEE Secur. Privacy Mag. **4**, 21–29 (2006)
21. Omariba, Z.B., Masese, N.B., Wanyembi, G.: Security and privacy of electronic banking. Int. J. Comput. Sci. Issues **9**(4), 432–446 (2012)
22. Fang, X., Zhan, J.: Online banking authentication using mobile phones. In: International Conference on Future Information Technology, pp. 1–5 (2010)
23. Gandhi, A., Salunke, B., Ithape, S., Gawade, V., Chaudhari, S.: Advanced online banking authentication system using one time passwords embedded in Q-R code. Int. J. Comput. Sci. Inf. Technol. **5**(2), 1327–1329 (2014)
24. Abbott, J.: Smart Cards: How Secure Are They? SANS Institute InfoSec Reading Room (2002)
25. Das, M.L., Saxena, A., Gulati, V.P.: A dynamic ID-based remote-user authentication scheme. IEEE Trans. Consum. Electron. **50**(2), 629–631 (2004). [2] Misbahuddin, M., Aijaz Ahmed, M., Shastri, M.H.: A simple and efficient solution to remote user authentication using smart cards. In: Proceedings of IEEE Conference on Innovations in Information Technology, pp. 1–5 (2006)
26. Raddum, H., Nestås, L.H., Hole, K.J.: Security analysis of mobile phones used as OTP generators. In: IFIP International Federation for Information Processing, pp. 324–331 (2010)
27. Chefranov, A.: One-time password authentication with infinite hash chains. In: Novel Algorithms and Techniques in Telecommunications, Automation and Industrial Electronics, pp. 283–286 (2008)
28. Gurabi, M.A., Alfandi, O., Bochem, A., Hogrefe, D.: Hardware based two-factor user authentication for the internet of things. In: 2018 14th International Wireless Communications & Mobile Computing Conference (IWCMC), Limassol, pp. 1081–1086 (2018)
29. Narasimhan, H., Padmanabhan, T.: 2CAuth: a new two factor authentication scheme using QR-code. Int. J. Eng. Technol. **5**, 1087–1094 (2013)
30. Reynolds, J., Smith, T., Reese, K., Dickinson, L., Ruoti, S., Seamons, K.: A tale of two studies: the best and worst of YubiKey usability. In: 2018 IEEE Symposium on Security and Privacy (SP), San Francisco, CA, pp. 872–888 (2018)

A Survey on Biosignals as a Means of Human Computer Interaction

Tanuja Subba and Tejbanta Singh Chingtham

Abstract In recent years, the growth of human computer interaction (HCI) has expanded widely. It creates interactive interface, deals with the design and implementation for human and computer. HCI is a huge multidisciplinary subfield and one ever growing field is biosignals. The electrical signals measured from living beings is biosignals that gives data about human body structure-like biological and physical information. The study of biosignals based HCI depends on the use of different types of electrical signals and physiological information which acts as interaction modalities and for the mode of evaluation. This paper deals with the types of biosignals used for creating a suitable interface for HCI. Biosignals used are electromyogram, electrocuologram, electroencphelogram, and electrocardiogram.

Keywords HCI · Biosignals · Electromyogram · Electrocuologram · Electroencphelogram · Electrocardiogram

1 Introduction

With the increase in demand of communication that is easy, efficient and innovative the approach that came into highlight is human and computer interaction with the use of biosignals. The increase in interest of researchers, the use of biosignals for an interface between human and computer has made growth and success in this research field. The researchers in this field have been working since 1980s with a growth and development which is huge today. The research in this field has created a new development for certain group of people who needs constant support in their daily life. It has made a huge progress in developing and creating user command interface where user could give input and specify the input for a desired output. The study of biosignals makes this research more extensive as user can input its own biosignals and command the interface to do certain job which makes interfacing much easier and real time for not only for normal people but also for specially abled people. This

T. Subba (✉) · T. S. Chingtham
Department of CSE, Sikkim Manipal Institute of Technology, SMU, Majitar, Sikkim, India

H. K. D. Sarma et al. (eds.), *Contemporary Issues in Communication, Cloud and Big Data Analytics*, Lecture Notes in Networks and Systems 281,
https://doi.org/10.1007/978-981-16-4244-9_24

paper focuses on the study of various biosignals that are useful for developing and performing an analysis fora system of human and computer communication.

2 Background

2.1 HCI

HCI is the study of interrelation between human and computer. It deals with the interactive design, knowing the limits of the human body, mind and also the environmental factors that affect human performance along with how computers affect individuals, organizations and society. To show an individual the importance and benefits of HCI designing, a simple, resourceful and useable reciprocative interface is must [1]. It is a multidisciplinary field of study and many areas have contributed to its success. Figure 1 shows the wide range of HCI. The multidisciplinary nature of HCI makes it more flexible for wider range of input and output.

2.1.1 HCI Architecture

HCI focuses on creating an interactive design and evaluation of a model which is easy, efficient to use for interaction purpose. Architecture of HCI is defined by the number of inputs and number of output in an interface which shows the working of input and output together [2].

Unimodal HCI: The input of a HCI model is single-communication channel. Only one channel is used for communication between human and computer. There are many different channels as an input like eye movement, heart signal, brain signal, face gesture, speech recognition, etc. When only one among this modality is used for creating an interface in the system then it is called unimodel HCI system. Categories of modalities depend on the type of input for vision it is visual-based, for speech,

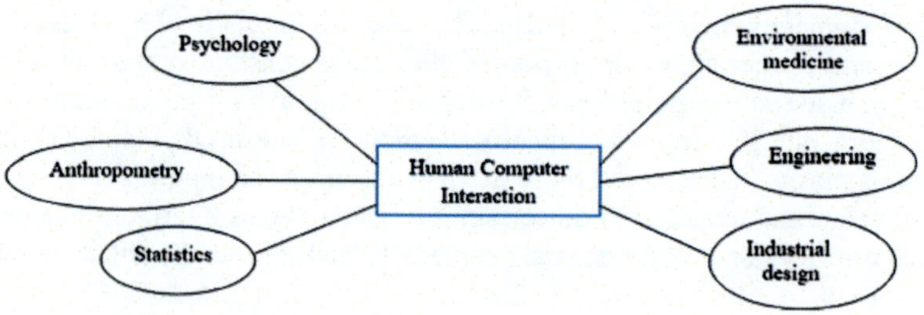

Fig. 1 Multidisciplinary field of HCI

it is audio-based and for sensing signal it is sensor-based. Therefore, there are three types of unimodel system.

- Visual/Vision-Based: Vision-based is noninvasive and provides a wide range of input. A computer vision technique is employed to capture the information perceived by the human eye and to process data from real cameras. Eye movement signal is captured by the sensor and is designed and evaluated to give a desired output.
- Audio-Based: A varied audio signal is an input for audio-based system where the information is acquired from the signal. It creates a more robust environment than a graphic based interface. Audio does not depend on visual and hearing it is independent of other source of interfacing. This type of technique can be implemented when research is conducted for visually challenged group.
- Sensor-Based: The use of sensors in HCI is growing rapidly as sensors are evolving it's becoming small, inexpensive and very responsive. It is not only used in HCI, but is popular in other areas also. Sensors can control activation, provide context-aware information, used when visual-based automation is done, collect human body signals for processing, etc.

Multimodal HCI: The input for HCI model is multiple channels. When more than one sense like speech, vision, hearing, human body movement signal and action are combined for input in a system for interfacing it is called multimodel channel communication. Combination of speech and gesture input for interfacing is one of the examples of this architecture. This type of technique is implemented when user wants more flexibility of input as well as output.

Multimodal HCI system is used:

- To diversify users, inputs and environmental situations.
- For developing interface which can be used by varied groups of people.
- To reduce error occurrence.
- To have flexible and enhanced use of input modes.
- To research for more existing types of mode for HCI interfacing.

2.2 Biosignal

Biosignal is a description of physiological phenomenon of any nature. It carries all the information about the living being. The analysis is done from the signal generated from the body. According to the physical nature of biosignals, it can be classified into electric, magnetic, chemical, mechanical, optical and thermal. To capture these kinds of signals, bioelectrodes are used which is the interface between biological structures and electronic systems. The activity in the human body is either sensed or stimulated. The electric potentials generated in the body are ionic potentials and to be measured externally this ionic potentials are to be converted into electronic potentials. The conversion is done by a device called electrodes which also acquires biosignals. There are many types of electrodes invasive and noninvasive. Surface electrodes which are placed on the skin are called noninvasive electrodes, but it measures very low biosignals, primarily used for electrocardiogram (ECG), electroencephalogram (EEG), electromyogram (EMG). Microelectrodes: it is invasive electrodes which are planted inside human body for better measurement of the activity of the signal it is very small in size electrodes, used in electrophysiology which records neural signals or electrical stimulation of nervous tissue. Internal electrodes: this type of electrodes is used for detecting fetal electrocardiogram during labor. Needle electrodes: it is also an invasive electrode which is penetrated in the skin to record the potentials difference. Here are some of the examples that capture signals using these electrodes EMG (electromyogram), ECG (electrocardiogram), EOG (electrooculogram), EEG (electroencephalogram), EP/RP (evoked-potential/event related potential). All signals can be measured using both invasive and noninvasive electrodes.

Biosignals offer an electrical interface from human body to the end user. This paper deals with multiple papers that have used human body signals and have created an interactive interface to communicate with the machine. Figure 2 shows the scheme of how biosignal can be used for HCI.

The factors that help understand HCI with its different emphasis are:

Human Cognition: A cognitive system which learns and adapts naturally and communicates like human. To understand human cognition all the senses like perception, visual cognition, auditory cognition, types of interfaces, motion cognition, memory, learning, language understanding, representations and mental models are important.

System Design for collaboration and communication: For human, collaboration and communication are important and designing an interface looking at this aspects are visual information, group dynamics.

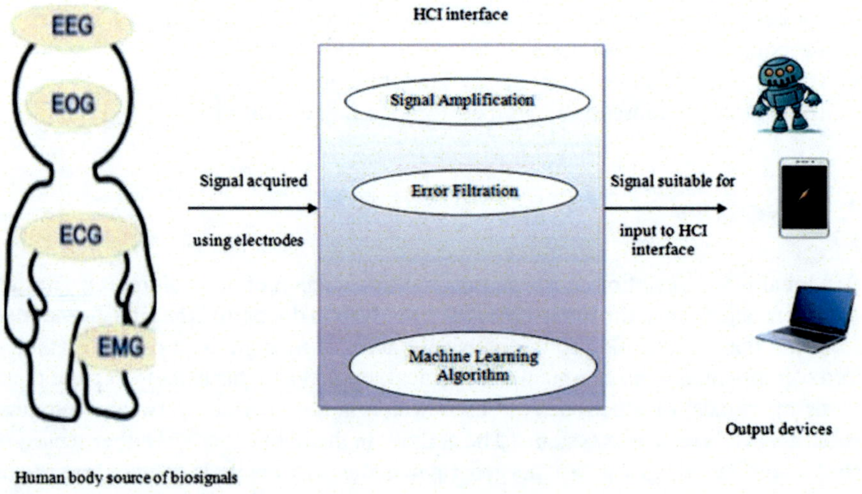

Fig. 2 Biosignal based HCI

Understanding how interfaces/technology affects users: In order to develop interactive interface understanding social and behavioral factors, ergonomics and environment study is important.

User-centered approaches to interaction design: Integration of types of user in prototyping a model, designing.

Usability evaluation: Understanding and testing users, modeling users.

Interaction styles: User centric interfaces, virtual environment, commands and natural language, circuit design etc.

Interaction devices: Vision based, biosignal based, sensory based, speech based, keyboard, hand movement based etc.

3 Related Works

The development on EOG based HCI [3] explains that human computer EOG based interfaces is one of the popular and widely researched areas considering its use good option for developing interface for specially abled people. Blinking and eye movement can be detected by conventional method. Researchers in this paper describe the quantitative estimation of horizontal position of eyeball. A system is designed for acquisition of EOG signal by the position of eyeball which also measures electromyogram (EMG) signal without extra electrodes. Sliding window algorithm, support vector regression (SVM) method is used to increase the performance of eye tracking and piecewise linear approximation (PLA) is integrated for reducing the complexity of the signal. This all are done for real-time application. In this paper, the researcher also have done the comparative study between conventional method which detects the eyeball position only in two levels, i.e., 0 and 1 and its developed EOG based interface which continuously keeps track of eyeball position for 0.2 s. It shows that enhanced system can drag and drop in mouse interface in less than 5 s. The developed system also increases the user flexibility by reducing number of electrodes.

The emotion-based detection techniques has been studied in [4] and explains the suitable signal to detect emotional expression is the use of electromyography (EMG). The acquisition of signal is done from face where the potential difference of each expression is measured. In this paper, researchers have classified expressions in for classes happy, neutral, sad and angry. It is a multimodel technique where the input channels are EOG and EMG. EMG is used to classify expression, and EOG is used to detect eye movement and saccades. A framework has been proposed in this paper to detect emotion states. Another similar paper which deals with EOG signal is [5], but is a unimodel system where the channel input is only one signal that is eye movement signal that controls human–computer interface (HCI) system. The researchers in this paper have studied the limitation and advantages of EOG signal that can be used to properly classify the output signal. They have developed a HCI device which is based on EOG signal and is wireless which could detect eye movement and blinking of an eye and the device could classify various activity of this movement. Eight direction of eye movement is been classified that is left, right, up, down, up-left, up-right, down-left and down-right, the direction of each signal are measured real time and is classified by extracting features from measured electrical signals.

Some researchers have also tried multimodel HCI technique which gives more accuracy in the result like [6]. They have researched and analyzed emotion detection using existing system also with the use of single-channel biosignal. In this paper two input source is considered ECG and EEG to improve the accuracy rate of emotion detection. A complex emotion recognition system was developed. The accuracy rate is high because EEG signal is used for autonomic nervous system and EEG to improve accuracy. With the use of multimodel, the system recognizes different kind of feelings like fear, sad, joy, angry, amusement and disgust. The interface was created for game with accuracy rate of average 98.06%.

The study on brain monitoring using noninvasive method [7] is one of the new approaches to acquire brain signal. A user-friendly method of measuring EEG signal made it more convenient for new researchers to study. The noninvasive technique uses surface electrodes or dry electrodes, active sensors to be used to minimize noise. This development has made study of continuous brain signal much easier than it used to be.

There is a growing demand for smart environment development to make life easier and improve quality [8]. Every device is turning into smaller version which makes it easier to carry around, and one of the applications is wearable devices which is capable of recording, monitoring, filtering, amplifying and classifying bio potential signals or environmental signals for various applications like health monitoring, physical activity monitoring, etc. wearable device can automatically measure human body, electric biosignals, classify them and perform any output desired command. Brain computer interface (BCI) is a growing field with many applications like wearable devices with BCI [8]. BCI is one of the future research topics. Another wearable based research is on electric power wheelchair [9] using EMG signals a source of input for motor disabilities group of people. The researchers considered a subject who was suffering from C4 and C5 spinal cord injury who had motor disabilities. The acquisition of EMG signal is done from shoulder elevation motions where subject on wheelchair could move forward or stop, turn left, turn right, and go forward with a high accuracy.

One of the reasons HCI is rapidly growing is because of the development of a wearable, portable and less costly devices which can help people with motor disabilities or specially abled people who is unable to communicate directly with the system. With the help of HCI these group of people will be able to interface with the computer or the system with alternative source of input and get desired output.

HCI implementation using biosignals as an input source for interfacing needs an understanding of human behavior, human needs, human limits and human cognition which is studied in [10]. It is a challenge for researchers as human factors depends on individuals and the environment. Together human and computer to reciprocate properly the proposed system should be highly efficient to accept wide range of input. A new HCI interface will replace collaboration by command, limitless freedom, and interface to become dynamic and responsive and cognitive system. Researchers have focused the implementation on cognitive science point of view which shows the study of human needs, behavior, limits and how all human behavioral factors are dependent on each other which works together to develop a real-time interface.

4 Summary of Different Biosignals

This survey paper highlights the use of bio electric signals that have been used by researchers for the implementation of HCI. Biosignal is one of the techniques that can not only create an interactive interface between human and computer, but it is one of the research areas where a researcher focuses on creating interface for differently abled people. Using these biosignals, researchers have created a successful interface.

4.1 EMG

Electromyogram is an electrical signals generated by muscles. Electromyography is a procedure that checks the health condition of muscles and the nerve cells that controls them it depends on automatic and physiological properties of muscles. The amplitude and the frequency of the signal are about 0–10 mV and 0–500 Hz [11], and most dominant is in between 50 and 150 Hz. Noise that affects this signal are all electronics equipment that is used to acquire this signal from human body which cannot be eliminated, ambient noise which is generated by electromagnetic radiation, motion artifacts caused by electrode interface and electrode cable and inherent instability of signal this kind of noise is unwanted and its removal is important. There are other factors that also affect EMG signal extrinsic and intrinsic. Extrinsic is caused by types of electrodes its structure, placement, and intrinsic is a physiological, anatomical and biochemical factors.

4.2 EOG

Electrooculogram is an electrical signals generated from eye movement by the change of corneo-retinal standing potential that exists between the front and back of the human eye. EOG can be one of the most efficients to create an interface for HCI as seen in [3, 12]. The amplitude and the frequency range of the signal vary from 0.5 μV to 5 mV and from dc up to 100 Hz [11]. The noise in EOG signal is by the interference from other electrical signals like electroencephalography, electromyography, blinks, speech, sensor noise, power line noise and electrical network.

4.3 ECG

Electrocardiogram is the electrical signal that originates from the activity of the human heart. ECG signal can diagnose different kind of heart disease by checking P wave, QRS wave and T wave signal. ECG is noninvasive techniques, i.e., using surface electrodes acquisition of heart signal is done. The amplitude and frequency range is about 1–10 mV and 0–20 Hz [11]. This procedure can also measure pulse, pulse rate and can diagnose many disease or abnormalities of human body. The noise that occurs in ECG data is power line interference these are the external noise and muscle movement artifacts such as EMG signal. Therefore, during acquisition of ECG the subject should be at rest.

4.4 EEG

Electroencephalogram is the continuous electrical signals activated from brain. A device contains multiple channel for data acquisition, and channel electrode is placed on the scalp of human head. EEG can acquire many signal from the active brain cell some are alpha, beta, theta and delta. These signals are active signal when user is sad, happy, excited or asleep. Just by thinking, a system could allow user to input data and work real-time. Brain states such as emotional level, stress level, etc. could be recognized using EEG data. The amplitude is about 2–100 μV [11] and according to active state of user there are four basic EEG frequency ranges from beta (14–30 Hz), alpha (8–13 Hz), theta (4–7 Hz) and delta (1–3 Hz). The noise in EEG signal is internal and external noise. Internal noise are strong than external as artifacts from other part of the body such as heart activity, muscle movement, eye movement, etc., which produces electrical signals, many times greater than the signal produced by brain. External noises are environmental, electrical sensors noise (Table 1).

5 Conclusion

Biosignals as a mechanism for HCI is a new and affordable approach creating a bridge between human and computer by its own biometrically generated bio electrical signals from human body. Interface developed with the input of biosignals can be user-friendly and it is also suitable for specially abled people. The interface will help such people to interact with machine in easy way. Biosignals can be unimodel and mulitimodel, which makes easier access for wide variety of users. EMG, EEG, ECG, EOG are the signals which can create real-time interface even though the amplitude of the signal is very low. From the survey, it can be found that developing an interface using biosignals could upgrade the quality of life of specially abled people.

Table 1 Summary of biosignals used for HCI

Biosignals used	Description
EMG [9, 13, 14]	• Electric powered wheelchair is modified with shoulder elevation EMG signal • Wheelchair is powered with 4 different output direction • EMG based mouse system is designed using 6 wrist motion • Fuzzy Min–Max Neural Network is used for classification of signal
	• The average accuracy rate for recognition of wrist motion is 97% • A facial EMG signal is used to identify 6 basic emotional states with the accuracy of 69.5%
	• k-nearest neighbor classifier is used for classification
EOG [3, 15]	• EOG based HCI system is developed to find eyeball position in two level 0 and 1 • Dragging and dropping operation is performed for mouse interface • Drag and drop of mouse system is done in less than 5 s • HCI interface is designed in using EOG signal with deep learning
	• Pattern Recognition Neural Network classifier is used for signal classification with 91% accuracy • A study between male and female subjects was performed which proves that performance and recognition accuracy is higher in case of male subjects
ECG [6, 16]	• Motion recognition interface is developed combining ECG and EEG • 6 different emotions were classified with average of 98.06% • QRS complex of ECG signal is used to detect emotion • Three different classifiers are used K-nearest neighbor (KNN), fuzzy KNN and regression tree
	• The average accuracy is 70.23% in general using KNN

References

1. Dix, A.: Human-Computer Interaction. Encyclopedia of Database Systems, pp. 1327–1331. Springer US, 2009
2. Adkar, P.: Unimodal and multimodal human computer interaction: a modern overview. Int. J. Comput. Sci. Inf. Eng. Technol **2**(3), 1–8 (2013)
3. Yang, J.-J., Woo Gang, G., Kim, T.S.: Development of EOG-based human computer interface (HCI) system using piecewise linear approximation (PLA) and support vector regression (SVR). Electronics **7**(3), 38 (2018)
4. Perdiz, J., Pires, G., Nunes, U.J.: Emotional state detection based on EMG and EOG biosignals: a short survey. In: 2017 IEEE 5th Portuguese Meeting on Bioengineering (ENBENG). IEEE, 2017
5. Wu, S.-L., et al.: Controlling a human–computer interface system with a novel classification method that uses electrooculography signals. IEEE Trans. Biomed. Eng. **60**(8), 2133–2141 (2013)
6. Jerritta, S., et al.: Emotion detection from QRS complex of ECG signals using Hurst exponent for different age groups. In: 2013 Humaine Association Conference on Affective Computing and Intelligent Interaction. IEEE, 2013
7. Mihajlović, V., Patki, S., Xu, J.: Noninvasive wearable brain sensing. Sensors, 2017 IEEE. IEEE

8. Udovičić, G., Topić, A., Russo, M.: Wearable technologies for smart environments: a review with emphasis on BCI. In: 2016 24th International Conference on Software, Telecommunications and Computer Networks (SoftCOM). IEEE, 2016
9. Kim, J.S., Jeong, H., Son, W.: A new means of HCI: EMG-mouse. In: IEEE International Conference on Systems, Man and Cybernetics, vol. 1, pp. 100–104. IEEE, Oct 2004
10. Torok, A.: From human-computer interaction to cognitive infocommunications: a cognitive science perspective. In: 2016 7th IEEE International Conference on Cognitive Infocommunications (CogInfoCom). IEEE, 2016
11. Schmidt, A.: Biosignals in human-computer interaction. Interactions **23**(1), 76–79 (2015)
12. Merino, M., Rivera, O., Gómez, I., Molina, A., Dorronzoro, E.: A method of EOG signal processing to detect the direction of eye movements. In: 2010 First International Conference on Sensor Device Technologies and Applications, pp. 100–105. IEEE, July 2010
13. Moon, I., et al.: Wearable EMG-based HCI for electric-powered wheelchair users with motor disabilities. In: Proceedings of the 2005 IEEE International Conference on Robotics and Automation. IEEE, 2005
14. Jerritta, S., et al.: Emotion recognition from facial EMG signals using higher order statistics and principal component analysis. J. Chin. Inst. Eng. **37**(3), 385–394 (2014)
15. Teng, G., et al.: Design and development of human computer interface using electrooculogram with deep learning. Artif. Intell. Med. **102**, 101765 (2020)
16. Shin, D., Shin, D., Shin, D.: Development of emotion recognition interface using complex EEG/ECG bio-signal for interactive contents. Multimedia Tools Appl. **76**(9), 11449–11470 (2017)

A Survey on Application of Machine Learning in Property and Casualty Insurance

Amlan Jyoti Dey and Hiren Kumar Deva Sarma

Abstract In today's world, data is playing the key role in every spheres of life like in business, retail, healthcare, travel, finance, and insurance sector. Machine learning (ML) is a subset of artificial intelligence (AI) that helps in data-driven decision making, which focuses mainly on analytical decision making and interpreting patterns and structures in data. Property and casualty (P&C) insurers are already reaping benefits with ML, and it has enough scope ahead with intelligent exploration. When we talk about ML, data come as its closest association because without data ML is of no use, and when we talk about data, cloud comes as its closest association as storing of data requires huge space which cloud can only provide. In the advanced analysis in P&C sector, the main challenge is that the models and algorithms used are not sufficient to support insurers. ML then becomes the final option to overcome this challenge. This article presents a survey on application of ML in various processes involved in the insurance sector.

Keywords Artificial intelligence · Machine learning · Compound annual growth rate · Property and casualty · Gross Direct Domestic Premiums Written

1 Introduction

Every industry in the world is having tough competition in order to survive, and insurance industry is not an exception. The business landscape is evolving rapidly. New risks are coming from cyber and climate change, new incumbents, emerging distribution channels, rising customer expectations, etc. These factors are compelling insurance companies to evolve new ways to come out from existing stagnant business processes. Traditional insurance giants who once dominated the industry are changing their strategies to deliver new product lines and new on-demand digital

A. J. Dey
Technology Specialist, NIIT Technologies Ltd, Noida, India

H. K. D. Sarma (✉)
Department of Information Technology, Sikkim Manipal Institute of Technology, Majitar, Sikkim, India

H. K. D. Sarma et al. (eds.), *Contemporary Issues in Communication, Cloud and Big Data Analytics*, Lecture Notes in Networks and Systems 281,
https://doi.org/10.1007/978-981-16-4244-9_25

platform to serve customers faster and remain competitive. To achieve the goals, the insurance industry must adopt new technologies to make their existing processes more efficient. Machine learning (ML) is an emerging technology, which can play an important role to achieve this goal if applied judiciously.

ML needs huge dataset in order to perform better. ML plays with any dataset be it structured, semi-structured, or unstructured. The insurance industry is an ideal candidate for the application of these technologies, due to the availability of huge amounts of both structured and unstructured data. The complexity of insurance products and processes, and the ascendant need to gain deeper insights into customers and risks. In property and casualty (P&C), by analyzing the dataset, ML can predict claims operation, customer behavior, and risk associated with claims. When compared to human, ML predicts more accurate result. Be it claim management, coverage changes, customer behavior, or risk management, ML predicts more accurate result in very efficient manner helping insurers to take appropriate actions, thus saving time and cost. There are some important areas in P&C where insurers can reap benefits from ML. Examples include rapid product development with dynamic pricing, individually developed loss predictions for claims, pricing, and reserving, distribution optimization, automated underwriting and marketing triage, underwriting risk portfolio optimization, etc. Moreover, by focusing on underwriting accuracy and claims collection efficiency, firms can boost their operating performance and thereby making profits to the organizations. All of these would be close to impossible to achieve without leveraging the power of ML [1–3].

2 Property and Casualty (P&C) Insurance Overview

Property insurance and casualty insurance (also known as P&C insurance) are types of coverages that cover both insured and insured's property. Property insurance covers, e.g., home or car one owns. On the other hand, casualty insurance as the name suggests means liability coverage to help protect anyone, if found responsible for an accident that causes injuries either to the other person or damage caused to the other person's belongings. Property and casualty insurance are usually bundled together into one insurance policy, for example, homeowners insurance, auto insurance, condo insurance, renters insurance, power sports insurance, landlord insurance, etc. [4–8].

Homeowners insurance covers damages and losses to a house owned by an individual along with other assets in the house. It also covers liability coverages against accidents in the house or property. **Auto insurance** is a contract between insured and the insurance company that protects against financial loss in the event of an accident or theft. **Condo insurance** covers a few standard coverages which includes to repair one's condo unit and belongings if they are stolen or damaged by certain perils, such as fire or vandalism. **Renters insurance** is an insurance policy that covers loss or damage to the property of someone who is renting a property. If something happens to the rental property, then landlord has insurance to cover

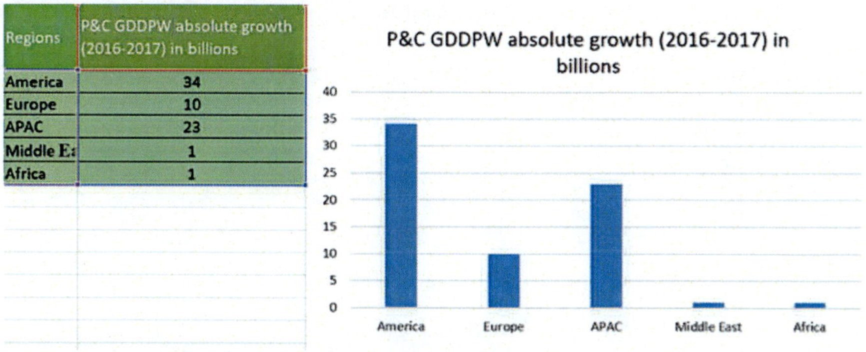

Fig. 1 P&C insurance in different regions of the world

the structure itself. **Landlord insurance** gives insurance coverage to a property that generates rental income. This type of policy typically helps to protect the building of the policyholder in the event of the damage caused by hail or fire, for example [2, 3].

3 Global Presence of P&C Market

Figure 1 depicts the regions in the world that deals with P&C insurance along with their **Gross Direct Domestic Premiums Written** (GDDPW) absolute growth for the year 2016–2017 [9–11].

4 An Overview of Machine Learning Capabilities

Machine learning is a subfield of artificial intelligence (AI), which is used for predictive analytics or predictive modeling. For different types of predictions, anomaly detection, clustering, and categorization, various types of machine learning algorithms are used. ML algorithms are mainly used for predicting values, detecting anomaly, clustering, and categorization.

Predicting Values: Predicting values means forecasting future outcome by studying the relationship between variables, e.g., predicting future sales figure, future product demand, etc. For prediction, various types of algorithms are used. They are linear regression, Poisson regression, Bayesian linear regression, neural network regression, decision forest regression, etc.

Anomaly Detection: Anomaly detection means identifying and predicting unusual data points. For anomaly detection, various types of algorithms used are one-class support vector machine and principal component analysis-based anomaly detection.

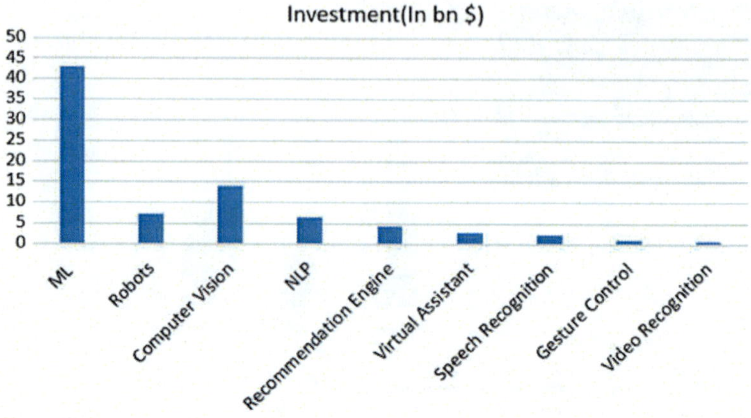

Fig. 2 Investments in various technologies

Clustering: Clustering means separating similar data points into groups, e.g., customer segmentation, customer taste, etc. For example, K-means algorithm is mostly used in clustering.

Categorization: Categorization means classification of an information. It may be two-class classification or multi-class classification.

In two-class classification, the algorithms used are two-class support vector machine, two-class Bayes point machine, two-class logistic regression, etc. Two-choice answer like true or false, and yes or no may be good example of two-class classification.

In multi-class classification, the algorithms used are multi-class logistic regression, multi-class-neural network, multi-class decision forest, etc. Question with multiple possible answers are examples of multi-class classification.

As per the source Statista, machine learning tops AI Dollars, May 10, 2019, an amount of $42.9B was invested in ML (ML platform + ML application) segment alone in the first calendar quarter of 2019, leading all other AI investment segments against the total of over $82B which was invested in all AI segments as shown in Fig. 2 [9–12].

5 Ml in P&C Insurance

The application of machine learning has grown tremendously especially in P&C insurance since last couple of years. It is used extensively in various segments of P & C insurance. Figure 3 shows application of ML for various segments of P&C insurance []. The result shows that business/underwriting is the segment where application of ML can be maximum.

Use Cases

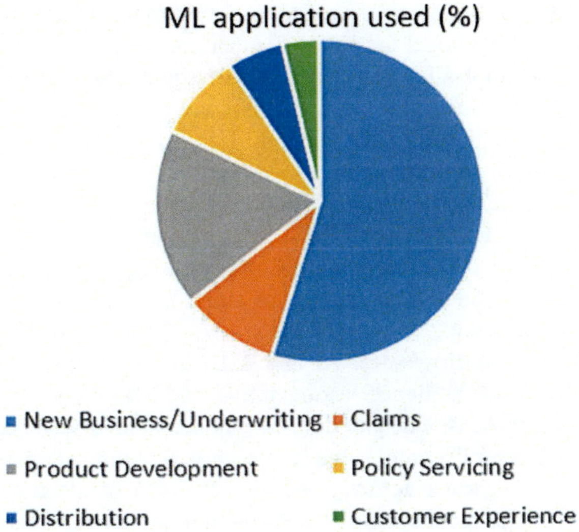

Fig. 3 Application of ML in P&C insurance

Below are some of the use cases where ML technology has been used extensively in order to gain benefits by the insurers.

Claims processing: The insurance companies are doing lot of research work on claims processing for accurate prediction of claims settlement. The more accurate the prediction is, lesser is the financial loss for the company.

Data and textual analysis: Here, the AI with ML enabled engines scan personal information records along with corporate audits to identify the possibility of risk. P&C firms sometimes customize their pricing policy or even may deny insurance to an entity that it feels as a high-risk proposition. Sometimes, machine learning techniques depict text to deliver a specific result, like to identify sentiment in a tweet. There are many ML algorithms used in text classification. The most frequently used are the Naive Bayes family of algorithms (NB), support vector machines (SVM), and deep learning algorithms.

Fraudulent claims: Machine learning and predictive models can help insurers with a better understanding of claim closure pattern. Insurance companies in general lose an estimated US$30 billion a year to fraudulent claims. ML can help them identify potential fraudulent claims faster and more accurately.

ML algorithm logistic regression is an ideal way for finding the claim as either fraud or non-fraud.

Upselling/Cross-Selling: ML plays an important role in improving cross-selling and up-selling opportunities. ML uses data-driven recommendations to customers who will get the right offers at the right time, and, therefore, purchases more products.

Policy comparisons: A global commercial lines insurer is collaborating with ML for policy comparisons. Policy language and endorsements are analyzed to support the product development process, match policy language to customer needs, and ultimately improve speed to market with new products. Linear and nonlinear models of ML are ideal for policy comparison.

Auto damage assessment: A large national insurer is upgrading its mobile app for auto insurance policyholders via a capability to assess images of damage to a vehicle after an auto accident. Image recognition and ML algorithms do an initial damage assessment.

Chatbots: Customer support is a massive cost center for any business, and this applies to insurance as well. With the help of AI & ML, this can be achieved through Chatbots. The concept of Chatbots is again text analysis. Therefore, the application of Naive Bayes family of algorithms (NB) plays an important role in Chatbots functioning.

Marketing/sales: AI and ML can help hyper-customize marketing initiatives based on audience segmentation and sentiment analytics. The various segments where AI & ML can do wonders are predictive targeting, predictive lead scoring, customer lifetime, value forecasting, recommendation, churn prediction, etc.

Insurance actuary: AI and ML can create more accurate exposure analysis models, specifically in trend-based areas such as catastrophe modeling.

RPA in distribution: Robotic process automation (RPA) in conjunction with ML can significantly improve various automation process like agent support, form filling, performance evaluations, etc. Many vendors such as Blue Prism, UIPath, Automation Anywhere, Edge Verve, Kofax Kapow, Pega Systems, Help Systems, Ant Works, Softomotive, and Argos Laboratory have developed a robust ecosystem, with some implementing their own powerful ML algorithms.

Renewals/reinsurance: AI and ML can assist in price optimization, negotiating the best value based on risk.

There are many tools in the reinsurance space especially in reinsuring weather derivatives such as agriculture, windstorm, hurricane, and other CAT (Catastrophe) lines of business where ML has significant contribution.

6 Business Houses Practising AI and ML in P&C Sector

AI and ML opens up a world of new opportunities for P&C insurance. Below are some of the examples of application of AI and ML in P&C by some of the business giants in insurance sector.

State Farm: State farm has used AI and ML in auto sector. They used the technology of computer vision to identify distracted drivers; they used image classification on two main photo regions: the head region and the bottom-right quarter where the driver's hand normally appears.

Liberty Mutual: In January 2017, Liberty Mutual has plans to develop automotive apps with AI capability and products aimed at improving driver safety. The insurance company is already working on it to help drivers involved in a car accident to quickly assess the damage to his car in real time using a smartphone camera.

Allstate: Allstate is all set to develop a virtual assistant called ABle (the Allstate Business Insurance Expert) to help business grow by data analysis.

Allstate is also working on ML technology to shrink customer-waiting times. It has developed a new AI-enabled customer support mechanism, which can process an impressive 25,000 queries per month, reducing lost business volume.

Ant Financial: Ant Financial, a company part of the Alibaba group, is a Chinese fintech company. The company created a software called Ding Sun Bao, which can analyze claims for vehicle damage and handles claims using ML vision. Ant Financial also offers a solution for automobile insurance with the help of AI and ML technologies called auto insurance points.

Tractable: Tractable is a UK based company which has created a tool, which can help insurance agencies automate the claims process using machine vision.

Geico: Geico offers a virtual assistant and named Kate, which it claims can help customers answer questions they have about their auto insurance using what seems to be natural language processing.

Progressive: Progressive company has developed a Chatbot called Flo, which can help customers using natural language processing and cloud-based API insurance data to alter payment schedules, file insurance claims, and request auto insurance quotes.

7 Challenges in Implementing Machine Learning

Below are some of the challenges faced by insurers while implementing ML in practice.

Training: ML based intelligent systems must be trained in a domain, e.g., claims or billing for an insurer, which requires some extra training. Insurers find it hard to support for training the AI and ML based model.

Data quality: The quality of data used to train predictive models is very important. The more accurate the data is the better is the accuracy of prediction.

Profit prediction: It is not very easy to predict profit that machine learning can bring to a project. Budgeting is a tough job in projects with ML application. Therefore, it is very difficult to predict the return on investment.

Data security: The huge amount of data used for ML algorithms needs to be protected and thus created an additional security risk for insurance companies. With such an increase in collected data and connectivity among applications, there is a risk of data leakage and security breaches.

Time consuming implementation: ML implementation is time consuming because of the rigidity of business. Model creation with proper dataset is time consuming, and end user training is also needed.

Lack of talent: ML is an emerging technology, and very few specialists are available in this field.

Dataset preparation for mode: Although data is available in the market, but preparing dataset for the model is a tedious job. ML specialist must be familiar with various dataset related problem and should be familiar with various types of algorithms, so that appropriate algorithm should be used for a model.

Infrastructure: ML needs huge dataset. Legacy system cannot hold such a bulk dataset and results in failure of ML implementation. So proper infrastructure should be there to support ML operations.

8 Conclusion

ML is an emerging technology and is moving extremely fast making its ways into various fields of the business. P&C insurance industry is also trying to reap the benefit from it like other fields of business. Thus, the fact that insurance companies are actively using data science analytics, which uses statistics, is not giving amazing result. So, the aim of applying data science analytics applying ML models in the insurance is to optimize marketing strategies, to improve the business, to enhance the income, and to reduce costs. In this paper, we presented several ML techniques to analysis the insurance claims efficiently and compare their performances using various metrics. The applications of Machine Learning (ML) are spreading very fast in every segments of P&C insurance. It is already in use in claim processing, Chatbots for customer service, automobile industry, underwriting, and various textual analysis, thereby enhancing customer experience and saving time and cost. Thus, ML is replacing the paper-based work in insurance sector especially in P&C sector with digital transformations. It is unclear just how the technology's role in the ecosystem will evolve further, but it is certain that its impact on the insurance sector will only grow leaps and bounds replacing the paper-based processes.

References

1. https://www.ijitee.org/wp-content/uploads/papers/v8i6s4/F11180486S419.pdf
2. https://www.mckinsey.com/~/media/McKinsey/Industries/Financial%20Services/Our%20Insights/2019%20global%20insurance%20trends%20and%20forecasts/2019-Global-Insurance-Pools-trends-and-forecasts-P-and-C-and-health-insurance-vF
3. https://medium.com/activewizards-machine-learning-company/top-10-data-science-use-cases-in-insurance-8cade8a13ee1
4. https://emerj.com/ai-sector-overviews/machine-learning-at-insurance-companies/
5. https://www.chisel.ai/blog/property-and-casualty-insurance-companies-are-in-love-with-ai
6. https://emerj.com/ai-sector-overviews/ai-auto-insurance-current-applications/
7. https://www.forbes.com/sites/louiscolumbus/2020/01/19/roundup-of-machine-learning-forecasts-and-market-estimates-2020/#ef4b7805c020
8. https://www.birlasoft.com/articles/ai-and-machine-learning-in-p-and-c-insurance-technology
9. Source: Statista, Machine Learning Tops AI Dollars, May 10, 2019.
10. https://www.orbisreports.com/global-pc-insurance-software-market/
11. https://iireporter.com/ais-tremendous-potential-for-pc-insurance-sma-study/
12. https://www.willistowerswatson.com/en-IN/Insights/2018/11/Finding-the-future-for-AI-in-P-and-C-insurance

Exploratory Data Analysis on IPL Data

Santanu Mohapatra, Angana Goswami, Ashi Singh, Vikash Kumar Singh, Biswaraj Sen, and Kalpana Sharma

Abstract Exploratory data analysis for Indian Premier League (IPL) data is widely covered in the field of data analytics and machine learning problem and specifically based for match prediction and team prediction IPL. IPL is a global tournament with over 1200 players participating in the auction every year, and it draws the attention of many cricketing fans around the globe. So, the proposed system predicts the team and winner with an accuracy of over 50% and engages many cricketing fans. The goal of this paper is to develop a system for predicting the Dream11 team and the possible team to win the match every day. The common attributes to be included while choosing a team involves a player's batting average, batting strike rate, bowling's economy rate, match ratings, and his experience; similarly, while predicting match winner, the machine will see different factors like toss outcome, venue of the match, team track record in that venue, and team to play that match which are stored in the datasets and generated according to the conditions given to the machine and the criteria's generated for the team selection. The objective of the paper is to develop a model which will give the accuracy of more than 50%, and it satisfies all the necessary conditions for team selection. And, the best part is that it will create a buzz around many cricketing fans around the globe and keep them engaged throughout the course of the tournament and it helps in increasing its fan base.

S. Mohapatra · A. Goswami · A. Singh · V. K. Singh (✉) · B. Sen · K. Sharma
Department of Computer Science & Engineering, Sikkim Manipal Institute of Technology, Sikkim Manipal University, Sikkim, India
e-mail: vikash.s@smit.smu.edu.in

S. Mohapatra
e-mail: santanu_201700005@smit.smu.edu.in

A. Goswami
e-mail: angana_201700067@smit.smu.edu.in

A. Singh
e-mail: ashi_201700186@smit.smu.edu.in

B. Sen
e-mail: biswaraj.s@smit.smu.edu.in

K. Sharma
e-mail: kalpana.s@smit.smu.edu.in

H. K. D. Sarma et al. (eds.), *Contemporary Issues in Communication, Cloud and Big Data Analytics*, Lecture Notes in Networks and Systems 281,
https://doi.org/10.1007/978-981-16-4244-9_26

Keywords IPL · EDA · Data analysis · Prediction · Regression

1 Introduction

Data analysis is the process of systematically applying various statistical techniques and logical functions to the given data to describe and illustrate, condense and recap, and evaluate data and draw necessary information to build the model [1]. An essential component of ensuring data integrity is the accurate and appropriate analysis of research findings. In statistics, exploratory data analysis is an approach to get accurate and vast study on the datasets using visualization tools. A statistical model can be used or not, but primarily, EDA is for seeing what the data can tell us beyond the formal modeling or hypothesis testing task [2]. Various statistical techniques are used for vast observation on data:

The statistical mean refers to the mean or average that is used to derive the central tendency of the data in question. It is determined by adding all the data points in a population and then dividing the total by the number of points. The resulting number is known as the mean or the average [3].

"The median is a simple measure of central tendency. To find the median, arrange the observations in order from the smallest to the largest value. If there is an odd number of observations, the median is the middle value." If there is an even number of observations, the median is the average of the two middle values [4].

"The mode of a set of data values is the value that appears most often." If X is a discrete random variable, the mode is the value X at which the probability mass function takes its maximum value.

"Correlation analysis is a method of statistical evaluation used to study the interdependency of a relationship between two, numerically measured, continuous variables (e.g., height and weight) [5]."

Collect all the data related to IPL. Those datasets are comprised of ball-by-ball data, season data, player's data, and team's data. There is a possibility that data may be organized in a random way, in that case, arrange the data according to their properties individually. Repeated data or null data can be eliminated [6].

Identify the different features of players for each team to be selected for the pools. A new cleaned dataset will be obtained after dimensionality reduction, and applying supervised learning technique to that cleaned datasets finalizes 30–40 players on each team for next season from the pool of players [7]. Once the pool of players is obtained, predict the Dream11 players in each team for every match. This can be done for different pool games in IPL seasons and also helps in advising the team management the best possible playing eleven for every match, as it helps in studying all the aspects regarding the team selection focusing all the characteristics of different players and finalizing the set of players who will be the part of Dream11 team [8].

The proposed system predicts the team and winner with an accuracy of over 50% and engages many cricketing fans. The common attributes to be included while choosing a team involves a player's batting average, batting strike rate, bowling economy rate, and match ratings. While predicting the match winner, the machine will see different factors like toss outcome, venue of the match, team track record in that venue, and team to play that match. Data are stored in the data frames, and according to the conditions given to the machine, the criteria are generated for the team selection [9].

2 Literature Survey

2.1 ERP and Application Integration

"Marinos Themistocleous, Zahir Irani, and Robert M. O'Keefe, August 2001: This paper proposes to identify, analyze, and present the problems of ERP systems, as well as examines new approaches for AI." In doing so, a multi-choice questionnaire has been designed and was distributed to ERP specialists over the Internet. Responses show that ERP systems amplified the need for integration, as existing systems have to be incorporated with ERP applications. AI securely incorporates functionality from disparate applications and has shown to lead to the development of new strategic business solutions for enterprises [10].

2.2 Exploratory Analytics in Learning Analytics

"David Gibson and Sara de Freitas, March 4, 2015: From exploratory analytics in learning analytics by David Gibson and Sara de Freitas, this article summarizes the methods, observations, challenges, and implications for exploratory analysis drawn from two learning analytics research projects. The cases include an analysis of a game-based virtual performance assessment and an analysis of data from 52,000 students over a 5-year period at a large Australian university." The complex datasets were analyzed and iteratively modeled with a variety of computationally intensive methods to provide the most effective outcomes for learning assessment, performance management, and learner tracking. The article presents the research contexts, the tools and methods used in the exploratory phases of analysis, and the major findings and the implications for learning analytics research methods [2].

2.3 Dream11 Literature and Working

Wikipedia study, August 2018: By studying this literature survey for the Dream11 fantasy cricket app, it clarifies basic concepts for the app creation and its working fundamentals. The concepts regarding team selection and app execution are being referred in the proposed data analysis for IPL data. The toughest challenge these apps can face is that it needs a complete understanding, and the data need to be up-to-date in order to give the accuracy in the prediction which by itself makes the creation or development of these apps very challenging. Each player who is eligible for selection are given with some points, and the fans are given with fixed points and based on that they select the team which according to the fans is a major drawback because they need to compromise a player due to less available points [11].

The Dream11 system that focused on designing was a bit different from the app already running wild on the Internet [12]. The proposed system focused on initially creating a pool of all the players eligible for selection based on their past performances. This system does not have any point criteria for the team selection, if you are a good player, then you are directly eligible for the team selection, and system can assure that this technique will provide the best playing 11 than the existing Dream11 app because the proposed system considers all the aspects to predict the playing 11 of the match including past performances, average, strike rates, favorite opposition of that particular player, and favorite venue. And, it also helps to predict the match outcomes compared to the actual results, and it gives the accuracy of more than 50% which is not there in the existing Dream11 app [13]. And also in the existing Dream11 app, this system also gives the option to select only one wicketkeeper in the match out of the given; it does not matter what the records of those players is [13]. But, in our proposed analysis, more than one wicket keeper can be selected, one as a keeper and another one as a fielder based on the player's past performance so that they can contribute to the victory of the team. It does not matter how many wicketkeepers; it only focuses on giving the best possible playing 11 [14].

3 Problem Definition

IPL is one of the most popular T20 leagues around the globe. Since 2008, this event changes the complexion of T20 leagues, and sometimes, it refers as the toughest T20 league. IPL produces so many great players, and hopefully, it continues to do the same in the future. As the challenges and toughness of this tournament increases, it also becomes a headache for the franchises that which player is good enough to be included in their IPL squad based on their performances. IPL draws interest of millions of fans worldwide, keeps them engaged throughout the tournament, and draws a lot of viewership like that. There have been twelve seasons of the IPL tournament. Therefore, there is a need to develop a system which will keep all the records of the player's data and past performances and suggest the team management the best possible playing 11 for every match and based on the previous data and venue records as well as player's individual records and predict the outcome of every match. This paper takes the problem as to develop a system for exploratory data analysis on IPL data which will predict the Dream11 for every match, as well as predict the outcome of every match.

4 Fundamental Design

In the proposed system, univariate analysis technique is used to quantize within a single variable and take all the possible noting. For example, consider the team data and note their plottings and expressions. The same can be done with the player's data and season's data. In the proposed system, the bivariate analysis technique is used to quantize within two variables and take all the possible noting. For example, consider the data between players and their respective team, plotting their relationships considering the player_id and the team_id, and eliminating the null values. Same can be done for other datasets like the relationship between player data and the ball-by-ball data and the relationships between team data and season data.

Principal component analysis (PCA) is a statistical procedure that uses an orthogonal transformation to convert a set of observations of interdependent variables or correlated variables and reduce its dimensionality in such a way that it does not alter the given data and still gives a high accuracy using the reduced data [15]. It is a feature extraction technique. It is a dimension reduction tool. Consider the data such as players and teams, check all of their corresponding values to their respective tuples, and eliminate all of those rows that can have the repeated data whose elimination from those sets cannot alter other data; i.e., it contains the subset.

Collect all the IPL data and categorize them based on their characteristics. Compare the correlation among different data. Taking all the datasets, predict the initial state. Remove the constraints (if any). Check the accuracy of players' dataset such as batsman, bowlers, and all-rounders [16]. Compare the roughly predicted pool with the pool generated by the system. For example, with the IPL data available, a batsman strike rate can be predicted based on the number of balls he faced and his total scores, and the average of a batsman can also be predicted, and also the economy rate of the bowler can be predicted [17].

From the above design depicted in Fig. 1, the following steps are used to get the Dream11 team prediction and match outcome prediction:

Collect all the data related to IPL such as player's data, team's data, season's data, and ball-by-ball data. Data may be collected in a random way, so group the datasets based on their properties. Process and manipulate the datasets. Clean the datasets by removing all the null values and perform the technique of dimensionality reduction under principal component analysis.

Classify the types that the data possesses, e.g., nominal and numerical. Initially, use matplotlib library function to see the plotting of various IPL datasets and obtain their relationships graphically. Apply supervised learning technique to obtain certain predictions about the known input data and to support decision making. Use correlation of statistics methods to check the interdependency between the sets of IPL data. Use univariate and bivariate analysis to obtain relationships among various sets of the given IPL data. Use regression methods and obtain the finalized pool of players for each team. Among the pool of players, use the classification technique to obtain the Dream11 for each team in every match. And, we have the datasets of the previous

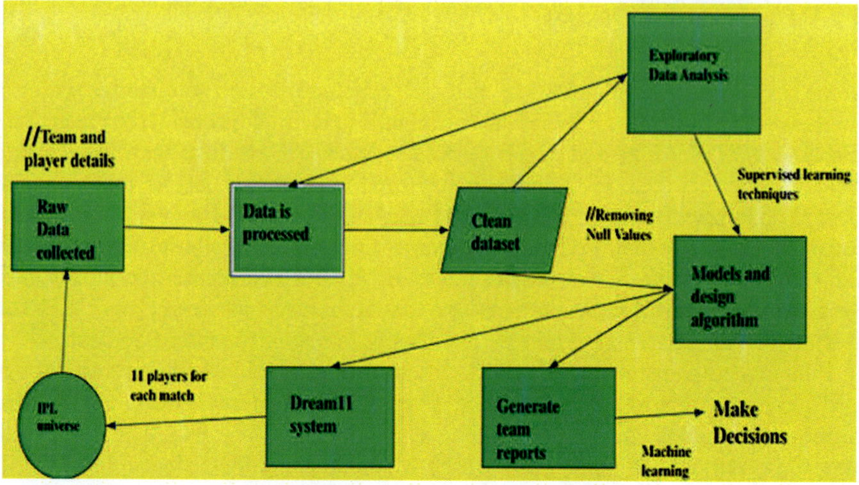

Fig. 1 Block diagram of Dream11 team prediction system

results of various IPL matches, and we will compare those results with the prediction that we want to get the accuracy of our prediction model.

5 Implementation

The data given were extracted by eliminating their null values from the huge set of data. The method used is principal component analysis (PCA), i.e., feature selection and feature extraction. Then, based on different franchises of the Indian Premier League (IPL), the players are categorized, and after extraction, the pool of players according to their teams are generated. The data are overwritten based on the latest statistics, and the heat map for the null-valued data is generated. Now, various plots are obtained such as distribution of teams, distribution of players, nation of players, and year-wise statistics. Then, out of the 30-man squad for each team, 15 players are finalized based on their previous performances. Now, two teams are taken out of eight teams for a match, and based on the 30 players combined from both the teams, using various machine learning techniques, top 11 players are selected to form a Dream11 team for that match. Each team must have a minimum of seven Indian players and a maximum of four overseas players. The team must consist of six batsmen and five bowlers or five batsmen and five bowlers and an all-rounder. One player should be selected as a wicketkeeper in the team, and one player should be assigned with the role of captaincy based on their past performances.

Now, a dataset has been taken from Kaggle which contains the past results of each match in each season and their various statistics. These data have all the records of all the matches conducted, and their results are used for predicting the outcome of every match, and it comprise various attributes such as season, city, date, team1, team2, toss_win, toss_decision, result, and match_winner. From these data, the results can be predicted and accuracy of the results can be checked. The prediction was checked in binary values consisting of "1" and "0". Predicting the outcome of every match is based on previous statistics. The value "1" under the attributes represents the team wins the match, "0" under the attributes represents the team loses the match, and "1" under both the teams attributes represents results cannot be determined. The various classifiers which are used are logistic regression classifier, decision tree classifier, SVM classifier, and random forest classifier. Using four classifiers is a good idea to obtain the highest accuracy to get a best-fit model. The data are tested and validated before proceeding which includes the player's data, team data, and previous results data. The benefit of this approach is that a clear picture can be seen about how the model reacts to previously unseen data. This approach is useful to handle large amounts of data and the data having null values as this approach can test each attribute. The most basic method is the train/test split. The principle is simple, "simply split the IPL data randomly into roughly 70% used for training the model and 30% for testing the model. The validation dataset is different from the test dataset that is also held back from the training of the model, but is instead used to give an unbiased estimate of the skill of the final-tuned model when selecting between final models." And in the data, first identify the null values, eliminate the attributes containing null values, and then take the data that are validated for making the model. The result is that the model shows more than 50% accuracy—"best-fit model."

6 Results and Discussion

Initially, the systems provide with six datasets such as "match," "player," "ball-by-ball," "team," "player_match," and "season," and by performing PCA, univariate, and bivariate analysis, we get the reduced data and eliminate those player's data who are no longer playing in the international matches to give the best results which will solely focus on the current cricket stars around the globe. Then, the system gets a pool of player's datasets who are eligible for selection by performing these analyses, and then, the system extracts the datasets from Kaggle and overwrites on our datasets so that the missing data can be interpreted, and similarly, the system generates the actual match outcome results from the year 2016 to 2019 to be used for prediction of the match (Fig. 2).

The finalized datasets for the pool of players are valid for the team selection, but they require some analysis.

The match outcome datasets are to be compared with our prediction to calculate the accuracy of the prediction.

Now, predict the Dream11 team for a given match. Let us take one match between Delhi Capitals and Chennai Super Kings. Criteria for selection will be top six batsman or top five batsman and one all-rounder based on their past performances and top five bowlers.

Country	Player	Team	Runs	BattingAv	StrikeRate	Wickets	EconomyF	Is_Keeper	Is_Captain
India	MS Dhoni	Chennai S	4432	42.21	137.85	0	0	1	1
India	Suresh Ra	Chennai S	5368	33.34	137.11	25	7.39	0	0
India	Ambati R	Chennai S	3300	28.7	125.95	0	0	0	0
Australia	Shane Wa	Chennai S	3575	29.91	139.54	92	7.93	0	0
South Afri	Faf du Ple	Chennai S	1853	31.41	126.74	0	16	0	0
India	Murali Vij	Chennai S	2587	26.4	122.84	0	8.17	0	0
India	Kedar Jadl	Chennai S	1079	22.96	126.49	0	0	0	0
India	Ravindra J	Chennai S	1927	24.09	122.66	108	7.58	0	0
India	Rituraj Ga	Chennai S	0	0	0	0	0	0	0
West Indi	Dwayne B	Chennai S	1483	23.17	128.29	147	7.8	0	0
India	Karn Shar	Chennai S	317	15.1	115.69	54	7.82	0	0
South Afri	Imran Tah	Chennai S	20	5	74.07	79	7.88	0	0
India	Harbhajar	Chennai S	829	15.07	138.17	150	7.05	0	0
New Zeal	Mtchell S	Chennai S	32	32	139.13	4	6.71	0	0
India	Shardul T	Chennai S	36	7.2	171.43	36	9.04	0	0
India	KM Asif	Chennai S	0	0	0	3	12.5	0	0
India	Deepak C	Chennai S	71	11.83	161.36	33	7.63	0	0
India	N Jagadee	Chennai S	0	0	0	0	0	1	0
South Afri	Lungi Ngi	Chennai S	0	0	0	11	6	0	0
India	Monu Sin	Chennai S	0	0	0	0	0	0	0
England	Sam Curra	Chennai S	95	23.75	172.73	10	9.79	0	0
India	Piyush Ch	Chennai S	584	11.92	111.45	150	7.82	0	0
Australia	Josh Hazl	Chennai S	0	0	0	0	0	0	0
India	R Sai Kish	Chennai S	0	0	0	0	0	0	0
India	Shikhar D	Delhi Capi	4578	33.17	124.78	4	8.25	0	0
India	Prithvi Sh	Delhi Capi	598	23.92	141.04	0	0	0	0

Sheet1

Fig. 2 Finalized pool for the team selection

Combining selections from both the teams (Fig. 3).

This dataset gives a briefing of the results of the previous matches and can be used to compare with the results of the predicted matches to obtain the accuracy of the results. From the data which were obtained for batsman, bowlers, and all-rounders, we concatenate those three datasets to obtain the Dream11 team, and it is the best possible playing 11 that can be predicted for the match. And, it meets all the team selection criteria as in this team there are eight Indian players and three overseas players (Fig. 4).

id	season	city	date	team1	team2	toss_winr	toss_deci	result	dl_applie	winner	win_by_r	win_by_w	player_of	venue	umpire1	umpire2	umpire3
1	2018	Mumbai	07-Apr	Mumbai I	Chennai S	Chennai S	field			Chennai Super Kings							
2	2018	Mohali	08-Apr	Delhi Dar	Kings XI P	Kings XI P	field			Kings XI Punjab							
3	2018	Kolkata	########	Royal Cha	Kolkata K	Kolkata K	field			Kolkata Knight Riders							
4	2018	Hyderaba	09-Apr	Rajasthan	Sunrisers	Sunrisers	field			Rajasthan Royals							
5	2018	Chennai	########	Kolkata K	Chennai S	Chennai S	field			Chennai Super Kings							
6	2018	Jaipur	11-Apr	Rajasthan	Delhi Dar	Delhi Dar	field			Rajasthan Royals							
7	2018	Hyderaba	########	Mumbai I	Sunrisers	Sunrisers	field			Sunrisers Hyderabad							
8	2018	Bangalore	13-Apr	Kings XI P	Royal Cha	Royal Cha	field			Royal Challengers Bangalore							
9	2018	Mumbai	14-Apr	Mumbai I	Delhi Dar	Delhi Dar	field			Delhi Daredevils							
10	2018	Kolkata	14-Apr	Kolkata K	Sunrisers	Sunrisers	field			Sunrisers Hyderabad							
11	2018	Bangalore	15-Apr	Rajasthan	Royal Cha	Royal Cha	field			Rajasthan Royals							
12	2018	Mohali	15-Apr	Kings XI P	Chennai S	Chennai S	field			Kings XI Punjab							
13	2018	Kolkata	16-Apr	Kolkata K	Delhi Dar	Delhi Dar	field			Kolkata Knight Riders							
14	2018	Mumbai	17-Apr	Mumbai I	Royal Cha	Royal Cha	field			Mumbai Indians							
15	2018	Jaipur	18-Apr	Rajasthan	Kolkata K	Kolkata K	field			Kolkata Knight Riders							
16	2018	Mohali	19-Apr	Kings XI P	Sunrisers	Kings XI P	bat			Kings XI Punjab							
17	2018	Pune	20-Apr	Chennai S	Rajasthan	Rajasthan	field			Chennai Super Kings							
18	2018	Kolkata	21-Apr	Kolkata K	Kings XI P	Kings XI P	field			Kings XI Punjab							
19	2018	Bangalore	21-Apr	Delhi Dar	Royal Cha	Royal Cha	field			Royal Challengers Bangalore							
20	2018	Hyderaba	22-Apr	Chennai S	Sunrisers	Sunrisers	field			Chennai Super Kings							
21	2018	Jaipur	22-Apr	Mumbai I	Rajasthan	Mumbai I	bat			Rajasthan Royals							
22	2018	Delhi	23-Apr	Kings XI P	Delhi Dar	Delhi Dar	field			Kings XI Punjab							
23	2018	Mumbai	24-Apr	Sunrisers	Mumbai I	Mumbai I	field			Sunrisers Hyderabad							
24	2018	Bangalore	25-Apr	Royal Cha	Chennai S	Chennai S	field			Chennai Super Kings							
25	2018	Hyderaba	26-Apr	Sunrisers	Kings XI P	Kings XI P	field			Sunrisers Hyderabad							
26	2018	Delhi	27-Apr	Delhi Dar	Kolkata K	Kolkata K	field			Delhi Daredevils							

matches - Copy

Fig. 3 Result datasets of previous matches

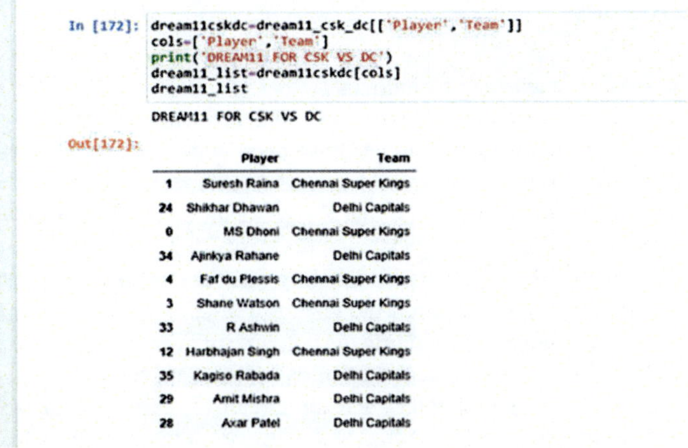

```
In [172]: dream11cskdc=dream11_csk_dc[['Player','Team']]
          cols=['Player','Team']
          print('DREAM11 FOR CSK VS DC')
          dream11_list=dream11cskdc[cols]
          dream11_list

DREAM11 FOR CSK VS DC
```

Out[172]:

	Player	Team
1	Suresh Raina	Chennai Super Kings
24	Shikhar Dhawan	Delhi Capitals
0	MS Dhoni	Chennai Super Kings
34	Ajinkya Rahane	Delhi Capitals
4	Faf du Plessis	Chennai Super Kings
3	Shane Watson	Chennai Super Kings
33	R Ashwin	Delhi Capitals
12	Harbhajan Singh	Chennai Super Kings
35	Kagiso Rabada	Delhi Capitals
29	Amit Mishra	Delhi Capitals
28	Axar Patel	Delhi Capitals

Fig. 4 Predicted Dream11 team

Figure 5 gives the accuracy for the predicted test set. Four classifiers are taken to obtain the best classifier to be used based on their ratio of the accuracy obtained.

Accuracy of logistic regression classifier on test set is 0.5461.

Accuracy of decision tree classifier on test set is 0.5395.

Accuracy of SVM classifier on test set is 0.5789.

Accuracy of random forest classifier on test set is 0.5929.

This gives the prediction accuracy of more than 50%. So, this model can be optimized and classified as a best-fit model.

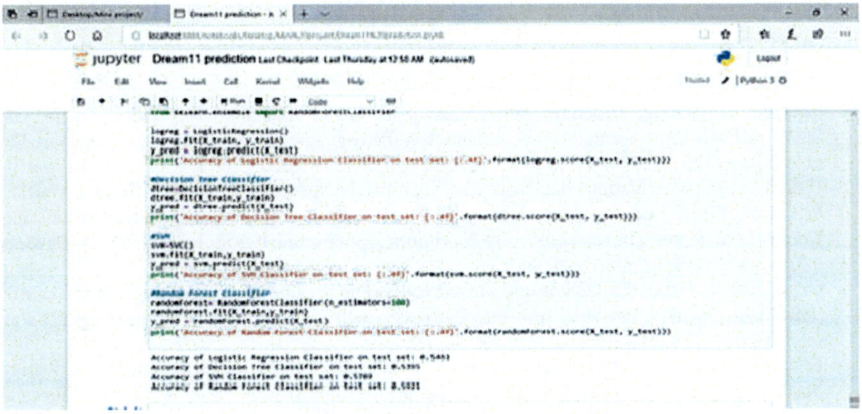

Fig. 5 Accuracy of various classifiers for prediction

7 Conclusions

"These days, machine learning techniques are being widely used to solve real-world problems by storing, manipulating, extracting, and retrieving data from large sources. Supervised machine learning techniques have been widely adopted; however, these techniques prove to be very expensive when the systems are implemented over a wide range of data. This is due to the fact that a significant amount of effort and cost is involved because of obtaining large labeled datasets. Thus, active learning provides a way to reduce the labeling costs by labeling only the most useful instances for learning." In this paper, the required model for prediction of the Dream11 team is achieved. And, it also helps in predicting the outcome of every match. It was observed that this machine works with an accuracy of over 50%, and so, it is a best-fit model. From the model, it can be concluded that the Dream11 team that can be predicted gives a result almost the best possible 11 for every match. Machine learning and exploratory data analysis algorithms perform more efficiently for a classification task when they are combined together. For the prediction of the correct output class, the combined learner selects the class to which the highest probability has been assigned among all the learners. Further, it can be concluded that feature selection is important but only as long as it does not decrease the efficiency of the learners by discarding too many attributes on the basis of their relevance. Principal component analysis is also used to handle huge dataset, and some of the attributes may not be relevant as they contain null values, and so, feature extraction process can be used to eliminate those attributes. And, finally taking the finalized data, those data can be tested, and applying various machine learning techniques, the finalized team and outcome of the match are predicted.

References

1. Shekhar, S., Xiong, H.: Exploratory Data Analysis. Published in encyclopedia of GIS 2008 retrieve from https://www.semanticscholar.org/paper/Exploratory-Data-Analysis-Shekhar-Xiong/72fd91ed74ab310e474071e2bd8f9a0359a0c8f0?p2df
2. Hair, J.F., Jr., Multivariate data analysis, a global perspective. Kennesaw State University William C. Black Louisiana State University Barry J. Babin University of Southern Mississippi Rolph E. Anderson Drexel University from https://pdfs.semanticscholar.org/6885/bb9a29e8a5804a71bf5b6e813f2f966269bc.pdf
3. Data Analysis Using Regression and Multilevel/Hierarchical Models Andrew Gelman Jennifer Hill June 13, 2012 retrieve from http://lac-essex.wdfiles.com/local--files/meetings1213/gelman_1.pdf
4. Batch Normalization: Accelerating Deep Network Training by Reducing Internal Covariate Shift Sergey Ioffe Christian Szegedy Google, 1600 Amphitheatre Pkwy, Mountain View, CA 94043 retrieve from http://proceedings.mlr.press/v37/ioffe15.pdf
5. Statistical Topological Data Analysis using Persistence Landscapes Peter Bubenik Department of Mathematics Cleveland State University Cleveland, OH 44115-2214, USA retrieve from https://www.jmlr.org/papers/volume16/bubenik15a/bubenik15a.pdf
6. An Introduction to Categorical Data Analysis, Department of Statistics University of Florida Gainesville, Florida retrieve from https://www.isical.ac.in/~arnabc/discrete/ICDA2e.pdf

7. Meta-Learning Update Rules for Unsupervised Representation Learning Luke Metz, Niru Maheswaranathan, Brian Cheung, Jascha Sohl-Dickstein retrieve from https://arxiv.org/abs/1804.00222
8. Cook, H.E., Johnson, P.D., Matti, J.C., Zemmels, I.: Methods of Sample Preparation and X-Ray Diffraction Data Analysis, X-Ray Mineralogy Laboratory, Deep Sea Drilling Project, University Of California, Riverside1, University of California, Riverside retrieve from https://pdfs.semanticscholar.org/0e78/699f3c5f17b50ea71c79dfd058f84a76b448.pdf
9. Bayesian data analysis: what it is and what it is not Prof. Andrew Gelman Dept. of Statistics Columbia University Talk for Columbia University Department of Computer Science, 15 Dec 2003 retrieve from- https://stat.columbia.edu/~gelman/presentations/cs.pdf
10. https://www.edx.org/course/foundations-of-data-science-prediction-and-machine#::text=Machine%20learning%20is%20a%20way,on%20are%20regression%20and%20classification. Machine learning strategies
11. Challenging Common Assumptions in the Unsupervised Learning of Disentangled Representations Francesco Locatello, Stefan Bauer, Mario Lucic, Gunnar Rätsch, Sylvain Gelly, Bernhard Schölkopf, Olivier Bachem retrieve from https://arxiv.org/abs/1811.12359
12. Social Influence as Intrinsic Motivation for Multi-Agent Deep Reinforcement Learning Natasha Jaques, Angeliki Lazaridou, Edward Hughes, Caglar Gulcehre, Pedro A. Ortega, DJ Strouse, Joel Z. Leibo, Nando de Freitas retrieve from https://arxiv.org/abs/1810.08647
13. Dropout: A Simple Way to Prevent Neural Networks from Overfitting Nitish Srivastava Geoffrey Hinton Alex Krizhevsky Ilya Sutskever Ruslan Salakhutdinov Department of Computer Science University of Toronto 10 Kings College Road, Rm 3302 Toronto, Ontario, M5S 3G4, Canada retrieve from https://jmlr.org/papers/volume15/srivastava14a/srivastava14a.pdf
14. https://towardsdatascience.com/a-one-stop-shop-for-principal-component-analysis-5582fb7e0a9c
15. https://www.datacamp.com/?utm_source=adwords_ppc&utm_campaignid=1242944157&utm_adgroupid=58673831928&utm_device=c&utm_keyword=%2Bdatcamp&utm_matchtype=b&utm_network=g&utm_adpostion=1t1&utm_creative=340731356773&utm_targetid=kwd361360284674&utm_loc_interest_ms=&utm_loc_physical_ms=9040190&gclid=EAIaIQobChMIytbUrYaP5wIV1I6PCh2l6ALOEAAYASAAEgIPg_D_BwE
16. https://dataaspirant.com/2017/01/30/how-decision-tree-algorithm-works
17. https://medium.com/@tapan.soni/dream11-android-app-user-research-case-study-2f874644d083

Online Music Recommendations Using User's Zodiac Sign

Sudipta Chakrabarty, Anushka Bhattacharya, Md. Ruhul Islam, and Hiren Kumar Deva Sarma

Abstract Normally, music recommendation system is the software that is used to create user's personalized list of music. The paper proposed an approach of music recommendation system based on user's zodiac sign. According to the date of birth of a person, there must be one specific zodiac sign, and each zodiac sign may have some basic features, like lucky number, suitable color, suitable stone, behavior, habit, and so on. According to the Indian classical music, there are 72 Melakarta or parent ragas and these ragas are subdivided into 12 chakras or cycles, and each cycle consists of six individual ragas. All the cycles have also some specific features like, color, stone, position, etc. In this work, the key feature is the stone. So after finding the suitable stone of one user, it has been implementing the key feature and mapping the required raga cycle. Any song from that raga cycle is suitable for that person; therefore, the work offers to create the song list generation of one specific raga cycle which contains six ragas.

Keywords Music recommendation system · Indian classical music · Raga cycle · Zodiac sign · Melakarta raga

1 Introduction

The background of the work is fully based on the concept of zodiac. Zodiac consists of twelve signs like Aries, Taurus, Gemini, Cancer, Leo, Virgo, Libra, Scorpio, Sagittarius, Capricorn, Aquarius, and Pisces. Normally, late March is the starting point of the Astrological year when the planet Sun enters Aries, and after that, the signs change

S. Chakrabarty (✉)
Department of Computer Applications, Techno India, Salt Lake, Kolkata, West Bengal, India

A. Bhattacharya
Department of Computer Science, Bengal Institute of Technology, Kolkata, West Bengal, India

Md. R. Islam
Department of Computer Science, SMIT, Majhitar, Rangpo, Rangpo, East Sikkim, India

H. K. D. Sarma
Department of Information Technology, SMIT, Majhitar, Rangpo, Rangpo, East Sikkim, India

H. K. D. Sarma et al. (eds.), *Contemporary Issues in Communication, Cloud and Big Data Analytics*, Lecture Notes in Networks and Systems 281,
https://doi.org/10.1007/978-981-16-4244-9_27

at one month interval. Some of the signs grouped together to create four elements like—Fire (Agni), Earth (Pruthvi), Air (Vayu), and Water (Jal). Fire includes Aries, Leo, Sagittarius signs, Earth includes Capricorn, Taurus, Virgo signs, Air includes Libra, Aquarius, Gemini signs, and Water includes Cancer, Scorpio, Pisces signs. Zodiac consists of another two components like houses and planets. Houses denote conditions of human life, environments, etc., whereas planets denote human mind and their mental tendency. In the fifth house of the planet Sun, music has a strong appeal. In the first and third house of the Venus planet, music is used as a pleasant element. In the first house of the planet Neptune, music is implemented as devotional element. The human who has the zodiac sign Libra in two different planets like Mercury and Neptune, they have the great musical senses. The humans are normally artistic and have musical talents both from Taurus and Libra zodiac sign of Neptune planet. The persons having either from zodiac sign Cancer or Libra have also very strong musical appeal. Therefore, Zodiac signs and music recommendations are tightly related with each other [1].

Melakarta (Parent) ragas are the fundamental ragas, and other ragas may be generated from these ragas. These ragas are also called as sampoorna or perfect ragas as they consist of all the seven notes in both the ascending and the descending mode of the octave. According to the Indian classical music, there are 72 Melakarta or parent ragas, and these ragas are subdivided into 12 cycles, and each cycle consists of six individual ragas. Both Sadja (Sa) and Panchama (Pa) are common for all 72 Melakarta ragas; while Risabha (Re), Gandhara (Ga), and Madhyama (Ma) notes remain the same for each of the six ragas in each cycle, either Dhaivata (Dha) or Nisada (Ni) or both vary for every raga. These 12 cycles are—(1) Indu, (2) Netra, (3) Agni, (4) Veda, (5) Bana, (6) Ritu, (7) Risi, (8) Vasu, (9) Brahma, (10) Disi, (11) Rudra, and (12) Aditya, respectively. Each raga cycles have also some specific features like, color, stone, position, etc. In this work, the key feature is the stone. Figure 1 depicts the 12 raga cycles with the names of suitable stones.

This is a software product-based Web application which provides the facility to generate personalize playlist of songs online according to zodiac sign. In this work, administrator has the privilege to check user details and to insert, update, and delete data's, add another admin, and update his login credentials. User can visit the Web site without logging in, as well as register login and view zodiac details. Moreover, they can listen to songs online with the play option. Also, user can view songs by categorizing them. It is a music recommendation system using zodiac sign; hence, it has to be needed to build a Web-based application where a customer can generate the current year zodiac details and a particular song for that. There are three parts of the Web application as product specification

Product function overview: This is a Web Application project which provides the facility to generate songs online for listening suitable for each zodiac sign. In this Web site, administrator has the privilege to check user details and to insert, update, and delete data's, add another admin, and update his login credentials. User can visit the Web site without logging in, as well as he/she can register, login, and view zodiac details. Moreover, they can listen to songs online with the play option. Also, user can view songs by categorizing them.

Fig. 1 12 raga cycles with the names of suitable stones

User characteristics: The system is user friendly. The end user can be divided into two categories:

Administrator: Access to master forms for the purpose of insertion, updating, and deletion of databases and viewing the user details, authorize another admin, and update his login credentials. Administrator is the super-user of this Web site, having following functionalities:-

i. Can view all zodiac details.
ii. Has the privilege to insert, update, and delete the database of customers and songs.
iii. Can also authorize a new admin.
iv. Categorize songs based on ragas.

User: Access to his corresponding details. User is the customer who views the Web site and has following functionalities:-

i. Can register and login into the Web site.
ii. Can view and update his/her profile.
iii. Listen to songs online using the play option.

General Constraints and Assumptions: The system is user friendly. The end user can be divided into two categories:

i. The medium of instruction will be English only.
ii. All the software and hardware are assumed to be available with the developers.

2 Related Works

Indian classical music (ICM) is one unexplored area in the world music. One paper proposed an approach that finds out the complexity of any particular song based on one statistical measure [2]. Paper [3] proposes one music software that edits or generates versatile music combine with both vocal and instrument using a specific tempo-based aggregation concept. The paper presents to create a lot of music rhythms based on the memetic algorithm [4]. The paper [5] proposed to compute the music similarity based on coefficient of concurrent deviation. An approach generates the song similarity percentage through correlation coefficient [6]. The paper [7] represents to create personalized song list using user's listening habit and age factor of users for performing online. Again one paper represents the matching similarity between songs by their pitch values through coefficient of variance [8]. The paper [9] proposes a music recommender using different time slots by neural network. The paper [10] introduced an approach that identifies the music rhythm density and rhythm complexity. Paper [11] represents another time slot-based music recommendation system using raga-time database. Paper [12] describes a lot of music survey research and applications in the field of musicology. Paper [13] represents that a song of a particular raga can be represented through unified modeling language. Petri net is one framework of the musical pattern which is described in paper [14]. One paper introduces the mechanism that efficiently chooses the most fitted parent rhythms of a set of rhythm chromosomes for creating offspring rhythm using genetic algorithm optimization technique in the context-awareness pervasive music rhythm learning education pervasive education for computational musicology [15]. Another paper presents a method of automatic raga recognition [16]. A Petri net is a modeling tool, and the paper [17] is used for music pattern analysis for Indian classical music where object-oriented methodology is the basis for analysis of musical patterns. In an effort, the authors have introduced a quality music model used for ICM using genetic algorithm [18].

3 Proposed Work

The primary aim of this work is to create a Web-based application for music depending on user's zodiac predictions of the current year. The basic components of the Web application are as follows:

Step1: Open the Web application "music recommendation system using zodiac sign."

Step2: The following options are available:-

I. Login as customer
   a. View zodiac details
   b. Play song

II. Login as admin
   a. View, add, and update zodiac database
   b. View, add, and update songs
   c. View, add, and update ragas
   d. View user details
   e. Add a new admin
   f. Play song
   g. View and update profile

III. Login as guest
   a. View album
   b. Play song
   c. Register.

The basic workflow of the work as described below :

Step 1: Open the Web application.

Step 2: Then, complete the user log-in, if the user is a new user, then make sign-up process by filled up the user's details like name, address, phone number, email ID, gender, password, confirm password, profile picture upload, etc.

Step 3: After completion of log-in process, insert the date of birth (DOB) of the user, and then the application will display the zodiac sign as well as their lucky number, lucky color, and suitable stone of that particular user.

Step 4: Then, matching the stone with the raga cycle as every raga cycle consists of one specific stone also.

Step 5: Finally, all ragas (six ragas) from that cycle are suitable to listen for the user.

Figure 2 depicts the use-case diagram of admin and use-case diagram of user of the Web application "music recommendation system using zodiac sign."

Figure 3 depicts the activity diagram of user and activity diagram of admin of the Web application "music recommendation system using zodiac sign."

4 Result Set Analysis and Discussion

To establish the proposed work, it has been considered a lot of date of births of the users, and then, the Web application will display the Zodiac sign of the users as well

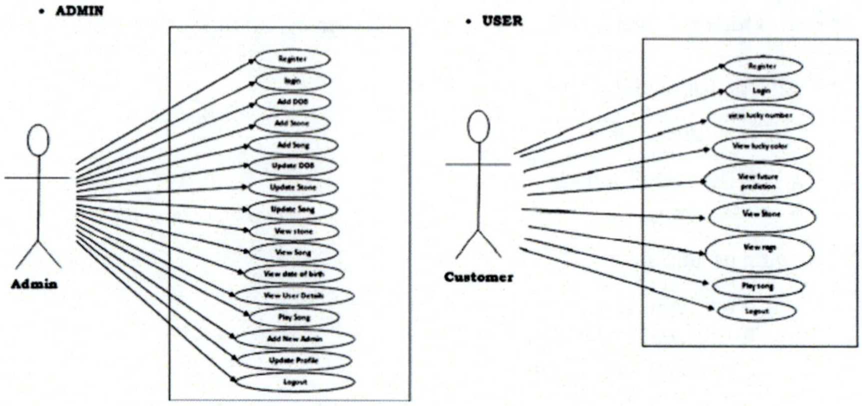

Fig. 2 Use-case diagram of "music recommendation system using zodiac sign"

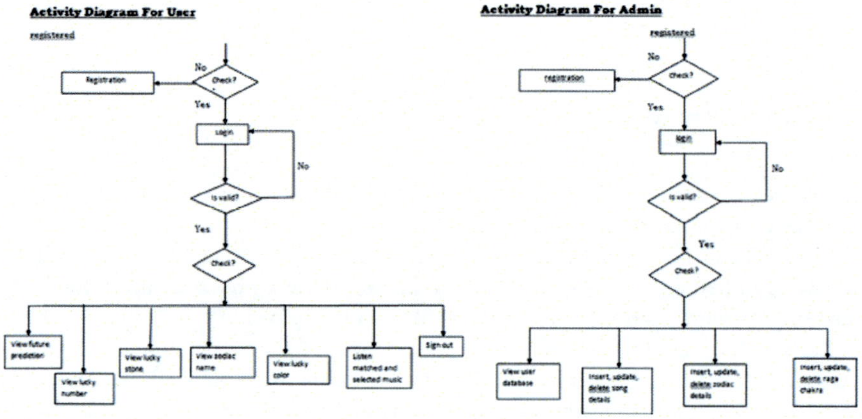

Fig. 3 Activity diagram of "music recommendation system using zodiac sign"

as their lucky number, lucky color, and suitable stone (Key feature) of that particular user. Then, matching the stone with the raga cycle as every raga cycle consists of one specific stone also. Finally, all ragas (six ragas) from that cycle are suitable to listen for the user. Figures 4 and 5 depict the sign-up and sign-in process of the Web application.

After completion of log-in process, insert the date of birth (DOB) of the user, and then, the application will display the zodiac sign as well as their lucky number, lucky color, and suitable stone of that particular user. Figure 6 depicts the various attributes as well as raga cycle name and the appropriate raga name with the raga song, respectively.

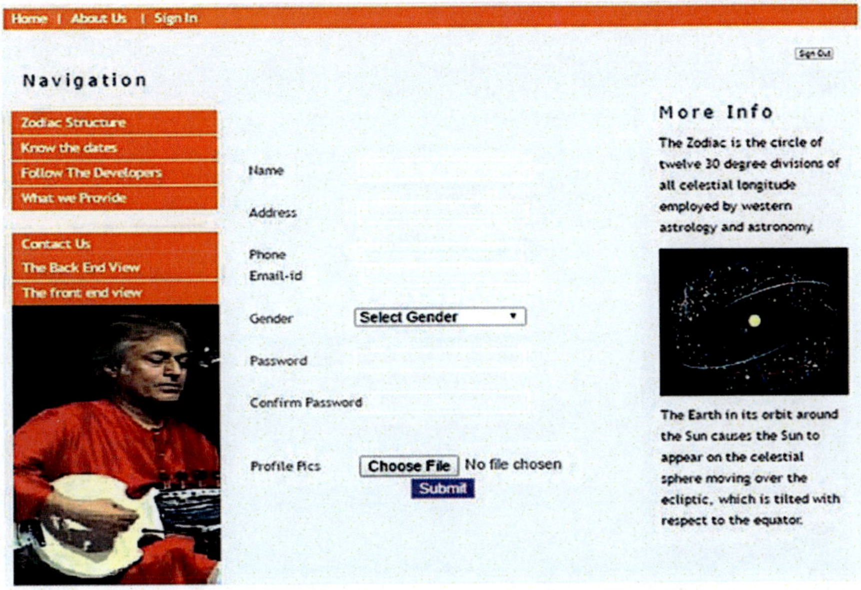

Fig. 4 Sign-up of "music recommendation system using zodiac sign"

Then, matching the stone with the raga cycle as every raga cycle consists of one specific stone also. Finally all six ragas from that cycle are suitable to listen for the user. Figure 7 depicts the appropriate raga and playing it for the particular user.

5 Conclusion

This is a software product-based Web application which provides the facility to generate personalize playlist of songs online according to zodiac sign. In this work, user provides their date of birth (DOB), and according to DOB, there is a specific zodiac sign of each person, and there is a specific suitable stone of that person. Here, suitable stone is the key feature of the work. Therefore, based on the specific stone, there is a specific suitable raga cycle and from that raga cycle any one raga from the six ragas are suitable for listen of that person.

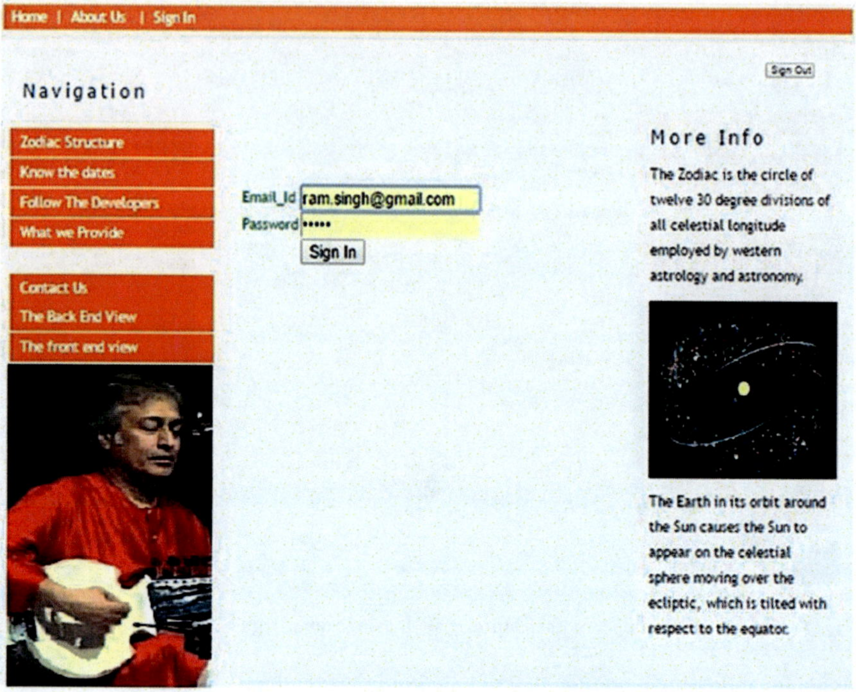

Fig. 5 Sign-in of "music recommendation system using zodiac sign"

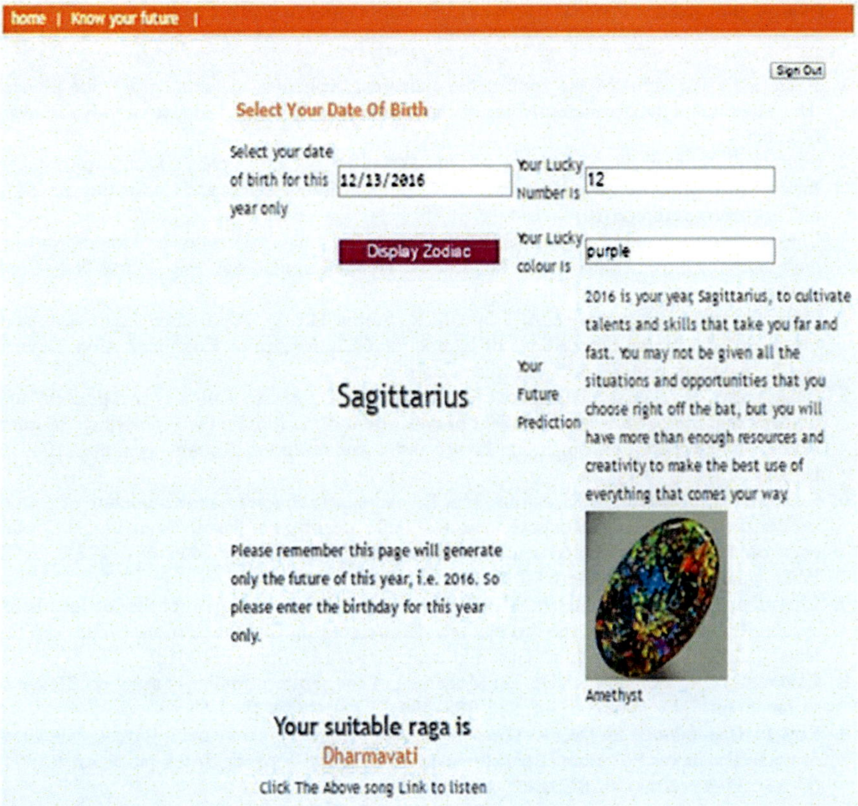

Fig. 6 Display the various attributes as well as raga cycle name

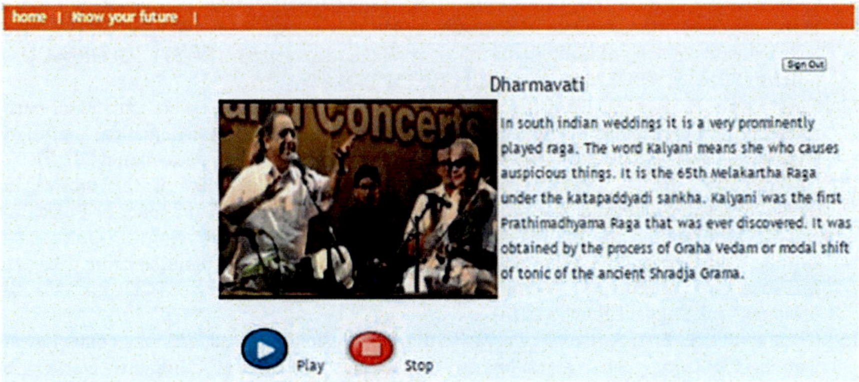

Fig. 7 Display the required raga and playing it

References

1. Bharati, J.: The signs of the zodiac, the elements, triplicities, quadruplicities, the planets, their aspects and the decanates. Notes on the elements of Astrology. Bharatiya Vidya Bhavan, Mumbai
2. Sudipta, C., Samarjit, R., Sarma, H.K.D.: Measuring song complexity by statistical techniques. In: the Proceedings of second International Communication Devices and Computing, pp. 687–695 Springer, Berlin (2020)
3. Chakrabarty, S., Samarjit, R. Sarma, H.K.D.: Intelligent music abstraction tool for improvising the quality of music composition. Int. J. Adv. Trends Comput. Sci. Eng. Warse 3641–3648 (2020)
4. Chakrabarty, S., Bhattacharjee, A., Islam, M.R., Sarma, H.K.D.: Algorithmic improvisation of music rhythm. In: the Proceedings of International Conference on Communication, Devices and Networking, pp. 323–335., Springer (2019)
5. Chakrabarty, S., Banik, S., Islam, M.R., Sarma, H.K.D.: Music similarity mapping through fundamental frequencies by coefficient of concurrent deviation. In: the Proceedings of International Conference on Computing, Power and Communication Technologies, pp. 910–915. IEEE (2019)
6. Chakrabarty, S., Islam, M.R., Sarma, H.K.D.: An approach towards the modeling of pattern similarity in music using statistical measures. In: Proceedings of the 5th International Conference on Parallel, Distributed and Grid Computing (PDGC), pp. 436–441. IEEE, JUIT, Waknaghat, Himachal Pradesh (2018)
7. Chakrabarty, S., Banik, S., Islam, M.R., Sarma, H.K.D.: Context aware song recommendation system. In: Proceedings of the 3rd National Conference on Communication, Cloud, and Big Data (CCB), in SMIT, Majhitar, East Sikkim, pp. 157–165. Springer (2018)
8. Chakrabarty, S., Islam, M.R., De, D.: Modelling of song pattern similarity using coefficient of variance. Int. J. Comput. Sci. Inf. Sec. 388–394 (2017). (ISSN 1947-5500)
9. Roy, S., Chakrabarty, S., De, D.: Time-based raga recommendation and information retrieval of musical patterns in Indian classical music using neural network. IAES Int. J. Artif. Intell. (IJ-AI). 33–48 (2017). (ISSN: 2252-8938)
10. Sudipta, C.,Gobinda, K., Islam, M.R., De, D.: Reckoning of music rhythm density and complexity through mathematical measures. In: Proceedings of the advanced computational and communication paradigm LNEE lecture note, pp. 387–394. Springer (2017)
11. Chakrabarty, S., Roy, S., De, D.: Chapter 12: time-slot based intelligent music recommender in Indian music. In: Siddhartha, B., Hrishikesh, B., De, S., Goran, K. (eds.) Handbook of Research on Intelligent Analysis of Multimedia information (Hardcover), pp. 319–351. IGI Global, USA (2016). ISBN 13: 9781522504986, ISBN 10: 1522504982
12. Chakrabarty, S., Roy, S., De, D.: A foremost survey on state-of-the-art computational music research. In: Proceedings of the recent trends in computations and mathematical analysis in engineering and sciences, pp. 16–25. International Science Congress Association (2015)
13. Chakrabarty, S., Roy, S., De, D.: Behavioural modelling of ragas of Indian classical music using unified modelling language. In: Proceedings of the 2nd International Conference on Perception and Machine Intelligence, pp. 151–160. ACM (2015)
14. Roy, S., Chakrabarty, S., De, D.: A framework of musical pattern recognition using petri nets. In: Proceedings of Emerging Trends in Computing and Communication (ETCC), pp. 245–252. Springer-Link Digital Library (2014)
15. Chakrabarty, S., Roy, S., De, D.: Pervasive diary in music rhythm education: a context-aware learning tool using genetic algorithm. In: Proceedings of Advanced Computing Networking and Informatics, pp. 669–677. Springer, Berlin (2014)
16. Sudipta, C., Roy, S., De, D.: Automatic raga recognition using fundamental frequency range of extracted musical notes. In: The Proceedings of Eight International MultiConference on Image and Signal Processing (ICISP 2014), pp. 337–345. Elsevier (2014)

17. Roy, S., Sudipta, C., Pradipta, B., De, D.: Modelling high performing music computing using petri nets. In: Proceedings of International Conference on Control, Instrumentation, Energy and Communication (CIEC), pp. 757–761. IEEE (2013)
18. Sudipta, C., De, D.: Quality measure model of music rhythm using genetic algorithm. In: Proceedings of International Conference on Radar, Communication and Computing (ICRCC), pp. 125–130. IEEE (2012)

Forecasting of Onion Prices

R. Sujay, V. K. Parvati, S. R. Biradar, Vijeta V. Kerur, and Shrikrishna Sharad Huilgol

Abstract The Indian agrarian market for onion commodity is elastic in form. Because the commodity market is uncertain, farmers, government institutions and traders need precise instruments for price prediction. An 'additive model' with a seasonal variation are adapted on an annual, weekly and daily span to assist forecast MSP (Minimum Support Price). By changing the trend, the predictive model addresses missing information and outliers. The findings acquired from the predictive model are very helpful and can be applied in a comparable inventory market-like real-time setting.

Keywords Time series forecasting model · Indian commodity · Price prediction · MSP · Minimum support price

1 Introduction

There are numerous commodities that are cultivated according to Agricultural Market in India. These commodities are cultivated in various parts of India. This generates uncertainty among peasants about other people's rates and accessibility of the cultivated goods. As separate commodity markets have evolved for the same commodity, distinct rates are offered. Indian market's complicated structure and the demand–supply proportion is a significant task for govt, distributors, and peasants. The cost fluctuation of commodities was primarily related to weather and producers have no benefit in obtaining the highest cost for the cultivated product [1, 2].

The Indian government therefore sets the MSP (Minimum Support Price) to preserve this divergence in order to preserve market stability. The minimum price of support is dependent on the proportion of demand and supply. It is therefore essential to embrace a model that can assist public, landowners, distributors, and customers to prevent and schedule cost fluctuations appropriately.

R. Sujay · V. K. Parvati · S. R. Biradar (✉) · V. V. Kerur · S. S. Huilgol
Department of Information Science and Engineering, SDM College of Engineering & Technology, Dharwad, Karnataka 580002, India

Affliated to Visvesvaraya Technological University (VTU), Belagavi, India

H. K. D. Sarma et al. (eds.), *Contemporary Issues in Communication, Cloud and Big Data Analytics*, Lecture Notes in Networks and Systems 281,
https://doi.org/10.1007/978-981-16-4244-9_28

Research relying on historical rates in the past century has resulted in predicting the cost for the future that aided in the decision-making of strategies or choices. One of the comparable investigations was to estimate or estimate oil prices depending on the Dynamic Model Average (DMA) by gathering prior gas rates [3]. A model was developed to forecast the cost of gold by gathering information from separate nations that analysed the volatile gold market [4]. Even comparable prototypes of cost forecast were created using various parts of the time series.

A predictive model is suggested in this document to suit the fluctuation of the season, uneven and tendency along with lacking information in it [5, 6]. The model forecasts potential year rates that can be used as an element of decision-making for landowners, public and traders. The information was gathered from the website Horticultural Producers' Cooperative Marketing (HOPCOM) (http://hopcoms.com). The chosen commodity is onion because there is a huge difference between the proportion of production and consumption. The cost ranges per quintals were drawn for each Indian market. The years listed are from 2014 to 2018 to obtain the cost distribution.

2 Proposed Forecast Model

Dragan's analysis [7] used the model Autoregressive Integrated Moving Average (ARIMA) that defines the modifications and potential trends in tomato cost parameters with descriptive statistics. The ARIMA model was poor in understanding the annual transition resulting in a large gap in forecast of prices. Boateng [8] indicated this in his article by drawing up a regular pattern system for fruit rates recognized as SARIMA (Seasonal-ARIMA) and discovered that this system improves predictability.

This document describes the model used for forecasting and managing price errors that are hard to manage in the ARIMA model. The model suggested is responsive to the annual pattern and it is possible to make judgments appropriately. It is possible to use the same model for various elements such as location, area, or product.

The model of forecasting is divided into three primary parts. Trend, seasonality, and incidents are these elements. The engineering formula is shown below,

$$y(t) = g(t) + s(t) + h(t) + e_t$$

The $g(t)$ is for the flow feature with non-periodic differences in the above equation, whereas the $s(t)$ reflects regular differences such as quarterly or monthly or annual changes. The $h(t)$ feature depicts activities such as holidays that happen one or more moments as an uneven case. The ε_t feature is the error function that occurs because of the independent variability or modifications that deemed to be a normal distribution.

The trend feature estimates confusion if any and gives the forecasting model a steady valuation. In the forecasting model, the seasonality function tests for any environmental feature that may be implemented. In commodity cases, the annual characteristic is the process in which the specific plant is cultivated. The feature of the case helps to prevent unseasonal status differences.

3 Results

The data is collected from the Agricultural Marketing Information Network (AGMARKNET) portal which is a Government of India portal on agricultural marketing. The data is collected from agricultural markets and State Marketing machineries. The portal is handled by the Directorate of Marketing and Inspection with the technical assistance of NIC (National Indian Council). The data is received from 3245 markets spread across India as reported on 31 March 2015. The data for onion commodity has been collected from 1st of January 2009 to 30th of April 2019. The data represents prices in Indian Rupees per Quintal quantity (Fig. 1).

The predictive model was introduced to the onion because it is one of India's main consumer commodities. The model predicts rates for the 2019–2021 year to come. Together with smaller and greater trends, costs were anticipated. The greater pattern and reduced curve show the variety within which onion rates may rise or fall.

This variety helps to know the price of commodities over the years. The findings forecast are a lot of tendencies towards the actual that can be seen in Fig. 2 (Predicted prices per quintal).

The model is time-constant and can be used appropriately to forecast the potential pattern. About the month, the seasonality has a large order variance between the information items, whereas at the start there is less ambiguity. Figure 3 demonstrates the predicted cost with the real onion value for the close lifetime.

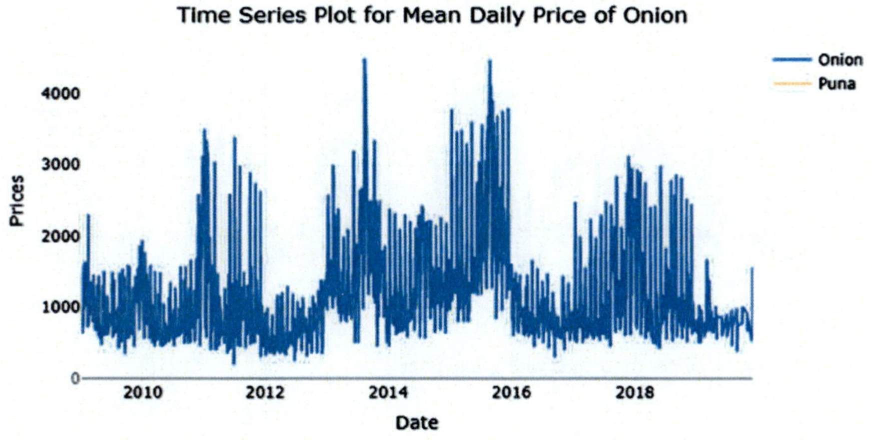

Fig. 1 Onion prices over the years

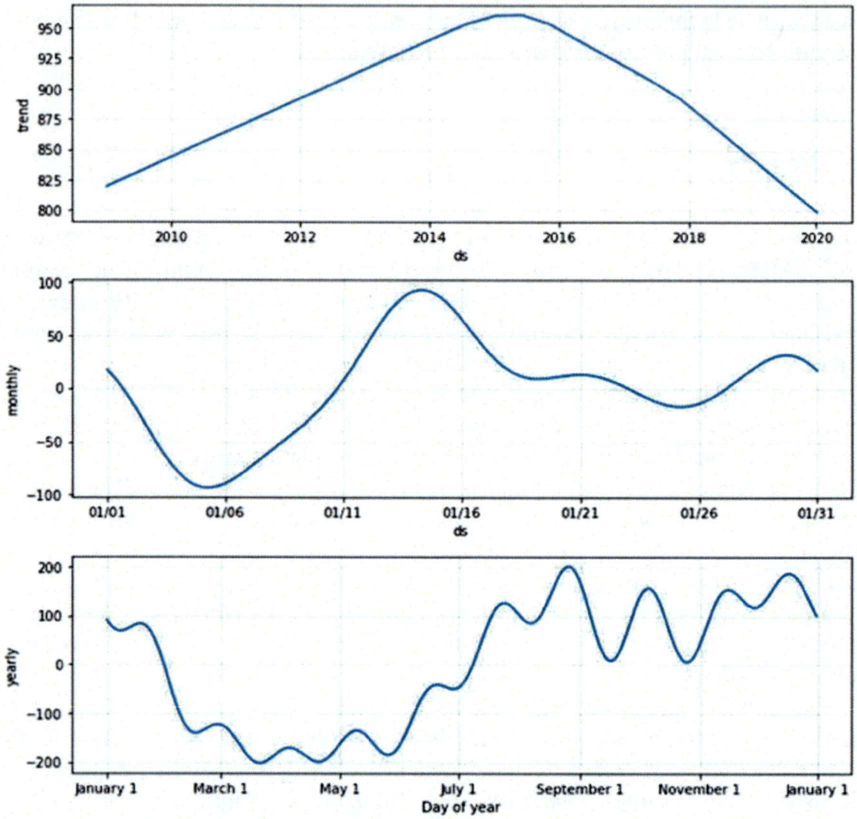

Fig. 2 Predicted Onion prices over the years, months and days

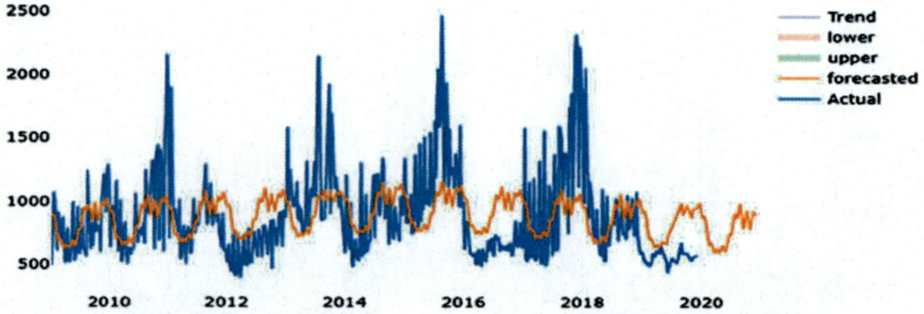

Fig. 3 Forecasted prices for onions with actual price

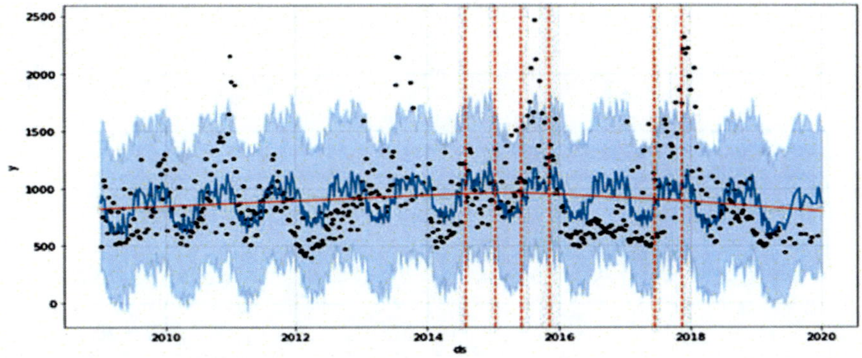

Fig. 4 Changepoints depicting change in trend

The model also detects the modifications in attitude over the years. This enables the pattern along the modifications to be adapted appropriately. The model's shift marks use regularization. In attempt to plan forward progress, the shift marks are inferred during the first 80% and avoid over-fitting triggered by changes depicted in Fig. 4.

The model is time-constant and can be used appropriately to forecast the potential pattern. With respect to the month, the seasonality has a large order variance between the information items, whereas at the start there is less uncertainty.

References

1. Farmer, J.D., Foley, D.: The economy needs agent-based modelling. Nature **460**(7256), 685 (2009)
2. Reddy, A.A., et al.: Analysis of pearl millet market structure and value chain in India. J. Agribus. Develop. Emerg. Econ. **8**(2), 406–424 (2018)
3. Drachal, K.: Forecasting spot oil price in a dynamic model averaging framework—have the determinants changed over time? Energy Econ. **60**, 35–46 (2016)
4. Sharma, S.S.: Can consumer price index predict gold price returns? Econ. Model. **55**, 269–278 (2016)
5. Taylor, S.J., Letham, B.: Forecasting at scale. Am. Stat. **72**(1), 37–45 (2018)
6. Reddy, A.A.: Price forecasting of tomatoes. Int. J. Veg. Sci. 1–9 (2018)
7. Ivanisevic, D., et al.: Analysis and prediction of tomato price in Serbia. Ekonomika poljoprivrede **62**(4), 951 (2015)
8. Boateng, F.O., et al.: Modeling of Tomato Prices in Ashanti Region, Ghana, Using Seasonal Autoregressive Integrated Moving Average Model.

Complete Test Set Generation for Control Flipping Faults in Reversible Circuits

Mousum Handique, Amrit Prasad, and Hiren Kumar Deva Sarma

Abstract The newly developed fabrication process reduces the size of the chips and increases the speed of a system. Simultaneously, power dissipation is also increased, which causes a major issue in semiconductor devices. Therefore, reversible computation is gaining interest in the low-power circuit design and the fast computation system. Moreover, the mechanism of reversible computation is widely used in quantum computing nanotechnology and also in digital communication systems. These inspire the researchers to give more attention to the reversible circuit. Thus, reversible circuit becomes more alternative than the conventional circuit. Testing these circuits is an essential aspect of ensuring these circuits' high reliability and integrity performance. This paper introduced a new fault model labeled as the control flipping fault model (CFF) in a Multiple-Control Toffoli (MCT)-based reversible circuit. Specifically, we target the positive control flipping faults (PCFFs) under the proposed control flipping fault model. The reported works present a scheme to detect fault to construct a complete test set (CTS), which is capable of detecting all the PCFFs in a reversible circuit. The paper also presents an experimental evaluation in order to show the efficiency of the CTS that covers 100% faults.

Keywords Reversible circuit · Control flipping fault model · Positive control flipping fault · MCT-based circuit

1 Introduction

Reducing energy dissipation and in turn energy efficiency is one of the challenging objectives in recent technologies under development, as the size of chips is becoming smaller and faster, and dissipate more energy in the form of heat. In 1961, Landauer [1] showed that irreversible circuits dissipate at least $kTln2$ Joules of energy because of

M. Handique (✉) · A. Prasad
Assam University, Silchar, Assam 788011, India

H. K. D. Sarma
Sikkim Manipal Institute of Technology, Rangpo, Sikkim 737136, India

H. K. D. Sarma et al. (eds.), *Contemporary Issues in Communication, Cloud and Big Data Analytics*, Lecture Notes in Networks and Systems 281,
https://doi.org/10.1007/978-981-16-4244-9_29

loss of information per bit. k signifies the Boltzmann constant, and T represents the temperature prevailing in the system. In 1973, Charles H. Bennett [2] observed the benefits of reversible computation in terms of information lossless, while reversible computation is executed. More precisely, the reversible logic computation can retain the information when operations are executed, and the system runs in a backward direction. Based on these properties, reversible logic is used as a circuit design alternative. The reversible gates are used to implement the reversible logic operation, and only the linear sequence formation of reversible gates is the structure of reversible circuits. The reversible circuit operations are bijective, and also, there is no direct concept of fan-out and feedback connections in reversible circuit [3].

Fault represents the incorrect state during the computation that leads to the functional error of a system. The presence of a single fault may affect the output performance in every computing device. In testing, fault detection is the first phase that detects all possible faults, and the next phase determines the exact location of faults. With the importance of the reversible circuit, testing is also an important parameter for the integrity performance of the reversible circuits. The physical description of faults is described by the fault models. The fault model is the mathematical description that helps the designer predict the system's erroneous state and efficiently construct the test pattern for evaluating the faults. The various fault models have been introduced in reversible circuits. An elaboration of these fault models may be found in [4].

The test generation process is required to determine the faults in a faulty circuit. A test set may be defined as a set of test sequences. A complete test set (CTS) can detect various possible faults available in a presented circuit [5]. The generation of the test set is straightforward in reversible circuits as compared to the traditional circuit because the reversible circuit maintains high controllability and observability [5].

In this paper, we have presented a proposed fault model labeled as control flipping fault model (CFF), which is more relevant in k-CNOT or Multiple-Control Toffoli (MCT)-based circuit structure. A fault detection scheme in order to generate the CTS to detect various possible positive control flipping faults in the presented reversible circuit has been proposed. At last, the experimental results are demonstrated based on the performance evaluation of MCT or k-CNOT-based reversible circuits to verify the proposed fault detection scheme.

The paper has been organized as mentioned below. Section 2 includes the introduction of logic function and gates in the reversible circuit, MCT, or k-CNOT-based circuits. This section also discussed structural fault models that are applied in reversible circuits and covers some of the prior works relevant to the proposed work. The description of the newly introduced fault model and the CTS generation process for detecting PCFFs is explained in Sect. 3. The experimental results based on the PCFFs and concluding remarks are presented in Sects. 4 and 5, respectively.

2 Preliminaries

2.1 Logic Function and Gates in Reversible Circuits

Reversible function $f : \mathbb{B}^n \Rightarrow \mathbb{B}^m$ is bijective in nature (i.e., one-to-one and onto mapping), and it maintains the equal inputs (n) and outputs (m) where each possible output vector can uniquely retrievable an input vector. Therefore, the reversible logic function can uniquely determine the inputs and outputs from each other.

Reversible logic operations are used to implement the reversible logic function, and these operations are considered as gate operations to construct the circuit. Each of the reversible gates has an n number of inputs that establish one-to-one mapping with the n number of outputs. In general, NOT gate, FEYNMAN or Controlled-NOT (CNOT) gate [6], TOFFOLI gate [7], and Multiple-Control TOFFOLI gate or k-CNOT gate [7] are the classical gates, and these gates are most commonly used to design the reversible circuits. The operation of the NOT gate in a reversible circuit is similar to a conventional circuit's operation. NOT gate constructed with a single input and output and input A simply inverts the output $X=\overline{A}$ as depicted in Fig. 1a. CNOT gate constructs with 2-input $\times$ 2-output lines. The first input A is connected to the positive control connection (•), and the output remains the same as it input $X=A$. The second input B is connected to the target connection ($\oplus$) and the output $Y=A \oplus B$. The input line B inverts if the logic value 1 is set to the input line A. The symbolic representation and gate operations are illustrated in Fig. 1b. TOFFOLI gate consists of 3-input and 3-output. The control lines' output X and Y are the same as the input of the control lines A and B, respectively. The output of the target connection Z inverts the input of the target connection C if both control input lines A and B are set as logic value 1, shown in Fig. 1c. The MCT gate consists of k number of control lines $k_1, k_2, \ldots, k_{n-1}$ where some of the control connection may be positive (•)/negative (∘) and one target connection k_n as demonstrated in Fig. 1d. In MCT gate, the target connection's output is only inverted if the logic value is 1/0 for the positive/negative control connections.

2.2 Reversible Circuits

In the reversible circuit, the gates are arranged in a linear cascade manner [8], and the structure of the reversible circuit is not directly allowed any fan-out and feedback connection [3]. Figure 2 illustrates the MCT or k-CNOT-based reversible circuit. Here, the initial input $\langle 0\ 1\ 0\ 1\rangle$ propagates to primary output $\langle 1\ 1\ 1\ 1\rangle$. The output vector $\langle 1\ 1\ 0\ 1\rangle$ is generated by the gate G_1 lies between the levels L_0 and L_1 after applying the initial input $\langle 0\ 1\ 0\ 1\rangle$ in level L_0. If any intermediate level of the gates generates other test vector, then the primary output vector also changes instead of $\langle 1\ 1\ 1\ 1\rangle$. Therefore, each reversible gate produces a unique test vector of the circuit.

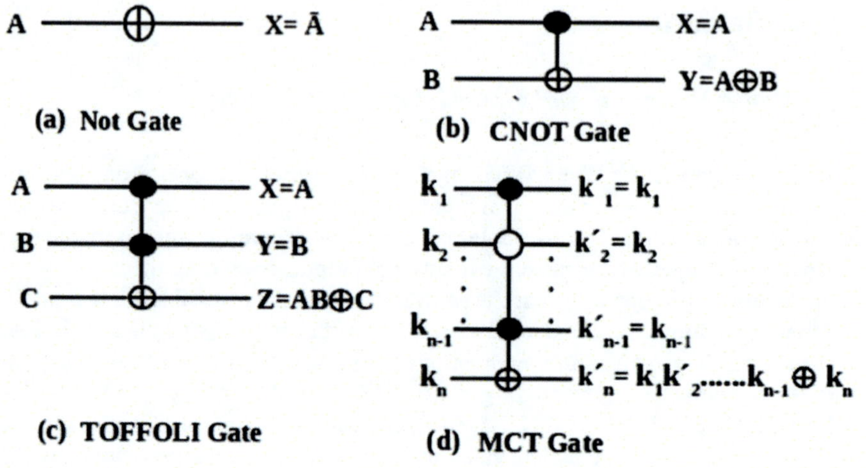

Fig. 1 Classical reversible gate

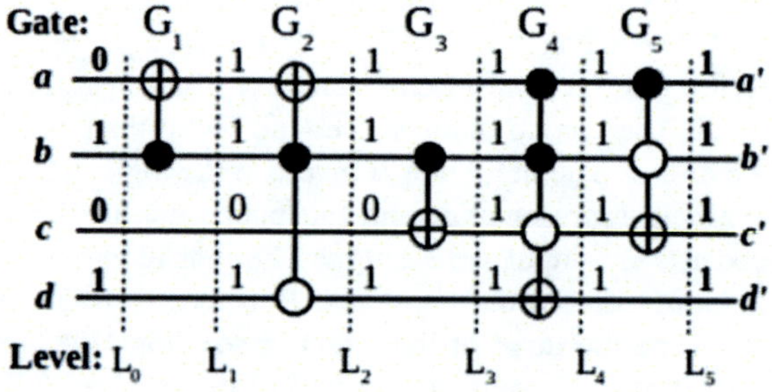

Fig. 2 Illustration of MCT circuit of composing various *k*-CNOT gates

2.3 Fault Models in Reversible Circuits

Description of the various levels of abstraction at which the fault model can be grouped. In other words, fault models can be described at various levels of abstraction in the circuit design hierarchy. These levels of abstraction in circuits are behavioral, functional, structural, and geometrical [9]. The proposed works concentrate only on the fault models that describe the structural level of abstraction in circuits. There are some fault models such as stuck-at [5] and bridging [10] fault models are common to both conventional and reversible circuit testing. Specifically, missing gate [11, 12], partial missing gate [12], and the crosspoint [13] fault models are applied only in reversible circuits that describe the gate level faults.

2.4 Related Work

Several existing works on fault detection approaches of reversible circuits testing have been described in the literature. A deterministic test generation algorithm is proposed to construct the CTS in [5] to detect the stuck-at faults, and the construction of CTS is formulated using ILP. The work in [10] discussed a test generation problem for testing the bridging faults that used in the MCT circuit. The proposed approach showed the generated CTS cardinality is $(d \log_2 n)$ for $n \times n$ reversible circuit with levels d. The authors in [14] have proposed an exact ATPG algorithm with the concept of a set cover method for generating the minimal test set for detecting the bridging faults in a reversible circuit designed with Toffoli, Fredkin and Peres gates. The authors in [15] developed a testing scheme capable of generating the lesser number of test vectors for identifying the single and partial missing gate faults in mixed control MCT-based circuits. The work in [16] presented a fault detection and fault localization scheme for newly developed faults, namely as gate appearance and control appearance in MCT-based circuit. This work showed that only one test vector is needed to detect the gate appearance faults. For evaluating the control appearance fault, this scheme is required n test vectors with n-input lines available in the circuit.

In this literature review, we observed that several approximate and heuristic testing approaches are being proposed to identify the faults in reversible circuits. However, the physical implementation technologies for reversible circuits are still in progress. In this work, we proposed and investigated a control flipping fault model (CFF) based on the MCT-based circuit design structure. This work also presented the complete test generation scheme for evaluating the positive control flipping faults under the proposed fault model.

3 Proposed Model

In this section, the proposed control flipping fault model (CFF) has been introduced. Here, the control flipping fault model is considered as a permanent fault due to incorrect interconnection in designing can be prescribed. Generally, controls are of two types positive ($\bullet$) and negative ($\circ$) control in MCT or k-CNOT-based circuit structure. Flipping of any control to another type can be termed as control flipping faults. Based on the k-CNOT circuit structure, CFF can be divided into three types: (i) Positive control flipping fault (PCFF), (ii) Negative control flipping fault (NCFF), and (iii) Mixed control flipping fault (MCFF). In this work, we specifically target only PCFF for constructing the CTS for all possible PCFFs.

3.1 Positive Control Flipping Fault

Positive control flipping is a fault that occurs in the control flipping fault model. Due to the designing error, when a positive control(s) is flipped to the negative control, such erroneous flipping type is called positive control flipping fault (PCFF). In the k-CNOT circuit, if the PCFF occurrence is involved with only one control, then it is termed as a single positive control flipping fault (SPCFF) as illustrates in Fig. 3a. If the PCFF is involved with two or more control connections, then it is termed multiple positive control flipping fault (MPCFF), which is shown in Fig. 3b.

3.2 Detection Logic for Positive Control Flipping Fault

In PCFF, one or more positive control is flipping to the negative control. The flipped control connection directly affects the target connection. Therefore, the detection logic of PCFF is performed such that the logic value of the target connection will be change after the control flipping. Here, unflipped control connection(s) is set with the logic value 1, and the logic value 1 or 0 (don't care) is applied to all other the affected flipped control connection(s), unconnected line(s), and the target connection. The proposed work applies the logic value 1 to all the flipped and unflipped connection(s) and unconnected connection(s) and the target connection.

Example 1 Figure 4a shows the fault-free *ham*3_*tc* benchmark reversible circuit. Here, we demonstrated the effect of SPCFF and MPCFF as a comparison with the fault-free circuit. Initially, the test vector ⟨0 1 1⟩ is applied to the gate G_1 in level L_0, and it propagates to the gate G_5 at level L_5. The primary output vector would be ⟨1 0 0⟩ in the fault-free circuit. In Fig. 4b, SPCFF occurs at control connection 'b' in gate G_1, and primary output is effected in the circuit due to the presence of the SPCFF. If we apply the same initial test vector ⟨0 1 1⟩ at level L_0 to the circuit, then primary output gate G_5 in level L_5 generates the test vector ⟨0 1 1⟩, which is faulty output of circuit. MPCFF occurs at both line 'b' and 'c' in gate G_1 which is depicted in Fig. 4c. The presence of MPCFF effects the primary output gate G_5 in level L_5 that produces the faulty test vector ⟨0 1 1⟩.

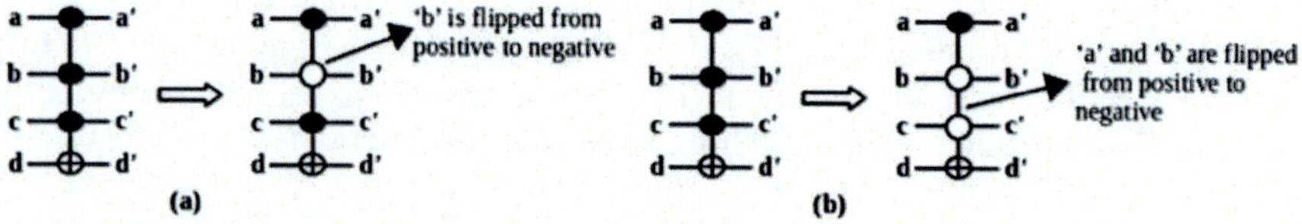

Fig. 3 **a** SPCFF at line 'a,' **b** MPCFF at line 'b' and 'c'

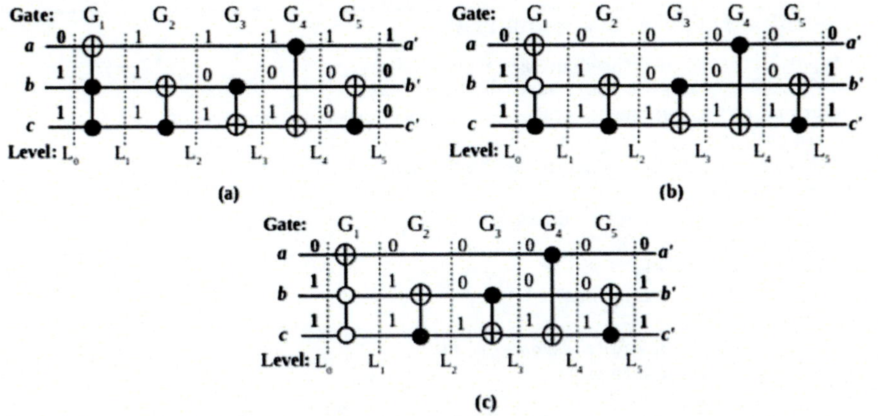

Fig. 4 *ham*3*_tc* reversible circuit: **a** fault-free circuit, **b** SPCFF at line 'b' in gate G_1, **c** MPCFF at line 'b' and 'c' in gate G_1

3.3 Generation of Complete Test Set for PCFF

To obtain the CTS in a presented MCT-based circuit, we consider the concept of fault detection logic for the PCFFs. Here, we consider the logic value 1 that is assigning to all the signal lines of the circuit, and this test sequence applies to each gate G_i. The possible test vector of the CTS retrieves from the initial input level by the backpropagation. Algorithm 1 presented the construction of CTS.

Example 2 The CTS generation algorithm is illustrated using *ham*3*_tc* benchmark circuit in Fig. 5. Here, gate G_j is denoted at the jth gate in the circuit, where $1 \leq j \leq N$, where N denotes the number of gates. The array $OutputList[\,]$ initially empty and the array $TP[\,]$ stores as $TP[000, 001, 010, 011, 100, 101, 110, 111]$ for n=3. The values of $RevReadGate[\,]$ stores {'c, b', 'a, c', 'b, c', 'c, b', 'b, c, a'} all information of controls and target connection that are present of gate G_j in reverse order. Let $RevReadGate[3]$ stores {'b, c'} information of gate G_3. If we apply the $TP[6]$=$\langle 1\ 1\ 0\rangle$ from primary output to the gate G_3 level, then $TP[7]$=$\langle 1\ 1\ 1\rangle$ is generated. Now, the test pattern $TP[7]$=$\langle 1\ 1\ 1\rangle$ propagates to the initial input level, and the $OutputList$ produces $TP[7]$=$\langle 1\ 1\ 0\rangle$. Therefore, the test vector $\langle 1\ 1\ 0\rangle$ is capable of detecting all PCFFs at gate G_3. In similar way, $OutputList$ is $\langle 0\ 1\ 1\rangle$, $\langle 1\ 0\ 1\rangle$, $\langle 1\ 0\ 1\rangle$, and $\langle 1\ 0\ 0\rangle$ for the gate G_1, G_2, G_4, and G_5, respectively. Finally, complete test set for the *ham*3*_tc* circuit is {011, 101, 110, 100} for all PCFFs.

Lemma 1 *The generated CTS covers all the possible PCFFs in MCT or k-CNOT-based circuit.*

Proof As per the operation of k-CNOT gate, the target connection output only inverts when all the controls are assigning the logic value 1, whereas the logic value 1 or

Algorithm 1: CTS generation for PCFF

Input: Generated tfc file using given circuit structure.
$TP[\,]$: stores binary encoding pattern of n input lines.
$OutputList[\,]$: stores input variables with their corresponding binary value.
$RevReadGate\,[\,]$: stores the target and control information in reverse.
G_j: store the control(s) and target variable for each gate at a time.
L_j: cardinality of G_j.
Output: Complete test set for detecting all PCFF

```
OutputList[ ] ← ∅
for i ← 0 to len(TP) − 1 do
    OutputList ← mapping(TP[i])
    Flag=0
    for j ← 0 to len(RevReadGate) − 1 do
        G_j ← RevReadGate[j]
        L_j ← len(G_j)
        if All elements in OutputList.values is 1 AND L_j >= 2 then
            Flag=1
        if L_j==1 then
            OutputList[G_j[L_{j−1}]] = NOT OutputList[G_j[L_{j−1}]]
        else
            Temp = OutputList[G_j[0]]
            for k ← 1 to L_{j−2} do
                Temp = Temp AND OutputList[G_j[k]]
            OutputList[G_j[L_{j−1}]] = OutputList[G_j[L_{j−1}]] EXOR Temp
    if Flag==1 then
        Print(OutputList.values)
```

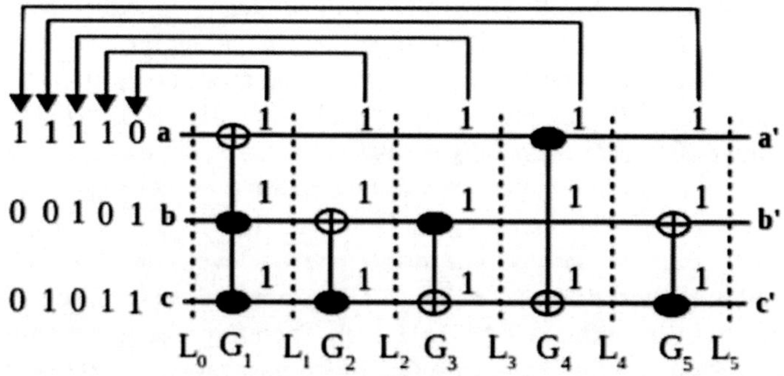

Fig. 5 Demonstration of Algorithm 1 for the *ham3_tc* circuit

0 (don't care) applies in target and unconnected lines. However, when one or more positive controls are affected by flipping control, then the logic value of flipping control inverts at the target connection. As a result, the output logic value of the target connection remains the same as the input. Therefore, one binary logic value

changes at the target connection when positive control is flipping. In this proposed method, the logic value 1 is assigning to the lines of each available gate G_j and stores in test pattern TP[i] and the backpropagation of TP[i] toward the input level to form the complete test set for each k-CNOT gate. Hence, a constructed test set at the initial input level is the CTS for all the PCFFs in an MCT- based circuit.

4 Experimental Results

The presented algorithm for PCFFs has been run on an Intel Pentium (R) CPU-2020 @ 2.40GHZ × 2 system running on Windows 10 (64-bit) with 4 GB RAM and implemented in Python 3.4. Several benchmark circuits [17, 18] are considered for analyzing the proposed CTS generation algorithm. Based on this analysis, the experimental results are shown in Table 1. Here, n and N are denoting as the number of input lines and gates presented in columns 2 and 3, respectively. Column 4 is presenting all possible faults for both SPCFFs and MPCFFs in a presented circuit. The cardinality of the generated CTS to detect the faults is shown in column 5. The last column of Table 1 mentioned the computational time for generating the CTS. From the experimental results, it is observed that in most of the circuits, the computational time increases when the number of gates is increasing as compared to the number of input lines in k-CNOT-based circuits.

Table 1 Experimental results for PCFFs detection and its computation time (s)

Benchmarks circuit	n	N	Total no. of faults (SPCFF+MPCFF)	CTS (SPCFF+MPCFF)	CPU time (s)
Peres_9	3	2	4	2	0.01562
3_17_14	3	6	9	4	0.01562
3_17_13	3	6	9	4	0.01563
ex-1-166	3	4	5	3	0.0162
fredkin_6	3	3	9	3	0.0155
ham3_tc	3	5	7	4	0.0163
4b15g_1	4	14	28	9	0.0161
4b15g_2	4	15	35	7	0.0158
hwb4-11-21	4	11	17	8	0.0156
hwb4d1	4	17	41	7	0.0176
mspk_4b15g_2	4	15	24	9	0.0156
mspk_hwb4_13	4	13	19	9	0.0165
ham15-70	15	70	278	53	11.325
ham15-109-214	15	109	427	94	14.418
ham15tc1	15	132	2816	73	24.392

5 Conclusion

This paper introduces a new fault model called a control flipping fault model (CFF), and the physical justification of this kind of fault model is considered a designing fault. In this method, we are efficiently extracting the test vectors for evaluating all single positive control flipping fault (SPCFF) and multiple positive control flipping fault (MPCFF) under the control flipping fault model. Our work may be extended to detect other faults like negative control and mixed control flipping faults in MCT-based circuits as future work. Moreover, this work may establish the correlation of other existing fault models of reversible circuits.

References

1. Landauer, R.: Irreversibility and heat generation in the computing process. IBM J. Res. Dev. **5**(8), 183–191 (1961)
2. Bennett, C.H.: Logical reversibility of computation. IBM J. Res. Dev. **17**(6), 525–532 (1973)
3. Nielson, M.A., Chuang, I.L: Quantum Computation and Quantum Information. Monograph Collection (Matt - Pseudo) (2000)
4. Rice, J.: An overview of fault models and testing approaches for reversible logic. In: 2013 IEEE Pacific Rim Conference on Communications, Computers and Signal Processing (PACRIM), pp. 125–130. IEEE (2013)
5. Patel, K.N., Hayes, J.P., Markov, I.L.: Fault testing for reversible circuits. IEEE Trans. Comput.-Aided Des. Integr. Circuits Syst. **23**(8), 1220–1230 (2004)
6. Feynman, R.P.: Quantum mechanical computers. Found. Phys. **16**(6), 507–531 (1986)
7. Toffoli, T.: Reversible computing. In: International Colloquium on Automata, Language, and programming, pp. 632–644. Springer, Berlin (1980)
8. Maslov, D.: Reversible logic synthesis. University of New Brunswick, Ph.D. diss. (2003)
9. Jha, N.K., Gupta, S.: Testing of Digital Systems. Cambridge University Press (2003)
10. Rahaman, H., Kole, D.K., Das, D.K., Bhattacharya, B.B.: Optimum test set for bridging fault detection in reversible circuits. In: Asian Test Symposium, ATS'07. 16th, pp. 125–128. IEEE (2007)
11. Hayes, J.P., Polian, I., Becker, B.: Testing for missing-gate faults in reversible circuits. In: 13th Asian Test Symposium, vol. 2004, pp. 100–105. IEEE (2004)
12. Polian, I., Fiehn, T., Becker, B., Hays, J.P.: A family of logical fault models for reversible circuits. In: 14th Asian Test Symposium (ATS'05), vol. 2005, pp. 422–427. IEEE (2005)
13. Zhong, J., Muzio, C.J.: Analyzing fault models for reversible logic circuits. In: 2006 IEEE International Conference on Evolutionary Computation, pp. 2422–2427. IEEE (2006)
14. Nagamani, A., Abhishek, B., Agrawal, V.K.: Deterministic approach for bridging fault detection in peres-fredkin and toffoli based reversible circuits. In: 2015 IEEE International Conference on Computational Intelligence and Computing Research (ICCC), pp. 1–6 (2015)
15. Mondal, B., Bandyopadhyay, C., Rahaman, H.: A testing scheme for mixed-control based reversible circuits. In: 6th International Symposium on Embedded Computing and System Design (ISED), pp. 96–100. IEEE (2016)
16. Mondal, B., Bandyopadhyay, C., Rahaman, H.: Detection and localization of appearance faults in reversible circuits. In: 7th International Symposium on Embedded Computing and System Design (ISED), pp. 1–5. IEEE (2017)

17. Maslov, D.: Reversible logic synthesis benchmark page. http://webhome.cs.uvic.ca/dmaslov/ (2015)
18. Wille, R., Große, D., Teuber, L., Dueck, G.W., Drechsler, R.: Revlib: an online resource for reversible functions and reversible circuits. In: Proceedings of 38th International Symposium on Multiple Valued Logic (ismvl 2008), pp. 220–225

Prediction of MOSFET Count in Processor Integrated Circuit Using Machine Learning Approach

Sourav Ghosh, Pranati Rakshit, Shankar Debnath, Dwaipayan Roy, Sanjit Paul, and Madhura Chakraborty

Abstract MOSFET count in an integrated circuit of processor has shown a large change over past years. Varying from different companies, the count has shown a gradual increase. The introduction of VLSI technology has been responsible for this increase and subsequently it is increasing on newer products. On observation from previous data of MOSFET count in an IC can predict the future if the increasing trend remains the same. This research is based on the prediction and comparison of various results that are obtained from different algorithms used to predict the value. The data hence obtained may be meaningful to the manufacturing companies to have an estimate on the upcoming value required and to plan accordingly. On analysis, the count not only depends upon the changing years but also depends upon other parameters like MOS processes, surface area, size of the register in bits. We have used gradient boosting regessor, K nearest neighbor regressor and linear regression for prediction purpose. We got good r-squared value which is near to 1 for all the regressor and for all the different companies.

Keywords MOSFET · Transistor count · Moore's law · Machine learning · Prediction

1 Introduction

In digital and analog circuits, MOSFET is the most common among semiconductors and power devices. It had truly been a primary compact transistor. It was miniaturized, mass-produced within a range of application that revolutionized industry includng world economy, and playing a serious role to the manufacturing of computers, digitalization, silicon age, information revolution and knowledge age. Miniaturization

S. Ghosh · P. Rakshit (✉) · S. Debnath · D. Roy · S. Paul
Department of Computer Science and Engineering, JIS College of Engineering, Kalyani, Nadia, West Bengal 741235, India
e-mail: pranati.rakshit@jiscollege.ac.in

M. Chakraborty
Department of Electronics and Communication Engineering, JIS College of Engineering, Kalyani, Nadia, West Bengal 741235, India

H. K. D. Sarma et al. (eds.), *Contemporary Issues in Communication, Cloud and Big Data Analytics*, Lecture Notes in Networks and Systems 281,
https://doi.org/10.1007/978-981-16-4244-9_30

and MOSFET scaling have driven the rapid growth to electronic semiconductor technology since 1960s, and has enabled high-density integrated circuits (ICs) such as microprocessors and memory chips. It is possibly considered to be the foremost important invention in the field of electronics, revolutionizing economy, modern culture, lifestyle and society.

Transistor count is the number of transistors used in an IC. Nowadays, all new generation ICs use MOSFETs, it refers to the number of MOSFETs (metal–oxide–semiconductor field-effect transistors, or MOS transistors) on an IC chip. It is a common measure of IC complexity (though majority of the transistors of new generation microprocessors are there in the cache memory that consists mostly of same memory cell circuits replicated many times). The growth of transistor count in general follows Moore's law [1] that observes the transistor count to double every two years approximately.

The idea is to generate a model that predicts the count in numbers. On analysing the trends, the count is found not only to depend upon the year of introduction but also depend upon the surface area, architecture, MOS processes of the integrated circuit. A comparative study is done over few machine learning algorithms that are used to predict the values using all the dependent parameters.

This paper consists of the following sections. Section 2 is representing the literature survey. The methodology of work is described in third section. In fourth section, we discuss on the results we have obtained. Finally, the conclusion is presented in fifth section.

2 Literature Survey

This type of prediction work on the count of MOSFETs in the design of IC is almost nil except the well-known Moore' law.

Moore's law: It states the count of MOSFET in an IC changes twice every two years. Moore's law was not initially a law but was a statement made by Gordon Moore observing the growth in the number in a few years. He had made this prophecy that later turned out to be a golden rule for designing integrated circuits. This prophecy was made due to shrinking transistor dimensions.

But this count can not only depend upon the year of introduction. The surface area, architecture, MOS processes of the integrated circuit may be the responsible factors to predict the number of counts of MOSFETs in IC [2]. Machine learning approach can be applied to do the analysis and predict the same.

According to microprocessor architects, advancement of semiconductor has slowed industry-wide past 2010, less the pace as predicted by Moore [3, 4]. In 2015, Gordon Moore saw observed rate of progress may reach saturation: "I see Moore's law dying here in the next decade or so". However, from 2018, leading semiconductor manufacturers have fabrication processes of IC in mass production with 7 and 10 nm features which claimed to pace with Moore's law.

For the last two decades, the count of transistors in an IC has been growing exponentially [5]. If we put our view on the count of transistors, it is proportional to time (Year). There are no such theoretical aspects available by which we can say what will be the number of transistors in a circuit after some year except one hypothesis called Moore's law [6–8]. But it has some limitations. To overcome the limitations, we took a statistical and probabilistic approach (Machine Learning approach) to solve this problem [9–11]. In our approach, we use various machine learning algorithms (Linear regression, KNN regressor, Gradient Boosting regressor) to get better R-square value as the count of transistors increases concerning time in a non-linear fashion. After a lot of hard work and research, we got very accurate results almost near equal to real values. As a result of our work, it will help to predict price, power consumption, effective heat generation of a future IC and help industry for the betterment of their product.

3 Methodology

3.1 Feature Engineering

This project predicts the number of MOSFET transistor count of an integrated circuit based on the values of 5 dependent variables as per collected data:

- Size of a register (Number of bits)
- Date of introduction in the market
- Surface area of the integrated circuit
- Processes to be execute
- IC manufacturer company

Since the collected data contains null values, and no mathematical operation can be performed on null values, so those null have been filled with the respective mean values of that column.

The relationship between independent and output (dependent) variable can be determined by correlation value which ranges from -1 to 1. If the correlation value is 0 then there is no relation between the two variables. A positive value indicates direct proportional relation and a negative value indicates inverse proportional relation. Higher is the value stronger is the relationship. The different correlation values as measured as per collected data are as shown in Table 1.

The number of transistor counts also depends on the manufacturing companies such as INTEL, IBM, APPLE, AMD, etc., as the different manufacturer uses different architectural designs. The dependency between each of the independent variable with the output variable is scatter plotted. A scatter plot is a plot where a data item is represented by a dot in Figs. 1, 2, 3 and 4.

The training and testing dataset is separated with 20% of total dataset as testing and the remaining as training data. The training and testing dataset are chosen randomly.

Table 1 Correlation between dependent and independent variable

Feature name	Correlation with transistor count	Remarks
Date of introduction	0.507861	Positive trend
Register size (in bits)	0.403127	Positive trend
Area of IC	0.46327	Positive trend
Number of process	– 0.184418	Negative trend

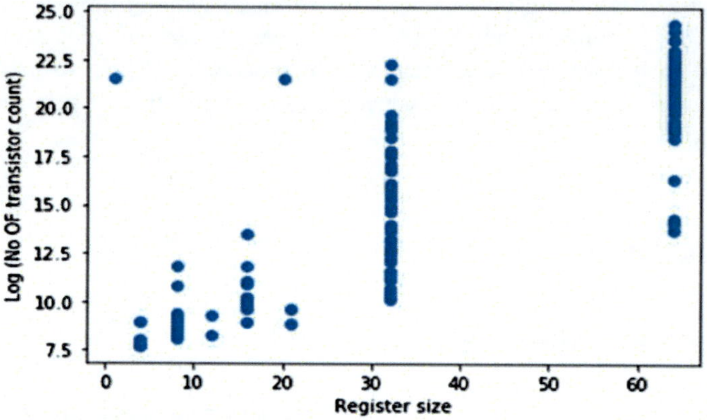

Fig. 1 Transistor count versus register size (bits)

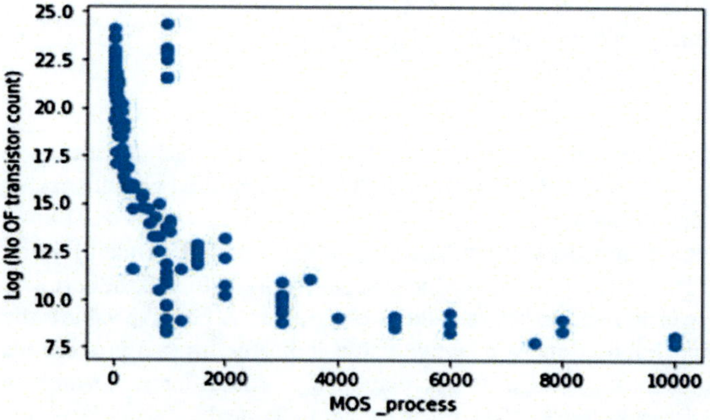

Fig. 2 Transistor count versus process

To confirm the absence of multi-collinearity in selected features correlation coefficient between every pair of the independent variable is calculated. The results according to our collected data as shown in Table 2.

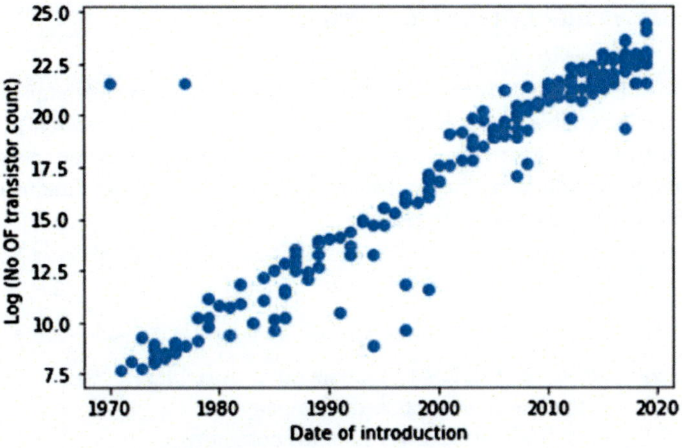

Fig. 3 Transistor count versus year of introduction

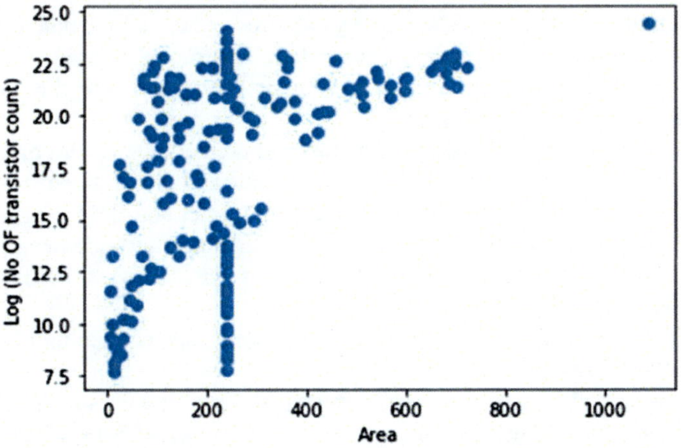

Fig. 4 Transistor count versus area

Table 2 Correlation table of independent variables

Feature name	Date of introduction	Register size (in bits)	Area of IC	Number of process
Date of introduction	1.000000	0.881416	0.546301	− 0.785295
Register size (in bits)	0.881416	1.000000	0.582218	− 0.727400
Area of IC	0.546301	0.582218	1.000000	− 0.420432
Number of process	− 0.785295	− 0.727400	− 0.420432	1.000000

3.2 Multivariate Linear Regression

Regression defines a mathematical relationship which helps to understand the impact of one or more variables (independent) over other variables (dependent).

The multivariate linear regression [12] model takes more than one predictor to predict the output values. It fits the best straight line in multiple dimension. In this experiment, our equation will look like:

No of transistor $= a_0 + a_1 *$ bit $+ a_2 *$ date_of_introduction $+ a_3 *$ process $+ a_4 *$ area $+ \sum_{n=0}^{8} c_n p_n$

where,

a_1, a_2, a_3, a_4	are the coefficients of independent variables.
c_n	is the coefficient value of company *n*.
p_n	is the independent variable for company *n*.
a_0	is the intercept on number of transistor (value of transistor when all the independent variable is 0).

This equation is applied to the training data from where all the unknown parameters are calculated, and a model is generated. This model is used to predict the outcome of testing data and predicted results are compared with the original one using different statistical methods.

Since a manufacturing company uses different methodologies, our project segregates the different companies to enhance the results and primarily considered three manufacturing units to work with, namely: AMD, Intel and IBM. The correlation matrix is calculated concerning dataset of each manufacturing companies and the new relationships between independent and output (dependent) variables are shown from Tables 3, 4, 5, 6, 7 and 8.

Three different training datasets are obtained with each containing rows relating to a particular company.

The same model is applied to different testing dataset of three different manufacturing companies, and different outcome is observed for each case.

After comparing, the training and testing dataset are again splitted randomly to observe a change in result and this process continued for more times until the achievable efficiency of the model is highest.

Table 3 Correlation between dependent and independent variable (AMD dataset)

Feature name	Correlation with transistor count	Remarks
Date of introduction	0.567677	Positive trend
Register size (in bits)	0.336333	Positive trend
Area of IC	0.855772	Positive trend
Number of process	0.494309	Positive trend

Table 4 Correlation table of independent variables with respect to AMD

Feature name	Date of introduction	Register size (in bits)	Area of IC	Number of process
Date of introduction	1.000000	0.822168	0.387450	0.190732
Register size (in bits)	0.822168	1.000000	0.330121	− 0.069187
Area of IC	0.387450	0.330121	1.000000	0.480153
Number of process	0.190732	− 0.069187	0.480153	1.000000

Table 5 Correlation between dependent and independent variable (Intel dataset)

Feature name	Correlation with transistor count	Remarks
Date of introduction	0.584697	Positive trend
Register size (in bits)	0.503678	Positive trend
Area of IC	0.656252	Positive trend
Number of process	− 0.258674	Negative trend

Table 6 Correlation table of independent variables with respect to Intel

Feature name	Date of introduction	Register size (in bits)	Area of IC	Number of process
Date of introduction	1.000000	0.851282	0.533967	− 0.760123
Register size (in bits)	0.851282	1.000000	0.563298	− 0.640117
Area of IC	0.533967	0.563298	1.000000	− 0.401992
Number of process	− 0.760123	− 0.640117	− 0.401992	1.000000

Table 7 Correlation between dependent and independent variable (IBM dataset)

Feature name	Correlation with transistor count	Remarks
Date of introduction	0.652208	Positive trend
Register size (in bits)	0.344577	Positive trend
Area of IC	0.800306	Positive trend
Number of process	− 0.375099	Negative trend

3.3 Regression with K-Nearest-Neighbors

Since multivariate linear regression is a parametric model, that is, it assumes the relationship in the form of straight line, $f(x)$, the outcome of it, will not be accurate

Table 8 Correlation table of independent variables with respect to IBM

Feature name	Date of introduction	Register size (in bits)	Area of IC	Number of process
Date of introduction	1.000000	0.917130	0.819958	− 0.932070
Register size (in bits)	0.917130	1.000000	0.548571	− 0.999163
Area of IC	0.819958	0.548571	1.000000	− 0.581785
Number of process	− 0.932070	− 0.999163	− 0.581785	1.000000

if the relationship is not linear regression with *k* nearest neighbor which is more flexible in this matter [13].

Similar dataset is used with this algorithm. The training and testing data is splitted in 80–20 fashion. The training and testing dataset are chosen randomly.

The correlation between dependent and independent variables is same as Table 1 in Sect. 3.1, and the correlation matrix of independent variables remains the same as in Table 2 in Sect. 3.1.

In our project, value of *k* used is 2, which means the model will look for two neighboring multidimensional points for its prediction. Values of *k* can be used which will give different *R*2 scores. According to this project, optimal value of *k* is 2. Figure 5 shows the different values of accuracy level based on different values of k.

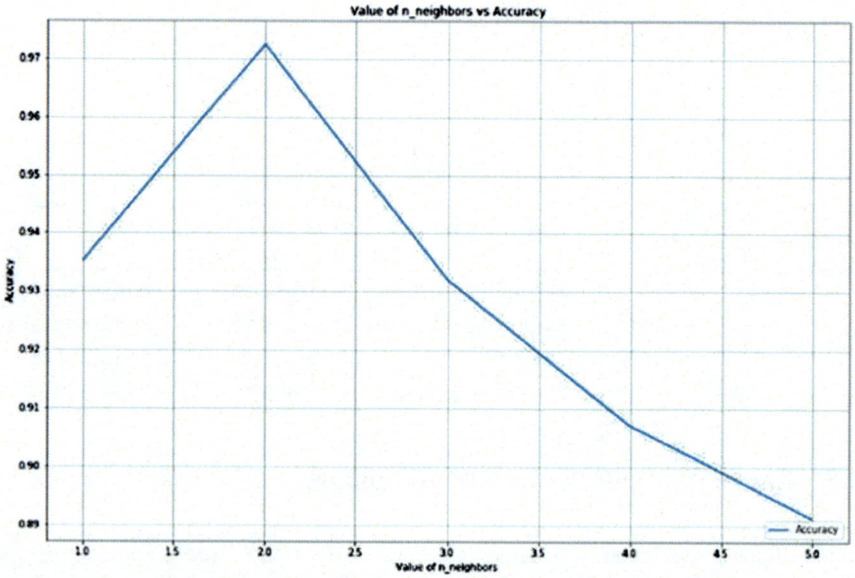

Fig. 5 Value of *n* neighbor versus accuracy

After training the model, the testing dataset is taken as input, and the model is used for generalization. When an unknown feature vector (Combination of different values of independent variable) is given, the model will calculate 2 nearest known distance points (known feature vector) by calculating Euclidean distances. Then, average of two output is predicted as the outcome of unknown feature vector.

This model is also used in respective datasets of the individual companies to enhance our results as is done in linear regression. The tables related to correlation between different variables of different manufacturing companies can be referred from Tables 3, 4, 5, 6, 7 and 8 in Sect. 3.2.

3.4 Gradient Boosting Regressor

Gradient boosting algorithm provides even more flexible and accurate outcomes than other regressor model used in this project [14, 15].

Similar dataset is used with this algorithm. The training and testing data contain 80 and 20% of the total dataset. The entire dataset is chosen randomly.

The correlation between independent and dependent variables is same as Table 1, and the correlation matrix of the independent variables remains the same as in Table 2 in Sect. 3.1.

This algorithm consists of multiple models (e.g., decision trees, neural networks), and each model provides an alternate solution to the problem, whose predictions are combined in a way to produce final model output.

Decision tree is chosen as base model in this project with learning rate of 0.01 to control overfitting of curve. Predicted results are compared with original transistor count, and accuracy is observed for different learning rates.

4 Result and Discussion

After the methods were applied, the following results were obtained:

The dataset was further classified on the basis of the manufacturing company. Since there was limited data hence the companies with more number of data was taken into consideration and the companies are Intel, IBM, AMD.

On applying the regression algorithm on Intel, the following results were obtained:

i. Linear regression model was applied which gave an $r2$ score of 0.9215. It had a mean square error value of 1.774603232249939e + 17. Figure 6 shows the difference between predicted data and original data.

ii. K Neighbors regressor was applied which gave an $r2$ score of 0.9988. It had a mean square error value of 3,552,547,500,002,232.0. Figure 7 shows the difference between predicted data and original data.

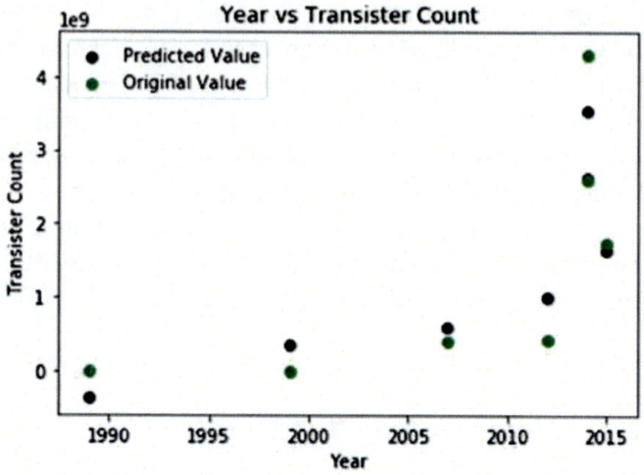

Fig. 6 Result obtained for linear regression on Intel's dataset

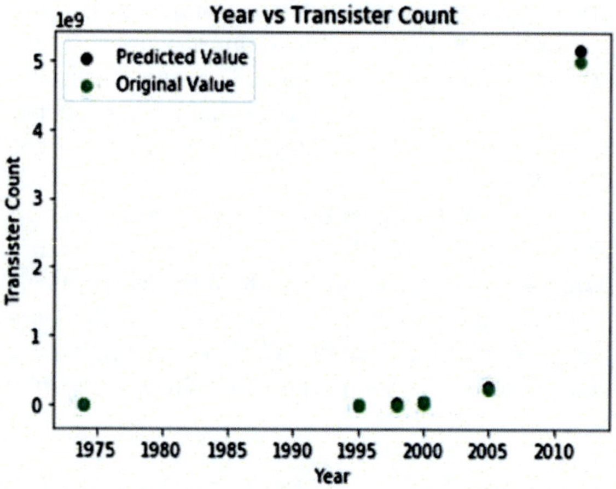

Fig. 7 Result obtained for K neighbors regressor on Intel's dataset

iii. Gradient boosting regressor was applied which gave an $r2$ score of 0.9936. It had a mean square error value of 2.3974345827528424e + 16. Figure 8 shows the difference between predicted data and original data.

On applying the regression algorithm on IBM, the following results were obtained:

i. Linear regression model was applied which gave an $r2$ score of 0.9938. It had a mean square error value of 4.272497573205097e + 16. Figure 9 shows the difference between predicted data and original data.

ii. K Neighbors regressor was applied which gave an $r2$ score of 0.9961. It had a mean square error value of 5,021,125,000,000,000.0. Figure 10 shows the difference between predicted data and original data.

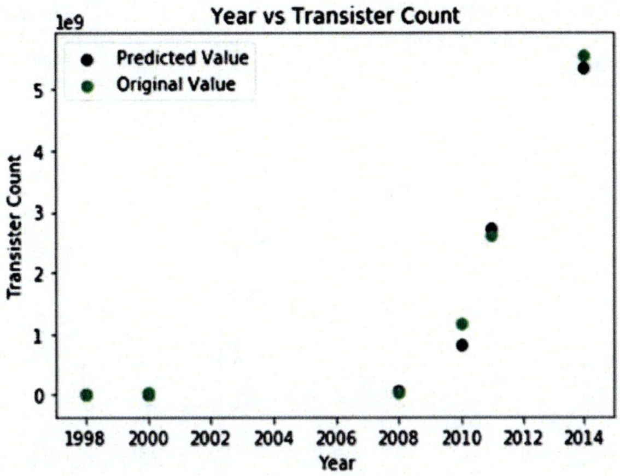

Fig. 8 Result obtained for gradient boosting regressor on Intel's dataset

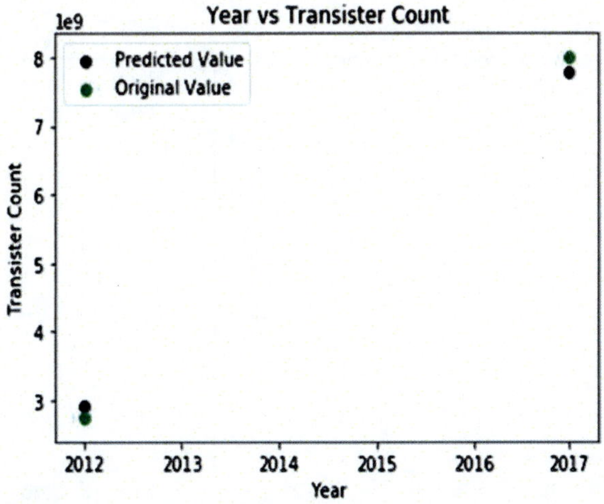

Fig. 9 Result obtained for linear regression on IBM's dataset

iii. Gradient boosting regressor was applied which gave an $r2$ score of 0.9995%. It had a mean square error value of 5,478,343,784,667,719.0. Figure 11 shows the difference between predicted data and original data.

On applying the regression algorithm on AMD, the following results were obtained:

i. Linear regression model was applied which gave an $r2$ score of 0.9548. It had a mean square error value of 1.0032699000754845e + 18. Figure 12 shows the difference between predicted data and original data.

ii. K Neighbors regressor was applied which gave an $r2$ score of 0.9999. It had a mean square error value of 17,405,000,000,000.0. Figure 13 shows the difference between predicted data and original data.

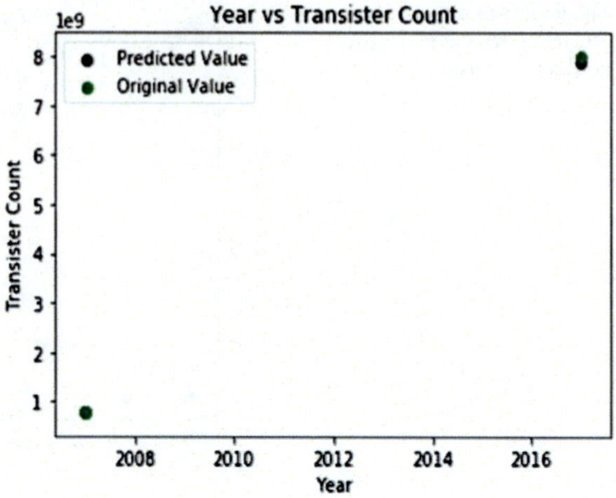

Fig. 10 Result obtained for K neighbors regressor on IBM's dataset

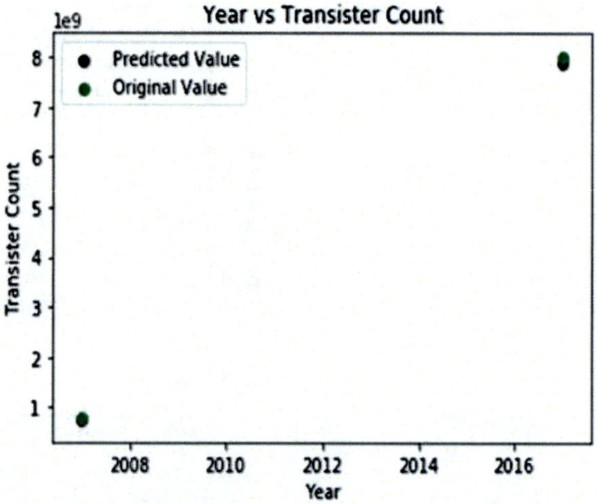

Fig. 11 Result obtained for gradient boosting regressor on IBM's dataset

iii. Gradient boosting regressor was applied which gave an $r2$ score of 0.9984. It had a mean square error value of 8,902,009,028,097,026.0. Figure 14 shows the difference between predicted data and original data.

The result is summarized in Tables 9, 10 and 11.

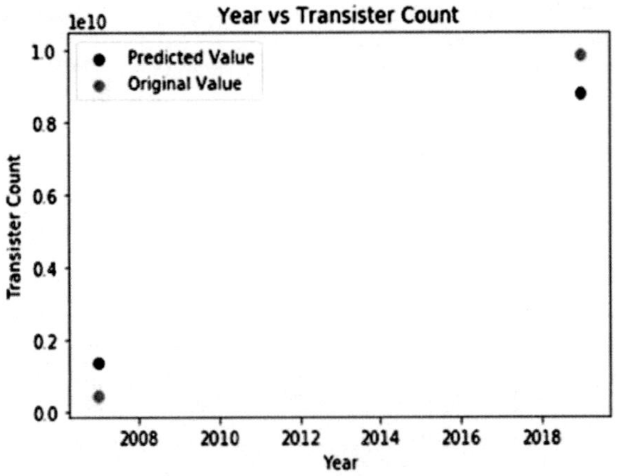

Fig. 12 Result obtained for linear regression on AMD's dataset

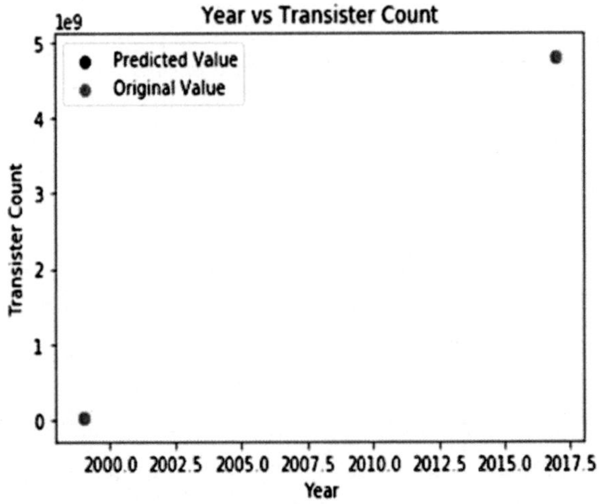

Fig. 13 Result obtained for K neighbors regressor on AMD's dataset

5 Conclusion

For the predictive model, dataset consisted of data till 2019. After achieving the results, a prediction was done to predict the transistor count in the year 2020 and 2021 using our model:

In the year of 2020,

- In case of Intel (Consideration is MOS Process = 14 mn and area is 100 mn^2), the transistor count as per our predictive model is approximately 1.82 billion (1,825,000,000) and in reality it is actually 1.75 billion.
  Here, the difference reality and prediction is 0.08 billion that means we have only 4% error in our model and 0.96 r2 score in case of Intel.

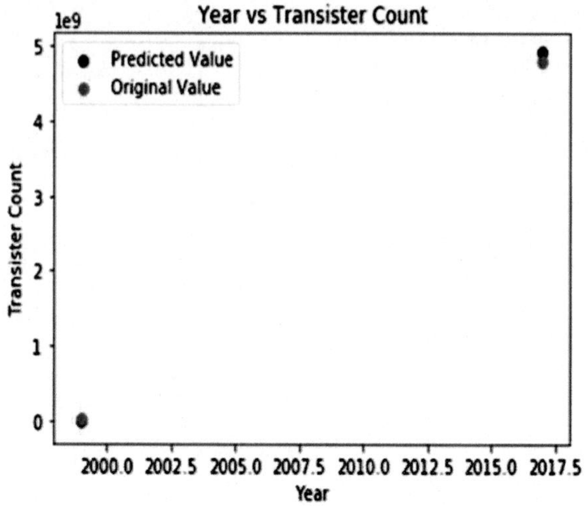

Fig. 14 Result obtained for gradient boosting regressor on AMD's dataset

Table 9 Linear regression on individual industry (Intel, IBM, AMD)

	Intel	IBM	AMD
*r*2 score	0.9215	0.9938	0.9548
MSE	1.774603232249939e + 17	4.272497573205097e + 16	1.0032699000754845e + 18

Table 10 K neighbors regressor on individual industry (Intel, IBM, AMD)

	Intel	IBM	AMD
*r*2 score	0.9988	0.9961	0.9999
MSE	3,552,547,500,002,232.0	5,021,125,000,000,000.0	17,405,000,000,000.0

Table 11 Gradient boosting regressor on individual industry (Intel, IBM, AMD)

	Intel	IBM	AMD
*r*2 score	0.9936	0.9995	0.9984
MSE	2.3974345827528424e + 16	5,478,343,784,667,719.0	8,902,009,028,097,026.0

- In case of AMD (Consideration is MOS Process = 14 mn and area is 100 mn^2), the transistor count as per our predictive model is 4.8 billion (4,800,000,000) and in reality also it is actually 4.8 billion which was highly accurate.
  In the year of 2021,

- In case of Intel (Consideration is MOS Process = 14 mn and area is 90 mn^2 and 64-bit processor), the transistor count as per our predictive model is near 1.09 billion (1,091,000,000).
- In case of AMD (Consideration is MOS Process = 14 mn and area is 90 mn^2 and 64-bit processor), the transistor count as per our prediction is near 2.42 billion (2,427,150,000).

It is our observation that the transistor count depends more on the area rather than the year of manufacturing.

References

1. Mack, C.A.: Fifty years of Moore's Law. IEEE Trans. Semiconductor Manuf. **24**(2) (2011)
2. El-Aawar, H.: Increasing the transistor count by constructing a two-layer crystal square on a single chip. Int. J. Comput. Sci. Inf. Technol. (IJCSIT) **7**(3) (2015)
3. Moore, G.E.: Progress in digital integrated electronics. IEEE Solid-State Circuits Society Newslett. **11**(3) (2006)
4. Moore, G.E.: Lithography and the future of Moore's Law. SPIE **2440** (1995)
5. Frank, D.J., Dennard, R.H., Nowak, E., Solomon, P.M., Taur, Y., Wong, H.S.P.: Device scaling limits of Si MOSFETs and their application dependencies. Proc. IEEE **89**(3) (2001)
6. Ahmad, K., Schuegraf, K.: Transistor wars: rival architecture face off in a bid to keep Moore's Law alive. IEEE Spect. **48**(11) (2011)
7. Brock, D.C.: Understanding Moore's Law: Four Decades of Innovation. Chemical Heritage Foundation, Philadelphia (2006)
8. Mollick, E.: Establishing Moore's law. IEEE Ann. Hist. Comput. **28**(3), 62–75 (2006)
9. de Oliveira Conceição, C.M., da Luz Reis, R.A.: Transistor count reduction by gate merging. IEEE **66**(6) (2019)
10. Possani, V.N., Reis, A.I.: Transistor count optimization in IG FinFET network design. IEEE **36**(9) (2017)
11. Xue, J., Al-Khalili, D., Rozon, C.N.: Tree-based transistor topology extraction algorithm for library-free logic synthesis. In: Proceedings of IEEE International Conference on Semiconductor Electronics, pp. 5, Dec 2004
12. Montgomery, D.C., Peck, E.A., Geoffrey Vining, G.: Introduction to Linear Regression Analysis, vol. 821. John Wiley & Sons (2012)
13. Zhang, M.-L., Zhou, Z.-H.: ML-KNN: a lazy learning approach to multi-label learning. Pattern Recogn. **40**(7), 2038–2048 (2007)
14. Friedman, J.H.: Stochastic gradient boosting. Comput. Stat. Data Anal. **38**(4), 367–378 (2002)
15. Feng, Ji., Yang, Yu., Zhou, Z.-H.: Multi-layered gradient boosting decision trees. Adv. Neural. Inf. Process. Syst. **31**, 3551–3561 (2018)

PSO Optimum Design-PID Controller for Frequency Management of Single Area Multi-Source Power Generating System

V. Kumarakrishnan, G. Vijayakumar, K. Jagatheesan, D. Boopathi, B. Anand, and V. Kanendra Naidu

Abstract In this article, Particle Swarm Optimization (PSO) tuned Proportional-Integral-Derived (PID) controller is proposed for frequency management of a single area multi-source power generation unit. The power system comprises of both renewable and non-renewable energy sources which includes thermal, solar and wind power generating units. In this work, Integral (I), Proportional -Integral (PI), and PID controllers are utilized as a subsidiary controller to regulate frequency deviation of the power system during unexpected load variation. The gain values of the controller are tuned by applying conventional (trial and error) scheme and PSO technique. Conventional method tuned Controller tuned using conventional method shows that PID controller provides superior response over I and PI controller response. Subsequently, PSO is implemented to tune PID controller gain values. To demonstrate the superiority of the PSO-PID controller, the output response is compared to the conventional tuned PID controller result. It is obvious from the comparison, the PSO-PID controller is provides fast settling time with minimal frequency overshoot and undershoot at various loading conditions.

Keywords Particle Swarm Optimization · Proportional-Integral-Derivative · Frequency regulation · Frequency deviation · Settling time

V. Kumarakrishnan (✉) · K. Jagatheesan · D. Boopathi
Department of EEE, Paavai Engineering College, Pachal, Tamil Nadu, India

G. Vijayakumar
Department of EEE, Muthayammal Engineering College, Rasipuram, Tamil Nadu, India

B. Anand
Department of EIE, Hindusthan College of Engineering and Technology, Coimbatore, Tamil Nadu, India

V. Kanendra Naidu
School of Electrical Engineering, College of Engineering, Universiti Teknologi MARA, Shah Alam, Selangor, Malaysia

H. K. D. Sarma et al. (eds.), *Contemporary Issues in Communication, Cloud and Big Data Analytics*, Lecture Notes in Networks and Systems 281,
https://doi.org/10.1007/978-981-16-4244-9_31

1 Introduction

In the modern world, electrical power demand is constantly increasing. In order to satisfy the ever growing power demand of the consumers, power generating units is required to increase their generation capacity. As the power demand grows, the complexity of the system follows the trend as their proportional to each other. Hence greater effort is required to ensures that system efficiency in delivering reliable power is maintained. LFC scheme is utilized to further control and sustain the power quality via frequency regulation [1, 2]. Researchers have investigated various types of controllers and optimization techniques for the LFC. A micro grid power system containing Renewable Energy Sources (RESs) and SMES power storage unit utilizing a PID controller is investigated by [3]. Author [4] investigated LFC problem by implementing PSO-PID while in [5] Ant Colony Optimization (ACO)-PID is investigated to control the frequency deviation in a interconnected nuclear based generating unit [5]. A power generating unit comprising of thermal, PV and wind-based sources is analyzed for LFC problem based on Bacterial Foraging Optimization (BFO) tuned PID controller [6].

Double Chains Quantum Genetic Algorithm (DCQGA) using Active Disturbance Rejection Controllers (ADRC) is examined for a multi area power system [7]. A wind based power plant is integrated to an existing power system network and the LFC for this system is investigated using Genetic Algorithm (GA) tuned classical controller [8]. Firefly algorithm is applied to optimize many critical controlling schemes as discussed in [9]. Authors in [10] employed ACO-PID controller and the results shows that ACO-PID controller performs better compared to classical controller. Similar performance is noticed in [11] when PSO-PID controller is able to quell the frequency deviation for thermal power system with boiler and GRC [11], Moth Flame Optimization (MFO) based PI controller for LFC is implemented for two area power system in [12]. A hybrid Bacteria Foraging Optimization incorporating PSO(hBFOA-PSO) is implemented in [13] for LFC in thermal power plant. Quasi-Oppositional-Harmony Search (QOHS) is used to optimize PID parameters of an autonomous power system for LFC in [14]. Firefly algorithm is implanted [15] to tune PID gain to achieve optimum values for reheate thermal power system.

Artificial Bee Colony (ABC) tuned PID is inspected for multi area thermal units for LFC problem [16] while in [17] authors investigated LFC of a four area thermal unit incorporating GRC and GRD by adopting fuzzy logic based PI controller. Wind power plant integrated power system based on GA-PID is investigated by [18]. Flower pollination tuned classical controller is investigated in [19] as the wind–PV interconnected autonomous power grid, [20] implemented Quantum Inspired Evolutionary Algorithm (QIEA)-PID controller of a power network for LFC. PSO technique has been implemented for various medical application like Dengue fever classification [21], Patient monitoring [22], and also in [23] to identify faulty design in building. Apart from PSO, many other optimization methods are discussed by researchers [24, 25]. This literature review gives pertinent information about the importance of optimization technique for LFC and the various techniques utilized by researchers to ensure that frequency deviation is minimized.

Contribution of the Research Work

The main contribution of this research work is as follows:

- With the current importance placed on increasing the penetration of renewable energy, the wind and photovoltaic based plants integrated with conventional thermal unit is investigated in detail in this study.
- Proposed research work designs a PID regulator which is implemented to regulate the power fluctuation in system and also to maintain the power quality. The interconnected thermal generating station with major RE sources (solar and wind) is developed.
- To achieve optimum gain values from the PID controller, the PSO technique is used to optimize the controller gain values in order to generate better performance during emergency conditions.
- To demonstrate the superiority of the PSO-PID controller, the output response is compared to the conventional tuned PID controller result.

2 System Strategy for Investigation

A hybrid power generating power system (Thermal, wind form and PV) is developed for analyzing system stability during sudden loading conditions by applying LFC schema. The proposed Simulink model is shown in Fig. 1 [3]. The transfer function of the various power system component is provided in Table 1. The parameters of the system is given in Appendix 1.

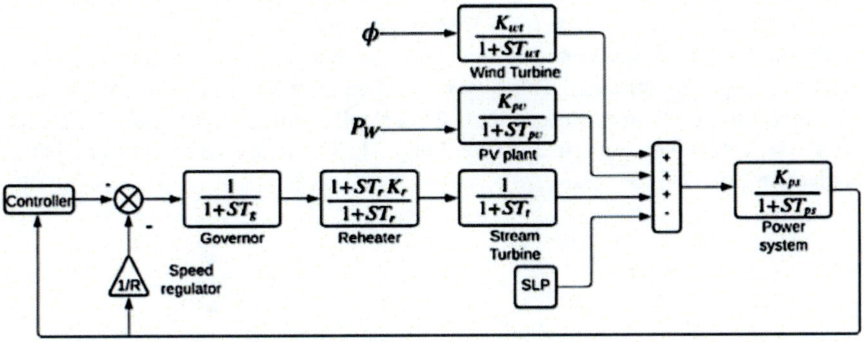

Fig. 1 Simulink model of proposed single area—multi-sources power generating system

Table 1 Mathematical functions in Simulink model

Components	Mathematical function
Governor	$\frac{1}{1+ST_g}$
Re heater	$\frac{1+ST_r K_r}{1+ST_r}$
Turbine	$\frac{1}{1+ST_t}$
Wind turbine	$\frac{K_{WT}}{1+ST_{WT}}$
PV Plant	$\frac{K_{PV}}{1+ST_{PV}}$
Power system/generator	$\frac{K_{ps}}{1+ST_{ps}}$
Speed regulator	$\frac{1}{R}$

3 Controller design

PID controller is the most effective industrial controller used to compensate error. PID controller is designed by incorporating three different controllers like Proportional, integral and derivative controller [25]. Mathematical expression of PID controller is mentioned in Eq. 1. Control signal generated by the PID controller is represented in Eq. 2 [4]. Typical assembly of PID controller is displayed in Fig. 2.

$$G_{\mathrm{PID}}(s) = K_{\mathrm{P}} + K_{\mathrm{I}}s + \frac{K_{\mathrm{D}}}{s} \tag{1}$$

$$u(t) = K_{\mathrm{P}}e(t) + K_{\mathrm{I}}\int_0^t e(t)\mathrm{d}t + K_{\mathrm{D}}\frac{\mathrm{d}e(t)}{\mathrm{d}t} \tag{2}$$

where K_{P} = Proportional gain, K_{I} = Integral gain, K_{D} = Derivative gain, $u(t)$ = Control signal to system, and $e(t)$ = Error signal from output.

By the suitable choice of controller regulation method, stability improves and the oscillations quelled instantly. This means quick settling time and minimal overshoot and undershoots. In order to tune the PID controller, Integral Absolute Time Error (ITAE) objective function is used in this study. The mathematical expression of ITAE is given in Eq. 3 [4].

Fig. 2 PID basic structure

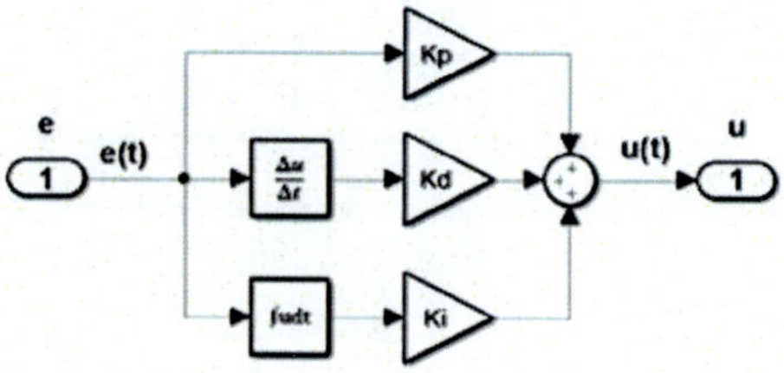

$$J = \mathrm{ITAE} = \int_{0}^{tsim} t.|e(t)|\mathrm{d}t \quad (3)$$

4 System Performance Analysis

Proposed research work includes of two different controller gain tuning methods which are conventional (trial and error) method and particle swarm optimization technique method.

Conventional PID Controller

Conventional method is a classical method to tune the PID controller gain values. Gain values were adjusted by trial and error patterns in this system. First tune integral gain value, while remaining gain values are kept as zero; then K_I is kept constant, and proportional gain value is tuned after getting optimal value K_I, K_P is kept constant. Repeat the same process for K_D [5]. Finally all the controller gain values are acquired, the obtained numerical gain values clearly mentioned in Table 2. I, PI, and PID controller indices curve are displayed, respectively, in Figs. 3, 4, and 5.

Table 2 Conventional method tuned controller gain values

Gain value/controller	K_I	K_P	K_D
I	0.44	–	–
PI	0.44	3.4	–
PID	0.42	2.4	0.1

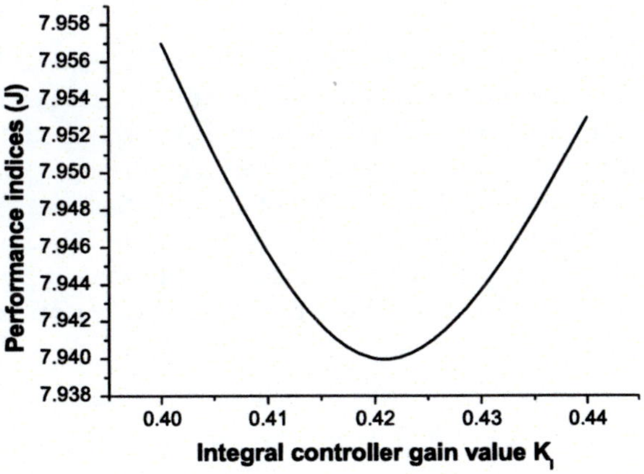

Fig. 3 Performance indices curve of I controller

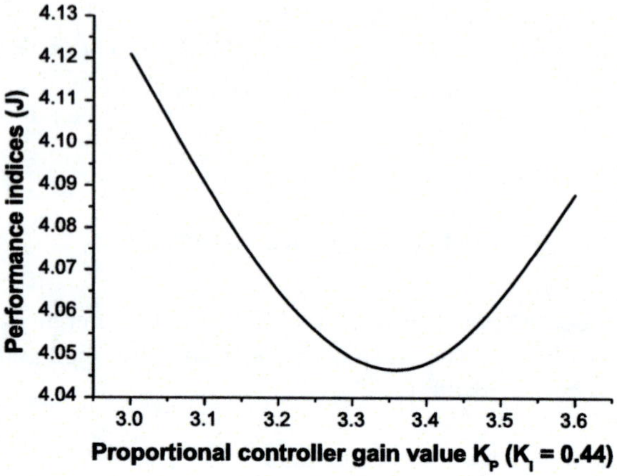

Fig. 4 Performance indices curve of PI controller

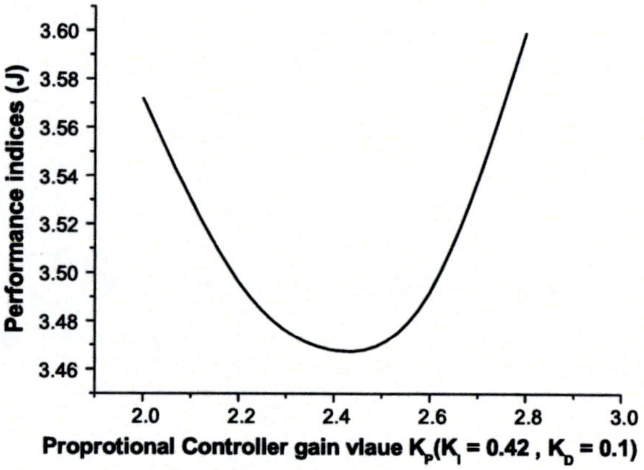

Fig. 5 Performance indices curve of PID controller

Particle Swarm Optimization Technique Based PID Controller

The population-based optimization strategy is the particle swarm optimization technique. PSO is a population based optimization strategy developed in the year of 1995 by Dr. Kennedy and Dr. Eberhart with the help of birds flying behavior and fish schooling [4]. The flow chart of PSO technique is shown in Fig. 6.

To acquire the optimum gain values for the PID controller to the proposed system, PSO technique is employed and the corresponding gain values are obtained based on the ITAE cost function. The acquired values of controller gain are dispatched in Table 3.

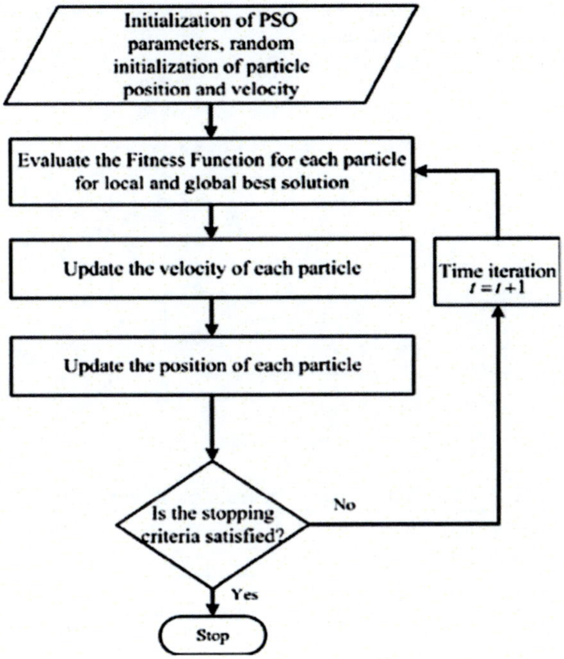

Fig. 6 Functional flow diagram for PSO

Table 3 PSO technique tuned controller gain values

Controller/gain value	Proportional gain	Integral gain	Derivative gain
PID	2.6889	0.7704	2.8844

5 Simulation and Result Discussion

The proposed power system Simulink model is simulated by MATLAB 2014a version for 60 s time duration for both conventional and PSO technique method by applying 1% SLP. First, conventional tuned method I, PI, and PID controller performance were compared. Corresponding simulation result are shown in the Fig. 7. The time domain specific parameters of the frequency response is tabulated in Table 4.

Based on the performance response comparison of conventional I, PI, and PID controllers, it is clearly seen in Fig. 8, the graph indicates that PID controller performs well compared to other controllers in terms of quick settling time and minimal frequency deviations. Bar chart comparison of settling time of conventional I, PI and PID controllers are displayed in Fig. 8. This figure confirms that PID controller is superior compared to other controllers.

To confirm the efficiency of the PSO-PID controller, response of classical PID controller result is compared against response of PSO-PID controller. Graphical representation of comparison is presented in Fig. 9.

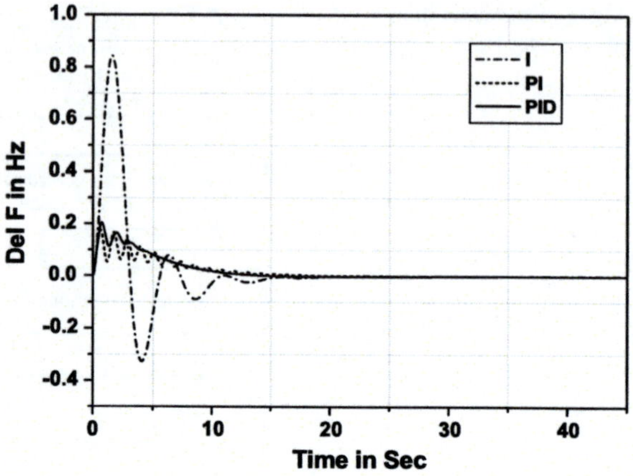

Fig. 7 Conventional I, PI, and PID controller result performances evaluation

Table 4 Time domain specific parameters of conventional PID controller result

Time domain specific parameter/controller	Relaxation time (s)	Overshoot (Hz)	Undershoot (Hz)
I	42	0.85	0.32
PI	38	0.175	0.25
PID	30	0.2	0.002

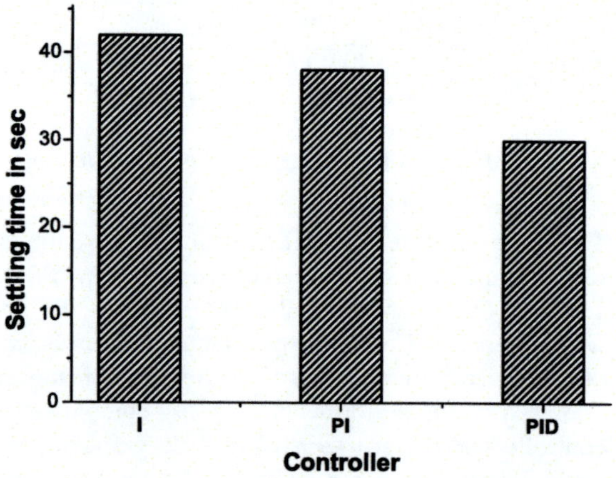

Fig. 8 Relaxation time comparison of conventional I, PI, and PID controllers in bar chart

Time domain specific parameters of conventional tuned PID and PSO-PID is dispatched in Table 5.

Based on bar chart comparison show in Fig. 10, it is evident that PSO-PID controller performs well during emergency conditions.

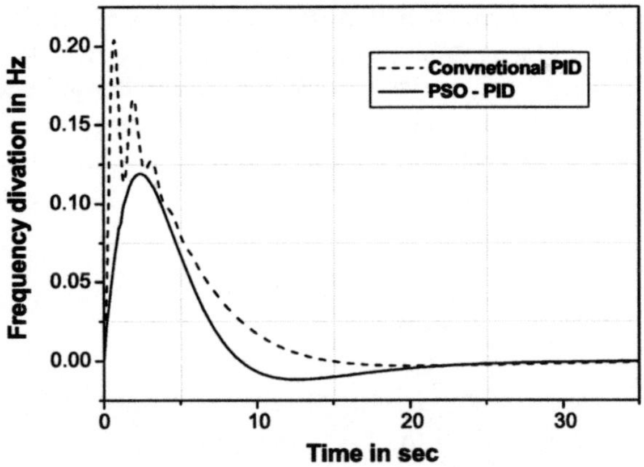

Fig. 9 Conventional PID and PSO-PID controller response comparison

Table 5 Conventional PID and PSO-PID controller time domain specific parameters

Time domain specific parameters/tuning method	Relaxation time (s)	Overshoot (Hz)	Undershoot (Hz)
Conventional PID	30	0.2	0.002
PSO-PID	27	0.12	0.011

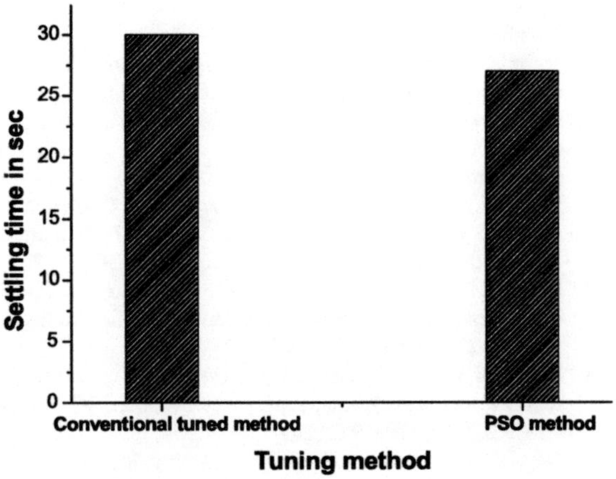

Fig. 10 Bar chart comparison of conventional tuned PID and PSO-PID controller relaxation time

6 Conclusion

This article presents frequency control of a single area multi-source power generation system with renewable energy sources. LFC based on PSO-PID is investigated to ensure good system performance during various loading conditions. The conventional tuned I, PI, and PID controller's performances are comparatively analysed and it is clearly identified that PID performs better than I and PI controller. The traditional tuned PID controller is subsequently compared with PSO-PID controller and it can be seen from results that PSO-PID controller provides good performance in terms of quick settling time and minimal peak overshoot and undershoot. Hence, it can be concluded that the PSO-PID controller performs well during sudden load varying condition.

Appendix [3, 24]

$T_{PS} = 20$ S, $K_{PS} = 120$, $T_g = 0.08$ S, Tr = 10 S, $K_r = 0.5$ (p.u), Twt = 0.5 S, $T_{pv} = 1.5$ S, $K_{wt} = 1$ (p.u), $K_{pv} = 1$ (p.u), $T_t = 0.3$ S, $R = 2.4$.

References

1. Elgerd, O.I.: Electric energy systems theory: an introduction (1982)
2. Nagrath, I.J., Kothari, D.P.: Power System Engineering. Tata McGraw Hill Publishing Company limited, New Delhi (1994)
3. Mohamed, E.A., Gouda, E., Mitani, Y.: Impact of SMES integration on the digital frequency relay operation considering high PV/wind penetration in micro-grid. Energy Procedia **157**, 1292–1304 (2019)
4. Jagatheesan, K., Anand, B., Samanta, S., Dey, N., Ashour, A.S., Balas, V.E.: Particle swarm optimisation-based parameters optimisation of PID controller for load frequency control of multi-area reheat thermal power systems. Int. J. Adv. Intell. Paradig. **9**(5–6), 464–489 (2017)
5. Dhanasekaran, B., Siddhan, S., Kaliannan, J.: Ant colony optimization technique tuned controller for frequency regulation of single area nuclear power generating system. Microprocessors Microsyst. **73**, 102953 (2020)
6. Koley, I., Bhowmik, P.S., Datta, A.: Load frequency control in a hybrid thermal-wind-photovoltaic power generation system. In: 2017 4th International Conference on Power, Control & Embedded Systems (ICPCES), pp. 1–5. IEEE (2017)
7. Huang, Z., Chen, Z., Zheng, Y., Sun, M., Sun, Q.: Optimal design of load frequency active disturbance rejection control via double-chains quantum genetic algorithm. Neural Comput. Appl. 1–21 (2020)
8. Boopathi, D., Saravanan, S. Jagatheesan, K., Anand, B.: Performance estimation of frequency regulation for a micro-grid power system using PSO-PID controller. Int. J. Appl. Evol. Comput. (IJAEC) **12**(2), 36–49 (2021)
9. Dey, N.: Applications of Firefly Algorithm and Its Variants. Springer Singapore (2020)
10. Jagatheesan, K., Anand, B., Omar, M.: Design of proportional-integral-derivative controller using ant colony optimization technique in multi-area automatic generation control. Int. J. Electr. Eng. Inform. **7**(4), 541 (2015)
11. Haroun, A.G., Li, Y.Y.: A novel optimized hybrid fuzzy logic intelligent PID controller for an interconnected multi-area power system with physical constraints and boiler dynamics. ISA Trans. **71**, 364–379 (2017)
12. Mohanty, B., Acharyulu, B.V.S., Hota, P.K.: Moth-flame optimization algorithm optimized dual-mode controller for multiarea hybrid sources AGC system. Optim. Control Appl. Methods **39**(2), 720–734 (2018)

13. Panda, S., Mohanty, B., Hota, P.K.: Hybrid BFOA–PSO algorithm for automatic generation control of linear and nonlinear interconnected power systems. Appl. Soft Comput. **13**(12), 4718–4730 (2013)
14. Shankar, G., Mukherjee, V.: Load frequency control of an autonomous hybrid power system by quasi-oppositional harmony search algorithm. Int. J. Electr. Power Energy Syst. **78**, 715–734 (2016)
15. Naidu, K., Mokhlis, H., Bakar, A.H.A., Terzija, V.: Comparative performance analysis of firefly algorithm for load frequency control in automatic generation control of interconnected reheat thermal power system (2014)
16. Naidu, K., Mokhlis, H., Terzija, V.: Performance investigation of ABC algorithm in multi-area power system with multiple interconnected generators. Appl. Soft Comput. **57**, 436–451 (2017)
17. Arya, Y., Kumar, N., Sinha, S.K.: Fuzzy logic based load frequency control of multi-area electrical power system considering non-linearities and boiler dynamics. Int. Energy J. **13**(2) (2012)
18. Hussain, I., Ranjan, S., Das, D.C., Sinha, N.: Performance analysis of flower pollination algorithm optimized PID controller for wind-PV-SMES-BESS-diesel autonomous hybrid power system. Int. J. Renew. Energy Res. (IJRER) **7**(2), 643–651 (2017)
19. Jagatheesan, K., Samanta, S., Choudhury, A., Dey, N., Anand, B., Ashour, A.S.: Quantum inspired evolutionary algorithm in load frequency control of multi-area interconnected thermal power system with non-linearity. In: Quantum Computing: An Environment for Intelligent Large Scale Real Application, pp. 389–417. Springer, Cham (2018)
20. Chatterjee, S., Hore, S., Dey, N., Chakraborty, S., Ashour, A.S.: Dengue fever classification using gene expression data: a PSO based artificial neural network approach. In: Proceedings of the 5th International Conference on Frontiers in Intelligent Computing: Theory and Applications, pp. 331–341. Springer, Singapore (2017)
21. Dey, N., Ashour, A.S., Bhattacharyya, S. (eds.): Applied Nature-Inspired Computing: Algorithms and Case Studies. Springer Singapore (2020)
22. Dey, N. (ed.): Advancements in Applied Metaheuristic Computing. IGI Global (2017)
23. Gopal, M.: Control Systems: Principles and Design. Tata McGraw-Hill Education (2002)
24. Malik, S., Suhag, S.: Salp swarm algorithm tuned control scheme for mitigating frequency deviations in hybrid power system and comparative analysis. Int. J. Comput. Digit. Syst. **10**, 2–10 (2020)
25. Chatterjee, S., Sarkar, S., Hore, S., Dey, N., Ashour, A.S., Balas, V.E.: Particle swarm optimization trained neural network for structural failure prediction of multistoried RC buildings. Neural Comput. Appl. **28**(8), 2005–2016 (2017)
26. Chakraborty, S., Samanta, S., Biswas, D., Dey, N., Chaudhuri, S.S.: Particle swarm optimization based parameter optimization technique in medical information hiding. In: 2013 IEEE International Conference on Computational Intelligence and Computing Research, pp. 1–6. IEEE (2013)

Spatial Analysis-Based Study on Impact of COVID-19 on Power Demand and Supply Prospective in Hilly State of Sikkim, India

Kamal Sapkota, Shabbiruddin, and Karma Sonam Sherpa

Abstract This study aims to investigate the power system scenario in ongoing COVID-19 pandemic, and various challenges being faced by the different stakeholders and individuals who are involved to manage the cause. It investigates how this sector responded to other sectors availing essential services like healthcare, security, data center, etc., during health crisis situation. In order to categorize the impact, region-wise detection of COVID-19 cases for the whole nation has been analyzed for developing the geoprocessing map model using advance technology like Q-GIS 2.18.0. Author also tries to examine some of the major challenges and disruptions faced on supply chain to new and ongoing renewable energy projects particularly solar, wind and hydropower projects in India. Moreover, this paper also tries to investigate first ever 9 min "Light off Event" in India, discussed some of its major consequences that could arise if not handled properly. The role of Power System Operation Corporation Limited (POSOCO) during the event in retaining the grid frequency and grid voltage profile within their recommended band has also been discussed. This unprecedented event has been studied and analyzed by taking case study on Sikkim, India and explores the different challenges being faced at state level to manage smooth operation of power supply system.

Keywords COVID-19 · Q-GIS · Renewable · POSOCO · Geo processing

1 Introduction

As the energy sector constitutes a backbone to shape the GDP of almost all the countries in the world. So, most importantly, energy is an important input to every goods and service sector in the system. For this reason, stable and reasonable energy prices are desirable for reigniting, sustaining and expanding economic growth of a

K. Sapkota · Shabbiruddin (✉)
Department of Electrical and Electronics Engineering, Sikkim Manipal Institute of Technology, Sikkim Manipal University, Majitar, Sikkim, India

K. S. Sherpa
Sikkim Manipal University, Gangtok, Sikkim, India

H. K. D. Sarma et al. (eds.), *Contemporary Issues in Communication, Cloud and Big Data Analytics*, Lecture Notes in Networks and Systems 281,
https://doi.org/10.1007/978-981-16-4244-9_32

nation. On the top of it, public health safety is a primary concern to maintain the demand, supply as well as growth of the economy. In the past history, the growth and development were interrupted many a times due to outbreak of diseases such as Circa of China in 3000 B.C., Plague of Athens in 430 B.C. to the last recent Zika Virus epidemic in 2015 causing thousands of death worldwide [1].

Novel coronavirus disease (COVID 19) is caused by the severe acute respiratory syndrome coronavirus-2 (SARS-CoV-2) which had started from China in the month of December 2019 and has overwhelmed the health care systems across the world [2]. This novel coronavirus (COVID-19) was first identified in Hubei Province of Wuhan City in China during late December 2019 as reported by WHO on December 31, 2019 and the outbreak started to begin after that [3]. This virus spreads through close contact of one already infected with the virus making it a highly infectious disease, if not cautiously dealt with, may spread on a community level. The patient will have no symptoms at all or have common symptoms like fever, cold, cough, bone pain and breathing problems, and ultimately leading to pneumonia after this viral infection [4]. Being a novel disease, till date COVID-19 doesn't have any clinically proven antiviral vaccine or medicine. Basically, this disease spreads from person to person via small droplets from the nose or mouth when expelled by an infected person when they cough, sneeze or speak. These droplets are relatively heavy and don't travel far and quickly sink into the ground [5].

In the present study, author has tried to present the scenarios of COVID-19 in Indian nation and details case study has been done in Sikkim state. Author was motivated by investigation on first-ever light-off Event declared due to outbreak of novel coronavirus diseases in India. In order to provide repaid information to society, Q-GIS-based Geoprocessing thematic map model has been developed as a novel work to perform study on COVID-19 outbreaks in India and focus has been given to state of Sikkim.

The main objective of the present study is to perform an investigation on impact of COVID-19 in power demand and supply system in India. Author has tried to explore the impact incurred on ongoing renewable energy projects in Sikkim as well as in India due to COVID-19 pandemic. An investigation has been done on 10 min light-off event in India, and its cause, consequence and outcome has also been studied. Some observation and suggestion have also been made on such unprecedented event if strikes in the future. Considering the available data, Q-GIS-based thematic map model has also been prepared to study the region-wise impact on COVID-19 in India and in Sikkim.

Q-GIS-based Geoprocessing map model has been prepared to study the outbreak of COVID-19 for India as well as for state of Sikkim. From the available literature, it is to be known that the study on impact of COVID-19 on power supply system incorporating field of Q-GIS software has rarely been done in state of Sikkim. So, the author has tried to implement Q-GIS software on COVID-19 outbreak and has conducted study on its impact on power demand and supply chain and ongoing renewable energy projects in Sikkim.

Q-GIS is a research tools for viewing, analyzing, editing and storing geological or spatial map throughout the world [6]. GIS is a powerful research tool for successful identification of sites for renewable energy system development at potential locations. GIS consists of two types of fundamental data structure, i.e., Raster and vector. Raster resembles to be a rectangular grid also known as pixel containing specific information of any given location, whereas vectors usually stores data in terms of coordinates and represented by geographic figure like points, lines and polygons [6].

1.1 Spatial Analysis and GIS

In order to prepare a thematic map for COVID-19 outbreak in any particular region of interest Q-GIS 2.18.0 software, first step involve is preparation of shape file map (.shp file format) on Q-GIS software. An administrative map has been prepared as shape file (.shp file) considering all the states and union territories in case of India and for state of Sikkim considering all districts. In this process, a physical map for particular region of interest, i.e. Sikkim and India has been being scan, georeferenced and then digitized on Q-GIS software. In the next section, the COVID-19 current raw data has been downloaded from web base portal like COVID-19 Pandemic India, World Health Organization or Worldometer. This data has been feed on digitized administrative map for both the case through attribute table. While the data from State IEC Bureau, Sikkim the COVID-19 data has been obtained. Finally, the COVID -19 thematic maps have been developed for India as well as for Sikkim.

Around 36,744,349 persons are infected; 1,066,819 deaths and 27,663,555 persons recovered as on October 7, 2020 affecting 216 countries and territories around the world and 2 international conveyances [7]. Out from several countries; USA, India, Brazil, Russia and Colombia are mostly affected. In India, 6,757,131 coronavirus cases have been detected, out of which 104,555 deaths and 5,744,693 persons have been recovered as on October 7, 2020 [8]. Maharashtra being the mostly affected state as indicated by red color in Fig. 1, considering scenarios in Indian nation. The state with highest recorded COVID-19 cases is shown by red color and state with minimum cases is shown by green color. Detailed case study has been done on COVID-19 outbreaks in Sikkim state of India. Sikkim being the study area for current work has also witnessed the COVID-19 cases since after detection of first case on May 23rd 2020. A Geoprocessing model has been prepared considering COVID-19 cases district wise in Sikkim shown in Fig. 2. Presently, 3234 people are infected, 2534 cured and discharged with 49-person death and 81 migrated in Sikkim as on October 7, 2020 [9]. While in Sikkim, East district being the most affected region with 2321 total confirmed cases as on October 7, 2020 [8, 9].

Many countries throughout the world have restricted the movement of the people to minimum and imposed strict quarantine to control the spread of this highly communicable disease. Identification of this disease at an early stage is very crucial to control the rapid spread and if otherwise, may take many years for this pandemic to subside [10]. As most of the countries imposed lockdown by slowing down their manufacturing products, social restriction, travel ban, and by emphasizing work from home policy that forced most of the people to stay inside the house, which heavily

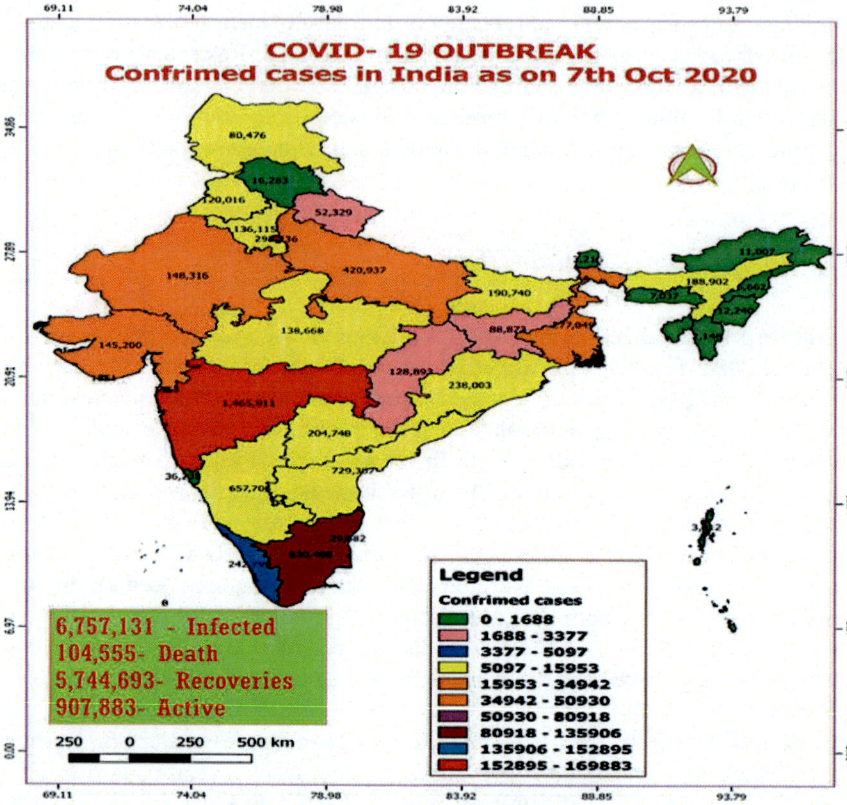

Fig. 1 Thematic map showing COVID-19 confirmed cased in India

affected the day to day life of people, normal business and operations. Industries were moving toward the minimum manual operation with limited productions. Due to this, the working condition has changed as a result of which energy demand has changed that eventually changed energy demand profile of a nation. Power system equipments and networks may hamper which may ultimately affect the transformers, protections, network equipments, etc. [1]. Due to drastic changes in power demand-supply ratio, electric grid failure may be the ultimate result which would be catastrophic as the power sectors availing services to emergency responses like healthcare, security, data center, etc., during this pandemic [11]. Moreover, distribution companies (Discoms) have unfavorably affected due to amid lockdown since people being consigned at their home posing lower efficiency of power system supplying to domestic consumers, shortfall of revenue collection from higher tariff due to subsidized consumers paying lower tariffs with higher aggregates and including technical and commercial losses [12].

Before declaration of one-day nationwide strike called "Janata Curfew" in India by Hon'ble Prime Minister as expression of gratitude to frontline worriers during this

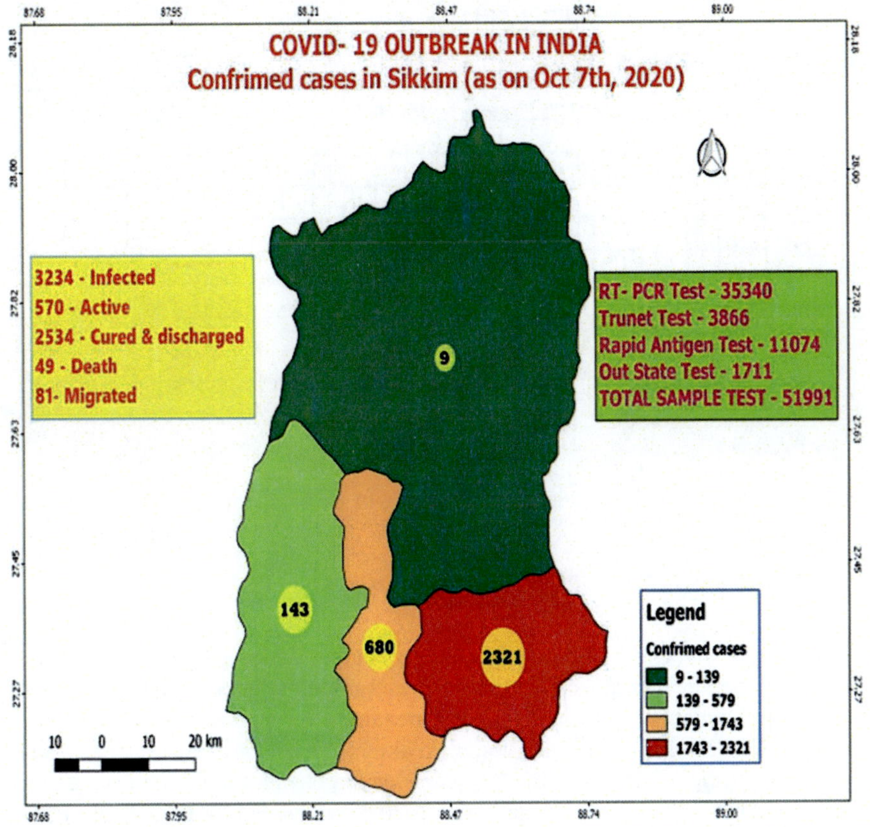

Fig. 2 Thematic map showing COVID-19 test and confirmed cases in Sikkim, India

ongoing health crisis, the energy consumption across the country was recorded as 3500 GWh [13] as shown in Fig. 3. Early 3000 GWh was the reduced demand after declaration of strike, i.e., on March 22, 2020, on April 1, 2020 the demand reduced to nearly 2500 GWh or at the lower stretched. Hence, on an average of about 1000 GWh daily energy consumption reduced as compared to year 2019 represented on Fig. 3 [1, 13].

As per the data from Global Statistics Yearbook 2019, there was steady increase in energy demand from 2014 to 2018 in G20 countries up to 17,000 TWh but it is forecasted that consumption demand will not exceed 18,000 TWh in 2020 due to outbreak of novel coronavirus [14]. Industrial and manufacturing activities will again resume post-COVID-19, and consumption may again increase up to the mark [15].

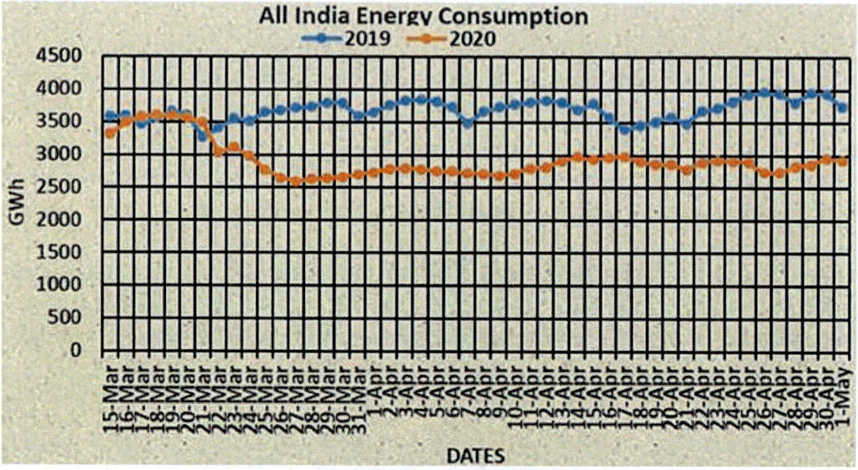

Fig. 3 All India energy consumption (GWh) in year 2019 and 2020 [13]

1.2 Impact on Renewable Energy Projects

Global supply chain ratio of renewable energy sources during COVID-19 pandemic is considerably slowing down due to logistic and shipment import delay, which may miss the deployment deadline and may face penalties. As per director IRENA, due to threats of COVID-19, it will have major breakthrough on renewable energy supply chain on stopping net low or zero CO_2 emission target and transition of sustainable energy goal [16]. Global solar installation 2020 slowed down by 10 GW from 121–152 to 108–143 GW than previously forecasted target as per the analysis firm [11]. Many others under construction projects particularly solar power have been disrupted which was mostly manufactured in East Asia including China [17]. As per report by Global Wind Energy Council's (GWEC), during year 2019 about 70% of the wind power project was installed and completed in India along with China, U.S., U.K. and Spain and these project are greatly suffering due present scenario of COVID-19 pandemic and further predicted that wind energy projects in 2020 will certainly be disrupted due to this ongoing health crisis [18]. In India, 3 GW wind and solar PV Wood Mackenzie project have been delayed due to this pandemic situation, since the solar PV modules were imported from China and have extensively disrupted their installation procedure [8, 19]. Therefore, this situation definitely gives a lesson that every nation should be well prepared in terms of energy availability and securities in advance. Due to less supply chain of raw materials, the world has become less reliant on fossil fuel. Thus, renewable energy may be successful in fulfilling the historic decline on energy demand [20, 21]. In March 2020, China has successfully increased the solar and wind power generation by 8.6% and 18.1%, respectively, by chasing to reduce in thermal generation [22]. Reduction in fossil fuel consumption has significantly reduced carbon emission, and its impact on the environment.

Due to ongoing COVID 19 outbreak, there has been huge impact on the ongoing hydropower projects and industries in India since past few months, due to which supply chain has been greatly hampered, causing delays in many ongoing projects [23]. In Sikkim, many such ongoing projects like 500 MW Teesta IV project in North Sikkim have been disrupted causing difficulty in its completion on given time frame, putting nearly 6.0 crores cost run over and other financial compensation during this situation. Apart from this, NHPC took over the 500 MW (125 MW X4) project from debt-ridden Lanco Teesta Hydro Power Ltd (LTHPL) as Teesta VI HEP on Teesta River also under halt. Nearly Rs 119.43 crores of generation loss is anticipated on three power stations, i.e., Chamera-II Power Station, Kishanganga Power Station, Loktak Power due to restriction on vehicular movement during this ongoing COVID-19 pandemic [20, 24].

2 Background

The Hon'ble Prime Minister of India appealed to the citizens of the country at 09:10 h of April 3, 2020 to switch off their lights and light lamps or candles for 9 min at 21:00 h on April 5, 2020 to show solidarity and confidence toward the nation's collective fight against the novel coronavirus [13]. The event was declared as ten-minute nationwide light-off event in India. Hospitals, police booths and medical manufacturing facilities as well as street lights were not supposed to be switched off. People had also been advised not to shut off other appliances such as fans, televisions, etc. The impact could be a serious to national power grid stability system if was not handled properly by all power utilities. Since point of equilibrium could be lost if load demand keeps on deceasing, and system frequency keeps on increasing which could be catastrophic and may collapse grid permanently. So, in order to successfully manage this unprecedented event, an advance methodology was adopted for calculation of demand of each state, region and all India level at National Load Dispatch Centre (NLDC) or Regional Load Dispatch Centre (RLDCs), whereas State Demand at RLDCs is worked out as the sum of all the intra-state generation and the net drawl of the state. This involves the algebraic summation of intra-state generation and all the tie lines connecting to the respective state, whereas regional demand was calculated as the summation of demand of state within the region. All India demand is calculated as summation of demand of the entire region [13].

POSOCO at NLDC level in coordination with State Load Dispatch Centers (SLDCs) of each state of India started preparing for the event immediately after decision made by Govt. of India. During the course of this preparation, several web-based meetings and interactions were conducted with SLDCs including official representatives of SLDC-Sikkim, Independent Power Producer (IPP) and Transmission Licensees on high priority basis. Every stakeholder was assigned specific responsibilities and compliance of event to take place to the extent possible measures. However, the most challenging aspect in managing the event was to match the sharp rate of change of demand with matching change in generation across the country

[13, 25]. Around 12,000–14,000 MW was the anticipated reduction in load during 9-min event in all India demand on April 5, 2020 [26].

3 Event Summary at National Load Dispatch Centre (NLDC) Level

During the event, 31,089 MW was the total maximum reduction in power demand in national level in India as shown on Table 1.

From 20:45 h, this demand started reducing and recorded 85,799 MW at 21.10 h. After the completion of event from 21:10 h, the all India demand started rising again and reached 1, 14, 400 MW at 22.10 h as shown in Fig. 4. The range of grid frequency was 50.26–49.70 Hz where 50.259 Hz maximum at 20:49 h and 49.707 Hz was the minimum at 21:08 h [26].

In order to tackle the situation, at 20:45 h hydro generation throughout the country was maximized and recorded to 25,559 MW. At 21:10 h, hydropower generation was reduced to 8016 MW. So, the total hydro generation reduction recorded was 17,543 MW matching with demand reduction of 31,089 MW with hydro. Between 21:10 and 21:27 h, this generation was again ramped up from 8016 MW to 19,012 MW to meet the increased demand after the post-event. Reduction of total 10,950 MW generation was achieved through Thermal (6992 MW), Gas (1951 MW) and Wind generation (2007 MW) during 20:45–21:10 h [26]. The event was successfully managed and handled without any untoward incident keeping the power system parameter within the permissible limit.

3.1 Grid Frequency Control Measures

The point of equilibrium between power demand and power generation is the system frequency if ignore losses on power transmission. The grid frequency is mainly

Table 1 The All India and Region wise demand details for the event [25]

Time (h)	Demand (MW)						Reduction w.r.t. All India demand at 20:45 h
	NR	WR	SR	ER	NER	All India	
20:45	31,791	32,474	35,012	15,815	1796	116,887	0
20:50	31,339	32,113	35,109	15,452	1761	115,775	– 1113
20:55	30,148	31,462	35,019	14,928	1693	113,251	– 3637
21:00	26,683	28,091	32,688	12,752	1453	101,667	**– 15,220**
21:10	22,061	24,010	29,034	9679	1015	85,799	**– 31,089**
21:15	24,956	26,992	30,665	11,879	1303	95,795	**– 21,092**

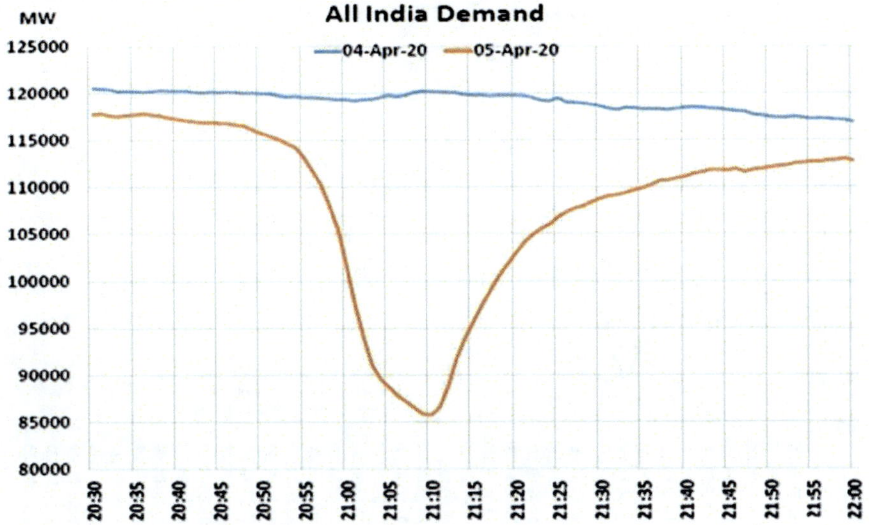

Fig. 4 All India power demand (MW) during light-off event [13]

influenced by the large power stations when there is imbalance between demand and supply, which in turn leads to line outage and tripping on the grid [22]. So, retaining frequency within the recommended band would be a difficult task [27]. During peak hours on event day, i.e., from 18:10 to 20:00 h of hydro generation were reduced by reducing governor droop setting and keep in balance for providing flexible load during start of event, i.e., during 21:00 h as shown in Fig. 4. So, during this period thermal and gas generation managed to cater load during this peak time. Accordingly, thermal and gas generation was reduced to their minimum level of 60% after the completion of peak hour around 22:55 h and subsequently hydro power was increased and rolling their generation of 0–10% of their rated generation to maintain the load-generation balance by taking care of the system frequency. After stabilizing the system parameter, hydro units withdrawal was done in consultation with RLDC/ SLDCs.

All India frequency was kept at lower side of Indian Electricity Grid Code (IEGC) band, i.e., 49.90 Hz from 20.30 onwards in view of anticipated frequency rise due to demand reduction at 21:00 h. Similarly, all India grid frequency kept at higher side of IEGC band at 50.15 Hz around 21:09 h in order to anticipate drop in frequency due to restoration of load. Hence, the grid frequency will vary from 49.50 to 50.50 Hz between 20:45 and 21:30 h [1, 28] as Fig. 5.

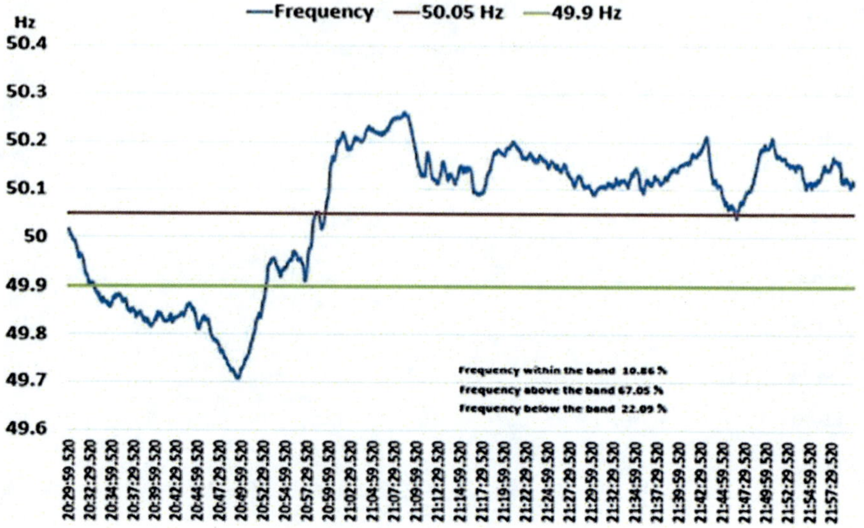

Fig. 5 Grid frequency (Hz) at NLDC during light-off event [1]

3.2 *Grid Voltage Control Measures*

Since COVID 19 containment measures nearly 220–240 lines at kV voltage level and above are kept in open condition for voltage control. System voltages have been within the IEGC band. As per studies during the above event, the voltage levels would be within control keeping the generation scheduling. The list of 400 kV and above busses where the voltage level is expected to rise by more than 0.01 pu. STATCOMS and SVCs shall be in voltage control mode with reference voltage [26].

3.3 *Event Summary at State Load Dispatch Centre (SLDC) at Gangtok, Sikkim*

To ensure the proper security and reliability of the grid during 10 min' light-off event, POSOCO issued an advisory after a conference call on 4th April with all the SLDCs including SLDC-Sikkim and heads of all prime hydro power plants in the country. All precautionary measures were discussed via video conference to all stakeholders headed by NLDC to face the system challenges. Sikkim's demand at 20:50 h was 45.20077 MW which was reduced to 27.1892 MW at 21:10 h. So, during the event, 18.011 MW was the total recorded reduction in demand [29] as shown in Fig. 6.

The demand started picking up and reached 38.62957 MW at 21:21 h. Sikkim falls under Eastern Region (ER). ER demand at 20:45 h was 15,815 MW which was

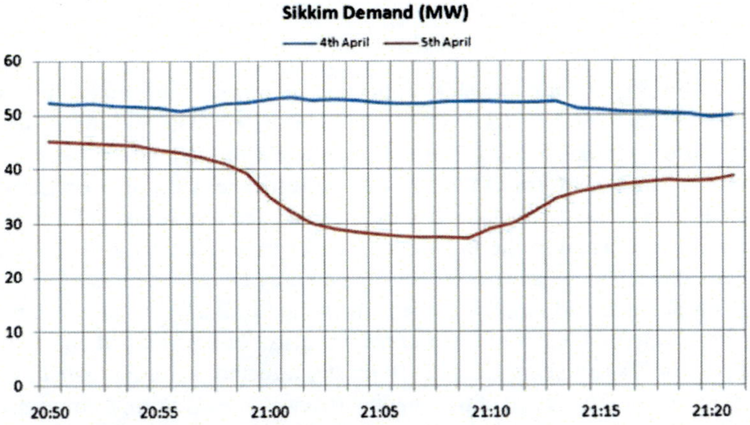

Fig. 6 Sikkim power demand (MW) during light-off event [29]

reduced to 9679 at 21:10 h. During the event, 6136 MW was the total reduction in ER demand. The demand started picking up and reached 15,231 MW at 22:00 h [13].

Grid frequency during the event in Sikkim remained in the range of 49.72 Hz and 50.24 Hz as recorded at 20:51 h and 21:09 h, respectively, as shown in Fig. 7. Since all of the Sikkim State-owned hydro generating station were off bar, no ramping of these station was carried out during this period. The load-generation balance was carried out successfully.

Hydropower stations responded quickly to the directions issued by SLDC and participated well in system management. The event was managed smoothly without any untoward circumstances while the power system parameters remained within the tolerable limit.

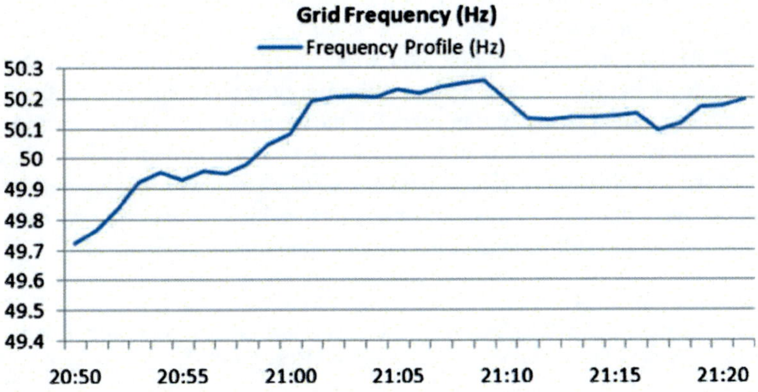

Fig. 7 Grid frequency (Hz) at SLDC-Sikkim during light-off event

4 Conclusion

This paper investigates the technical influences on power management system due to nationwide lockdown in COVID-19 pandemic along with its impact on socio economic life style and work pattern of the people throughout the world. Using open source cross platform Q-GIS 2.18.0 software thematic map of COVID-19 confirmed cases along with total active cases, death and recoveries in India has also been developed to analyze the individual state of India that is the mostly affected due to this pandemic. Almost 60% confirmed cases being accounted from Maharashtra, Tamil Nadu and Andhra Pradesh in India. Similarly, the thematic map for Sikkim for studying total confirmed cases has also been developed and found the East district of Sikkim being COVID-19 mostly affected region followed by South District. Because of this, most of the ongoing renewable and hydropower projects are negatively affected. International or interstate free movement was banned and most of power project equipment and accessories were to be imported from China and other Asian countries. The demand-supply profile has been hit badly after imposing lockdown and may continue to persist depending upon future lockdown scenario. So, it is very challenging to handle and manage the power system to ensure stable and continuous supply. This study further investigates different measure undertaken to retain grid frequency and system voltage within their recommended band during "10-min light off event" in India. Technically hydro power plant has provided maximum flexibility and reliability on grid operation and management. Hence, it is evident that Indian utilities and stakeholders have successfully managed to overcome the challenges during this unprecedented event. This study has clearly recognized and analyzed some steps to be undertaken in power management system in a healthy manner to tackle such type of unprecedented events and may be useful to all stakeholders and power utilities throughout the world if such events are raised in the near future. The author has incorporated Q-GIS software for studying location-wise COVID-19 outbreak, and its impact on renewable energy ongoing project in India, same software can be utilized for finding better opportunities regarding renewable energy development for both solar and wind energy. Since, in Sikkim very less number of work has been done on renewable energy exploration. Hence, the field of Q-GIS can be applied to for finding potential location for sitting solar or wind power plant in Sikkim. Moreover, this health crisis also has also given an opportunity to look forward for adoption of renewable sources of energy with raw materials for its generation freely abundantly availed from nature making it cost-effective during the operation period and fulfilling the necessity of global climate demand.

Acknowledgements Author is highly acknowledged with State Load Dispatch Centre (SLDC) DPH road, Gangtok for providing necessary information and data on "9 minutes light off event" for the Indian state Sikkim. Further, author is acknowledged with Health and Family Welfare Department and State IEC Bureau, Sikkim for providing relevant health bulletin data on COVID-19 to carry out this study.

References

1. Shafiullah, G.M., Jamal, T., Reddy, S.K., Subramaniam, U.: COVID-19: Impact Analysis and Recommendations for Power and Energy Sector Operation. Available on https://www.researchgate.net/publication/341204513
2. Novel coronavirus COVID-19: current evidence and evolving strategies. J. Bone Jt. Surg. Am. 1–11 (2020)
3. WHO: Novel Coronavirus (2019-nCoV) Situation Report-1. World Health Organization. World Health Organization. https://www.who.int/docs/default-source/coronaviruse/situation-reports/20200121-sitrep-1-2019-ncov.pdf?sfvrsn=20a99c10_4
4. Jin, Y.H., Cai, L., Cheng, Z.S.: A rapid advice guideline for the diagnosis and treatment of 2019 novel coronavirus (2019-nCoV) infected pneumonia (standard version). Mil. Med. Res. **7**(1), 4 (2020)
5. World Health Organization (WHO): Q&A on coronaviruses (COVID-19). Extracted from who.int
6. Church, R.L.: Geographical information systems and location science. Comput. Operat. Res. **29**(6), 541e562 (2002)
7. Worldometer coronavirus. Extracted from https://www.worldometers.info
8. COVID-19 pandemic India. https://en.wikipedia.org/wiki/COVID-19_pandemic_in_India
9. State IEC Bureau, Sikkim. www.covid19sikkim.org
10. Haleem, A., Javaid, M., Vaishya, R.: Effects of COVID-19 pandemic in daily life. Elsevier Publ. Health Emergency Collect. **10**(2), 78–79
11. Smith, D.C.: COVID-19 and the energy and natural resources sectors: little room for error. J. Energy Nat. Resour. Law. ISSN: 0264-6811, April 2020
12. Rout, A., Mainali, B., Chetan, S.S., Solanki, S., Bhati, G.S.: Assessing the Financial Sustainability of rural grid electrification pathway: a case study of India. Sustainable Production and Consumption. Available online 5 August 2020. https://doi.org/10.1016/j.spc.2020.08.001
13. Report on Pan India Light Off Event 9 pm 9 min on 5th April 2020. Power System Operation Corporation Ltd. (POSOCO)
14. Reuters: Five charts that show the global economic impact of COVID-19. Retrieved March 24, 2020, from https://www.weforum.org/agenda/2020/03/take-five-quarter-life-crisis/GoogleScholar, March 31, 2020
15. Qarnain, S.S., Muthuvel, S., Bathrinath, S.: Review on government action plans to reduce energy consumption in buildings amid COVID-19 pandemic outbreak. Materialstoday Proc. (in Press). https://doi.org/10.1016/j.matpr.2020.04.723
16. Macola, I.G.: What is the impact of Covid-19 on the power sector? Power Technol. (2020). https://www.power-technology.com
17. US Renewable Energy Projects and COVID-19 Impacts. https://www.natlawreview.com. April 11, 2020
18. Global Wind Energy Council. https://gwec.net, April 10, 2020
19. Wood Mackenzie, Energy Research and Consultancy. https://www.woodmac.com. April 10, 2020
20. Outlook The News Scroll. https://www.outlookindia.com, May 29, 2020
21. Renewable are the Only Winners in Historic Decline in Energy Demand. https://finance.yahoo.com/news/renewables-only-winners-historic-decline-040040629.html
22. National Energy Administration of China: Electricity Consumption of China in the First Quarter of 2020. http://shoudian.bjx.com.cn/html/20200417/1064308.shtml
23. Hydropower and the Impact of COVID-19. International Water Power and Dam Construction. https://www.waterpowermagazine.com, May 18, 2020
24. Sarkar, D.: The Economic Times-Power. https://economictimes.indiatimes.com, April 30, 2020
25. Report on Lights Switch Off Event on 5th April 2020. State Load Despatch Centre (SLDC), Odisha Power Transmission Corporation Limited
26. Preliminary Report on Pan India Lights Switch Off Event on 5th April 2020. Power System Operation Corporation Limited (POSOCO), National Load Despatch Center

27. Zhenhua, X.: Tuning method for governor control parameters of hydropower generator in isolated grid considering primary frequency performance and small-signal stability. Global Energy Interconnection **1**(5), 568–575 (2018)
28. Grid Security—Need for Tightening of Frequency Band and Other Measures. Central Electricity Regulatory Commission New Delhi, CERC Staff Paper March 2011
29. Report on Pan India Load Switch Off Event on 5th April 2020. Power Department Sikkim State Load Despatch Centre, Government of Sikkim

Ecological and Anthropogenic Factors Influencing Presence of Tiger: A GIS-Based Study in Sikkim

Aranya Jha, Polash Banerjee, and Ajeya Jha

Abstract Located in Eastern Himalayas, Sikkim is a mountainous state with an area of just 7096^2 kms. Altitudinal gradation here ranges from 300 msl to over 8000 msl and which provides possibilities of a multitude of climates from tropical to Arctic. This climatic diversity spawns biodiversity which for such a small tract of land is incomparable. *Panthera tigris* is one of the best-known fauna and has been reported by Sikkim as a rare visitor. This study attempts to comprehend the presence and movements of tigers in Sikkim and is based on the blending of available data with Geographical Information System (GIS). GIS has been used to map the locations of tigers' recorded presence upon spatial layers based on 6 key input variables, namely temperature, rainfall, water bodies, proximity to habitation, population and roads. Findings confirm the general disposition of tiger to prefer relatively warmer climate, heavy rainfall, avoidance of human population and nearness to waterbodies. Presence of four big cats (Tiger, leopard, Snow Leopard and Clouded Leopard) in a small geographical landscape of Sikkim has considerable implications for conservation measures.

Keywords Habitat · Suitability analysis · Mountain · Big cats

1 Introduction

Spread over the laps of Himalayas, Sikkim is the second smallest state of India, occupying merely about 7096 square kilometres of area. It is a melting point of varying climates, ranging from tropical to subtropical, temperate, subalpine, and arctic type of climates made possible because of massive altitudinal changes ranging from 300

A. Jha · P. Banerjee
Department of CSE, SMIT, SMU, Majitar, Sikkim 737136, India
e-mail: aranya_201800077@smit.smu.edu.in

A. Jha (✉)
Department of Management Studies, SMIT, SMU, Majitar 737136, India
e-mail: ajeya.jha@smit.smu.edu.in

H. K. D. Sarma et al. (eds.), *Contemporary Issues in Communication, Cloud and Big Data Analytics*, Lecture Notes in Networks and Systems 281,
https://doi.org/10.1007/978-981-16-4244-9_33

msl to approximately 8000 msl. Mount Kahngchendzonga, the third highest mountain peak in the world is located here. These climatic variations nurture biodiversity of immense magnitude. Sikkim harbours over 500 avian species [2]; 156 mammalian species [3]; 78 Reptilian species [31]; over 700 butterflies [24]. It is a region that falls at the confluence of two biogeographic realms—Palearctic and Oriental. Fauna, therefore, also is representative of these two realms. On one hand, we find species such as Indian Bison (*Bos gaurus)* and *Leopard (Panthera pardus*) as also Tibetan Lynx (*Lynx lynx isabellinus*) and Snow leopard (*Uncia uncia*). In total Sikkim has 10 cat species out of which there are 4 big cats (Tiger, Leopard, Snow Leopard, Golden Cat) and 6 small cats (Clouded Leopard, Leopard Cat, Marbled Cat, Fishing Cat, Jungle Cat, Tibetan Lynx).

Tiger is one of the best-known fauna across the world always well-known for its ferociousness and strength, and it is a universal emblem of courage, power and dominance. It is an apex predator and its home range today comprises of Eurasia, although almost 93% of the region it once occupied is bereft of it now. It has several subspecies. Royal Bengal Tiger (*Panthera tigris tigris*) is the only subspecies reported from India. Any geographic area harbouring tiger is of immense significance because the tiger is a keystone species, and its presence signifies a rich and vibrant habitat. Presence of Tiger in Sikkim has been based on extremely rare reporting. On the basis of evidence (not actual sightings) presence of tiger in Sikkim has been reported by Avasthe and Jha [3]. Such methodology has been used earlier by other researchers [14]. Based on evidence (not actual sightings) of the presence of tigers in Sikkim has been reported by [3, 14].

The present study has been undertaken to better comprehend the presence of tigers in Sikkim and is based on the blending of available data with the newly emerged Geographical Information System (GIS).

Remote sensing and geographic information systems are now recognized as powerful technologies for habitat suitability mapping for a species [46, 49]. These methods have an added advantage that they provide newer and deeper insight by assimilating existing information and knowledge for sharing a compatible spatial referencing system [5, 21]. More and more researchers are now turning towards these technologies for exploring physical dimensions of wildlife habitats and geospatial modelling [26].

Literature Review on the presence of Tiger in Sikkim and adjoining areas provides important insight on this super-cat's existence in this region. Sikkim is bounded by Bhutan in its east, Tibet (China) in the north, Nepal in the west and Indian state of West Bengal in the south.

Bhutan: The Jigme Singye Wangchuck National Park in Bhutan recently employed camera traps along with capture-recapture data analysis [15] to detect and provides details on the presence of tigers and leopards. Both are listed in Schedule I of the Forest and Nature Conservation Act (1995), of Bhutan [52]. It has been found that the habitat of tigers in Bhutan ranged from an altitude of 200 msl in the south to 3000 msl in the north [17, 50]. Apart from tigers and leopards, Snow Leopards were also found in Bhutan in regions above 3000 msl. Surveys had been conducted in two separate areas of Jigme Dorje National Park [28, 29] which confirmed their presence. Snow Leopards have repeatedly been reported in Bhutan [20, 27, 38].

West Bengal: Some of the eco-regions of North Bengal are the natural habitats of Tiger [36]. As early as 1923, tigers have been reported from Tonglu and Kalpokhri [4]. Both places are located at an altitude of 3,200 m both surrounded by rhododendron forests. As many as 19 tigers have been reported in 2002 from Neora Valley National Park [14, 35]. Presence of tiger in the regions of West Bengal neighbouring Sikkim is well established.

Nepal: Tiger has been reported from eastern Nepal [33].

As mentioned earlier, Tiger has been reported from Sikkim also [1, 3, 7, 30, 45]. There are reported studies on the presence of wildlife in various regions across the globe that emphasizes on the conservation of the wildlife [16, 37, 40, 41, 51]. Studies related to the eastern zone of India, and the state of Sikkim in specific, with respect to the biodiversity and wildlife including lesser cat and golden cat are reported in [6, 8, 16, 25].

How frequent is reporting of tigers in Sikkim? Which locations have been sighted at? What ecological and anthropogenic factors influence its presence in Sikkim? These are some research questions that have been envisaged in this paper.

2 Materials and Methods

The study is exploratory and is based on field work, and the literature survey and discussions with forest department officials to identify locations where the presence of tiger (even if no actual sightings) have been noted, based on their pug marks and kills. The information is based on evidence spread over about two decades (1999–2019). Once the locations were identified and reconfirmed, their latitudes and longitudes were noted from Google Earth and thereafter the GIS has been used to map the locations of tigers' recorded presence upon spatial layers based on 6 key input variables, namely temperature, rainfall, water-bodies, proximity to habitation, population and roads (Table 1). The climate-related rasters were accessed from WorldClim2 online resource that provides high-resolution average climatic data of every month. The climatic raster data were averaged over the entire year using Raster Calculator

Table 1 Details of the environmental descriptor layers

Environmental descriptor layer	Unit	Resolution	Source
Average temperature	°C	30 s	Worldclim2 [18]
Average rainfall	mm/month	30 s	
Proximity to water bodies	m	30.7 m	©OpenStreetMap
Proximity to human habitations	m	30.7 m	
Proximity to roadways	m	30.7 m	
Population density	Persons/km^2	250 m	[12]

in ArcGIS framework. The vector layers of water bodies, townships and roadways accessed from OpenStreetMaps were clipped with the vector layer of Sikkim administrative map accessed from SAGA GIS online resource site. The clipped layers of water bodies, townships and roadways were used to prepare proximity rasters using proximity tools in the ArcGIS framework. Population density raster was accessed from Google Earth Engine data catalogue and clipped with Sikkim administrative boundary.

These environmental descriptor layers were prepared to help in correlating locations with model input variables. It was hypothesized that (a) tigers will prefer higher temperatures and avoid relatively low temperatures; (b) tigers may follow a trail where rains are relatively heavier; (c) tigers may be found near water bodies as water is a vital need for tigers; (d) tigers may avoid proximity to habitation; (e) tigers will prefer regions away from high population and (f) tigers will avoid roads built for human movement to evade traffic.

In this paper, *Tiger* unambiguously refers to *Panthera tigris tigris* commonly known as Bengal Tiger.

3 Result and Discussion

3.1 Locations of the Recorded Presence of Tigers in Sikkim

Records of Tiger sighting in Sikkim have always been rare. In the twentieth century, there were just three sightings of tigers (citation). However, when based on evidence the presence of tiger has been noted recently, there are many more instances confirming the presence of Tiger in Sikkim. Over the years, there have been several pieces of evidence suggesting the presence of tigers in the state of Sikkim Locations from where the presence of tiger has been reported are shown in Table 2, and the same data has been shown in GIS.

From Fig. 1, it appears that tiger enters Sikkim from West Bengal or Bhutan and moves northward and then again turns towards the east to enter Bhutan. This observation is also reported by Avasthe and Jha [3].

3.2 Location of Tiger on Temperature Layer

Location of tigers in Sikkim on temperature layer has been shown in Fig. 2. As *Panthera tigris tigris* is primarily a creature of tropical forests, it is expected to be more comfortable at higher temperature locations than the lower ones. A large part of Sikkim falls under freezing temperatures. This is not to underestimate the adaptability of this species to freezing temperatures at high altitudes, also as has been noted [13, 44, 49]. From Fig. 2, it becomes clear that tigers in Sikkim confine themselves to

Table 2 Locations from where the presence of tiger reported

Location	Latitude	Longitude
Rachela	27.08088	88.67556
Talkharka	27.17897	88.72361
Regu	27.19671	88.71861
Phadamchen	27.23738	88.76778
Zuluk	27.25185	88.7775
Gnathang	27.31902	88.82806
Bhusuk	27.35069	88.6725
Lake Menmecho	27.33022	88.84639
Changu	27.37456	88.76333
Kabi	27.40585	88.6175
Phensung	27.42018	88.60194
Phodong	27.40798	88.57167
Tong RF	26.93481	88.27167
Chyakhung RF	27.15281	88.24389
Lema	27.655	88.724
Lachung	27.68906	88.74306
Dombang valley	27.63333	88.75

relatively higher temperatures and this endows credence to the hypothesis formulated in this respect.

3.3 *Location of Tiger on Rainfall Layer*

Figure 3 displays the tiger location on a rainfall layer. We have hypothesized that tigers will prefer regions with heavy rainfalls. From Fig. 3, this is evident. Very low rainfall areas are absent from the tiger route. We may conclude that tigers are comfortable at moderate and high rainfall regions but avoid low rainfall areas. This may be accepted with a cautionary note. Low rainfall areas are also very high altitude areas and which are extremely cold, and perhaps, it is altitude and freezing temperature combine that deters tigers.

3.4 *Location of Tiger on Water-Body Layer*

Tigers have been known to be highly water-dependent [32, 43]. This has been the base of our hypothesis that tigers in Sikkim are to be reported from regions relatively close to water bodies. This is evident from Fig. 4 depicting the tiger locations on

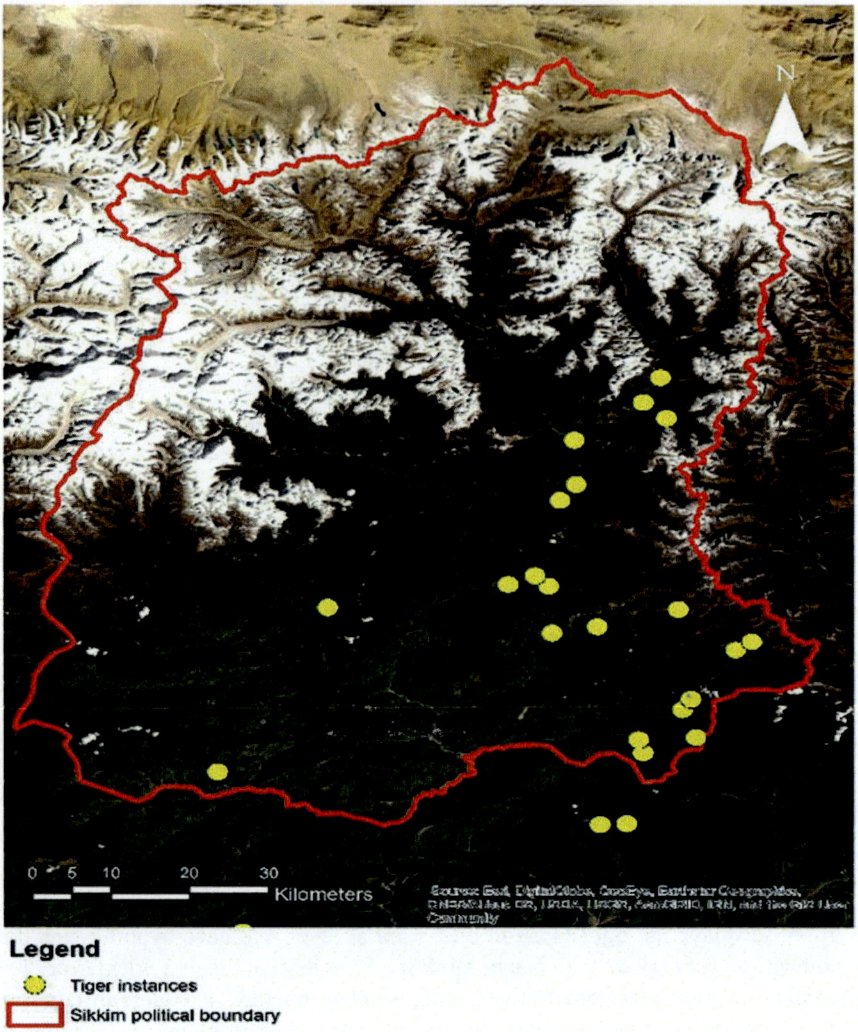

Fig. 1 Recorded presence of tiger in Sikkim

water-body layer. Although it must be said that the entire region is a thick network of mountain streams that owe their existence to heavy rains and mountainous terrain.

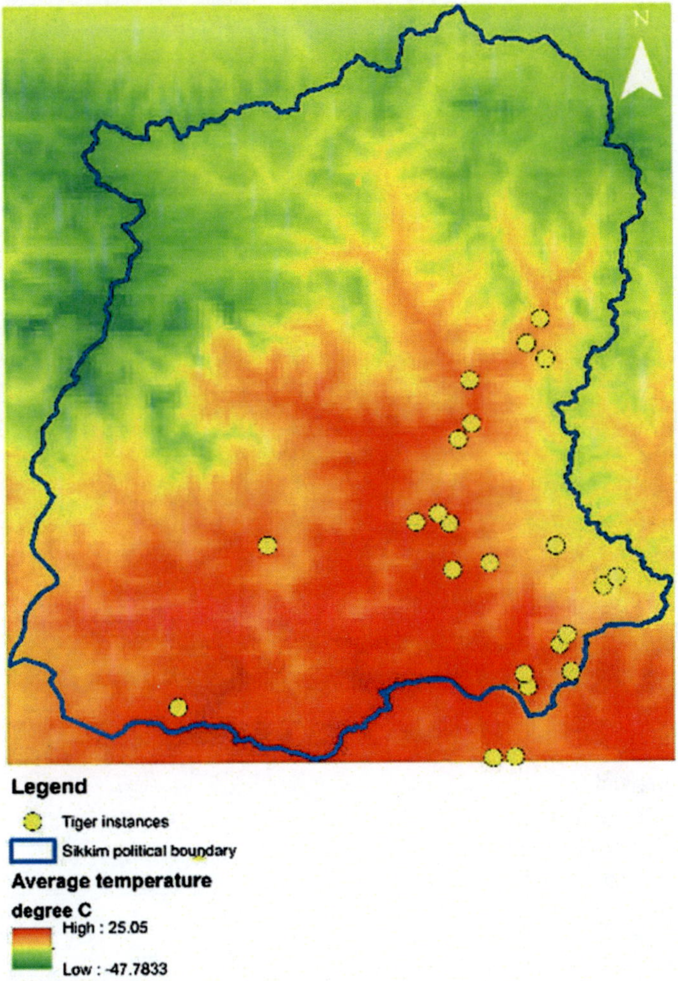

Fig. 2 Location of tigers on a temperature layer

3.5 *Location of Tiger on Proximity to Habitation Layer*

Decrease in wild species population due to increasing human population has been reported by several researchers [9–11, 19, 22, 23, 39, 42, 53]. Tigers like other animals have learnt to be wary of human habitation. Despite the availability of food (as cattle), their experience has been unpleasant and has resulted in injuries and death for them. This led to our hypothesis that tigers will avoid inhabited places. Figure 5 shows the tiger locations against a habitation layer and from it appears that this is not particularly true. What could be the likely explanation for this deviation? It is natural to expect the collection of evidence closer to habitation than in the deeper

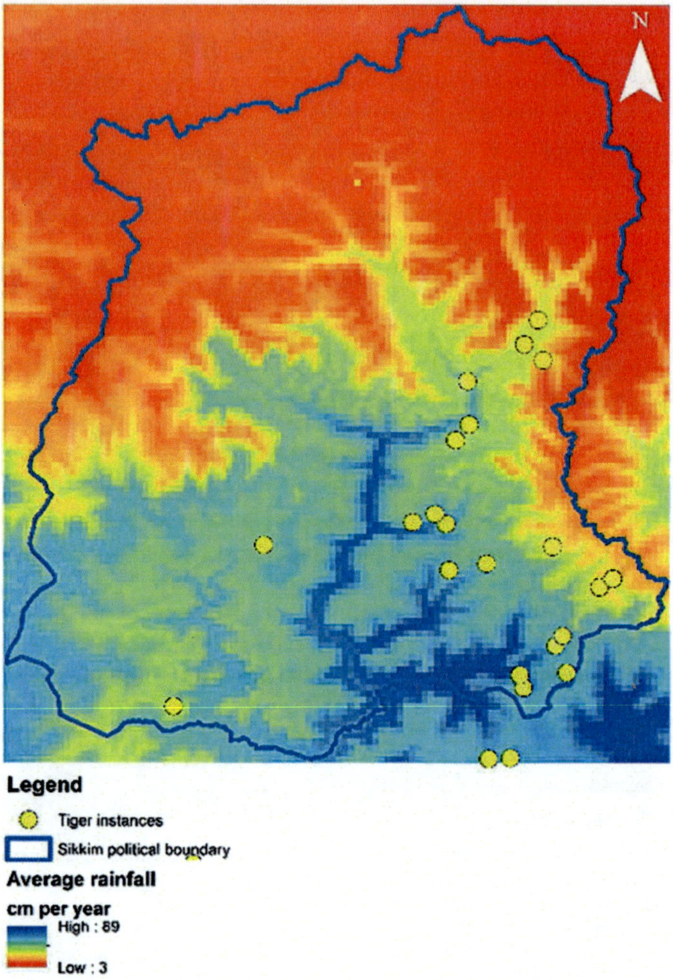

Fig. 3 Location of tigers on a rainfall layer

forest. And hence, this anomaly could be attributed a biased reporting and which tends to be heavily been sampled from inhabited regions than non-inhabited ones.

3.6 Location of Tiger on Population Density Layer

Figure 6 provides an insight into the correlation between tiger location and population density. It is expected as mentioned earlier that tigers will avoid human proximity and may rarely be found in densely or even moderately populated regions. Figure 6

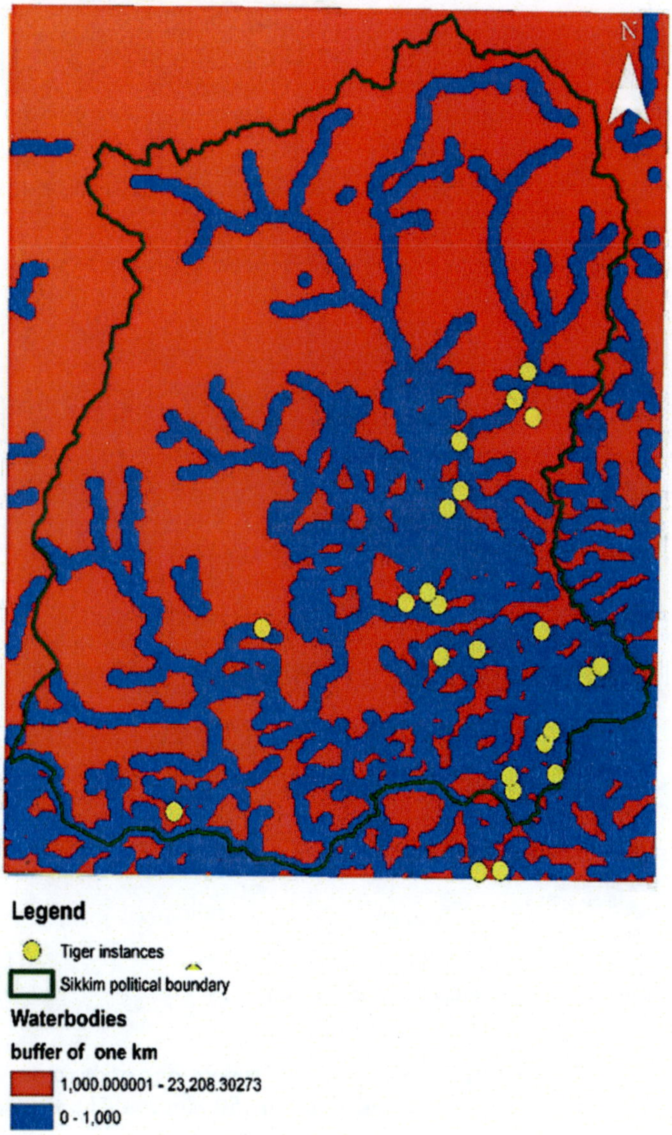

Fig. 4 Location of tiger on water-body layer

provides the salient details. It is to be noted that Sikkim has much lower population than the national average. From the figure, it is evident that tigers move through low-density areas and enter moderate population density areas only because their historical migratory route has been fragmented, and hence, they have no option

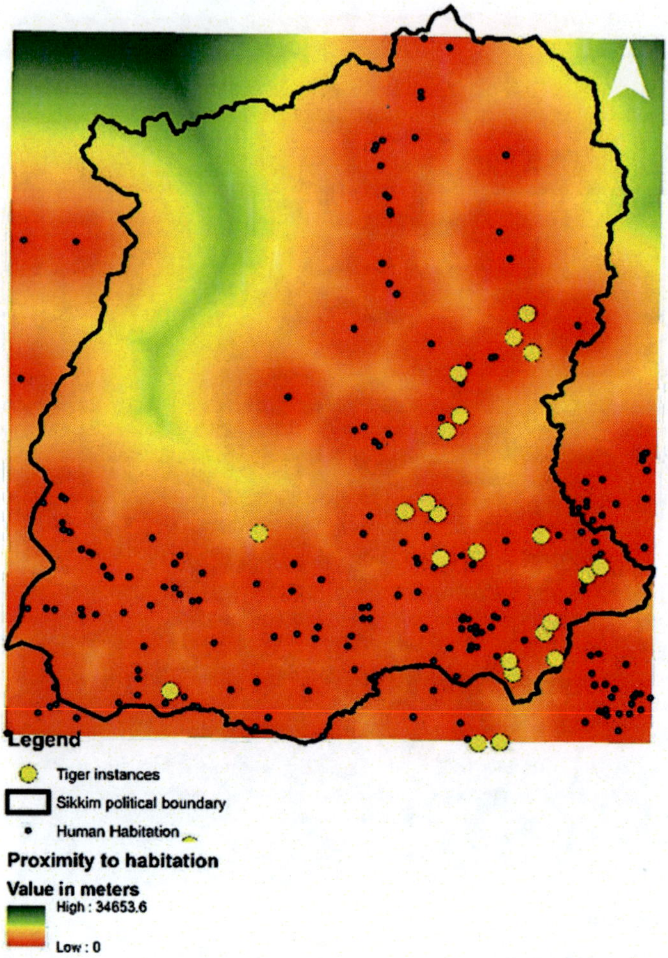

Fig. 5 Location of tiger on proximity to habitation layer

but to enter even moderately dense areas. This should ring bells to the conservationist else disappearance of tiger population from Sikkim could be attributed to the fragmentation of their traditional migratory pathway.

3.7 *Location of Tiger on Road Layer*

Researchers hypothesized that tiger locations will be away from the roads as human traffic is known to be deterrent. This is true for wild species in general [47, 48].

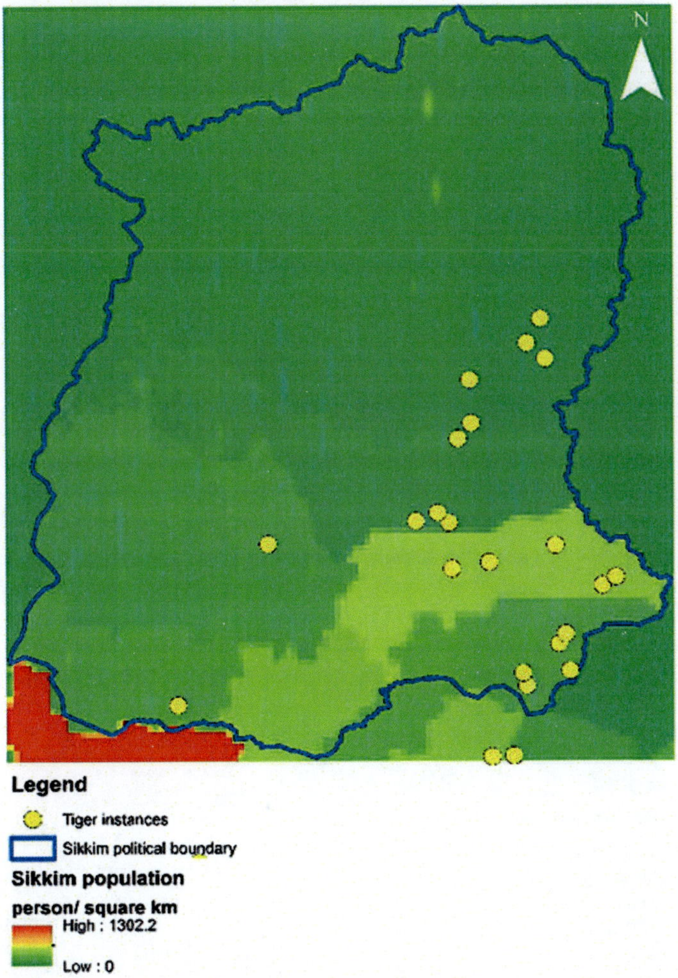

Fig. 6 Location of tiger on population density layer

Figure 7 reflects the result in this respect. Results appear to negate our hypothesis. But this again can be explained by the fact that our data is based on mainly opportunistic evidence than a random sample. Most evidence has been collected where the chances of them being gathered will be extremely high.

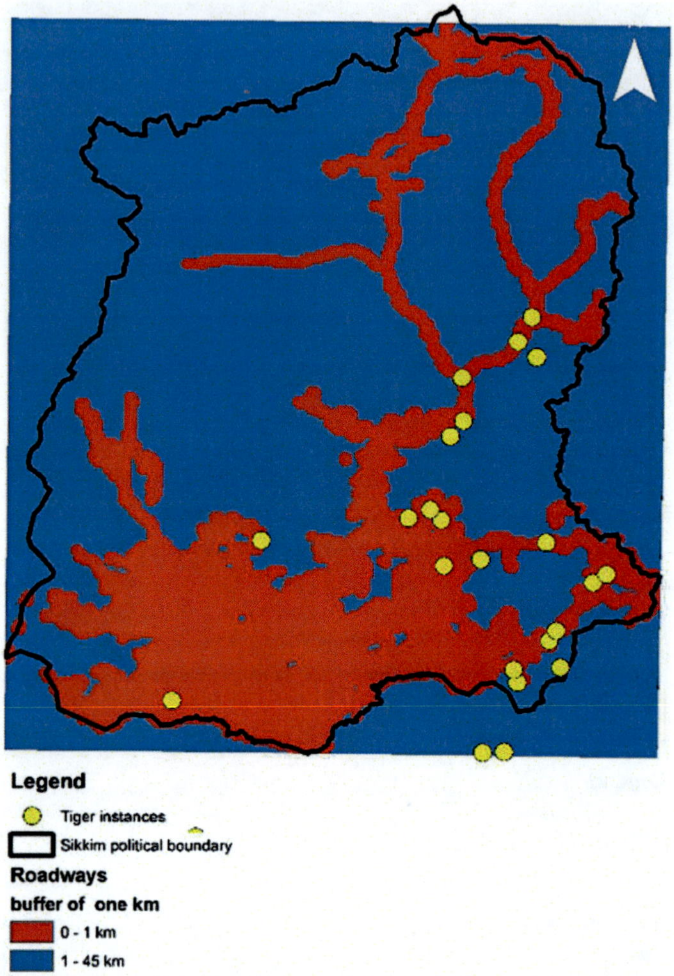

Fig. 7 Location of tiger on road layer

4 Conclusion and Implications

This study was undertaken to identify the locations where evidence of the presence of tigers has been found in the state of Sikkim and thereafter try to correlate these with six ecological and anthropogenic variables, namely temperature, rainfall, water bodies, proximity to habitation, population and roads. To make study meaningful, we had developed certain hypotheses based on the literature and theory.

Over the years, there have been quite many evidence suggesting the presence of tigers in the state of Sikkim (study area). The areas wherein these tigers' evidence were found or the tiger itself was spotted and had been recorded. Thereafter the latitude and longitude of the location were identified and the areas were plotted on the map of Sikkim.

Based on some of the geographical factors of Sikkim, an analysis was performed and few assumptions were noted down.

It was found that the tiger sightings took place in areas with slightly warmer temperature which is natural because it is already known that tiger prefers a warmer climate. It was also observed that the places with high rainfall density were the ones wherein most of the tigers were spotted. Another important assumption was that the tigers were seen in places having a lesser human population (i.e. person/km^2) which is obvious.

With the help of the data, few Euclidean distances were also obtained. It was seen that the tigers stayed closer to the water bodies most probably to quench their thirst and stay close to the prey stock coming for drinking water. Now surprisingly, the Euclidean distance of roadways showed that the tiger had been spotted near the road. But it may be possible that only the tiger that strayed closer to the roads were seen, there may be no evidence for the one's deeper in the forest. Similar to the roadways, it was also seen that the tigers were located closer to human habitation. Maybe to easily fulfil their food requirement or maybe the evidence or tiger citations had only been noted from areas closer to human habitation and from the deeper part of the forests or mountains.

Implications of this study are substantial for conservation fraternity. Presence of four big cats in one location is a rare and extraordinary phenomenon. It is known that Sikkim is home of Leopard, Clouded leopard and Snow Leopards—three big cats. Presence of tiger has been reported but has been so rare that considering it a species of the state has been a doubtful proposition (Though recently tiger was captured at Sikkim for the first time on camera and on the night of 6 December 2018, at 6:23 PM and 7:00 PM, and on December 28 near Goru Jurey inside Pangolakha Wildlife Sanctuary at an altitude of 9583 ft [34]. Sikkim thus acquires significant importance. A region, with less than 7000 km^2 area, harbouring 4 big cats is perhaps unique. This makes conservation of biodiversity at Sikkim even a greater challenge and opportunity. As tiger is migrating through Sikkim, its migratory route needs to be conserved on a war footing. Further, warm climate and water bodies that tiger seems to prefer human habitation too and hence species-specific and locality specific programmes need to be formulated and implemented. Future studies may look to radio-frequency identification (RFID) techniques for higher quality data rather than opportunistic data-based study as has been undertaken in this instance. Future RFID-based study can also explore predator–predator and predator–prey relationships of four big cats in Sikkim.

References

1. Agarwal et al: Mammalia, Zoological Survey of India. Fauna of West Bengal. Part 1, p. 439 (1992)
2. Ali, S.: The Birds of Sikkim. Oxford (1998)
3. Avasthe, R., Jha, A.: Mammals of Sikkim. WWF (1999)
4. Baldry, T.A.: The Tonglu Tiger. J. Darjeeling Natural History Soc. **1**(2), 80–82 (1923)
5. Banerjee, P., Ghose, M.K., Pradhan, R.: Analytic hierarchy process (AHP) based spatial biodiversity impact assessment model (SBIAM) of highway broadening in Sikkim Himalaya. Geocarto Int. **35**(5), 470–493 (2020). https://doi.org/10.1080/10106049.2018.1520924
6. Bashir, T., Bhattacharya, T., Poudyal, K., Sathyakumar, S.: Notable observations on the melanistic Asiatic golden cat (Pardofelistemminckii) of Sikkim, India. NeBIO **2**(1), 2–4 (2011)
7. Bhatnagar, Y.V., Mathur, V.B., McCarthy, T.: A regional perspective for snow leopard conservation in the Indian Trans-Himalaya. In: Unpublished Paper Presented at the National Workshop on Regional Planning for Wildlife Protected Areas, pp. 6–8, Aug 2001
8. Biswas, B., Ghose, R.K.: Progress Report 1 on Pilot Survey of the WWF-India/Zoological Survey of India Collaborative Project on the Status Survey of the Lesser Cats in Eastern India. Zoological Survey of India, Calcutta (1982)
9. Brashares, J.S., Arcese, P., Sam, M.K.: Human demography and reserve size predict wildlife extinction in West Africa. Proc. R. Soc. Lond. B Biol. Sci. **268**, 2473–2478 (2001)
10. Cardillo, M., Purvis, A., Sechrest, W., Gittleman, J.L., Bielby, J., Mace, G.M.: Human population density and extinction risk in the world's carnivores. PLoS Biol. **2**(7), e197 (2004)
11. Ceballos, G., Ehrlich, P.R.: Mammal population losses and the extinction crisis. Science **296**, 904–907 (2002)
12. Center for International Earth Science Information Network—CIESIN—Columbia University: Gridded Population of the World, Version 4 (GPWv4): Basic Characteristics, Revision 11. Palisades, NY: NASA Socioeconomic Data and Applications Center (SEDAC) (2018). https://doi.org/10.7927/H46M34XX
13. Cooper, D.M., Dugmore, A.J., Gittings, B.M., Scharf, A.K., Wilting, A., Kitchener, A.C.: Predicted pleistocene-holocene range shifts of the tiger (*Panthera tigris*). Divers. Distrib. **22**(11), 1199–1211 (2016)
14. Das, A.P., Bhujel, R.B., Lama, D.: Plant resources in the protected areas and proposed Corridors of Darjeeling, India. Biodiv. Conserv. Kangchanjunga Landscape. 57–79 (2008)
15. Dhakal, M., Karki (Thapa), M., Jnawali, S.R. et al.: Status of Tigers and Prey in Nepal. Department of National Parks and Wildlife Conservation and Department of Forests. Kathmandu, Nepal (2014)
16. Dhendup, T.: Status of Asiatic golden cat Catopuma temminckii Vigors & Horsfield, 1827 (Carnivora: Felidae) in Bhutan. J. Threatened Taxa **8**(4), 8698–8702 (2016)
17. Dorji, D.P., Santiapillai, C.: The status, distribution and conservation of the tiger *Panthera tigris* in Bhutan. Biol. Cons. **48**(4), 311–319 (1989)
18. Fick, S.E., Hijmans, R.J.: Worldclim 2: New 1-km spatial resolution climate surfaces for global land areas. Int. J. Climatol. (2017)
19. Forester, D.J., Machlis, G.E.: Modeling human factors that affect the loss of biodiversity. Conserv. Biol. **10**, 1253–1263 (1996)
20. Fox, J.L.:. Snow leopard conservation in the wild—a comprehensive perspective on a low density and highly fragmented population. In: Fox, J.L., Jizeng, D. (eds.) Proceedings of the Seventh International Snow Leopard Symposium (Xining, Qinghai, China, July 25–30, 1992), pp. 3–15. International Snow Leopard Trust, Seattle, Washington (1994)
21. Goodchild, M.F.: The state of GIS for environmental problem solving. In: Goodchild, M.F., Parks, B.O., Steyaert, L.T. (eds.) Environmental Modelling with GIS, pp. 8e15. Oxford University Press, New York, USA (1993)
22. Harcourt, A.H., Parks, S.A.: Threatened primates experience high human densities: adding an index of threat to the IUCN red list criteria. Biol. Conserv. **109**, 137–149 (2002)

23. Harcourt, A.H., Parks, S.A., Woodroffe, R.: Human density as an influence on species/area relationships: double jeopardy for small African reserves? Biodivers. Conserv. **10**, 1011–1026 (2001)
24. Haribal, M., Mulla, N.D., Chaturvedi, N.C.: The butterflies of Sikkim. J. Bombay Nat. Hist. Soc. **85**(2), 271–280 (1988)
25. https://www.downtoearth.org.in/news/wildlife-biodiversity/in-a-first-tiger-spotted-at-9-500-feet-in-sikkim-62755
26. Huang, Z.F., Liao, C.M., Chang, Y.: Applying GIS to design a habitat reserve for South-China tiger. Wildlife **11**, 18e19 (1998)
27. Hunter, D.O., Jackson, R.: A range-wide model of potential snow leopard habitat. In: Jackson, R., Ahmad, A. (eds.) Proceedings of the 8th International Snow Leopard Symposium, pp. 51–56, Islamabad, November 1995. International Snow Leopard Trust, Seattle and WWF-Pakistan, Lahore (1997)
28. Jackson, R., Fox, J.L.: Snow leopard and prey species workshop in Bhutan. Cat News **27**, 18–19 (1997)
29. Jackson, R., Wangchuk, P., Namgyal, T., Kumar, E.: Report on the second Bhutan SLIMS training workshop. Unpub. Report, International Snow Leopard Trust, Seattle (2000)
30. Jackson, R.M., Hunter, D.O.: Snow Leopard Information Management Handbook. International Snow Leopard Trust, Seattle (1996)
31. Jha, A., Thapa, K.: Reptiles and Amphibians of Sikkim. Shila Jha (2002)
32. Kafley, H., Gompper, M.E., Sharma, M., Lamichane, B.R., Maharjan, R.: Tigers (*Panthera tigris*) respond to fine spatial-scale habitat factors: occupancy-based habitat association of tigers in Chitwan National Park, Nepal. Wildlife Res. **43**(5), 398–410 (2016)
33. Kandel, P., Gurung, J., Chettri, N., Ning, W., Sharma, E.: Biodiversity research trends and gap analysis from a transboundary landscape, Eastern Himalayas. J. Asia-Pacific Biodivers. **9**(1), 1–10 (2016)
34. Lachungpa, D.: In a First, Tiger Spotted at 9500 ft in Sikkim, Down to Earth (2019)
35. Mallick, J.K.: Status of Red Panda Ailurus fulgens in Neora Valley National Park, Darjeeling District, West Bengal, India. Small Carnivore Conserv. **43**(30), e36 (2010)
36. Mallick, J.K.: *Panthera tigris*: range and population collapse in Northern West Bengal, India. Biodivers. Int. J. **3**(3), 110–119 (2019)
37. McCarthy, Weltzin, J. (eds.) Contributed Papers to the Snow Leopard Survival Strategy Summit. International Snow Leopard Trust, Seattle, Washington, USA. Available at http://www.snowleopard.org/sln/
38. McCarthy, T.M., Chapron, G.: Snow leopard survival strategy. International Snow Leopard Trust and Snow Leopard Network, Seattle, USA, 105 (2003)
39. McKinney, M.L.: Role of human population size in raising bird and mammal threat among nations. Anim. Conserv. **4**, 45–57 (2001)
40. Myers, N., Mittermier, R.A., Mittermier, C.G., da Fonseca, G.A.B., Kent, J.: Biodiversity hotspots for conservation priorities. Nature **40**, 853–858 (2000)
41. Olson, D., Dinerstein, E.: The Global 200. A representation approach to conserving the Earth's most biologically valuable ecoregions. Conserv. Biol. **12**(3), 502–515 (1998)
42. Parks, S.A., Harcourt, A.H.: Reserve size, local human density, and mammalian extinctions in U.S. protected areas. Conserv. Biol. **16**, 800–808 (2002)
43. Rathore, C.S., Dubey, Y., Shrivastava, A., Pathak, P., Patil, V.: Opportunities of habitat connectivity for tiger (*Panthera tigris*) between Kanha and Pench National Parks in Madhya Pradesh, India. PLoS One **7**(7) (2012)
44. Sangay, T., Rajaratnam, R., Vernes, K.: Wildife Camera Trapping in the Himalayan kingdom of Bhutan with Recommendations for the Future, pp. 87–98. Camera Trapping for Animal Monitoring. CSIRO Publishing, Collingwood (2014)
45. Sterndale, R.A.: Mammalia of India. First Indian Reprint, p. 540. Himalayan Books (1982)
46. Store, R., Jokimäki, J.: A GIS-based multi-scale approach to habitat suitability modeling. Ecol. Model. **169**, 1e15 (2003)

47. Taylor, B.D., Goldingay, R.L.: Roads and wildlife: impacts, mitigation and implications for wildlife management in Australia. Wildl. Res. **37**(4), 320–331 (2010)
48. Underhill, J.E., Angold, P.G.: Effects of roads on wildlife in an intensively modified landscape. Environ. Rev. **8**(1), 21–39 (1999)
49. Wang, S.W., Macdonald, D.W.: The use of camera traps for estimating tiger and leopard populations in the high altitude mountains of Bhutan. Biol. Cons. **142**(3), 606–613 (2009)
50. Wang, S.W.: A rare morph of the Asiatic golden cat in Bhutan's JigmeSingyeWangchuck National Park. Cat News **47**, 27–28 (2007)
51. Wang, S.W.: The impacts of wildlife damage andconservation policies on farmer attitudes in JigmeSingyeWangchuck National Park, Bhutan. M.S. Thesis, CornellUniversity. Ithaca NY, USA, 2004
52. Wang, S.W., Macdonald, D.W.: Livestock predation bycarnivores in JigmeSingyeWangchuck National Park, Bhutan. Biol. Cons. **129**, 558–565 (2006)
53. Woodroffe, R.: Predators and people: using human densities to interpret declines of large carnivores. Anim. Conserv. **3**, 165–173 (2000)

Health Informatics

Techniques in Detecting Diabetic Retinopathy: A Review

Parul Datta, Prasenjit Das, and Abhishek Kumar

Abstract In this research work, an exhaustive review has been done in the context of understanding the algorithms involved in building diabetic retinopathy systems. The study is important because with time India has become "Diabetic Capital" of the world. Due to diabetics, eyes of the people at large are getting impacted, and it plays a major role in blinding people and accelerating comorbidities. This study found that experts take into account specific features such as blood vessel area to detect abnormalities in eyes, and for this, they are primarily using fundus image processing algorithm in combination with statistical/machine/deep learning models. In this paper, we have conducted a review of the methods for automatically detecting and classifying diabetic retinopathy. This review points out that there are primarily three approaches that authors are applying for detecting diabetic retinopathy, and a meticulous view of algorithms is given for detecting diabetic retinopathy.

Keywords Diabetic retinopathy · Image processing · Segmentation · Fundus images

1 Introduction

Currently, the world is fighting a battle to tackle covid-19 pandemic. Due to pandemic health care services, all of the worlds have been curtained and doctors are attending a few patients as compared to the last year. This calls for innovation and rethinking on the way that we deliver our health care services, especially for people suffering from

P. Datta (✉) · A. Kumar
Chitkara University School of Engineering and Technology, Chitkara University, Baddi, Himachal Pradesh, India
e-mail: parul.datta@chitkarauniversity.edu.in

A. Kumar
e-mail: abhishek.kumar@chitkara.edu.in

P. Das
Chitkara University School of Computer Applications, Chitkara University, Baddi, Himachal Pradesh, India
e-mail: prasenjit.das@chitkarauniversity.edu.in

H. K. D. Sarma et al. (eds.), *Contemporary Issues in Communication, Cloud and Big Data Analytics*, Lecture Notes in Networks and Systems 281,
https://doi.org/10.1007/978-981-16-4244-9_34

blindness-related complications. As some of them are living in remote areas and due to multiple lockdown and travel restrictions, many of them cannot have access to eye care services.

In the context of eye diseases, the problem can be resolved by using remote eye care facilities and diagnostics. Rapid screening of many patients can be done using a safe environment for both doctors and patients. Mass screening for diabetic retinopathy can be done with help of intelligent algorithms and methods. This would help to reduce the burden of retinopathy for society. Hence, in this research work, an attempt has been made to evaluate and review algorithms and systems that can support remote mass level screening of retinopathy. However, before we move further, let us understand the term diabetic retinopathy briefly: When the veins and arteries of the eyes change their properties and some part of eyes are impaired due to high sugar levels in the blood, the disease is known as diabetes mellitus (DM). This condition is also referred to as diabetic retinopathy (DR). The main symptoms of this problem include loss of vision, blurred vision, spots, or dark strings or empty areas in vision. Research shows that people who are suffering from Type I and Type II diabetics are most prone to such problems, and their vulnerability increases many folds in case they get infected with a virus such as a corona. Due to diabetic condition, the retinal cells may cause irregularity in blood vessels, morphological changes including changes in diameter, blood vessel volume, etc.

If automatic algorithm needs to detect and track such changes in eyes, then the first step will be to separate the arteries and different parts of the eyes, so that a deep analysis of the eye can be done. The computer-aided examination of the eyes will give the stage at which the DR will be. The imaging modality suitable for remote examination is fundus images. Sample fundus image may be observed in Fig. 1 [1]. However, it should be noted that fundus images quality may vary due to many reasons such as non-alignment of the camera, untrained photographer, eye movement, camera focusing, etc. These factors affect the quality of fundus images. Hence, a preprocessing step may be required to upgrade the quality of fundus images.

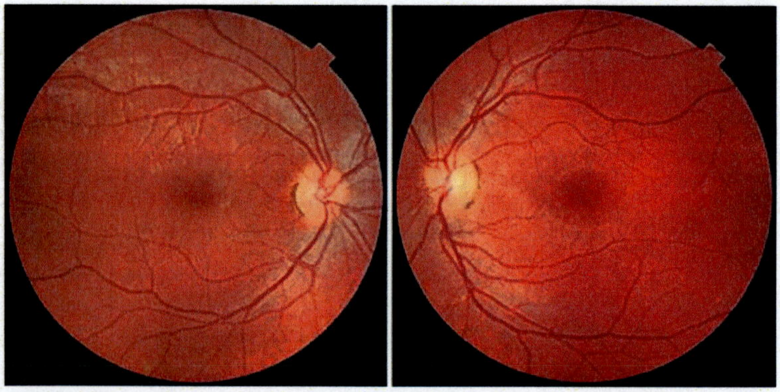

Fig. 1 Sample fundus images [1]

There might be a need for conversion of image to color space other than RGB to get better segmentation results. Image quality can be improved by contrast enhancement and denoising techniques. Gabor, mean, median, Gaussian filters, etc., can be used for image enhancement. A background subtraction and edge enhancement techniques also play a vital role in preprocessing the image.

The computed-assisted diagnose must find out medically relevant information from images, so that the detection of the appropriate eye disease can be done. The image processing algorithm may find damaged capillaries, detect leakage in the blood vessels, or may find some cloud-like objects in the eyes. Figure 2 shows healthy eye and diabetic eye. All these objects found refer to one of the medical condition written below [2]:

- **Hemorrhages**: These are the chunks of blood formed by the seepage of the blood from the smashed capillaries.
- **Microaneurysms**: These are the bulges filled with blood in the artery wall of the eyes.
- **Soft exudates**: These are the small whitish/gray cloud-like lesions within the eye.
- **Hard exudates**: These are the bright yellow colored objects on the retina.

For final decision making on the objects, the computer algorithm is found. Following are the levels of DR that are medically approved [2]:

- **Normal**: This means that the eye does have any abnormality in blood vessels, and there are no unwanted artifacts such as exudates found in the image.
- **Mild DR**: Only microaneurysms are seen.
- **Moderate DR**: Number of microaneurysms and hemorrhages are less than twenty in each quadrant.
- **Severe DR**: Number of microaneurysms and hemorrhages are more than twenty in each quadrant.

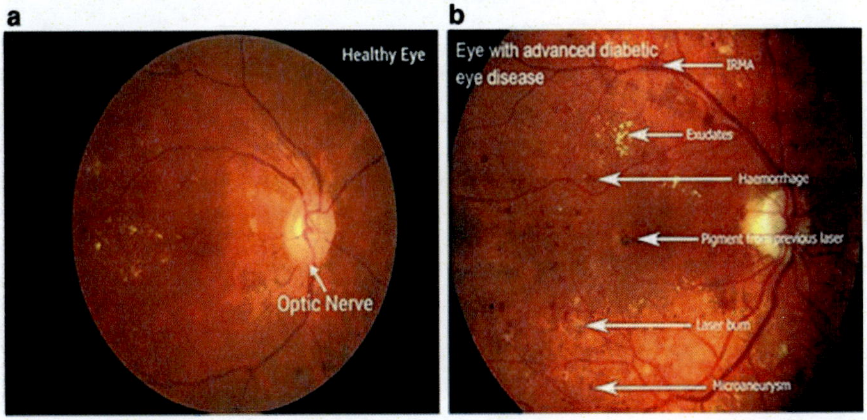

Fig. 2 **a** Healthy eye, **b** diabetic eye [3]

1.1 Availability of Datasets

Some of the datasets available online for the research work related to DR are:

1. **DAIRTDB**: This data collection has two levels, DIARTDB0 consists of 130 color imagery in which 20 are unaffected and 110 are affected with DR. DIARTDB1 consists of 89 fundus images in which 84 are affected with DR and 5 are normal.
2. **STARE**: STARE is University of California project consisting of 400 images.
3. **DRIVE**: This dataset has been created from the DR screening program in the Netherlands. It contains 40 images, and out of 40, 33 images are unaffected and 7 images are affected with DR.
4. **KAGGLE**: High-resolution fundus images are available in Kaggle dataset [4].
5. **MESSIDOR**: It contains 1200 colored fundus images [5].

It can be observed from this list, that data related to Indian demographics is not publicly available even though India has an existing active epidemic of diabetics. This paper is organized as follows: Sect. 2 summarizes some of the DR detection algorithms in a tabular form. Section 3 concludes this review work.

2 Literature Review

For maintaining good readability of this paper, a table is presented with the details of the work done by contemporary authors (Table 1).

2.1 Research Gaps and Possible Solutions

From the current literature survey, following research gaps can be listed:

(1) The biggest challenge faced by all the people around the world is related to the access of eye care facilities due to corona virus pandemic. Most of the health care services are running on low levels of operations. Due to this, there is urgent need for building remote eye care services that can be operated in situations such as covid-19 pandemics.

(2) The remote eye care service center will require stake of multiple technologies, which include mobile, global position system, and imaging systems. The amalgamation of such technologies and integration with location-based services is the need of the hour. With such technologies, eye care facilitation centers are required to be build.

(3) Technically, the images will be captured from hand-handled devices such as mobile devices and send to the remote server, so that contact less medical care can be done. This poses new kind of challenges. The quality of images

Table 1 Tabular summary of literature review

S. No.	Paper	Year	Major findings
1	[6]	2020	IDRID, ROC, and local datasets were used in the research work. And CNN achieved a sensitivity of 98.2%, specificity of 98.45%, the accuracy of 98.56% and average AUC of 0.9
2	[7]	2020	The authors have used DenseNet161, kappa score of 0.9025, sensitivity 90%, and specificity 87%
3	[8]	2020	The authors have used the hierarchical ensemble of CNNs and were able to achieve 96.1% accuracy
4	[9]	2020	Twenty-four studies involving 235 subjects were included and ANN used for detection
5	[4]	2020	Multi-stage transfer learning method followed for detection. The detection method had sensitivity and specificity of 0.99
7	[10]	2020	Automated Hyperparameter Tuning Inception-v4 (HPTI-v4) model developed for constructing classifier
8	[11]	2018	The classification accuracy was 95.68% by using CNN and transfer learning. The work has been done using Kaggle dataset
10	[12]	2019	Weighted thresholds were used, and this led to a good level of performance of the classifier
12	[5]	2020	Synergic deep learning model was applied to classify fundus images (Messidor DR dataset). This method performed better as compared to the preceding methods

becomes an issue. The alignment between the left and right eye image may get disturbed. This would require preprocessing on the image before they can be clinically examined.

(4) The fundus images may suffer from noise, growth of artificial artifacts, and most important they may suffer from low resolution. This would lead to inaccuracies in the computing the width and length of the arteries for diagnoses of DR. This is also required to be addressed. Super-resolution algorithms may be applied to overcome these issues for denosing and improving the resolution quality of the fundus images.

(5) Active Learning: Survey of the databases also show that there are limited fundus eye databases related to the studies conducted on Indian patients. Hence, there is a need for collecting primary image data related to Indian demographics. Secondly, in many cases, it has been found that unlabeled databases are also publically available and manual labeling of these images has not been done. In such cases, also either manual labeling of images as per the disease is required or active learning models may be required to overcome the problem of rarity of experts in this field.

(6) Machine Learning and Transfer Learning: The current research also shows that large number of dataset are publically available in context of DR, but their size and volume of the datasets are small. Such dataset may require treatment for

getting their effective use in machine/ deep or transfer learning. The treatment may include use of data argumentation algorithms or use the concept of transfer learning on them.

3 Conclusions

From this review, it is apparent that most the work done in context of building systems that can detect DR work on fundus images. There are publically available many datasets on this problem, but limited data is available for doing research work in Indian context. This is in spite of the fact that India is the biggest capital of the world of diabetics. Secondly, this review points out that there are primarily three approaches that authors are applying for detecting the DR. The approaches include statistics, machine learning, and deep learning. And, almost all the research papers focus on specific modality of the DR. In other words the contemporary demonstrated are specific to either extraction blood vessels or examination of optical disk size, etc., none of the work considers examination all aspects of eye morphology.

References

1. Fundus photography overview. opsweb **22**(11), 1–5 (2015)
2. Salamat, N., et al.: Diabetic retinopathy techniques in retinal images: a review. Artif. Intell. Med. **97**, 168–188 (2019)
3. Coatsworth: Fundus Photography—Coatsworth Eye Clinic. Coatsworth Website (2018). [Online]. Available: https://www.coatswortheyeclinic.co.uk/fundus-photography/
4. Tymchenko, B., et al.: Deep learning approach to diabetic retinopathy detection. In: ICPRAM 2020—Proceedings of the 9th International Conference on Pattern Recognition Applications and Methods (2020)
5. Shankar, K., et al.: Automated detection and classification of fundus diabetic retinopathy images using synergic deep learning model. Pattern Recognit. Lett. (2020)
6. Murugan, R., et al.: An abnormality detection of retinal fundus images by deep convolutional neural networks. Multimed. Tools Appl. 24949–24967
7. Sheikh, S., Qidwai, U.: Smartphone-based diabetic retinopathy severity classification using convolution neural networks. In: Advances in Intelligent Systems and Computing (2020)
8. Singh, R.K., Gorantla, R.: DMENet: diabetic macular edema diagnosis using hierarchical ensemble of CNNs. PLoS ONE 15(2)
9. Wang, S., et al.: Performance of deep neural network-based artificial intelligence method in diabetic retinopathy screening: a systematic review and meta-analysis of diagnostic test accuracy. Eur. J. Endocrinol. (2020)
10. Shankar, K., et al.: Hyperparameter tuning deep learning for diabetic retinopathy fundus image classification. IEEE Access (2020)
11. Wan, S., Liang, Y., Zhang, Y.: Deep convolutional neural networks for diabetic retinopathy detection by image classification. Comput. Electr. Eng. (2018)
12. Dutta, S., et al.: Classification of diabetic retinopathy images by using deep learning models. Int. J. Grid Distrib. Comput. (2018)

Specular Reflection Removal in Cervigrams

Ayush Agarwal, Faraaz Ali, Aditya Kopparthi, Priya Ranjan, Kumar Dron Shrivastav, and Rajiv Janardhanan

Abstract Removal of specular reflection also called specularity or highlights from an image is very important before the image is processed or analyzed by computer for image processing, computer graphics and computer vision because it is considered as part of the image by the computer. Removal of specular reflection is important because when there are specular reflections in the image it causes errors in the results of many algorithms of computer vision and image processing. Due to specular reflections in images, shapes in images get distorted, and therefore, all shape detecting algorithms and algorithms based on shape detection, boundary detection results in erroneous outputs. Removal of specular reflection includes detection of the reflection or reflections in the given image and then changing those pixels with some other color pixel based on the other portions of the image which are not faulty. There exist many different techniques for the detection and removal of specular reflection. In this paper, some of such techniques and approaches to tackle the problem are discussed. We have also proposed a simple yet efficient algorithm for specular reflection removal.

Keywords Specular reflection · Cervical cancer image · Filtering · Accuracy

---

A. Agarwal · F. Ali · A. Kopparthi
Amity School of Engineering and Technology, Amity University Noida, Sector 125, Noida, Uttar Pradesh, India

P. Ranjan
SRM University Andhra Pradesh, Neerukonda, Mangalagiri Mandal Guntur District, Mangalagiri, Andhra Pradesh 522502, India
e-mail: ranjan.p@srmap.edu.in

K. D. Shrivastav · R. Janardhanan (✉)
Amity Institute of Public Health, Amity University Noida, Sector 125, Noida, Uttar Pradesh, India
e-mail: rjanardhanan@amity.edu

K. D. Shrivastav
e-mail: kdshrivastav@amity.edu

H. K. D. Sarma et al. (eds.), *Contemporary Issues in Communication, Cloud and Big Data Analytics*, Lecture Notes in Networks and Systems 281,
https://doi.org/10.1007/978-981-16-4244-9_35

1 Introduction

Specular reflection in the image has been a challenge in the field of computer vision and image processing from a very long period. It is defined as a type. They generally seem like surface features, but they are just artifacts that are caused due to changes in illumination from different angles [10]. Removing specular reflection can be seen as a problem of taking out information that is present in an image and then applying various transformation techniques to convert the information into meaningful representations. Specular reflection removal is quite important before image analysis and processing as applying processing algorithms like segmentation, recognition, etc., to an image containing specular reflection may lead to significant inaccuracies in the results. It also reduces the robustness of certain algorithms and as a result makes the algorithms less effective and reducing their applicability. They can also hide various defects in images that will remain hidden during inspection [15]. Specular reflection, especially on medical images like cervigrams, can result in misdiagnosis of cervical cancer or other types of diseases. So, it can be seen easily that reflections in images cause a lot of problems ranging from making algorithms ineffective to life-threatening. Therefore, it is crucial to develop effective algorithms and hardware solutions. One of the methods is to use a polarizing filter. But it is not convenient enough as not everyone possesses a polarizer every time they try to capture an image or take a photograph. Other methods have also been studied like changing the position of the light source [9] or changing the position of the camera instead of a light source [12] in order to separate the specular reflection components. But, both the above-mentioned methods require the change in the position of the light source or the camera in the image processing system, thereby producing a huge strain or limitation when a picture is being taken. Here, the speckle removal algorithms take the lead. In this paper, we discuss the various algorithms developed for specular reflection removal.

2 Methods

The papers selected were taken from various renowned publishing Web sites like IEEE, Springer, and Elsevier. The selected papers were then evaluated, and papers with significant results and good accuracy were further taken into consideration. The papers were selected from the past 20 years. The algorithms discussed in the paper mainly vary by virtue of their applications and implementations. Some algorithms perform better in situations when reflections are caused due to moisture on the surface, some perform better when reflections are caused due to the material of the object, some when reflections occur during medical imaging, while others perform better in general conditions. The images used in the paper are taken from Kaggle Intel/mobile ODT dataset [5].

3 Algorithms and Techniques

3.1 Light Field Imaging-Based Accurate Image Specular Highlight Removal

Wang et al. have developed a light field imaging-based image specular highlight removal algorithm [10]. Their algorithm works well in complex scenarios that occur in real life. This algorithm takes advantage of the light field of imaging technology (Lytro ILLUM). The image is first captured (with specularity present) by the light field camera. Then, the depth of the image is accurately analyzed and estimated, and the specular pixels are then classified into two categories, namely "Unsaturated" and "Saturated" using a straightforward and concise thresholding strategy. For depth estimation and to achieve refocusing, utilization of a depth estimation algorithm by combining/integrating both the correspondence and defocus cues has been done. 4D epipolar image (EPI) is exploited after deriving it from the LF data, making shears in order to operate refocusing. Then, the responses of the two cues are computed using a single contrast-based approach presented in the paper. Both the locally estimated cues are then combined with a measure of confidence, and global depth estimation is computed using MRFs to obtain the final result. At last, multiple views are subjected to color variance analysis, and the two categories are used to recover diffused color information by conducting local color refinement individually on the two categories. The method is then experimentally evaluated by comparing them with the existing methods based on the light field dataset along with the standford light field archive, thus verifying its effectiveness.

$$D_\alpha(x, y) = \frac{1}{|W_D|} \sum_{(x^f, y^f) \epsilon W_D} |\Delta \overline{I}_x(x^f, y^f)|$$

3.2 An Image Correction Method for Specular Reflection Removal Using a High-Speed Stroboscope

Tsuji [14] suggests a method for separating diffused reflection components from specular reflection components. The author's group developed a method of separating specular reflection components present in a high-speed video from diffused reflection components. A luminance variation (because of the flicker of a strobe)-based estimation technique has been used. But the issue with this method is that a new specular reflection component is produced by the strobe. An algorithm is suggested to remove this specular reflection produced by the strobe. The arithmetic algorithm consists of the two processes mentioned below:

- Specifying the highlight area
- Compensation in the highlight area.

The original color of the object cannot be defined accurately as the highlighted area is produced by strong specular reflection components caused by the light source. This method also solves the problem of intensity discontinuation due to image synthesis with the different light sources. Experimental results are then used to show the efficiency and validity of the algorithm.

3.3 Removal of Specular Reflection in Large-Scale Ocean Surface Images

Wang et al. used an image inpainting technique for removal of specular reflection in large-scale ocean surface images [16]. After detection of specular reflection in an image for removing reflection area, image inpainting technique proposed by Alexandru Telea based on fast marching method is used [13]. In this algorithm, filling of the specular reflection is done from the boundary pixels that decreases the size of area, and repetition of filling boundary decreases the size of the area of specular reflection to zero. A pixel at the boundary is changed depending upon the surrounding pixels of the image which are already correct. Effect of a surrounding pixel to the pixel to be corrected is based on the value of the surrounding pixel and its weight. Weight for a pixel is based on the distance between that pixel and the pixel to be corrected.

$$I(p) = \frac{\sum_{q \in B_\epsilon(p)} w(p, q)[I(q) + \nabla I(q)(p - q)]}{\sum_{q \in B_\epsilon(p)} w(p, w)}$$

where I is inpainting, and p is the point to be inpainted. B_ϵ is the surrounding region where q points exist. q points are in the surrounding region that do not contain specularity. $w(p, q)$ is the weight of the points in B_ϵ.

3.4 Removal of Specularities Using Color and Polarization

Nayar et al. proposed an algorithm for removal of specular reflection that uses polarization and color for separation of the two components of reflection, diffused and specular [8]. Most of the methods are either based on color or polarization. This algorithm applies new constraints on speculatrities by simultaneously using both color and polarization. Color of specular component is determined locally using polarization, thereby constraining the diffused color at a pixel to a linear subspace of a single dimension. Neighboring pixels whose color does not change much with the pixel are found using this subspace. Information of diffuse color from consistent neighbors in subspace determines the value of the diffuse color of the pixel. This method of specular reflection removal can be used for specularity removal from direct source illuminations as well as inter-reflection between points in the scene. Neighbors having same diffuse color for each point are required for the algorithm to be used. The magnitude of neighboring pixels may have a different magnitude of diffuse color, but the direction in color space must be the same. The advantage of this algorithm over other methods is that the diffuse component in the image which is under the specularity region is not considered to be constant in the whole region. Another great advantage is that the diffuse component under the highlight region can be a texture and algorithm performs well in removing specularity. Reflection depends upon the Fresnel ratio which is dependent on the angle of incidence and the material of the surface on which light is incident, and this algorithm also performs well in case of reflection of different Fresnel ration in a particular specularity region. So, if in a region of specularity there are reflections due to different materials, it can also be removed using this algorithm.

3.5 Specular Reflection Reduction with Multi-flash Imaging

Raskar et al. [3] proposed a technique to solve the problem of specular reflections. A camera with multiple flashes is used in this method. The camera will not be dependent on a single image of a scene rather it captures multiple images of a particular scene with different flash for each image. One flash at a time is kept in ON state, and all the other flashes are kept in OFF state. This technique uses the phenomenon that position of specular reflection changes with the position and angle of the incident light. So, there will be n number of images captured with n flashes, all at a different position that leads to specularity at a different position on each image. When captured images are combined, some highlights may overlap or may not overlap. This results in the three cases mentioned below:

- Some highlights remain distinct (no overlapping).
- Some highlights partially overlap.
- Some highlights overlap completely.

The proposed method works well with the first two cases but fails in the third case. It is to be noted that the boundaries/intensity edges around the highlights do not generally overlap even though specularities do overlap in input images. The principle idea of the approach is to take advantage of the variation of the gradient in n images, taken under n separate lighting conditions at the location of the given pixel, (x, y). If the mentioned pixel (x, y) is in the specular region in the case of no overlap or partial overlap, the gradient because of the specularity edges will be high in a minority or in only one of the n given images. A variable named I_k has been defined that represents an input image taken using a light source k, where k ranges from 1 to n. This formula separates out the maximum intensity components from the image. These maximum intensity components are specularities. This formula calculates the median of pixels at location (x, y) of every image.

3.6 Real-Time Specular Highlight Removal Using Bilateral Filtering

Yang et al. [17] suggest a method for real-time specular reflection removal based on the observation that maximum diffuse chromaticity changes quite smoothly in local patches. This property is used to then estimate the maximum diffuse chromaticity values by the usage of low pass filter. The low pass filter is applied directly to the maximum fraction of the color components present in the original image, such that the propagation of maximum diffuse chromaticity values is done from diffuse pixels to the specular pixels. Computation of diffuse pixel at each pixel is then done as a nonlinear function of the maximum diffused chromaticity that was estimated. If edge-preserving filters are used, this method can be extended for multi-color surfaces. Joint bilateral filtering can be used to smoothen the maximum chromaticity $\sigma_{\max}$. As for the smoothing guidance, maximum diffuse chromaticity $\Lambda_{\max}$ is used. The maximum diffuse chromaticity can be defined as

$$\Lambda_{\max} = \max(\Lambda_r, \Lambda_g, \Lambda_b)$$

Recent developments in the field of fast bilateral techniques make the method run nearly 200 times faster than the state of the art on the standard CPU. This technique is capable of processing high-resolution images at video rate and therefore is suitable for applications that work in real time. Also, the diffused reflections that are estimated will be locally smooth and do not result in noticeable artifacts.

3.7 Correction of Specular Reflection by Recursively Applying Smoothing Spatial Filter

Generally, the detected specular reflections are considered as noise, and we use smoothing spatial filter to either remove or smoothen this noise in the image. By this filter, the reflections can be blurred that is the filter covers places on the original image and divides by the sum of all kernel elements, then multiplied with input pixel intensity and kernel values [2], and the formula for this filter is

$$C = \frac{1}{N}\{M(1)I(1) + M(2)I(2) + \cdots + M(N)I(N)\}$$

$$C = \frac{1}{N}\sum_{k=1}^{N}\{M(k)I(k)\}$$

Here, the c is renewed average, and N is multiple of length and width [4]. The result will be an average value as it is divided by the multiple of length and width N. This formula calculates the median of pixels at location (x, y) of every image. To have a very smooth process, we have to consider mask elements as 1, and the above expression will be changed as

$$C = \frac{1}{N}\{I(1) + I(2) + \cdots + I(N)\}$$

$$C = \frac{1}{N}\sum_{k=1}^{N} I(k)$$

Therefore, the final result of this filtering will be an average value as it divides with N after adding all intensities of the pixels. Our main task is to remove or minimize the specular reflections, but after this process, the image will be blurred as it is not our main aim we have to update the intensity of pixels present in the specular reflection area which is found through perception neural net as it was shown before but not change the whole intensity of the pixel.

3.8 Automatic Detection of Specular Reflections in the Cervix Images

Here, the specularities are detected in cervix images using intensity, saturation, gradient information, and identification of highlighted region is done in a two-stage segmentation process. First, coarse regions which have the reflections are found, and second probabilistic modeling and segmentation are done to perfectly select the coarse regions. For the resulting region, a simple filling scheme is proposed. Highlights are most likely to found in the SR regions which are coarse regions, and these specularities normally have low saturation values S and very high intensity I values, so initial regions can be found by using thresholds on I and S [7].

$$I = \frac{R + G + B}{3}$$

$$S = 1 - \frac{\min(R, G, B)}{I}$$

Now, a precise selection of previously defined coarse region is required, for this, they used Gaussian mixture model in which each pixel is considered as the 2-d vector, these pixels are grouped as homogeneous regions, and each region is represented as Gaussian distribution [11].

$$(x) = \sum_{j=1}^{k} a_j \frac{1}{\sqrt{(2\pi)^d |\sum_j|}} \exp\left\{-\frac{1}{2}(x - \mu_j)^T \sum_j^{-1} (x - \mu_j)\right\} \quad (1)$$

For the final step, a filling scheme is used, which removes the strong gradients while making sure the original texture is not affected.

4 Algorithm Comparison

See Table 1 and Figs. 1, 2.

5 Results

See Table 2.

5.1 *Removing Specular Reflection Components for Robotic-Assisted Laparoscopic Surgery*

In this paper, they propose that at first all the chromatic information should be collected for the spatio-temporal volume so that this information can be used to remove the specular reflections on the epicardial surface of the heart while performing a robotic laparoscopic surgery while preserving the original image structure. By the above collected information, the pixel intensities can be shifted for separating the specular and diffuse image components. Mostly all the processes that remove the specular reflections normally have a little bit of image data lost in the process, but here, as there are thin nerves, we cannot use normal methods, and we have to rather use a method that can collect as much as the information of the image by extending chromaticity [6].

Here, I_c^{diff} is the diffuse component of reflectance and $\Lambda = \Lambda_r \Lambda_g \Lambda_b$.

It is clear that chromaticity of the diffuse pixels is always higher than that of the specular pixels. By making use of the image pixel distribution projection into D space, the diffuse reflectance component can be found as shown in the above formula and stored in a spatio-temporal volume.

5.2 *A Video Stream Processor for Real-Time Detection and Correction of Specular Reflection in Endoscopic Images*

In this paper, the detection and correction of the specular reflections are by an algorithm called inpainting algorithm. Correcting a frame represents removing all the

Table 1 Comparison table

Authors	Algorithm	Technique used	Advantages
Wang et al.	Light field imaging-based accurate image specular highlight removal	Based on light imaging technology	Works well in complex real-life scenarios
Tsuji et al.	An image-correction method for specular reflection removal using a high-speed stroboscope	Luminance variation-based estimation technique	Solves discontinuous intensity problem due to image synthesis with a light source
Wang et al.	Removal of specular reflection in large-scale ocean surface images	Image inpainting based on fast marching method	Works well with underwater images of oceans
Nayar et al.	Removal of specularities using color and polarization	Based on color and polarization of surrounding pixels	Removes direct reflection as well as inter-reflection. Works well in images with texture
Raskar et al.	Specular reflection reduction with multi-flash imaging	Uses multi-flash camera (hardware), the position of specularity changes with a change in angle of incidence of light	Removes distinct highlights. Removes partially overlapped highlights
Yang et al.	Real-time specular highlight removal using bilateral filtering	Based on the observation that maximum diffuse chromaticity changes quite smoothly in local patches	Capable of processing high-resolution images at video rate. Suitable for applications that work in real time.
Lehman et al.	Correction of specular reflections by recursively applying a smoothing spatial filter	Using smoothing spatial filter to remove noise	The new image will not be blurry as the whole intensity of the pixel is updated
Gonzalez et al.	Automatic detection of specular reflections in the cervix images	Probabilistic modeling and segmentation	As the filling scheme is used, original texture is not affected
Bertalmio et al.	Removing specular reflection components for robotic-assisted laparoscopic surgery	Using chromatic information and store it in spatio-temporal volume	The original image is distorted in the process as only image data is collected
Lin et al.	A video stream processor for real-time detection and correction of specular reflections in endoscopic images	We use single frame memory architecture	It takes very less memory and does not need high computational power

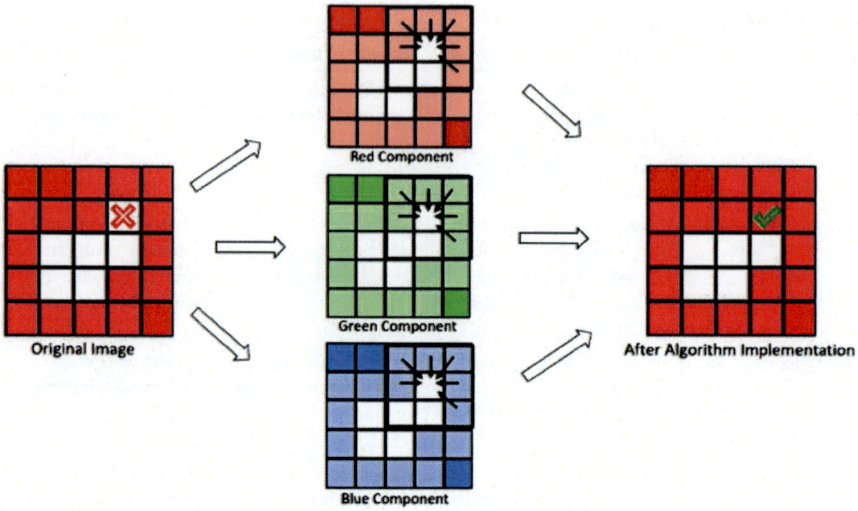

Fig. 1 Visual representation of algorithm

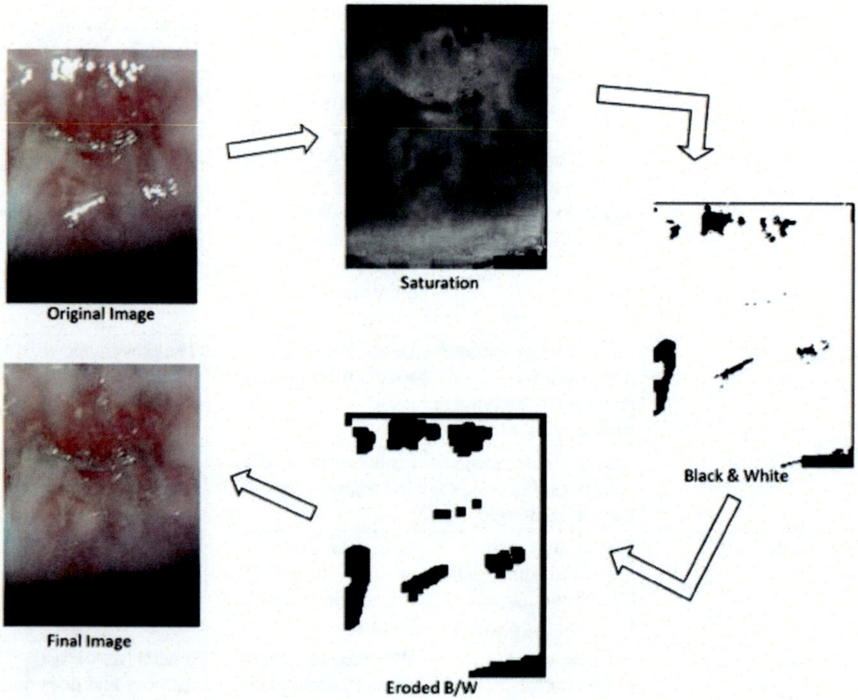

Fig. 2 Process of reflection removal

Table 2 Results

S. No.	Original image	Image after applying algorithm	EMD score
1	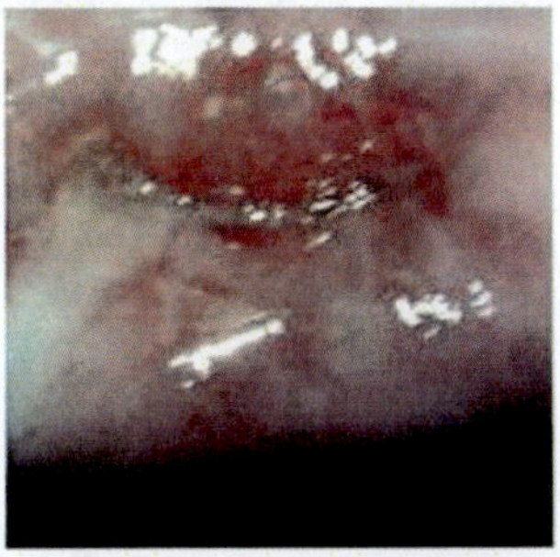	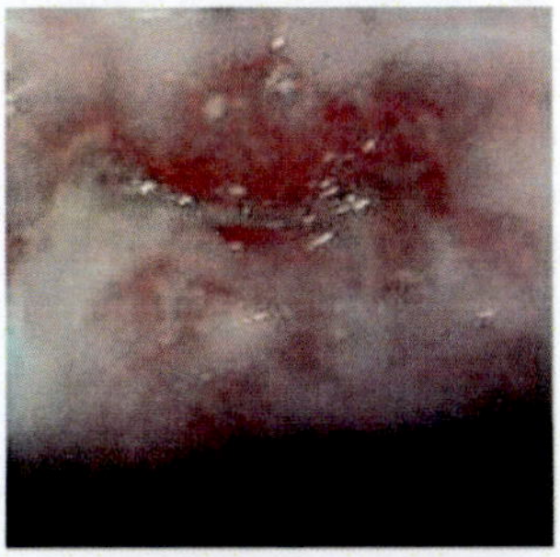	0.01679449
2	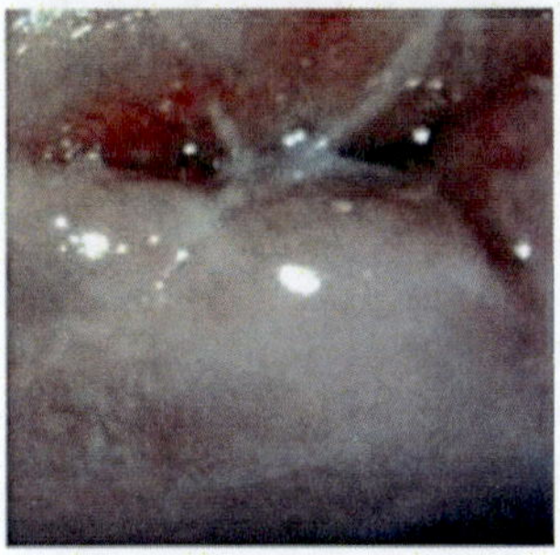	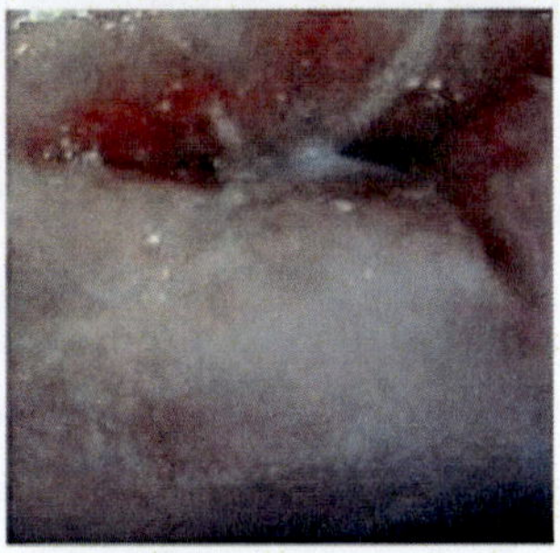	0.01771109
3	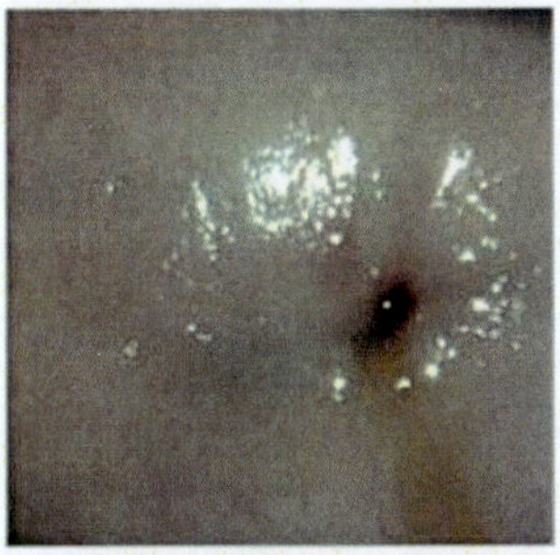	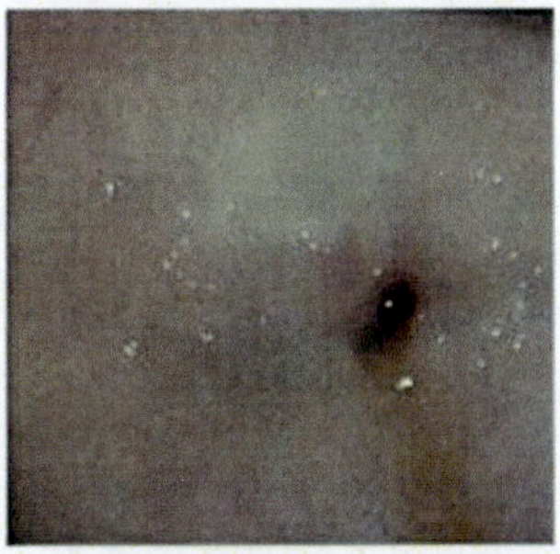	0.01355387
4	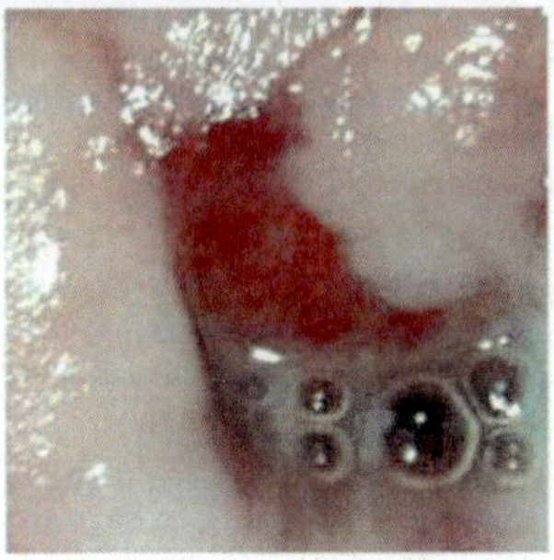	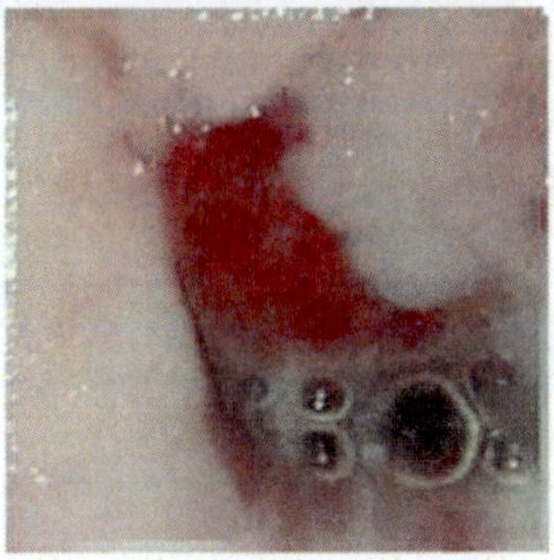	0.02003022

specularities that were determined before and replacing them for neighborhoods information, and the common way is to use algorithms like Navier-Stokes algorithm [1]. As it has many loops through a frame, it requires large amounts of memory and computational power, so we use single frame memory architecture. It operates line by line

1. A line of the frame is stored.
2. We collect three data information, they are pixel value before the specular region P_b , pixel value after specular region P_c, and the region's width
3. Now, we can kind linear show a as:

$$a = \frac{P_c - P_b}{w}$$

4. The leftmost pixel has index 0 in the specular region, and P_0 is given value $P_b + a$. The new value of all other pixels is

$$P_{i+1} = P_i + a$$

As it only operates in the horizontal direction, it is necessary to correct the vertical dimension. This is done with a smoothing window by changing modified pixels with the average of its neighbors, and it may have a little delay, but it can be considered negligible while the time of real operation.

6 Algorithm

Let I_m be an image from the set of images S. We first transform our RGB image to (hue, saturation, and value (HSV) format. HSV format was chosen as it is closer to human vision. Hue is the tint or color potion expressed as number from 0 to 360°. Saturation describes the amount of gray in a particular color. Saturation varies the color from white to primary color in hue. Value is the brightness/intensity value of the resulting color. Saturation varies the color from white to primary color in hue. Value is the brightness/intensity value of the resulting color. The saturation component s was extracted from $I_m(I_{ms})$ which was observed to better separate/extract bright specular reflection portion than the other components. The objective is to highlight the specular component region of the image I_{ms} and suppress the background or other unimportant regions. Thus, the image I_{ms} was then converted to black and white I_{bw} , which makes the specular region black and other regions/background white. It was observed in certain cervigrams that there was some sort of black ring or boundary to be found around the specularity (which is very thin). The boundary needs to be removed from the images as it can affect the performance of the algorithm. To tackle this, the size of the black colored region in I_{bw} was slightly increased so that the increase in the region will cover up the black boundary which is found around the specular reflections.

The removal procedure: Next, we describe the procedure to remove the specularities present in the original image. For this, a mask m of size $n \times n$ (the size of the mask can be changed according to the requirements) was created. The pixels P_{ij} in the black region of the I_{bw} were found out. A m $n \times n$ mask was created with the pixel P_{ij} at its center.

Then, repeat below-mentioned process for each color channel C_i of the image I_m in RGB color scheme:

- Traverse I_{bw} pixel by pixel, and wherever the non white pixels P_{ij} are found (white in I_m's context), calculate the mean of all the corresponding non-white pixels inside the mask in the original image $I_m(A_{ij})$ and channel C_i.
- Replace the pixel at (i, j) position in channel $C_i = (red, green, yellow)$ of image I_m with A_{ij}.

This process is repeated for every pixel in the image I_{bw} until every specularity is averaged out. The resultant image is I_{mr}. After running the algorithm, one can easily compute the degree to which the algorithm was effective. In order to compare the original image with the I_{mr} we got after running the algorithm, we used earth mover's distance (EMD) to compare the two images. Higher EMD score signifies better results achieved, i.e., more specular reflection removed.

Table 2 shows the EMD score along with the images for which the score was calculated.

$$A_{ij} = (P_{11} + P_{12} + \cdots + P_{1n} + P_{21} + P_{22} + \cdots$$
$$+ P_{2n} + \cdots + P_{n1} + P_{n2} + \cdots + P_{nn})/K$$

$$A_{ij} = \frac{\sum P_{ij}}{K}$$

where P_{ij} represents the non-white pixels in the mask $mn \times n$, and K represents number of non-white pixels.

7 Conclusion

In this paper, we have discussed various specular reflection removal algorithms. Some of these algorithms perform better in certain conditions, while others are effective in general conditions. Some are more accurate, while others are less accurate but require less computation power and perform their operation in very less time. Characteristics of various materials in the real-world make specular reflection inevitable. Currently, existing specular reflection removal methods are quite effective and achieve good separation of specular components, but are limited to their applicability conditions. Most of the techniques mentioned above rely on specific reflection model. This along with other problems like noise sensitivity makes these algorithms less effective and reduces the range of their applicability. More accurate and robust algorithms are required which can overcome the limitations of the current methods. Our future work will involve developing a more general and robust algorithm for specular reflection removal which overcomes the limitations of most of these algorithms along with maintaining its accuracy.

References

1. Bertalmio, M., Bertozzi, A., Sapiro, G.: Navier-Stokes, fluid dynamics, and image and video inpainting, vol. 1, pp. I–355. https://doi.org/10.1109/CVPR.2001.990497 (2001)
2. Fausett, L., Fausett, L.: Fundamentals of Neural Networks: Architectures, Algorithms, and Applications. Prentice-Hall International Editions, Prentice-Hall. https://books.google.co.in/books?id=ONylQgAACAAJ (1994)
3. Feris, R., Raskar, R., Tan, K.H., Turk, M.: Specular highlights detection and reduction with multi-flash photography. J. Braz. Comput. Soc. **12**, 35–42 (2006)
4. Gonzalez, R., Woods, R., Woods, R.: Digital Image Processing. Pearson/Prentice Hall. https://books.google.co.in/books?id=8uGOnjRGEzoC (2008)
5. Intel, M.: https://www.kaggle.com/c/intel-mobileodt-cervical-cancer-screening/data (2017)
6. Lin, S., Li, Y., Kang, S.B., Tong, X., Shum, H.Y.: Diffuse-specular separation and depth recovery from image sequences, vol. 2352, pp. 210–224. https://doi.org/10.1007/3-540-47977-5_14 (2002)
7. Lehmann, M.T., Palm, C.: Color line search for illuminant estimation in real-world scenes. J. Opt. Soc. Am. A Opt. Image Sci. Vis. **18**, 2679–2691 (2001). https://doi.org/10.1364/JOSAA.18.002679
8. Nayar, S.K., Fang, X., Boult, T.: Removal of specularities using color and polarization. In: Proceedings of IEEE Conference on Computer Vision and Pattern Recognition, pp. 583–590. https://doi.org/10.1109/CVPR.1993.341071 (1993)
9. Sato, Y., Ikeuchi, K.: Temporal-color space analysis of reflection. In: Proceedings of IEEE Conference on Computer Vision and Pattern Recognition, pp. 570–576. https://doi.org/10.1109/CVPR.1993.341073 (1993)
10. Shah, S.M.Z.A., Marshall, S., Murray, P.: Removal of specular reflections from image sequences using feature correspondences. Mach. Vis. Appl. **28**(3), 409–420 (2017). https://doi.org/10.1007/s00138-017-0826-6
11. Smith, A.R.: Color gamut transform pairs. SIGGRAPH Comput. Graph. **12**(3), 12–19 (1978). https://doi.org/10.1145/965139.807361
12. Swaminathan, R., Kang, S.B., Szeliski, R., Criminisi, A., Nayar, S.K.: On the motion and appearance of specularities in image sequences. In: Heyden, A., Sparr, G., Nielsen, M., Johansen, P. (eds.) Computer Vision—ECCV 2002, pp. 508–523. Springer, Berlin Heidelberg, Berlin, Heidelberg (2002)
13. Telea, A.: An image inpainting technique based on the fast marching method. J. Graph. Tools **9**(1), 23–34 (2004). https://doi.org/10.1080/10867651.2004.10487596
14. Tsuji, T.: An image-correction method for specular reflection removal using a high-speed stroboscope. In: IECON Proceedings (Industrial Electronics Conference). https://doi.org/10.1109/IECON.2011.6120050 (2011)
15. Wang, H., Xu, C., Wang, X., Zhang, Y., Peng, B.: Light field imaging based accurate image specular highlight removal. PLoS ONE **11**(e0156), 173 (2016). https://doi.org/10.1371/journal.pone.0156173

16. Wang, S., Yu, C., Sun, Y., Gao, F., Dong, J.: Specular reflection removal of ocean surface remote sensing images from UAVs. Multimed. Tools Appl. **77** (2018). https://doi.org/10.1007/s11042-017-5551-7
17. Yang, Q., Wang, S., Ahuja, N.: Real-time specular highlight removal using bilateral filtering. In: Daniilidis, K., Maragos, P., Paragios, N. (eds.) Computer Vision—ECCV 2010, pp. 87–100. Springer, Berlin Heidelberg, Berlin, Heidelberg (2010)

Kloman Metre: An EMD-Based Tool for Triaging Diseases Leading to Lung Infections Including COVID-19

Niranjan Chavan, Priya Ranjan, Uday Kumar, Kumar Dron Shrivastav, Hiren Kumar Deva Sarma, and Rajiv Janardhanan

Abstract The digital revolution can help developing countries to overcome the problem of limited healthcare infrastructure in developing nations such as India. The COVID-19 pandemic has shown the urgency of integration of digital technologies into healthcare infrastructure. In order to solve the issue of lack of trained healthcare professionals at public health centres (PHCs), researchers are trying to build tools which can help to tag pulmonary ailment within a fraction of second. Such tagging will help the medical community to utilize their time more efficiently. In this work, we have tried to assess the "lung health" of patients suffering from a variety of pulmonary diseases including COVID-19, tuberculosis and pneumonia by applying Earth Mover's Distance algorithm to the X-ray images of the patients. The lung X-ray images of patients suffering from pneumonia, TB and COVID-19 and healthy persons are pooled together from various datasets. Our preliminary data based upon 100 random images depicting each type of lung disease such as COVID-19, tuberculosis and pneumonia revealed that patients suffering from tuberculosis have the highest severity as per the values obtained from the EMD scale.

Keywords Lung health · Chest radiographs · Earth Mover's Distance · Triaging · Community level · Large-scale screening

N. Chavan · K. D. Shrivastav · R. Janardhanan
Health Data Analytics & Visualization Environment, Amity Institute of Public Health, Amity University Uttar Pradesh, Sector 125, Noida, India

P. Ranjan
SRM University, Neerukonda, Mangalagiri Mandal Guntur District, Mangalagiri, Andhra Pradesh 522502, India

U. Kumar
Delhi State Cancer Institute, Dilshad Garden, Delhi, India

H. K. D. Sarma (✉)
Department of Information Technology, Sikkim Manipal Institute of Technology, Majitar, East Sikkim 737136, India

R. Janardhanan
Laboratory of Disease Dynamics & Molecular Epidemiology, Amity Institute of Public Health, Amity University Uttar Pradesh, Sector 125, Noida, India

H. K. D. Sarma et al. (eds.), *Contemporary Issues in Communication, Cloud and Big Data Analytics*, Lecture Notes in Networks and Systems 281,
https://doi.org/10.1007/978-981-16-4244-9_36

1 Introduction

Among low- and middle-income countries (LMICs), India is considered to be an important nation to study the emerging burden of Non-Communicable Diseases (NCDs) as it indeed presents a unique landscape of disease burden. India is projected to experience more deaths from NCDs than any other country over the next decade, due to the size of the population and worsening risk factor profile, associated with recent dramatic economic growth. The country has deep and entrenched social and economic disparities, with affordable health care being beyond the reach of large sections of the society. The epidemiological evidence on the socio-economic status (SES) related patterning of NCDs remains limited in LMICs.

In the last two decades, a significant leap of thought has resulted in the translation of the fundamental concepts of precision medicine at a community level to understand the patterns and processes associated with the landscape of disease burden in the low- and middle-income countries having fractious and fractionated healthcare ecosystems. India endowed with heterogeneous genetic base, wide variations in socio-cultural norms, along with divergent geological relief structures provides a veritable landscape of disease burden resulting in the prevalence of healthcare disparities. The need of the hour is to create **fruganomic** community empowering solutions aimed alleviating the healthcare disparities.

During this pandemic situation (2020 onwards), clinicians are dramatically looking forward to fast diagnostic tools for COVID-19, characterized by a right balance between sensitivity and specificity, leading to acceptable predictive values in a context of a variable prevalence, depending on policies ranging from testing only symptomatic subjects to mass screening. Of note, any tool to be applied for this aim should have an excellent cost–benefit ratio for the healthcare service [1].

Although India has witnessed an epidemiological transition from communicable diseases to NCDs in the last two decades with cardiovascular diseases (CVD) and cancer accounting for a significant proportion of morbidities and mortalities, the unfinished agenda of communicable diseases has led to emergence of recent pandemics such as COVID-19 [2]. The World Health Organization (WHO) has recognized the outbreak of COVID-19 to be a Public Health Emergency of International Concern on 30 January 2020 and declared it as a pandemic on 11 March 2020 resulting in enforced nationwide locked downs in several countries across the world including India [3]. This has had a statutory impact on the economy of all the countries including India.

In this paper, a tool based on Earth Mover's Distance has been reported which has been developed for analysing X-ray images of lungs. The paper also presents some preliminary results along with relevant discussions.

The rest of the paper is organized as mentioned below. Section 2 presents background of the work. Motivation behind the work is presented in Sect. 3. The basics of Earth Mover's Distance are discussed in Sect. 4. Section 5 presents implementation approach and results along with an analysis of the results. The paper is concluded in Sect. 6.

2 Background

Tuberculosis (TB) is a leading cause of morbidity and death worldwide, with approximately two billion people infected and approximately two million annual deaths attributable to it [1]. COVID-19 infection primarily spreads through contact with an infected person and via respiratory droplets when people cough or sneeze. It shows the initial symptoms like viral pneumonia such as fever, cough, myalgia, fatigue and shortness of breath. In serious cases, complications like acute respiratory distress syndrome and cardiac failure. But about 5% can develop severe stages of pulmonary distress syndrome and can lead to death [4].

The existence of environmental factors such as air pollution is known to significantly contribute to premature mortality and disease burden globally, with the highest impact in low-income and middle-income countries such as the Indian sub-continent endowed with resource limited healthcare systems [5, 6].

A significant drawback of the existing interpretation algorithms based on artificial neural networks (ANN) is its black-box nature. When an image is fed to a neural network and it matches it to another image, it is not easy to understand why and how it came up with a match [7–9]. Further, the nonlinear dynamical behaviour of deep neural networks is prone to chaotic nature and fundamental underlying unpredictability [10]. Addressing to this issue is an extreme demand for computation in deep learning which is forcing researchers to explore other techniques [11, 12]. On the other hand, static and predictable algorithms like Earth Mover's Distance (EMDs) perform image match by computing perceptual similarity and provide more meaningful and interpretable solutions to matching problems. To re-emphasize, ANNs have a long and very well researched history of inherent instability and its automated decisions cannot be entrusted to make decisions critical to the survival of a patient afflicted with a severe case of a cardiac episode. India faces many challenges in delivering health care, especially in the rural domain due to a shortage of resources and skilled personnel. The fact that chronic respiratory diseases account for 8% of mortalities in India clearly undermines the magnitude of the public health challenge it imposes. The lack of suitable algorithms further compounds this to map out the clinical resource allocation especially in rural underserved population where doctor-patient ratio is low and access to affordable, and quality healthcare options is significantly marginalized.

The diagnosis of TB and pneumonia is performed by multiple tests such as sputum test, skin test, blood test, pulse oximetry or imaging tests (X-ray, CT scans). COVID-19 is diagnosed using RT-PCR test and imaging tests. Chest X-rays are the most commonly ordered diagnostic imaging tests, with millions of X-rays performed globally every year [13]. Whilst the chest X-ray is frequently performed, interpreting a chest X-ray is one of the most subjective and complex of radiology tasks, with inter-reader agreement varying from a kappa value of 0·2–0·77, depending on the level of experience of the reader, the abnormality being detected and the clinical setting [14–17]. Due to their wide availability and affordability, chest X-rays are performed all over the world, including remote areas with few or no radiologists. A more prudent step, therefore, will be to develop tools and applications that can be integrated easily into the current healthcare system. Additionally, the lack of well-structured databases for referencing and analysis hinders the progression of research from aiding and optimizing processes and clinical decision making with the help of artificial intelligence (AI).

Presence of diverse genetic base and socio-cultural norms in India presents a unique landscape of disease burden necessitating the need for niche specific databases for enhancing the accuracy of AI tools. Apart from the complexities as mentioned earlier, the development of these technologies come with substantial technical, ethical, confidentiality and clinical challenges. Despite the afore-mentioned hindrances, the socio-economic impact and benefits of AI-based automation of ECG analysis for LMICs like India will be significant. The potential healthcare application of AI-based platforms is vast, encompassing screening, disease detection and patient risk stratification along with niche specific optimal intervention strategies. Since the penetration of the mobile platforms and Internet is extensive across the Indian subcontinent, development of AI-enabled computational applications on these platforms would provide a valuable, precision public health tool for better management of lung disease epidemic alleviating a significant burden on the national exchequer.

3 Motivation Behind Building This Tool

This work is motivated by the following objective.

> To develop Digital Chest Radiograph Segregation system which will segregate the Covid-19, Pneumonia, and Tuberculosis associated lung diseases in both tertiary care settings and extended community.

Our fruganomic data intensive AI-enabled tool will not only facilitate the same which will aim at not only resolving the dogma of missed and misdiagnosis of lung diseases such as tuberculosis or pneumonia at tertiary care centres and extended community, but also individualize the risk assessment of patients with suspected myocardial infarction or to categorize patients into low- or high-risk groups.

In recent years, various computer-based tools have been developed which can be reliably used for computational disease tagging purposes. Healthcare professionals with the help of such tools can accurately computationally tag different disease conditions within a short time.

In the past, people have developed AI models which use X-ray images to predict COVID-19 in the patients [18]. In this paper, we have explored the possibility to predict the lung ailment by applying Earth Mover's Distance algorithm to the X-ray images of the patients. EMD mimics the human perception of texture similarity. EMD outperforms many other texture similarity measures when used for texture classification and segmentation [19]. All the previously developed models have chosen the process-based approach which has inherent instabilities built in. We have used a programmatic approach to detect the diseases using chest radiographs (lung X-ray) of patients. Our programmatic approach has the advantage of being free from any dynamic instability as compared to tools like deep learning.

The lung X-ray images of patients suffering from pneumonia, TB, COVID-19 and healthy persons were pooled together from various datasets [5, 18, 19]. There are 25 random images selected from each dataset. The focus of our current study is restricted to understanding the patterns and processes associated incremental burden of lung diseases in the Indian populace.

4 Earth Mover's Distance

The ground distance between two single perceptual features can often be found by psychophysical experiments. For example, perceptual colour spaces were devised in which the Euclidean distance between two single colours approximately matches human perception of their difference. This becomes more complicated when sets of features, rather than single colours, are being compared. This correspondence is key to a perceptually natural definition of the distances between sets of features. This observation led to distance measures based on bipartite graph matching [20, 21] defined as the minimum cost of matching elements between the two histograms.

Earth Mover's Distance (EMD) is a method to calculate the disparity between two multi-dimensional distribution in some space where a distance magnitude between single ones (ground distance) is given. Suppose the two distributions are there, one can be considered as the area with the mass of earth, and the other as a collection of holes in that same area. Then, the EMD is the measure of the least amount of work required to fill the holes with earth.

Here the unit of work is the force needed in transporting unit earth by a unit of ground distance. Hence, it can also be defined as the minimum cost that must be provided to convert one histogram into other. Measuring of EMD is based on a solution of transportation problem. For finding mathematical representation, firstly we formalised it as the following linear programming problem:

Let X be the first signature with n clusters, X_i s the cluster representative and W_{xi} is the weight of cluster.

Let Y be the second signature with m clusters,

Y_i is the cluster representative and W_{yi} is the weight of cluster.

Let D be the ground distance matrix, D_{ij} is the ground distance between clusters x_i and Y_j.

Let F be the flow matrix and F_{ij} is the between X_i and Y_j.

Then,

$$X = (X_1, W_{X1}), (X_2, W_{X2}), (X_3, W_{X3}) \dots (X_n, W_{Xn}) \tag{1}$$

$$Y = (Y_1, W_{Y1}), (Y_2, W_{Y2}), (W_3, Y_{Y3}) \dots \left(Y_n, W_{yn}\right) \tag{2}$$

$$D = [D_{ij}] \tag{3}$$

$$F = [F_{ij}] \tag{4}$$

Now, the WORK $(X, Y, F) = \sum_{i=1}^{n} \sum_{i=1}^{m} f_{ij} D_{ij}$.

Subject to constraints:

(i) $f_{ij} \geq 0$ where $0 \leq i \leq n, 0 \leq j \leq m$

(ii) $\sum_{i=1}^{m} f_{ij} \leq w_{ij}$ where $0 \leq i \leq n$

(iii) $\sum_{i=1}^{n} f_{ij} \leq w_{yj}$ where $0 \leq j \leq m$

(iv) $\sum_{i=1}^{m} \sum_{i=1}^{n} f_{ij} = \min \sum_{i=1}^{n} w_{xi}, \sum_{j=1}^{m} w_{yj}$

The constraint (i) enables mass moving from X to Y. (ii) and (iii) restricts the amount of mass that can be sent by the clusters in X to their weights and the clusters in Y to receive no more mass than their weights. And (iv) forces to move the maximum amount of mass possible. It is also known as the total flow. Once we solve the transportation problem, we will get the optimal flow F. Now the Earth Mover's Distance is defined as the work normalised by the total flow:

$$\text{EMD}(X, Y) = \sum_{i=1}^{n} \sum_{j=1}^{m} F_{ij} D_{(ij)} - \sum_{i=1}^{n} \sum_{j=1}^{m} F_{ij}$$

Metric is a function which defines a distance between each pair of elements in the set. Here usage of distance is not only limited to knowing how far two objects are from each other. It is also used for calculating the similarity between two datasets, for example, watermelon is similar to orange because it is spherical; but at a same time, watermelon is also similar to guava because they are same in colour. Both shape and colour are characteristics of the objects, and they can be expressed in number form, and the difference between them will be the "distance" between them.

There are multiple ways to calculate the similarity in datasets. Following metrics are available in EMDIST library.

Euclidean Distance

The Euclidean distance between two points in n-dimensional space is the straight length of a line connecting the two points and is the most common and a simple way of calculating the distance. In n-dimensional Euclidean space, we get for the distance $d(A, B)$ between two points $A = (A_1, A_2, \ldots, A_n)$ and B $= (B_1, B_2, \ldots, B_n)$

$$D_{\text{Eucl}}(A, B) = \sqrt{(A_1 - B_1)^2 + (A_2 - B_2)^2 + \cdots + (A_n - B_n)^2}$$

$$D_{\text{Eucl}}(A, B) = \sqrt{\sum_{i=1}^{n}(A_i - B_i)^2}$$

Manhattan Distance

The Euclidean distance formula is usually referred for theoretical distance measurements. But, in actual scenario, it is not feasible to in move in straight line from one point to other point. The Manhattan distance formula is much more useful because it allows calculating the distance between two data points on a uniform grid.

A generalized formula for the Manhattan distance is in n-dimensional vector space between two points, A = $(A_1, A_2, \ldots, A_n)$ and $B = (B_1, B_2, \ldots, B_n)$:

$$D_{\text{Manh}}(A, B) = \sum_{i=1}^{n}|A_i - B_i|$$

Dataset

X-ray images of subjects of Indian origin were accessed from multiple sources. The data of the patients was anonymised before subjecting it to algorithms for image processing (Table 1).

Table 1 Images of chest radiograph (X-ray) typifying a variety of images along with their sources

No of Images	Patient Condition	Dataset Source
25	Normal	https://www.kaggle.com/parthachakraborty/pneumonia-chest-x-ray, https://www.kaggle.com/raddar/chest-xrays-tuberculosis-from-india
25	Pneumonia diagnosed	https://www.kaggle.com/parthachakraborty/pneumonia-chest-x-ray
25	Tuberculosis diagnosed	https://www.kaggle.com/raddar/chest-xrays-tuberculosis-from-india
25	Covid-19 diagnosed	Dr. Uday Shankar. Rajiv Gandhi super speciality Hospital, Tahirpur

5 Implementation Environment and Results and Discussion

Open-source software "*R*" version 3.6.3 has been used as scientific programming environment as it is equipped with highly advanced image analysis tools and techniques with magnificent visualizations. For image processing and EMD calculations, imager and EMDIST libraries were installed, respectively, in the R studio. The computing machine used for the image analysis has the following features.

Processor: Intel® Core™ i7-5500U CPU @ 2.40 GHz × 4.

Operating System (OS) Name: Ubuntu 20.04.1 LTS.

OS Type:—64-bit.

Graphics: NVIDIA Corporation GM108M [GeForce 840 M].

Memory: 7.7 GB.

To maintain the homogeneity, images were converted to the grayscale for uniform basis and scaled down to 64 × 64 pixel due to limited computational resource availability. It should be noted that bigger the pixel size, larger the time required to process the images. EMD can be calculated using Euclidean or Manhattan approach. Here, we have used Euclidean approach for the analysis. The pairwise comparison of all images using EMD has been tabulated in the LibreOffice calc which gives us an interesting comparison about relative impact of COVID-19, tuberculosis and pneumonia.

Results and discussion

The pairwise EMD comparison for each image gave us EMD values. The image, which has least average EMD among the respective group, is considered as representative image of the particular group. The least EMD of each group is presented in Table 2. From the table, we can infer that different diseases are affecting the lungs at a different scale. Further, we compared the healthy representative image of with other group images. It has been noticed that comparison is effectively showing distinct values.

The computational analysis of greyscale chest/lung X-ray images of lung shows EMD values of X-rays of patients afflicted with tuberculosis are maximum followed by EMD values of X-ray of patients affected with pneumonia, whilst the least EMD values are observed in the X-ray images of lungs of COVID-19 patients. We also observed that the EMD comparison value of COVID-19 image is similar to the EMD value of pneumonia which means pneumonia and COVID-19 cause almost the same level of damage to the lungs.

Table 2 EMD values of chest radiographs (X-rays) different types of lung diseases

X-ray image group	Least EMD in each cluster	Total Images in cluster
Healthy images	1377	25
COVID-19 positive	1523	25
Pneumonia positive	1666.2	25
TB positive	1833.05	25

Pairwise EMD comparison of images in the group is shown in Figs. 1, 2, 3 and 4.

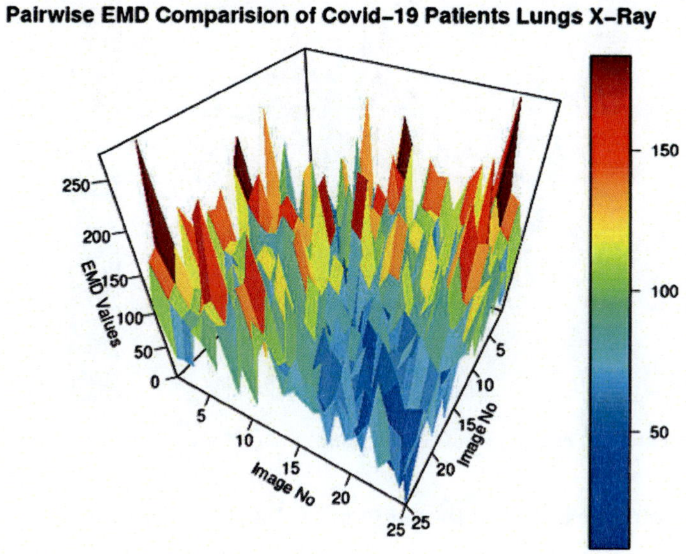

Fig. 1 Pairwise EMD comparison of COVID-19 patients' lung X-ray

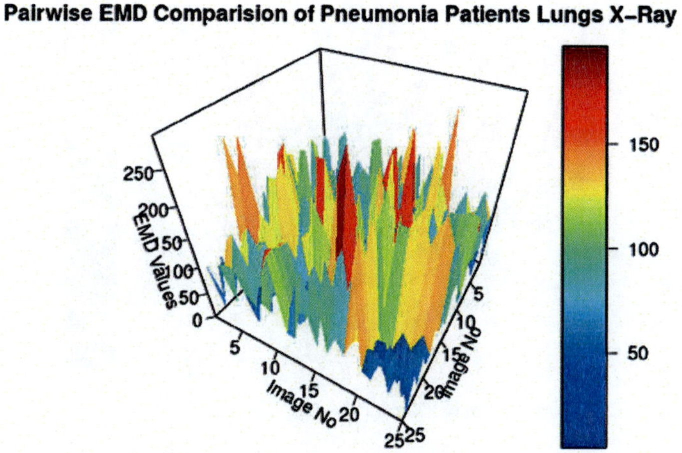

Fig. 2 Pairwise EMD comparison of pneumonia patients' lung X-ray

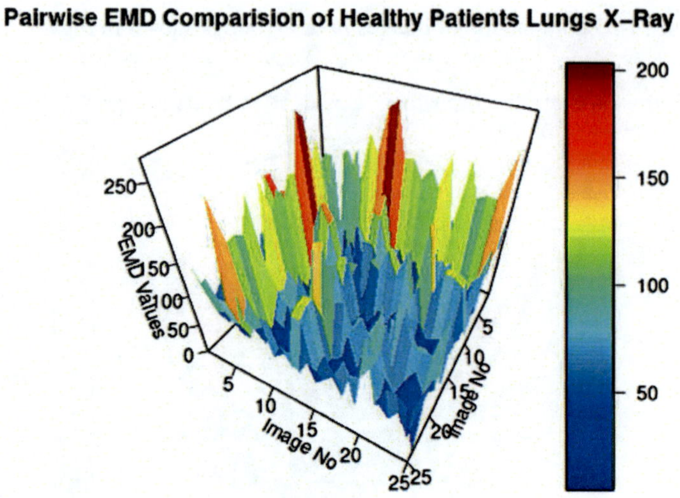

Fig. 3 Pairwise EMD comparison of healthy patients' lung X-ray

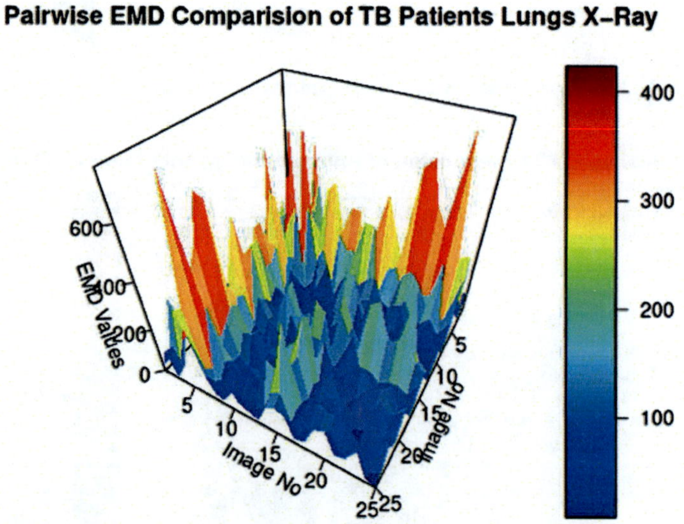

Fig. 4 Pairwise EMD comparison of TB patients' lung X-ray

6 Conclusion

We take our breathing and our respiratory health for granted, but the lung is a vital organ that is vulnerable to airborne infection and injury. Despite the fact that respiratory diseases are leading causes of death and disability in the world, it was only after the emergence of COVID-19 pandemic that the field of both malignant and non-malignant lung infections has attracted considerable attention. About 65 million people suffer from chronic obstructive pulmonary disease (COPD) and 3 million die from it each year, making it the third leading cause of death worldwide. About 334 million people suffer from asthma, the most common chronic disease of childhood affecting 14% of all children globally. Pneumonia kills millions of people annually and is a leading cause of death among children under 5 years old. Over 10 million people develop tuberculosis (TB) and 1.4 million die from it each year, making it the most common lethal infectious disease. Lung cancer kills 1.6 million people each year and is the most-deadly cancer. Globally, 4 million people die prematurely from chronic respiratory disease. At least 2 billion people are exposed to indoor toxic smoke, 1 billion inhale outdoor pollutant air, and 1 billion are exposed to tobacco smoke. The truth is that many of us are naïve to these stark realities.

Fortunately, most of the non-malignant respiratory diseases are preventable by improving the quality of the air. Controlling unhealthy air in the workplace along with strengthening of immunization programmes in low-and middle-income countries can prevent several types of pneumonia. The use of mobile health platforms to disseminate valuable health literacy modules on "Lung Health" and its statutory impact on the cardiometabolic system will go a long way in improving the respiratory health of the Indian populace.

The 2030 sustainable development agenda was adopted by world leaders in 2015 at a historic UN Summit in New York and came into force on 1 January 2016. The very ambitious agenda is a plan of action to achieve 17 Sustainable Development Goals (SDGs) and 169 targets by the year 2030, which include the economic, social and environmental dimensions of sustainable development. The goal 3 of SDGs is focussed to ensure healthy lives and promote well-being for all at all ages as one of the most important goals with active participation from all the stakeholders of the healthcare ecosystems. Improved health will bring people out of poverty and contribute substantially to sustainable development.

Non-Communicable Diseases, including cardiovascular diseases, cancers, chronic respiratory diseases and diabetes, are the biggest killers today. A fact affirmed by the poor clinical prognosis of patients afflicted with COVID-19 having underlying risk factors of NCDs. In fact, one of the ambitious goals of SDGs is to significantly reduce the mortality due to the ravages of NCDs by 30% by the year 2030. Although India has 18% of the world's population with a rising incidence of chronic respiratory diseases, it is endowed with a resource limited healthcare ecosystem incapable of tackling the healthcare needs of its populace in an equitable manner. This is further compounded by the fact that evidences from the hospital-based registries are not adequately complemented by evidences from the community at large necessitating the need to develop disruptive and cutting-edge interventions capable of collecting evidences in an affordable and accessible manner. To this, we believe that our AI-enabled algorithm proposed in the current study will not only predict the incidence of lung infections but also effectively triage the patients in the order of their severity.

Data analytical processes often require large amount of data for showing accurate analyses. Our preliminary analyses suggest that EMD can be used for tagging the chest radiographs X-ray images with diseases in more efficient way. We are trying to acquire bigger datasets for studies so the accuracy level of our tool will increase, and tool can also be used for tagging the untagged X-ray images. Complementing the chest radiographs (X-rays) with chest computed tomography (CT) scans and molecular biomarkers of tissue hypoxia at both angiogenic and fibrotic phases of the disease will significantly contribute towards alleviating the burden of respiratory diseases in a LMIC like India endowed with divergent genetic base as well as varied geological relief structures presenting a unique landscape of disease burden.

We believe that development of indigenous and fruganomic tools towards collection of niche specific datasets will significantly aid in the development of disease surveillance and forecasting strategies at both regional and national levels aimed at controlling the menace of cardio-pulmonary diseases across the Indian sub-continent.

References

1. Desikan, P.: Sputum smear microscopy in tuberculosis: Is it still relevant? Indian J. Med. Res. **137**(3), 442–444 (2013)
2. Leung, C.: Clinical features of deaths in the novel coronavirus epidemic in China. Rev. Med. Virol. e2103 (2020)
3. Chen, N., Zhou, M., Dong, X., Qu, J., Gong, F., Han, Y., et al.: Epidemiological and clinical characteristics of 99 cases of 2019 novel coronavirus pneumonia in Wuhan, China: a descriptive study. Lancet **395**, 507–513 (2020)
4. GBD 2015: Chronic Respiratory Disease Collaborators. Global, regional, and national deaths, prevalence, disability-adjusted life years, and years lived with disability for chronic obstructive pulmonary disease and asthma, 1990–2015: a systematic analysis for the Global Burden of Disease Study 2015. Lancet Respir. Med. **5** 691–706 (2017)
5. Cohen, A.J., Brauer, M., Burnett, R., et al.: Estimates and 25-year trends of the global burden of disease attributable to ambient air pollution: an analysis of data from the Global Burden of Diseases Study 2015. Lancet **389**, 1907–1918 (2017)
6. Landrigan, P.J., Fuller, R., Acosta, N.J.R., et al.: The Lancet Commission on pollution and health. Lancet **391**, 462–512 (2018)
7. Roman V.Y.: Unpredictability of AI. Arxiv-1905.13053 (2019)
8. Roman V.Y.: Unexplainability and Incomprehensibility of Artificial Intelligence. Arxiv-1907.03869 (2019)
9. Sohl-Dickstein, J., Weiss, E., Maheswaranathan, N., Ganguli, S.: Deep Unsupervised Learning using Nonequilibrium Thermodynamics. Arxiv-1503.03585 (2018)
10. Li, H.: Analysis on the Nonlinear Dynamics of Deep Neural Networks: Topo- logical Entropy and Chaos. Arxiv-1804.03987 (2018)
11. Thompson, N.C., Greenewald, K., Lee, K., Manso, G.F.: The Computational Limits of Deep Learning. Arxiv-2007.05558 (2020)
12. https://venturebeat.com/2020/07/15/mit-researchers-warn-that-deep-learning-is-approaching-computational-limits/. Published on July 15, 2020. Accessed on 14 Nov 2020
13. England, N.H.S.: Diagnostic Imaging Dataset Statistical Release. Department of Health, London (2017)
14. Moncada, D.C., Rueda, Z.V., Macías, A., Suárez, T., Ortega, H., Vélez, L.A.: Reading and interpretation of chest X-ray in adults with community-acquired pneumonia. Brazilian J. Infect. Dis. **15**(6), 540–546 (2011)
15. Hopstaken, R.M., Witbraad, T., Van Engelshoven, J.M.A., Dinant, G.J.: Inter-observer variation in the interpretation of chest radiographs for pneumonia in community-acquired lower respiratory tract infections. Clin. Radiol. **59**(8), 743–752 (2004)

16. Moifo, B., Pefura-Yone, E.W., Nguefack-Tsague, G., Gharingam, M.L., Tapouh, J.R.M., Kengne, A.P., Amvene, S.N.O.: Inter-observer variability in the detection and interpretation of chest x-ray anomalies in adults in an endemic tuberculosis area. Open J. Med. Imaging, **5**(03), 143 (2015)
17. Sakurada, S., Hang, N.T., Ishizuka, N., Toyota, E., Hung, L.D., Chuc, P.T., Lien, L.T., Thuong, P.H., Bich P.T.N., Keicho, N., et al.: Inter-rater agreement in the assessment of abnormal chest X-ray findings for tuberculosis between two Asian countries. BMC infect. Dis. **12**(1), 1–8 (2012).
18. India State-Level Disease Burden Initiative CRD Collaborators: The burden of chronic respiratory diseases and their heterogeneity across the states of India: the Global Burden of Disease Study 1990–2016. Lancet Glob Health. **6**(12), e1363–e1374 (2018). https://doi.org/10.1016/S2214-109X(18)30409-1
19. World Health Organization. Naming the coronavirus disease (COVID-19) and the virus that causes it. Available from: https://www.who.int/emergencies/diseases/novel-coronavirus2019/technical-guidance/naming-the-coronavirus-disease-(covid-2019)-and-the-virus-that-causes-it. Accessed: 15 Nov 2020
20. Peleg, S., Werman, M., Rom, H.: A unified approach to the change of resolution: Space and gray-level. IEEE Trans. Pattern Anal. Mach. Intell. **11**, 739–742 (1989)
21. Zikan, K.: The Theory and Applications of Algebraic Metric-Spaces. Ph.D. Thesis, Stanford University (1990)

Artificial Intelligence in Diagnosis of Polycystic Ovarian Syndrome

Debasmita Ghosh Roy and P. A. Alvi

Abstract The polycystic ovarian syndrome (PCOS) is one of the most developing endocrine disorders, which is a frequent cause of infertility. Around 8–10% of women globally are affected by this syndrome in their fertility period. In light of the heterogeneity of symptoms, the diagnosis of PCOS is a challenging task for clinicians. Artificial intelligence (AI) techniques are now an emerging methodology as a decision support system in the healthcare industry. In this paper, different AI techniques are used for the prognosis and diagnosis of PCOS. The dataset comprises of metabolic and biochemical features of 541 females under 47 years of age having a positive and negative class labels. The reduced dataset is used to diagnose PCOS and assess the prediction efficacy. Finally, the result showed that the SVM outperforms other AI techniques with the prediction efficacy of 96.92%.

Keywords Polycystic ovarian syndrome (PCOS) · Artificial intelligence (AI) · Correlation · Efficacy

1 Introduction

The technological advancement and human hands should be combined together effectively for the achievement of a good medical care, and AI has the potential to reshape health care completely [1]. The healthcare framework is used to handle a considerable measure of information, which is exceptionally hard to examine through the conventional approach. The accomplishment of AI in medical services guarantees prognosis, determination, and treatment of ailments. Machines will not supplant human clinicians, yet AI-based predictive models unquestionably help clinicians settle on better clinical decisions [2]. Guided by relevant clinical inquiries, powerful AI methods can clinically open significant data covered up in the vast measure of information, helping in clinical decision making. AI can utilize advanced algorithms

D. G. Roy (✉)
School of Automation, Banasthali Vidyapith, Jaipur, Rajasthan, India

P. A. Alvi
Department of Physics, Banasthali Vidyapith, Jaipur, Rajasthan, India

H. K. D. Sarma et al. (eds.), *Contemporary Issues in Communication, Cloud and Big Data Analytics*, Lecture Notes in Networks and Systems 281,
https://doi.org/10.1007/978-981-16-4244-9_37

to learn features from a huge data set for building a legitimate clinical dynamic framework. Consequently, feature selection is one of the prime assignments in AI that can improve clinical practices to educate fair patient consideration. The AI framework is also utilized to extract valuable data from an enormous patient population to make continuous inductions for health hazard alert and result expectations. Presently, AI research is expanding into several subdivisions, but among them, ML (subset of AI) is becoming increasingly prevalent in the healthcare industry. Optimists predict that the early stage of the disease can be diagnosed and can be treated effectively by AI and ML. One of the most prime applications of ML in healthcare industry is to develop predictive models for disease prediction and deploy them for maximum benefits in real clinical scenarios. Data and computation algorithms are a keystone for designing an accurate and useful predictive model. The computational algorithms are used to compute data, learn the underlying pattern, and offer knowledge utilizing in which decision and forecast about the real-world events can be made [3].

PCOS is a multifaceted disease that can significantly influence the reproductive health of women. It is the leading cause of ovulatory dysfunction, and 90% of infertility cases are identified with ovulatory dysfunction [4]. As per the National Institute of Health Office Disease, around 5 million women are influenced by PCOS around the world. It is more common in Indian women than white women, where symptoms are gentle [5]. The primary assessment of PCOS is a challenging task for clinicians due to varied symptoms. Thus, a physical evaluation is required for the analysis of the disease. These all are cost-inadequate and tedious methodology, which may cause mental stress to the patients. PCOS is an elaborate clinical presentation and manifestation of this disease, including menstrual irregularities, hirsutism, obesity, acne, sleep apnea, and baldness [6]. Numerous examinations featuring the effect of identity in PCOS have thought about the disorder's metabolic perspectives, including insulin resistance, glucose intolerance, lipid abnormities, coronary artery disease, type 2 diabetes, and malignancy. Besides, PCOS's effect is not limited in the reproductive stage, yet it is a drawn-out health hazard related to this condition. Due to the ill-defined symptoms, PCOS needs more attention to prevent long-term health hazards. Additionally, there is a necessity to build awareness concerning disease indications [7].

To address this issue, authors proposed a predictive model with a minimal number of features that helps to diagnose the PCOS at an early stage. The objective of the presented research work is to reduce the testing expenses for early stage diagnosis of PCOS and increase the social awareness and risk association related to this syndrome. The essential contribution of this paper is as follows:

- Diagnosis of PCOS with eight metabolic and biochemical information which helps to reduce the testing expenses.
- Among eight features, five features can create social awareness among adolescent girls having possibility of the syndrome.

Recently, the I-Hope model for PCOS detection is presented in [8], where the objective is to design the early prediction and detection of PCOS with a minimum set of potential markers. Also, the authors have identified the optimized features contributing toward PCOS and infertility. In this study, six ML algorithms such as linear regression (LR), k-nearest neighbors (K-NN), classification and regression tree (CART), random forest (RF), naïve Bayes (NB), and support vector machine (SVM) are used where RF gives an optimal 89% accuracy with optimized features. The study in [9] has analyzed PCOS genes' properties for the prediction of PCOS candidates. Three ML classifiers such as SVM, K-NN, and decision tree (DT) are used to identify PCOS genes. The optimal accuracy of 81% is obtained from linear kernel-based SVM. In [10], Gabor wavelet-based feature extraction method and a competitive neural network (CNN) technique have been used to determine the PCOS from ultrasound images. The accuracy achieved using CNN is 80.84%. In [11], three classifiers, such as neural network, DT, and NB, are utilized to classify PCOS, and a comparison of their performance in terms of accuracy was made. The result showed that NB outperforms other classification algorithms with 97.65% accuracy. Cheng et al. used the rule-based classifier and gradient boosted tree classifier to determine PCOS from ultrasound images. The rule-based classifier's accuracy of 97.6% is higher than the gradient boosted tree [12]. The study in [13] has implemented a hybrid algorithm, namely a neural fuzzy rough set (NFRS) and artificial neural network (ANN), to predict PCOS. A comparative study has been done using five feature selection methods such as information gain, neuro-fuzzy rough set, gain ratio, principal component analysis, and correlation. Three classification algorithms such as ANN, CART, and NB are used to classify PCOS with the above-mentioned feature reduction techniques. The obtained result revealed that ANN and NFRS outperform others. In [14], a study was performed for filtering the attributes of PCOS by two feature selection techniques such as Neuro-fuzzy rough set (NFRS) and information gain subset (IGS). ID3 and J48 are used to estimate the performance accuracy of the techniques mentioned above for PCOS classification. The two classifiers achieve an accuracy of 93.7% with the utilization of NFRS method. Joham et al. performed a cross-sectional analysis of a longitudinal cohort using a logistic regression technique for figuring out the factors associated with infertility and fertility treatment [15]. In [16], the primary screening of PCOS has been done by LR and Bayesian classifiers. The Bayesian classifier outperforms LR with 93.3% accuracy. The authors also identified the significant features for PCOS using the t test method.

The remaining part of the paper is organized as follows: The dataset and model description are discussed in Sect. 2, followed by Sect. 3 containing the design and development of models utilized in this article. The results are being provided in Sect. 4, and finally, the relevant concluding remark is made in Sect. 5.

2 Dataset and Model Description

2.1 Dataset

The dataset has been acquired from the Kaggle ML Repository, including patient's elementary information such as patient file number, metabolic, and biochemical information [17]. Total 541 samples are used with 39 features, and the output label is assigned as PCOS and non-PCOS. Out of 541 samples, 364 samples were non-PCOS and 177 simples were PCOS.

2.2 Feature Description

The features that are used in this study are age (20–48 yrs), weight (31–108 kg), height (137–180 cm), BMI (12–38.9), blood group ($A^+ = 11$, $A^- = 12$, $B^+ = 13$, $B^- = 14$, $O^+ = 15$, $O^- = 16$, $AB^+ = 17$, $AB^- = 18$), pulse rate (18–82 bpm), respiratory rate (16–28/min), Hb (8.5–14.8 g/dl), cycle pattern (regular = 2, irregular = 4), cycle length (0–12 days), marital status in yrs (0–30), pregnant (yes/no), no. of abortions (0–5), hip (26–48 in.), waist (24–47 in.), waist-to-hip ratio (0.76–0.98), hair loss (yes/no), weight gain (yes/no), Vit-D3 (0–6.01 K mg/mL), progesterone (PRG) hormone level, thyroid stimulating hormone (TSH), level (0.04–65), follicle stimulating hormone (FSH) level (0.21–5.05 k mIU/mL), luteinizing hormone (LH) level (0.02–2.02 K mIU/mL), FSH-to-LH ratio (0–1.37 k mIU/mL), skin darkening (yes/no), anti-mullerian hormone level (AMH), prolactin hormone level (0.05–85), random glucose test (60–350 mg/dl), hair growth (yes/no), pimples (yes/no), fast food intake (yes/no), regular exercise (yes/no), BP systolic, BP diastolic, follicle number in right ovary (0–20) and left ovary (0–24), average follicular size in left and right ovary (0–24), endometrium thickness (0–18 mm), and output label is PCOS (yes/no) where yes = 1 and no = 0.

2.3 Syndrome Level Prediction

The study aims at diagnosis and prognosis of PCOS from clinical and metabolic features. Moreover, designing a good prediction model is essential for PCOS diagnosis by supervised learning techniques of machine learning. The model is trained with input variables as a number of features, $X = (X_1, X_2, X_3, \ldots X_n)$, including the patient's personal information such as age, marital status, and metabolic and biochemical factors. The output label represented as Y shows whether the patient has syndrome possibility or not with allotted value $Y = \{1, 0\}$, where 1 signifies that the patient has PCOS and 0 shows that the patient does not have PCOS. Different ML algorithms, such as support vector machine (SVM), decision tree (DT), and naïve

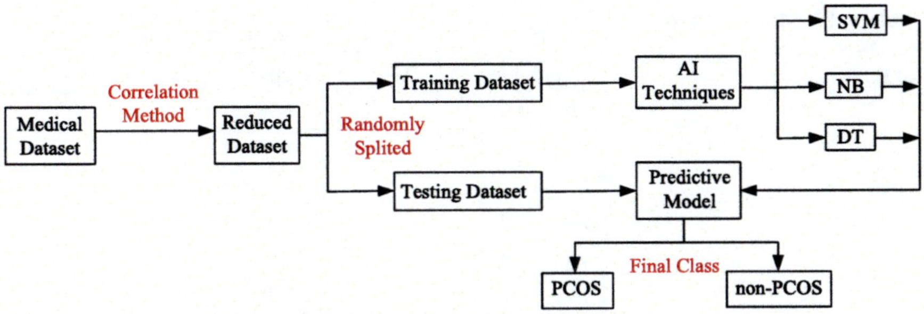

Fig. 1 Three AI techniques used for PCOS diagnosis

Bayes (NB) are used to implement the prediction model. Initially, for training the model, the trained data and test data are separated in the ratio of 7:3. Cross-validation technique is used to estimate the efficacy of ML algorithms on a constrained dataset. This process includes the data set randomly divided into K partitions, or folds, of approximately equal size. The partition is treated as a validation set, and the strategy is applied on the remaining $K - 1$ partitions of the model. Here, fivefold cross-validation is used for testing of classification accuracy [18]. Python language is utilized for the execution of the AI-based algorithms. The basic framework of the predictive model for PCOS diagnosis is shown in Fig. 1.

2.4 *Evaluation Parameters*

In case of classification problems, performance is measured using the computation of accuracy. So, accuracy has been considered for the determination of classifier performance across the data set.

3 Methodology for Model Implimentation

3.1 *Feature Extraction*

The main objective of feature extraction is to describe the dataset in a finest way with a smaller number of features than the original feature dataset. By utilizing this approach, the removal of unnecessary, undesirable, and irrelevant data is possible. In real-world health applications, the dataset is a combination of unimportant, missing, and noisy features. At that point, feature extraction technique is used for removing irrelevant data from the diseased dataset which accelerates the computation time and provides accurate outcome. Here, the correlation technique is used for extracting exciting features that are relevant to PCOS. Correlation is an estimate of a monotonic

Table 1 Significant features with positive coefficient values

Features	Coefficient values
Hemoglobin	0.8804
Follicle no. (L)	0.648223
Follicle no. (R)	0.603109
Skin darkening	0.475283
Hair growth	0.464245
Weight gain	0.440488
Cycle	0.401165
Fast food	0.375389

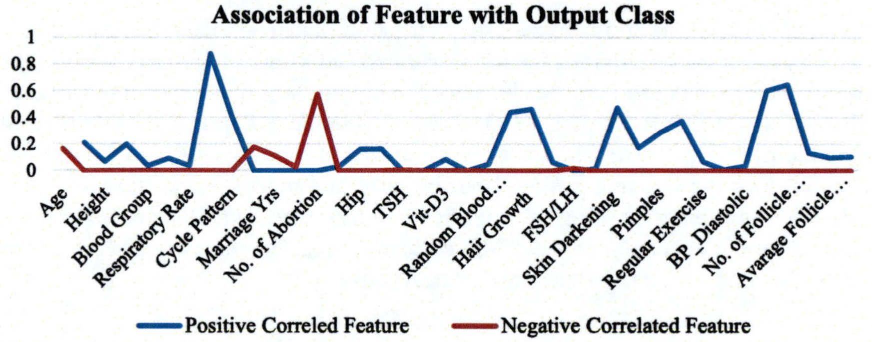

Fig. 2 Positive correlation and negative correlation feature with the output class

association between two parameters. In correlated data, the change in magnitude of one parameter is associated with a change in the parameter's magnitude, either positively or negatively. The correlation coefficient (cc) is scaled so that the range lies between −1 and +1. Various approaches have been suggested to translate the correlation coefficient into a description, such as weak, moderate, and strong [19]. In this manner, eight potential features such as hemoglobin, number of follicles in the left and right ovaries, skin darkening, hair growth, weight gain, cycle, and fast food are utilized to create a reduced dataset, which is listed in Table 1. The association of each feature with output class is graphically represented in Fig. 2.

3.2 Classification Techniques

(1) Support vector machine: SVM is a discriminative classifier used for classification and regression. It is a linear classifier that makes a hyperplane to order all inputs to a high measurement. The closest values to the classification

margin are known as support vectors. The prime goal of SVM is to maximize the limit among hyperplane and support vectors [20]. The algorithm of SVM uses various mathematical functions which are represented as a kernel. The kernel is used to gather information which is used as input and also converts the data according to desired framework. The individual SVM algorithm utilizes unique kernel function. These kernels are of various types, such as linear, nonlinear, sigmoid, polynomial, and radial basis function [21]. In this paper, a linear kernel is used for model designing. The output label is categorized into two classes, and a best-fit hyperplane is made, which isolates the output label. The algorithm steps are as follows:

- $Y = \{0, 1\}$
- The hypothetical SVM model equation is

$$h_{w,b} = g\left(w^T + b\right)$$

- Here, $g(z) = 1,\ \text{if}\, z \geq 0,\ g(z) = 0$ otherwise.
  Assume the vector w will always be normal to the hyperplane. The mathematical expression of the hyperplane is given by

$$w^T X + b = 0$$

The final optimization problem SVM solves to fit the finest variable given that
$y_i\left(w^T X_i + b\right) \geq 1\,\forall X_i$ is:

$$\min \frac{1}{2}||w||$$

For each vector X_i such that

- X_i belongs to class 1, then

$$w^T + X_i \geq 0$$

- X_i belongs to class 0, then

$$w^T + X_i \leq 0$$

X_i belongs to hyperplane (decision boundary), then

$$w^T + X_i = 0$$

(2) Decision tree: The classification algorithm of DT is clear, straightforward, and easy to convert into specific classification rules. Thus, this classification algorithm is broadly studied and applied. The structure of DT is like a flowchart,

utilizing a top-down recursive way. The tree's internal nodes are compared by attribute value, and the branches are judged under this node as per distinctive feature value. The ultimate conclusion can be reached from the leaf node [22]. The working process of the DT is as follows. Suppose the given dataset is

- $X = \{X_1, X_2, \ldots X_n\}$
- $Y = \{1, 0\}$
- Compute the total positive (P) and negative (N) instances.
- Calculate the entropy $E(S)$ for given dataset.

$$E(s) = \frac{-P}{(P+N)} \log_2\left(\frac{P}{P+N}\right) - \frac{N}{(P+N)} \log 2\left(\frac{N}{P+N}\right)$$

- Calculate entropy for each feature denoted as entropy (A).
- Take average information entropy for the present feature.

$$I(\text{Feature}) = \sum \frac{P_i + N_i}{P+N} \text{Entropy}(A)$$

- Calculate the information gain for the present feature.

$$\text{Gain} = \text{Entropy}(S) - I(\text{feature}).$$

- Choose the highest gain feature which is considered as a root node.
- Repeat these above steps until the last node should be the leaf node.

(3) Naïve Bayes: NB algorithm is one of the expected effective techniques in the field of ML. It is commonly known as a simple probabilistic classifier and accepts features' autonomy in a given class. This supposition can incredibly decrease the complexity of the improvement of the classifier. It is utilized in binary classes or multi-class classification problems [23]. Let us assume the input feature sequence as $X = (X_1, X_2, \ldots X_n)$ and the output label as $Y = \{0, 1\}$. The Bayesian rule for probability is

- $P(Y|X) = \frac{P(X|Y)P(Y)}{P(X)}$
  $P(Y|X)$ is discriminative, whereas $P(X|Y)$ is a generative and $P(X)$ is posterior. For an input X, compute the largest one from L probabilities output by a discriminating probabilistic classifier $P(Y_1|X) \ldots P(Y_L|X)$.
- Assign X to label Y^* if $P(Y^*|X)$ is the largest.
- Generative classification with MAP rule.
- Apply Bayesian rule to convert them into posterior probabilities

$$P(Y_i|X) = \frac{P(X|Y_i)P(Y_i)}{P(X)} \propto P(X|Y_i)P(Y_i).$$

Here, $P(X)$ indicates a common factor for all L probabilities. At that point, NB theorem is applied on given dataset and reduced data set having eight features. The decision making with MAP rule is as follows:

$$X = (X_1, X_2, \ldots, X_8)$$

$$P(\text{Yes}|X) \approx [P(F1|\text{Yes}), \ldots, P(F8|\text{Yes})]P(\text{PCOS} = \text{Yes})$$

$$P(\text{No}|X) \approx \left[P(F1|\text{No}) \ldots P(F8|\text{yes})\right]P(\text{PCOS} = \text{No})$$

A similar rule is also applicable for the original dataset.

4 Result Analysis

In this study, correlation method is applied on given dataset for identifying the most potential features with the help of cc. Generally, the cc value of highly influential features is kept between 0.9 and 1.00. In this dataset, the maximum positive cc value lies between 0.003 and 0.88. Thus, the threshold value is set as 0.3, and the coefficient value less than 0.3 indicates a negligible correlation.

After that, three ML techniques are applied to original and reduced data set. Their efficacy level evaluates the designed predictive model's performance, and the optimal efficacy is achieved from the SVM classifier with a linear kernel. The predictive model's accuracy using all classifiers with a different dataset size is listed in Table 2. Also, a comparison study of all classifiers' performance with different data size is depicted in Fig. 3.

Several articles have been mentioned about the utilizing of various methods for predicting PCOS, but optimal efficacy is obtained in [15, 16] where features are lifestyle factors and pelvic images. So, the direct comparison is meaningless in the articles as mentioned above. However, in [12], the obtained efficacy is 89%, where features are metabolic and clinical factors with 541 instances and the accuracy is lesser than to our result. Therefore, the SVM model is better assistive diagnostic aid for PCOS detection. Likewise, it is seen that for the prediction of syndrome probability just eight features are required rather than thirty-nine features. Thus, it can be said that PCOS is a disease with varied indications, whose exact risk cannot be anticipated by the usage of these essential features, rather the primary assessment can be easily achieved by these essential features.

Table 2 Efficacy of AI models

Algorithms	Original dataset (%)	Reduced dataset (%)
SVM	84	96.92
DT	80	89.52
NB	77	94.36

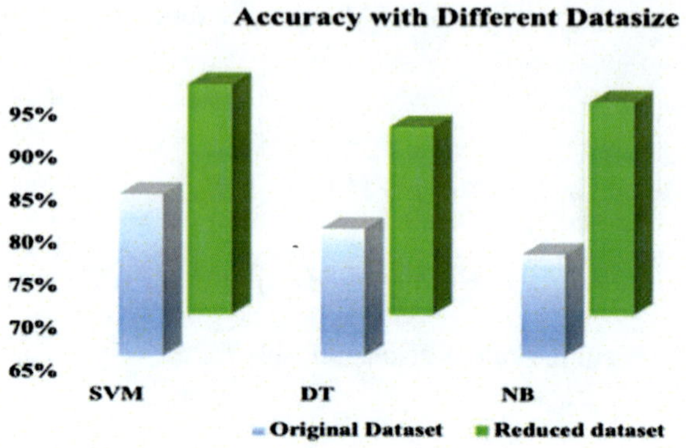

Fig. 3 Graphical representation of classifier performance with different data sizes

5 Conclusion

In present decades, PCOS is not just seen in women going through fertility age yet also found in adolescent girls. It has a significant risk factor for metabolic disorder and, thus, responsible for developing of type 2 diabetes. There are few potential complications which incorporate microalbuminuria, hypertension, hypercholesterolemia, and dyslipidemia that are fundamentally common with type 2 diabetics. These factors directly affect the human cardiac system, and it is a prime cause of mortality. Consequently, recognizing the girls at risk for PCOS and executing early treatment of PCOS may help prevent some of the long-term complications related to this syndrome. Also, two-thirds of the population encounters classic PCOS, and subsequently, laboratory testing is the prime criteria for analyzing this condition. In this paper, the SVM prediction model for PCOS determination acts as a clinical decision support system to help clinicians to make a treatment rule by keeping up with persistent patient health safety. Eight potential features are utilized for segregation among PCOS and non-PCOS patients. Among these eight features, five features are not associated with laboratory testing, but provide a primary indication of syndrome possibility that helps to create awareness among adolescent girls unaware of this disorder and risk association. The predictive model can spare clinician time, speed up the treatment strategy, and reduce undesirable laboratory testing expenses.

References

1. Fong, S.J., Dey, N., Chaki, J.: AI-enabled technologies that fight the coronavirus outbreak. In: Artificial Intelligence for Coronavirus Outbreak, pp. 23–45. Springer, Singapore (2020)
2. Jain, A., Bhatnagar, V.: Concoction of ambient intelligence and big data for better patient ministration services. Int. J. Ambient Comput. Intell. (IJACI) **8**(4), 19–30 (2017)
3. Santosh, K.C., Antani, S., Guru, D.S., Dey, N. (eds.): Medical Imaging: Artificial Intelligence, Image Recognition, and Machine Learning Techniques. CRC Press (2019)

4. Balen, H.A., Rutherford, J.A.: Managing anovulatory infertility and polycystic ovary syndrome. BMJ **335**(7621), 663–666 (2007)
5. Glintborg, D., Andersen, M.: An update on the pathogenesis, inflammation, and metabolism in hirsutism and polycystic ovary syndrome. Gynaecol. Endocrinol. **26**(4), 281–296 (2010)
6. Rotterdam: Revised 2003 consensus on diagnostic criteria and long-term health risks related to polycystic ovary syndrome (PCOS). Hum. Reprod. **19**(1), 41–47 (2004)
7. Wild, R.A., Carmina, E., Damanti-Kandarakis, E., Dokras, A., Escobar-Morreale, H.F.: Assessment of cardiovascular risk and prevention of cardiovascular disease in women with the polycystic ovary syndrome: a consensus statement by the Androgen Excess and Polycystic Ovary Syndrome (AE-PCOS) Society. J. Clin. Endocrinol. Metab. **95**(5), 2038–2049 (2010)
8. Denny, A., Raj, A., Ashok, A., Maneesh Ram, C., George, R.: i-HOPE: detection and prediction system for polycystic ovary syndrome (PCOS) using machine learning techniques. In: TENCON 2019-2019 IEEE Region 10 Conference (TENCON), pp. 673–678 (2019)
9. Zhang, X.-Z., Pang, Y.-L., Wang, X., Li, Y.-H.: Computational characterization and identification of human polycystic ovary syndrome genes. Sci. Rep. **8**(1), 1–7 (2018)
10. Dewi, R.M., Wisesty, U.N.: Classification of polycystic ovary based on ultrasound images using competitive neural network. J. Phys. Conf. Ser. **971**(1), 0120105 (2018)
11. Vikas, B., Anuhya, B.S., Chilla, M., Sarangi, S.: A critical study of polycystic ovarian syndrome (PCOS) classification techniques. Int. J. Comput. Eng. Manag. **21** (2018)
12. Cheng, J.J., Mahalingaiah, S.: Data mining and classification of polycystic ovaries in pelvic ultrasound reports. bioRxiv 254870 (2018)
13. Meena, K., Manimekalai, M., Rethinavalli, S.: Correlation of artificial neural network classification and NFRS attribute filtering algorithm for PCOS data. Int. J. Res. Eng. Technol. **5**(3), 519–524 (2015)
14. Meena, K., Manimekalai, M., Rethinavalli, S.: Implementing neural fuzzy rough set and artificial neural network for predicting PCOS. Int. J. Recent Innov. Trends Comput. Commun. **3**(12), 6722–6727 (2010)
15. Joham, A.E., Teede, H.J., Ranasinha, S., Zoungas, S., Boyle, J.: Prevalence of infertility and use of fertility treatment in women with polycystic ovary syndrome: data from a large community-based cohort study. J. Women's Health **24**(4), 299–307 (2015)
16. Mehrotra, P., Chatterjee, J., Chakraborty, C., Ghoshdastidar, B., Ghoshdastidar, S.: Automated screening of polycystic ovary syndrome using machine learning techniques. In: 2011 Annual IEEE India Conference, pp. 1–5 (2011)
17. https://www.kaggle.com/prasoonkottarathil/polycystic-ovary-syndrome-pcos
18. Kohavi, R., et al.: A study of cross-validation and bootstrap for accuracy estimation and model selection. IJCAİ **14**, 1137–1145 (1995)
19. Mukaka, M.M.: A guide to appropriate use of correlation coefficient in medical research. Malawi Med. J. **24**(3), 69–71 (2012)
20. Dey, N., Ashour, A.S., Borra, S. (eds.): Classification in BioApps: Automation of Decision Making, vol. 26. Springer (2017)
21. Deris, A.M., Zain, A.M., Sallehuddin, R.: Overview of support vector machine in modeling machining performances. Procedia Eng. **24**, 308–312 (2011)
22. Navada, A., Ansari, A.N., Patil, S., Onkamble, B.A.: Overview of use of decision tree algorithms in machine learning. In: 2011 IEEE Control and System Graduate Research Colloquium, pp. 37–42 (2011)
23. Langarizadeh, M., Moghbeli, F.: Applying Naive Bayesian networks to disease prediction: a systematic review. Acta İnform. Med. **24**, 364 (2016)

Comparative Assessment of Different Feature Selection Methods with Proposed Method in the Application of Diabetes Detection

Sourav Ghosh, Pranati Rakshit, Sayan Paul, Rishav Sen, Rinika Manna, and Sandeep Shaw

Abstract Machine learning, as we know, is a broad application of artificial intelligence (AI) which gives the capability to the machine to learn automatically and improve itself by own experience. Machine learning is doing excellent job in the medical field. As diabetes is one of the vital diseases which is basically a silent killer, it needs to be more addressed and detected or predicted as well. In this present work, several machine learning algorithms have been used to detect it. We are using various classification and prediction method to detect the diseases more accurately in real time. We have compared all the classification models like logistic regression, linear discriminant analysis, K-Nearest neighbor, decision tree, Naive Bayes, support vector machine to get the more accurate results. Univariate feature selection method also has been used and our new feature selection method has been developed. How this new feature selection method is working and how it over-performs the other feature selection methods has been shown in this work.

Keywords Classification · Diabetes · Logistic regression · Linear discriminant analysis · K-nearest neighbor · Decision tree · Naive Bayes · Support vector machine

1 Introduction

Diabetes mellitus is a metabolic disease that happens when the level of blood glucose known as blood sugar is too high. Blood glucose is the only energy source in our body and comes from food. Insulin is one of the hormones in our body which helps to move the sugar from the blood into our body cells to store energy. Our body does not produce enough insulin to serve the glucose from blood to cell sometimes. A low level of insulin attacks the immune system and destroys the cells in your pancreas that make insulin. There are some diabetic patients are asymptomatic mainly those, who during the early years of the disease marked with hyperglycemia and the children

S. Ghosh · P. Rakshit (✉) · S. Paul · R. Sen · R. Manna · S. Shaw
Department of Computer Science and Engineering, JIS College of Engineering, Kalyani, Nadia, West Bengal 741235, India
e-mail: pranati.rakshit@jiscollege.ac.in

H. K. D. Sarma et al. (eds.), *Contemporary Issues in Communication, Cloud and Big Data Analytics*, Lecture Notes in Networks and Systems 281,
https://doi.org/10.1007/978-981-16-4244-9_38

with absolute insulin deficiency may suffer from polyuria, polydipsia, polyphagia, weight loss, and blurred vision.

The only way to prevent this is education. Polish Dialectology Society guideline shows that education is the foundation of effective diabetes prevention. Through specific interventions, the educator can assist a patient with type 2 diabetes in minimizing the impact of environmental factors, preventing disease aggravation, and returning to their social roles.

To predict diabetes accurately, we have used several classifier methods such as—logistic regression, linear discriminant analysis, K-Nearest neighbor, decision tree, Naive Bayes, support vector machine. We have used several algorithms to get better predicted values. Several feature selection methods have been used to get predicted results more accurately. We used several feature selection like—chi-square and proposed method. We have made a comparison between the value of the classifier before the feature selection method and after applying feature selection methods.

.

2 Literature Survey

Sneha et al. [1] used optimal feature selection for early prediction on diabetes. Algorithms are used in this paper are decision tree algorithm, random forest, Naïve Bayesian, SVM. Here, they got the highest specificity of 98.20 and 98.00% on decision tree algorithm and random forest algorithm, respectively. The accuracy of the classification also been increased by generalizing the selection of the optimal features.

Singh et al. [2] predicted diabetes using medical data. There are three different types of supervised machine learning algorithms such as Naive Bayes, multilayer perceptron, decision tree-based, random forest. The methods such as 10-Fold cross validation (FVC), percentage split with 66% (PS) training dataset (UTD) and the rest as test dataset are used. Dataset used Pima Indians Diabetes Dataset. They got the highest accuracy of 84.93 using random forest (RF) with applying the all the test methods with pre-processing methods.

Saru et al. [3] analyzed and predicted diabetes using machine learning algorithm. In this paper, different algorithms used are random forest, decision tree, K-nearest neighbors, Naïve Bayes, and the dataset used Pima Indians Diabetes Dataset. They got the highest accuracy of **94.4%** from decision tree.

3 Methodology and Implementation

3.1 *Dataset Design*

Description of dataset

The datasets consist of several medical predictor (independent) variables and one target (dependent) variable outcome. Independent variables include the number of pregnancies the patient has had, their BMI, insulin level, age, and so on.

Attributes

Pregnancies: Number of times pregnant.
Glucose: Plasma glucose concentration a 2 hours in an oral glucose tolerance test.
Blood Pressure: Diastolic blood pressure (mm Hg).
Skin Thickness: Triceps skin fold thickness (mm).
Insulin: 2 h serum insulin (mu U/ml).
BMI: Body mass index (weight in kg/(height in m)^2).
Diabetes Pedigree Function: Diabetes pedigree function.
Age: Age (years).
Outcome: Class variable (0 or 1) 268 of 768 are 1, the others are 0.

3.2 *Classifier*

3.2.1 Logistic Regression

Logistic regression is a regression model in which the dependent variable should be categorical. That means binary dependent variable. The binary dependent variables only takes value "0" or "1". **B**inary variable represents outcome as impaired/unimpaired, clicked not clicked, good/bad etc. Logistic regression used in various fields like diseases diagnosis, fraud detection, spam detection. Logistic regression has sigmoid function which helps us to convert our prediction into probability which always lies between '0' and '1'.

3.2.2 Linear Discriminant Analysis

Linear discriminant analysis or LDA is one of the most common machine learning algorithms which is mainly used to solve the classification problems. The main goal of using LDA is that to find the direction which will provide us to separate maximum classes with the help of scatter matrices. In LDA, there are mainly two scatter matrices, 1. Scatter matrix between the data points on the plane and 2. Scatter matrix within the data points on that plane where distance between the two different classes' data points should be higher and the distance within the data points of same class should be minimum.

3.2.3 K-Nearest Neighbors

KNN is the most commonly used classification method to predict the result based on similar results. In this method, the number of neighbors is provided with the query. It works in this way:

1. There are some training samples with results to train the model and the **k** value is provided.
2. Then user provides some unknown value which is not present in the training samples to find out the result.
3. Then, the pre-trained model find similarity based on the Euclidean distance measurement.
4. Then based on the provided **k** value model predicts the result and gives us output.

For example, there are two classes in given training samples Class A and Class B and **k** value is provided as 5. In the output, we got the given query's 5 nearest neighbors and 4 of the neighbors belong to the class A and 1 in class B.

So the model gives us output for the given query is class A because we got 4 votes for class A and only 1 for class B.

3.2.4 Decision Tree

Decision tree is a supervised machine learning algorithm which is used to solve regression as well as classification problems. In this algorithm, we use tree like structure to solve the problems. In this decision tree model, there are basically two nodes one is decision node and other is leaf node. Leaf node represents the class label. We pass the entire dataset to the root node and then the classifier splits the data set into branches (decision node and leaf node or class). Then, an unknown value or data is provided to this model and then it gives the output of the class value for the input or test data based on this model.

3.2.5 Naive Bayes

Naive Bayes classifier belongs to probabilistic classifier family. Naive Bayes model is easy to implement and used for constructing classifiers. In this classification, we assumed the value of feature is independent of any other feature. It can not only handle the continuous data but also the discrete data and also it does not require much of a training data.

$$P(A|B) = (P(B|A)\,P(A))/P(B)$$

- *P*(*A*|*B*) is the posterior probability.
- *P*(*A*) is the prior probability.
- *P*(*B*|*A*) is the likelihood.
- *P*(*B*) is the marginal probability.

3.2.6 Support Vector Machine

Support vector machine belongs to supervised machine learning which is used to solve classification as well as regression problems. In this algorithm, we plot the data items. After that we have to separate the data points through the hyper plane, it is a plane which separates parallel the two classes. In the graph, there are two different classes. There is one in every two different classes which is nearer to the opponent class, and these nearer one is called as support vectors.

3.3 Feature Selection

3.3.1 Chi-Square Method (Univariate Feature Selection)

Univariate feature selection is one of the most common and basic feature selection techniques in the field of data analysis. In this feature selection method, basically one feature is selected from the given dataset and depending upon this feature it will provide us the results or class value. This type of feature selection helps us to create the histogram depending upon its feature and from that we can find out which features varies most in our analysis.

This method is one type of wrapper method in feature selection. It takes number of features as an input gives the best possible selected feature (multivariate) from a dataset. For an example if n is given as input then it produces a dataset with best possible n features.

In this case, we are using Pima Indian Diabetes Dataset. It selects best n features from that dataset and gives best selected features.

3.3.2 Proposed Method

In our proposed method, we are using all of the above-mentioned algorithms as follows:

We are using Pima-Indian-diabetes dataset here.

1. Initially, we remove all the null values from the dataset.
2. After that we applied a feature selection method to select the features from that dataset, such that for which features the accuracy of our model varies most.

3. In this case, the entire dataset contains total of 9 features, 1 of them is class for prediction. So other 8 are features.
4. Our model divides the data set in such a way that there are 80% of training data and 20% of test data.
5. Our method will perform different permutations and combinations to get the most essential features which provide the highest accuracy.

4 Results

After removing noisy and redundant data from our dataset, we perform different machine learning algorithms to find best fitted model for our dataset by comparing accuracy score of each model (Table 1).

Now, we performed univariate feature selection, and we got the maximum accuracy of 81.82%.

4.1 Feature Selection

The chi-square statistical method is commonly used for testing relationships between categorical variables. After performing chi-square test, we are getting the accuracy of 82.46%.

For **n = 4** or if we want best possible result for 4 features it gives:

Maximum Accuracy Score = 73.37%

For 4 features, proposed model gives the output dataset which contains four feature from main dataset which are Glucose, BMI, Diabetes Function, and Age.

For **n = 5** or if we want best possible result for 5 features it gives:

Maximum Accuracy Score = 82.46%

For 5 features proposed model gives the output dataset which contains five feature from main dataset which are Pregnancies, Glucose, Skin Thickness, Insulin and Age.

For **n = 6** or if we want best possible result for 6 features it gives:

Maximum Accuracy Score = 82.467532467532 46%

Table 1 After performing classifications without feature selection

Methods	Accuracy
K-nearest neighbors	70.77
Logistic regression	79.22
Linear discriminant analysis	79.22
Decision tree	73.37
Naïve Bayes	75.32
Support vector machine	63.63

For 6 features proposed model gives the output dataset which contains six feature from main dataset which are Pregnancies, Glucose, Skin Thickness, BMI, Diabetes Function, and Age.

We can see that we are getting best result for n = 5 which is 82.46%

Now, we are selecting the features using proposed method and compare the accuracy of each model which is referred in Table 3.

Now we are comparing each model with respect to univariate feature selection and proposed method which is referred in Table 4.

4.2 Analysis

Here are some graph results which we extract from our models. We plot the graph with respect to accuracy and different machine learning methods to visualize the table data. Figure 1 shows the comparison graph before feature selection and after applying univariate Selection [referred to Table 2]. Figure 2 indicates the comparison graph between before feature selection and after applying proposed method: [referred to Table 3]. The comparison graph between univariate feature selection and proposed method is shown in Fig. 3 [referred to Table 4].

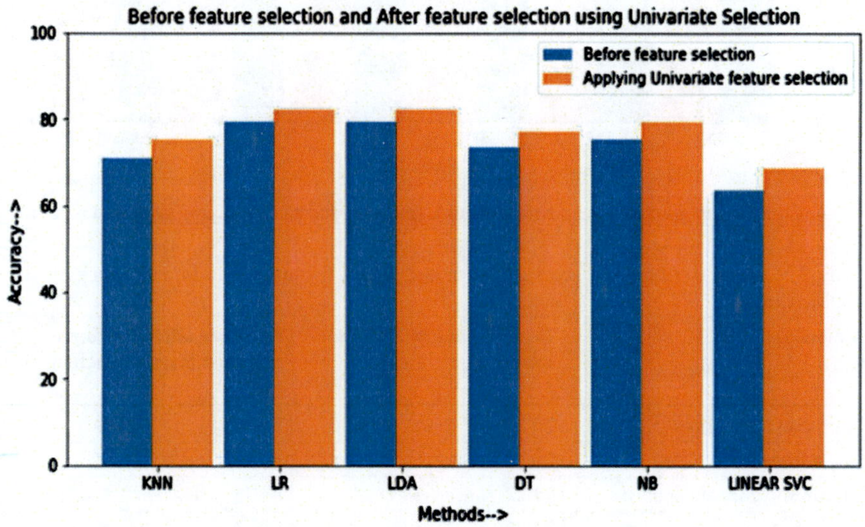

Fig. 1 Comparison graph before feature selection and after applying univariate selection

Table 2 Comparison between accuracy of model before feature selection and after applying univariate feature selection

Methods	Accuracy before feature selection	Accuracy after applying univariate feature selection
K-nearest neighbors	70.77	75.32
Logistic regression	79.22	82.46
Linear discriminant analysis	79.22	82.46
Decision tree	73.37	77.27
Naive Bayes	75.32	79.22
Support vector machine	63.63	68.83

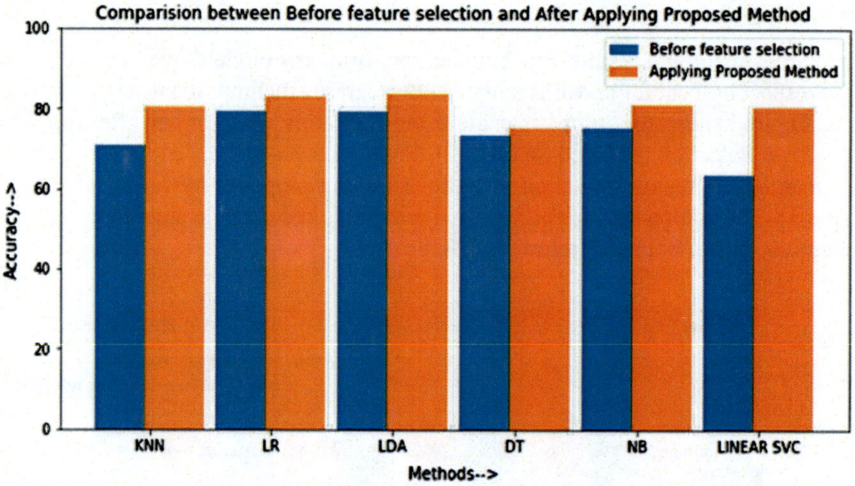

Fig. 2 Comparison graph between before feature selection and after applying proposed method

Table 3 Comparison between accuracy of model before feature selection and after applying proposed method

Methods	Accuracy before feature selection	Accuracy after applying proposed feature selection method
K-nearest neighbors	70.77	80.51
Logistic regression	79.22	83.11
Linear discriminant analysis	79.22	83.76
Decision tree	73.37	75.32
Naïve Bayes	75.32	81.16
Support vector machine	63.63	80.51

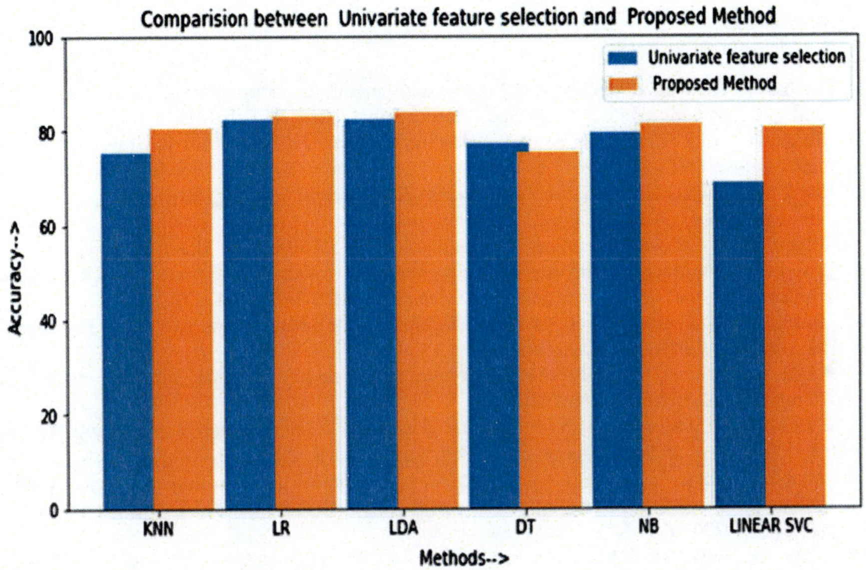

Fig. 3 Comparison graph between univariate feature selection and proposed method: [referred to Table 4]

Table 4 Comparison between univariate feature selection and proposed method

Methods	Accuracy after univariate feature selection	Accuracy after proposed method
K-nearest neighbors	75.32	80.51
Logistic regression	82.46	83.11
Linear discriminant analysis	82.46	83.76
Decision tree	77.27	75.32
Naive Bayes	79.22	81.16
Support vector machine	68.83	80.51

5 Conclusion

We have used various classification models to detect diabetes. To reduce the dimension as well as to increase the accuracy, we have used feature selection methods. Our novel feature selection approach also has been used to increase the accuracy. This new proposed method has given better result as compared to traditional chi-squared method.

References

1. Sneha, N.: Analysis of diabetes mellitus for early prediction using optimal features selection. Springer International Publishing Open Access First Online, 06 Feb 2019
2. Singh, D.A.A.G.: Diabetes prediction using medical data. J. Comput. Intell. Bioinform. **10**(1) (2017). ISSN 0973-385X
3. Saru, S.: Analysis and prediction of diabetes using machine learning. Int. J. Emerg. Technol. Innov. Eng. **5**(4) (2019)
4. Alam (2019) A model for early prediction on diabetes. Inform. Med. Unlocked **16**
5. Sisodia, D.: Prediction of diabetes using classification algorithms. In: International Conference on Computational Intelligence and Data Science (ICCIDS 2018), vol. 132.
6. Joshi, T.N.: Diabetes predictions using machine learning techniques. Int J. Eng. Res. Appl. **8**(1) (Part-II) (2018). ISSN: 2248-9622
7. Li, Y.: Analysis and study of diabetes follow-up data using a data-mining-based approach. Soft Comput. Anal. Biomed. Data **2018**
8. Christobel, Y., Sivaprakasam, P.: A new class wise K nearest neighbor method for the classification of diabetes dataset. Int. J. Eng. Adv. Technol. **2**(3), 396–400 (2013)
9. Anand, R., Kirar, V., Burse, K.: K-fold cross validation and classification accuracy of PIMA Indian diabetes data set using higher order neural network and PCA. Int. J. Soft Comput. Eng. **2**(6), 2231–2307 (2013)
10. Bang, H., Edwards, A.M., Bomback, A.S., et al.: Developmentand validation of a patient self-assessment score for diabetesrisk. Ann. Intern. Med. **151**(11), 775–783 (2009)
11. Shaw, J.E., Sicree, R.A., Zimmet, P.Z.: Global estimates of the prevalence of diabetes for 2010 and 2030. Diabetes Res. Clin. Pract. **87**(1), 4–14 (2010)
12. Coutinho, M., Gerstein, H.C., Wang, Y., Yusuf, S.: The relationship between glucose and incident cardiovascular events: a meta regression analysis of published data from 20 studies of 95,783 individuals followed for 12.4 years. Diabetes Care **22**(2), 233–240 (1999)
13. Akobeng, A.K.: Understanding diagnostic tests 3: receiver operating characteristic curves. Acta Paediatr. **96**(5), 644–647 (2007)
14. Fluss, R., Faraggi, D., Reiser, B.: Estimation of the Youden index and its associated cutoff point. Biom. J. **47**(4), 458–472 (2005)
15. Saxenal, K., Khan, Z., Singh, S.: Diagnosis of Diabetes Mellitus using K Nearest Neighbor Algorithm. Invertis University
16. Alam, T.M., Iqbal, M.A., Ali, Y., Wahab, A., Ijaz, S., Baig, T.I., Hussain, A., Malik, M.A., Razab, M.M., Ibrar, S., Abbas, Z.: A model for early prediction of diabetes. Inform. Med. Unlocked **16**, 100204 (2019)
17. Rashid, T., Abdullah, S., Abdullah, R.: An intelligent approach for diabetes classification. Predict. Description (2015). https://doi.org/10.1007/978-3-319-28031-8
18. Choi, S.B., Kim, W.J., Yoo, T.K., Park, J.S., Chung, J.W., Lee, Y., Kang, E.S., Kim, D.W.: Screening for prediabetes using machine learning models. Comput. Math. Methods Med. **618976**, 8 (2014). https://doi.org/10.1155/2014/618976

Author Index

H. K. D. Sarma et al. (eds.), *Contemporary Issues in Communication, Cloud and Big Data Analytics*, Lecture Notes in Networks and Systems 281,
https://doi.org/10.1007/978-981-16-4244-9